The Biological Basis of Cancer

Second Edition

Robert G. McKinnell
Ralph E. Parchment
Alan O. Perantoni
Ivan Damjanov
G. Barry Pierce

CAMBRIDGE
UNIVERSITY PRESS

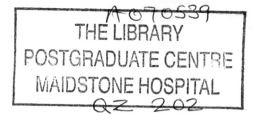
CAMBRIDGE UNIVERSITY PRESS
Cambridge, New York, Melbourne, Madrid, Cape Town, Singapore, São Paulo

Cambridge University Press
40 West 20th Street, New York, NY 10011-4211, USA

www.cambridge.org
Information on this title: www.cambridge.org/9780521844581

First edition first published 1998
Second edition first published 2006

Printed in the United States of America

A catalog record for this publication is available from the British Library.

Library of Congress Cataloging in Publication Data

The biological basis of cancer / Robert G. McKinnell . . . [et al.] – 2nd ed.
    p. ;   cm.
Includes bibliographical references and index.
ISBN-13: 978-0-521-84458-1 (hardcover)
ISBN-10: 0-521-84458-4 (hardcover)
ISBN-13: 978-0-521-60633-2 (pbk.)
ISBN-10: 0-521-60633-0 (pbk.)
1. Cancer.   2. Carcinogenesis.   3. Cancer cells.   I. McKinnell, Robert Gilmore.
[DNLM: 1. Neoplasms.   2. Neoplastic Processes.   QZ 200 B6147 2006]
RC261.M436   2006
616.99′4 – dc22   2005028814

ISBN-13   978-0-521-84458-1 hardback
ISBN-10   0-521-84458-4 hardback

ISBN-13   978-0-521-60633-2 paperback
ISBN-10   0-521-60633-0 paperback

The authors dedicate this book to two deserving groups. First, we recognize students who we confidently expect will make a greater impact on cancer than anyone has been able to do thus far. We are optimistic that these young scholars, committed to their research, will be rewarded by the eventual conquest of cancer. The other group is no less deserving. Let it be known that patients, who are the motivating force behind this book, despite enduring difficult treatment modalities, inspire us with their spirit – they have captured our respect and accordingly we dedicate this effort to them.

For Beverly Kerr McKinnell
who revealed extraordinary courage when learning that she had what Susan Sontag termed "the barbarian within"; that cancer claimed her life all too soon despite incredible bravery on her part and the care of many compassionate and dedicated people in the healing professions. She is missed terribly.

A note about this dedication: Beverly, wife of Robert G. McKinnell, died just months before this revised edition appeared. She and Robert are witnesses to the horrific reality that few families indeed escape the awfulness of cancer lethality.

Contents

Preface   *page* xv

Introduction: Letters illustrating clinical aspects of cancer • G. BARRY PIERCE   **1**

Colon cancer   2
Breast cancer   4
Acute leukemia   5
Lung cancer   6
Kidney cancer   7
Squamous cell cancer   8
Testicular cancer   9
Stomach cancer   10
Melanoma   11
Neuroblastoma   12
Summary   13

1 The pathology of cancer • G. BARRY PIERCE AND IVAN DAMJANOV   **14**

1.1   Introduction   14
1.2   Benign versus malignant tumors   18
1.3   The diagnosis of benign and
       malignant tumors   24
1.4   Tumor grading and staging   25
1.5   Classification and nomenclature   27
1.6   Metastasis   28
1.7   Tumor markers   30
1.8   How cancer kills   30
       1.8a   Organ failure   30

	1.8b	Obstruction of the gastrointestinal tract, ducts, and hollow organs	31
	1.8c	Cachexia and infection	33
1.9	Spontaneous regression		34
1.10	Dormancy		35
1.11	Initiation		36
1.12	Latency		36
1.13	Progression to the autonomous state		37
1.14	Selection and cellular heterogeneity		38
1.15	A developmental concept of cancer		40
1.16	Apoptosis		48
1.17	Summary		49

2 Invasion and metastasis • Robert G. McKinnell   51

2.1	Introduction		51
2.2	The metastatic cascade		54
	2.2a	Disruption of the basement membrane and lytic activity in the extracellular matrix	56
	2.2b	Cell detachment	59
	2.2c	Cell migration and motility	61
	2.2d	Invasion	64
	2.2e	Penetration of the vascular system	65
	2.2f	Cancer cells in the circulation	67
	2.2g	Arrest of circulating cancer cells (stasis)	67
	2.2h	Extravasation, growth of metastases, and metastasis of metastases	68
2.3	A multiplicity of genes are associated with metastasis		69
2.4	Soil and seed hypothesis of Paget		70
	Box: Stephen Paget: No "ploughman" was he!		71
2.5	Is metastasis limited to malignant cells?		72
2.6	How do we know a metastasis to the liver is not a primary neoplasm of the liver?		76
2.7	Why study metastasis?		77
2.8	Summary		78

3 Carcinogenesis • ALAN O. PERANTONI   **80**

3.1   Introduction   80
3.2   What is a carcinogen?   81
3.3   Carcinogenesis as a multistage process   82
3.4   Chemical carcinogenesis   84
    3.4a   Organic compounds   92
    3.4b   Inorganic compounds and asbestos   95
    3.4c   Naturally occurring chemicals   98
3.5   Radiation   98
    3.5a   Ultraviolet radiation   98
    3.5b   Ionizing radiation   100
    3.5c   Endogenous ionizing radiation   103
3.6   Radon   103
3.7   Viral carcinogenesis   105
3.8   Endogenous carcinogenesis   107
3.9   Metabolism of xenobiotics   109
    3.9a   Host defenses   109
    3.9b   Inducibility of xenobiotic
        metabolism   111
    3.9c   Metabolic activation of chemical
        carcinogens   113
    3.9d   Inactivation of chemical
        carcinogens   114
    3.9e   Systemic distribution of chemical
        carcinogens   114
    3.9f   Mechanisms for carcinogen
        suppression/chemoprevention   115
    Box: Elizabeth Cavert Miller with
    husband James   116
3.10   Modulation of carcinogenesis   117
3.11   Tumor promotion   120
3.12   Tumor progression   122
3.13   Alternative pathways for carcinogenesis?   123
3.14   Federal regulations   123
3.15   Summary   125

4 Genetics and heredity • ROBERT G. MCKINNELL   **126**

4.1   Introduction   126
4.2   Chromosomes and cancer   127
    4.2a   Aneuploidy   127

	4.2b	Euploidy does not preclude genetic change	129
	4.2c	Cancers with chromosomal aberrations	131
4.3	Chromosome damage, mutation, and vulnerability to cancer		135
4.4	Hereditary cancers		136
	4.4a	Retinoblastoma	136
	4.4b	Wilms tumor	137
	4.4c	Hereditary conditions that increase cancer risk	138
4.5	Familial cancer syndromes		139
	4.5a	Colon cancer	139
	4.5b	Breast cancer	141
	4.5c	Prostate cancer	142
	4.5d	Microarray technology as a way of examining many genes simultaneously	143
4.6	Summary		144

5 Cancer-associated genes • ALAN O. PERANTONI   **145**

5.1	Introduction		145
5.2	What is an oncogene?		145
5.3	Proto-oncogenes function in signal transduction, cell cycle regulation, differentiation, or programmed cell death (apoptosis)		148
5.4	Genetic approaches to delineate proto-oncogene function		150
	5.4a	DNA microarray analysis – global gene expression or genomic profiling	154
5.5	Classification of proto-oncogenes/oncogenes		155
	5.5a	Growth factors and their receptors	156
	5.5b	Nonreceptor tyrosine kinases	161
	5.5c	GTP-binding proteins: *ras* activation	162
	5.5d	Cytoplasmic serine/threonine kinases	163
	5.5e	Suppression of *ras* signaling	165
	5.5f	Nuclear signaling	165
	5.5g	Transcriptional activation	166
5.6	Regulation of DNA synthesis and the cell cycle		168

5.7   Other mechanisms for the
      regulation of signaling   171
5.8   Mechanisms of oncogene
      activation   173
5.9   Carcinogens and oncogene
      activation   178
5.10 Oncogene cooperation   179
5.11 Normal cells suppress tumor
      growth   180
5.12 Angiogenesis and tumor
      development   180
5.13 Tumor Suppressor genes   181
      5.13a   The *Rb* locus   183
      5.13b   *p53* suppressor gene   184
      5.13c   Other tumor suppressors   187
      5.13d   Apoptosis and its role in growth
              regulation   188
      5.13e   Senescence   191
5.14 Where pathology meets molecular
      biology   192
5.15 Summary   193

6 Cancer in nonhuman organisms • ROBERT G. MCKINNELL   195

6.1   Introduction   196
6.2   Plant growths   197
6.3   Invertebrate animals   200
      Box: Yoshio Masui   202
6.4   Cancer in selected ectothermic
      (cold-blooded) vertebrates   203
      6.4a   Fish   204
      6.4b   Amphibia   207
      6.4c   Reptiles   212
      Box: John C. Harshbarger   214
6.5   Cancer in selected warm-blooded
      vertebrates   215
      6.5a   Birds   215
      6.5b   Mammals   216
6.6   Summary – But try anyway!   220

7 Epidemiology • ROBERT G. MCKINNELL   221

7.1   Introduction   221

7.2   Cancer in fossil humans: A brief
        digression concerning
        paleopathology   226
7.3.   Epidemiology of selected human
        cancers   226
        7.3a   Lung cancer   227
        Box: Alton Ochsner   229
        Box: Richard Doll   231
        7.3b   Breast cancer   233
        7.3c   Skin cancer   236
        7.3d   Prostate cancer   239
        7.3e   Colorectal cancer   241
        7.3f   Cervical cancer: "The Beginning
                 of the End"   243
        7.3g   Hodgkin lymphoma   244
7.4   Occupational cancers   244
7.5   AIDS-related Kaposi's sarcoma   245
7.6   What is next?   246

8   Lifestyle: Is there anything more
    important? • ROBERT G. MCKINNELL   248

8.1   Introduction   248
8.2   Lung cancer is a preventable disease   249
8.3   Ultraviolet radiation and that
        "healthy tan"   251
        8.3a   How to minimize risk for skin cancer   252
        8.3b   The peculiar status of protection by
                 sunscreens   253
8.4   Diet, nutrition, and cancer   253
        8.4a   Dietary fiber and colorectal cancer   255
        Box: Denis Burkitt   256
        8.4b   Correlations between food substances
                 and cancer prevalence: Significance   258
        8.4c   Dietary fat and obesity   258
        8.4d   Vitamins and cancer   259
        8.4e   Selenium and calcium   260
        8.4f   Non-nutrient organic compounds in
                 food that may protect against cancer   260
        8.4g   American Cancer Society (2002)
                 Guidelines on Diet, Nutrition, and
                 Cancer Prevention   262
8.5   Exercise as it relates to cancer   263

8.6 A special note about breast cancer 263
8.7 Other lifestyle hazards 264
8.8 Summary 264

9 The stem cell basis of cancer treatment: concepts and clinical outcomes • RALPH E. PARCHMENT   **266**

9.1 Introduction 266
  9.1a Therapies remaining at the conceptual level 266
  9.1b Therapies being explored clinically: Differentiation therapy and cytostatic therapy 267
  Box: Leland Hartwell, R. Timothy Hunt, and Sir Paul Nurse 269
  9.1c Eradicating cancer cells – the aim of current cancer therapy 271
9.2 Absolute versus fractional cytoreduction 273
  Box: Howard Skipper 275
9.3 The meaning of "curing cancer" depends on whom you ask 279
9.4 The biological basis of multimodality therapy as optimal cancer treatment 282
9.5 Biological factors that contribute to treatment success 289
  Box: H. Rodney Withers 291
9.6 Biological factors that contribute to treatment failure 293
9.7 Treatment of intermediate-stage breast cancer as a clinical science success story 301
9.8 Summary 305

10 Oncology: The difficult task of eradicating caricatures of normal tissue renewal in the human patient • RALPH E. PARCHMENT   **307**

10.1 Surgical oncology 308
10.2 Radiation oncology 311
10.3 Chemotherapy 314
  10.3a Directly cytotoxic chemotherapy drugs 314

10.3b   Reactive chemicals as cytotoxic
         anticancer drugs   315
10.3c   Selective cytotoxicity as a screening tool
         to discover more cytotoxic drugs   315
10.3d   Indirect tumor cytotoxicity by nutrient
         deprivation ("antimetabolite therapy")   317
Box: George H. Hitchings, Jr.,
and Gertrude B. Elion   318
Box: Charles Brenton Huggins   329
10.3e   Trophic factor therapy to treat
         hematologic side effects of
         chemotherapy   341
10.3f   Therapy that exploits differentiation
         processes in malignancies   342
10.4   Pharmacological issues arising from
       tumor biology   349
10.5   Unknowns, the future, and the
       emergence of molecular oncology   353

Appendix: Description of selected tumors • G. BARRY PIERCE AND IVAN DAMJANOV   **355**

A.1   Adenocarcinoma of the breast   355
A.2   Adenocarcinoma of the prostate   358
A.3   Adenocarcinoma of the colon   359
A.4   Squamous cell carcinoma   361
A.5   Teratocarcinomas   365
A.6   Liver cell carcinoma   370
A.7   Lung cancer   370
A.8   Malignant melanoma   372
A.9   Retinoblastoma   374
A.10   Neuroblastoma   374
A.11   Wilms tumor (nephroblastoma)   375
A.12   Sarcomas   377
A.13   Lymphoma and leukemia   378

Glossary   381

References   401

Index   469

Preface

This version of *The Biological Basis of Cancer* is referred to as a "second edition" which, of course, it is. The term "second edition" does little to inform the reader to what extent material has been updated or rewritten. Major changes were mandated throughout by the rapid pace of cancer research and the translation of new findings into improved clinical care. Be that as it may, some fundamental aspects of cancer do *not* change. The letters in the Introduction are such. The letters are provided to introduce the student to the impact of cancer on the lives of ordinary people and the resultant need for cancer research. That impact and need are the same today as they were several years ago and are perhaps best reflected in the remarkable cathartic writings of author John Gunther (*Death Be Not Proud*) on the untimely loss of his seventeen-year-old son in 1947 from a brain tumor to the recent trials and triumphs over cancer of international cycling star Lance Armstrong (*It's Not About the Bike: My Journey Back to Life*). Similarly, the distinction of benign versus malignant, as well as the concept of tumor grading and staging and certain other aspects of cancer discussed in the chapter on pathology remain the same. Other material has been carefully revised or rewritten – there is an entirely new Chapter 8 devoted to measures recommended by cancer organizations, and the authors of this book, to hopefully (and likely) reduce risk for cancer. Chapter 10 presents a new therapeutic modality that was not established in 1998 – molecular therapy that specifically treats the molecular cause of cancerous growth. It also includes new clinical strategies not available to the oncologist in 1998: controlling blood supply to tumors and molecular profiling of patients to identify those few individuals who are extremely susceptible to particular drug toxicities. New to this revised edition are brief essays highlighting the contributions of major cancer researchers – this has been done to remind students that insights into cancer come from the endeavors of gifted scientists and physicians. The lives of the highlighted cancer researchers may even inspire and stimulate some students to go forth and do likewise. It is worth repeating that this book was originally designed for undergraduate students and beginning graduate students. Our intended audience has remained unchanged. However, students of any age with education in the biological sciences and time to pause and reflect will be able to master this material. We continue to hope that others, even nonbiologists with an interest in cancer, will find reading this book understandable and rewarding.

We are gratified that certain universities have chosen to use this text in their courses in the biology of cancer. We thank the instructors and seek from them and from their students suggestions on how the book can be improved in future editions. We acknowledge with appreciation the inclusion of this book in a list of references by the National Cancer Institute:

http://science-education.nih.gov/supplements/nih1/cancer/guide/references.htm

While errors are solely those of the authors, a number of readers have graciously scrutinized various stages of this text and by their efforts have minimized egregious mistakes. We are profoundly grateful for their efforts. We appreciate those who have served us in other ways to enhance this book. The list of readers and helpers includes the following: Helen Miller Alexander, University of Kansas, Lawrence, KS; Debra L. Carlson, Augustana College, Sioux Falls, SD; Elaine M. Challacombe, Owen H. Wangensteen Historical Library of Biology and Medicine, University of Minnesota, Minneapolis, MN; William L. Dahut, Medical Oncology Clinical Research Unit, National Cancer Institute, Bethesda, MD; Marie A. DiBerardino, Drexel University College of Medicine, Philadelphia, PA; John Harshbarger, George Washington University Medical Center, Washington, DC; Sherri A. Long, Health Partners and Department of Dermatology, University of Minnesota, Minneapolis, MN; Susan Kerr McKinnell, ADCS, University of Minnesota, Minneapolis, MN; Danica Ramljak, Medora Global Consulting, Inc., McLean, VA; Mark A. Sanders, Imaging Center, College of Biological Sciences, University of Minnesota, Saint Paul, MN; Michael Waalkes, NIEHS, Research Triangle Park, NC; William R. Waud, Cancer Therapeutics, Southern Research Institute, Birmingham, AL; and Michael J. Wilson, Veterans Administration Medical Center and Department of Laboratory Medicine and Pathology, University of Minnesota, Minneapolis, MN.

This tome could not have attained fruition had it not been for the extraordinary aid of James V. Curley and Chacko T. Kuruvilla, University of Minnesota Biomedical Library. The authors thank both for their superb help.

Ken Karpinski was of inestimable support during the production phase of this book. RGM is particularly in his debt and owes Ken much appreciation indeed.

Special thanks go to our editor Katrina Halliday at Cambridge University Press, Cambridge, England. She was always prepared to provide help and advice, always pleasant in doing so, and had inexhaustible patience for the authors during the production of this book.

We would be remiss indeed if we did not recognize the support of our families and their forbearance in tolerating our annoying habits relating to time management in the preparation of this book. There are no words that adequately express our appreciation.

The following paragraph appeared in the first edition of this book. It is repeated here word for word because we believe it to be pertinent as a brief statement of the debt we owe our students. It reads: Finally, we thank students who have and will take a course in the biology of cancer. In a conversation with a distinguished Scottish scientist, one of us commented that he wished he had more time for the laboratory. Perhaps, he thought, it would be better not to teach at all but rather devote full time to laboratory pursuits. The Scot, a Fellow of the Royal Society of London, responded, "Appreciate your students and give thanks that you must teach – there will be days upon days when research does not go well – that is the way of research. But, if you teach, you will leave the lab for lecture, and your flagging spirits will be rejuvenated by the enthusiasm and youthful concerns of your students. Then, with vigor renewed, return to the lab." The authors of this book say, "Thank you, students."

Homage to a colleague

Stem cells are much in the news for a variety of very good reasons. Those reasons need not be described in detail here because it is difficult not to be aware of their potential merit in the treatment of cancer and other chronic, debilitating, and lethal diseases. Stem cells merit the attention of scholars for another very important reason. Fully differentiated cells are post-mitotic and, as such, are unlikely to be an effective target of chemical, physical, or viral carcinogens. After all, if a cell is truly post-mitotic, how could it ever give rise to cell progeny, either normal or malignant? If not fully differentiated cells, what then is the target of carcinogenesis? Stem cells either proliferate into more stem cells that retain their pluripotential differentiative competence, or differentiate into lineage-specific mature cells. There has emerged in the last half century the concept that cancer is a disease of stem cells. Abnormal stem cells give rise to malignant cells which differentiate, as might be expected, along similar but aberrant pathways; i.e., they give rise to more abnormal stem cells and to cells that differentiate incompletely as cancer cells. The abnormal differentiation of cancer cells has been referred to as a "caricature" of the process of tissue renewal. The concept of cancer that originates in a caricature of the process of differentiation is not only of theoretical value but has certain very practical consequences – not the least of which is the notion that because there is a problem in differentiation of these progeny of stem cells, then treatment of some cancers by differentiation agents might give rise to cells that mature, become terminally differentiated, and ultimately are disposed of by apoptosis. An example of such a cancer is acute promyelocytic leukemia, which responds to all-trans-retinoic acid with maturation, terminal differentiation, and finally apoptosis.

The concept of the stem cell origin of cancer was a lifetime passion of one of the authors of this book. That author is Gordon Barry Pierce (see photos next page) known to many as "Barry." Barry's work is described briefly in Chapter 1. Barry, for health reasons, has finished his responsibilities with this book. We, the remaining authors, wish herewith to express our admiration for his lifetime of cancer research and to personally thank him for the many times he has aided us in our research and developed our intellectual drive to understand cancer. Barry, all of us wish for you and Donna many years of happiness, and please remember we are grateful for all that you have done for us individually and collectively.

Robert G. McKinnell
Ralph E. Parchment
Alan O. Perantoni
Ivan Damjanov

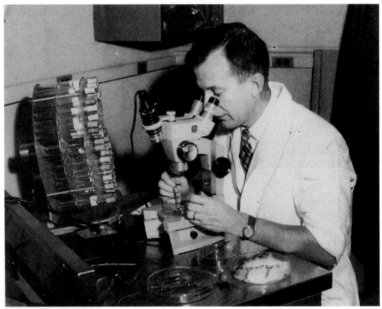

A youthful Barry Pierce working with a mouse embryonal carcinoma in 1958 (above). Portrait ca. 1992 (left). Photographs courtesy of Juan J. Rodriguez, *The International Journal of Developmental Biology*, Volume 37(1), 1993.

INTRODUCTION

Letters illustrating clinical aspects of cancer

G. Barry Pierce

Some may read this book because they or a loved one has cancer, students of science may read it because they are about to make career decisions, and students of cancer may read it because of a desire to be acquainted with aspects of cancer other than those with which they work. Although most students of cancer have never seen a malignancy in a human, they are nevertheless expert in one or more of the many fascinating and important aspects of the disease. They contribute to the understanding of DNA and its replication, control of gene expression, receptors and growth factors, developmental aspects, immunology, prevention, and treatment of cancer and a myriad of other important parts of the cancer problem. We are all impressed by the recent compounding of knowledge, but even more so by our lack of knowledge and understanding of how important facts generated in one aspect of investigation bear on another. Students, unaware of the clinical aspects of the disease, lack the information to make the correlations and see the important problems that result. In addition, they may be driven in their endeavors only by the intellectualism of their part of the problem. This book is designed to provide insight and understanding into the human aspects of cancer.

It is not our purpose here to present all-encompassing details of the clinical behavior of patients with cancer that are necessary for diagnosis, medical practice, and patient management. Neither is it our objective to detail clinical aspects that could be misunderstood or inappropriately applied. Rather, we present a flavor of the clinical aspects. The clinical problems are illustrated, and areas of ignorance and points of attack identified.

Cancer is a dreaded disease, and the impact of death and suffering are often lost when its research aspects are considered in isolation. Many believe research should be driven only by the intellectualism of the problem and not by practical concerns. But recall that the control of diabetes with insulin was driven by an emotionally charged clinical problem. It is often satisfying to be driven in research by not only an intellectual charge but also a practical goal, providing perspective is maintained.

It was decided to begin this book with a series of letters written by patients, their relatives, clinicians, and friends. The events described in these letters are real, but of course the names of individuals and particular circumstances have been changed. They illustrate the diversity of clinical conditions caused by cancer, stress the commonalities of the conditions, and the points of research attack.

Colon cancer: Appendix, Section A.3; Chapters 1, 2, 9, and 10

Dear Uncle Harvey,

I am writing to tell you about Dad. I know you would have come to see him if you could have. I will bring you up to date about what has happened. You remember our feeling of fear and panic when his bowels became obstructed two and half years ago. It did not seem reasonable – a man only fifty-five years old, a tough old rancher. Yet, they said he would die within a week if the obstruction of his colon was not "relieved" surgically. They removed part of his colon and anus, because the tumor had spread into the anus. Then they did a colostomy on his belly (it is like an artificial anus). The surgeon said the tumor had not spread to the liver or lymph nodes, and Dad felt pretty good, even though he hated that new opening with a passion.

About a year after surgery, he began to lose weight, but we didn't pay too much attention to it, because he could stand to lose a few pounds anyway. Then, he began to feel run-down and had some pain in his stomach, most on the right side up high. They did some tests and the doctor told us the tumor had spread after all, and the liver was enlarged with it. That was causing the pain. They gave him chemotherapy and he was pretty miserable. Actually, it made him mad because what hair he had left all fell out. Why would a man's hair fall out? So then he wore his old baseball cap from high school. Anyway, he got much better, but then when symptoms recurred, they treated him with chemotherapy again, but it didn't seem to do anything at all. They said his tumor was resistant. He was thin and sickly and had a good deal of pain. We couldn't get him to eat to get his strength up. The doctor was pretty good and gave him pain relievers. After that, it was just one thing after another – he got pneumonia that was cured with antibiotics, then Mom heard about a man, somewhere in Georgia, who had a new cancer cure. It was a secret cure, because he was afraid the American Medical Association would take it away from him. She wanted Dad to try the cure.

I went to the American Cancer Society and they said this man was a quack and all quacks have a secret. I explained this to Mom, but she said it was worth it, and besides, she had saved up $3,750 and they would use that. To make a long story short, the quack did not help, but Mom doesn't regret doing it even though she lost all her money.

Dad wanted to see all of my brothers and sisters. We talked and joked, but he was so frail and weak he could not spend much time with us. Would you believe he had lost seventy-five pounds? It broke our hearts. This was the guy who worked all day, partied all night, and wrestled the five of us to a standstill the next day.

Then he got really sick with a high fever and pneumonia, and they took him to the hospital. He died three days later. He had some kind of infection of the blood they couldn't treat. It was called septicemia. He had been in great pain, but there wasn't much we could do.

Mom is fine, but she feels guilty about Dad's death. I don't see what she could have done differently. I feel bad about some of his so-called friends who never came to see him, and I am angry about what happened to him. To see a strong, husky, independent man become bedridden and frail and have to be looked after hand and foot just didn't seem right.

Yours sincerely,

George

Author's note

This letter reveals much of what happens with people who have cancer of the colon (see Appendix A.3). At the onset, changes in bowel habits, passage of blood in the stools, or some nonspecific and vague pains in the abdomen may occur. It is not uncommon for the individual to present ("present" is medical jargon for appearance before a clinician for examination or treatment) with an obstruction, and it depends on the location whether the tumor can be removed and the bowel anastomosed (reconnected) end to end or whether a colostomy has to be made. As in this case, some patients find it extremely difficult to adapt to a colostomy, but fortunately organizations that can help patients deal with this or other cancer-related problems are available (in the United States, call the American Cancer Society at 1-800-ACS-2345 or the National Cancer Institute's Cancer Information Service at 1-800-4-CANCER). At the time of surgery the tumor had spread, but it took time for the metastases (Color Plate 11; Chapter 2) to grow and become clinically evident. Metastasis, wasting, and the death of the patient due to infection (in this case septicemia) are commonalities of the malignant phenotype. The loss of hair and diarrhea result from the killing of fast-growing normal cells of hair follicles and intestinal epithelium by the cytotoxic chemicals used in attempts to kill tumor cells (Chapters 1, 9, and 10). These chemicals lack specificity for cancer cells.

Note the strength of will and determination of human beings and their loyalties and lack of loyalties. Friends are often embarrassed and afraid to visit people with cancer, and as a result the patients are lonely.

Breast cancer

Dear Ed,

Well, it is the big C again. We thought we had it licked, but it is back. I am writing to you for some of your free legal advice.

You will remember that Joyce had that "thickening" in the breast. Her doctor told her it was nothing and not to worry about it. She went for her annual checkup ten months later and the doctor again said not to worry about it, that it was okay. Then, it began to enlarge, so she went to the doctor again, but he was away. His partner said she had a tumor and at her age it was probably cancer and should be operated on immediately.

When they removed the tumor, they also removed the fat from her armpit because this cancer usually spreads to the lymph nodes in the armpit first. They found it had spread to twelve of fifteen lymph nodes examined by the pathologist. The doctors said this was a very bad sign. She had X-ray treatments and chemo. She didn't like the chemo treatments and was glad to be done with them. They nauseated her and gave her diarrhea.

She was well for fourteen months, and then about six months ago she felt lousy and began to lose weight. We thought it was flu or something like that. Then she stepped off our back stoop and her hip broke. It turned out the cancer had spread and was quietly growing in her bones, of all places. It ate away the top of her leg bone, causing the fracture. Now she has pain in her back because the tumor has spread to her spine. She is losing weight and clearly cannot live much longer. I hate to say it, but this will be a blessing because of the suffering.

I am angry. I don't think that doctor knew what he was doing when he told her not to worry, that the lump was nothing. He let it grow and spread. I know that suing him won't help Joyce, but I don't think a man like that should be practicing medicine. If I sue him, I can say that publicly and maybe the warning will help other people. What should I do? Joyce doesn't want me to sue him, but if I don't I'll probably beat him up. I am desperate and angry, and I don't know what I am going to do. When you have been married to somebody for thirty-five years, it is difficult to reconcile what has happened. I will appreciate your advice.

Yours sincerely,

Ted

Author's note

As you will learn in this book, early tumors tend to behave less aggressively than later ones. The transition from less to more malignant is termed progression (Chapter 1). Usually the earlier the tumor is diagnosed and treated, the better the prognosis (outlook or predicted disease outcome) for the patient because the tumor has not progressed to the stage of invasion and metastasis. Delay in therapy can result from patient fears or, as in this case, misjudgment or ignorance on the part of the physician.

By the time definitive treatment was started in this patient, spread of the cancer to so many regional lymph nodes and distant organs had occurred that the patient's prognosis was hopeless. Note that some tumors often metastasize to particular sites (Chapter 2). In this case, breast cancer (Appendix A.1) has a predilection to spread to bone. Note the commonalities: growth leading to a rapidly enlarging mass, spread of cells (metastasis), and weight loss (cachexia).

Patients are fearful of cancer and cancer surgery, but all of these fears must be put aside because the only hope for the patient is early diagnosis and adequate initial therapy prior to progression of the disease. Chemotherapy and endocrine therapy for the patient with metastasis is palliative.

Acute leukemia: Appendix A.13; Chapters 9 and 10

Dear Aunt Molly,

I am sure you have heard that Jamie, our three-year-old son, was diagnosed with leukemia. I didn't write at the time the diagnosis was made because we were overwhelmed by what happened subsequently. He was such a lovely child – so pleasant, so full of energy, and we miss him so. Jamie developed an acute fever, and overnight became seriously ill, with bleeding from his gums and nose. He also had large bruises, but we knew he had not been hurt. The doctor suspected leukemia and said the illness resulted from "displacement of normal white blood cells from the bone marrow by the malignant leukemia cells." As a result Jamie could not fight infection. The leukemia cells also displaced the cells that stop bleeding. The doctors said it was very serious, but most children with acute lymphatic leukemia are saved with new types of treatment. They took a sample of blood and bone marrow, but although he did have acute lymphatic leukemia, he was not one of the lucky ones. He was a little better after chemotherapy and X-ray, and then he developed septicemia, was bleeding, and was so lethargic. Last night Jamie died. It is hard to believe. He had been sick only nine weeks but it seemed like ages. I hate to say it, but I feel relieved. He suffered so and he was so good. Now he doesn't have to suffer anymore. Please come and stay with me – I need your help.

Your loving niece,

Mary

Author's note

This letter speaks for itself. Malignant disease often has a bimodal incidence and occurs most commonly in the young and the elderly. Acute lymphatic leukemia (Appendix A.13) is a disease of the young, and its treatment is one of the triumphs of modern chemotherapy. The disease, which before modern treatment caused death in a matter of weeks, now allows cure in 50 percent or more of patients. However, in this case, the malignancy failed to respond to treatment (Chapters 9 and 10). There is no way of knowing which patients will respond to chemotherapy. Note commonalities of rapid growth of malignant cells that invade and displace normal tissue and interfere with its function, in this case replacement of normal marrow with malignant cells. This loss of white and red blood cells and platelet-forming cells makes the patient prone to opportunistic infections, anemia, and bleeding, respectively.

Lung cancer: Appendix A.7; Chapters 2, 7, and 8

Dear Cousin Janet,

We have bad news to report. You will recall that my dad had that nasty chronic cough. Well, it got worse and he developed pneumonia. The doctor took X-rays and treated him with one of the new antibiotics. He got better, but then six weeks later he developed pneumonia again in exactly the same spot. It made the doctor suspect lung cancer, because Dad was fifty-five and had smoked two packs a day for the last thirty years. They made him cough up sputum and found malignant cells in it. These were "small cell lung cancer cells," which are the worst kind. They are so malignant that the doctors refused to operate because by the time these tumors are discovered, they have already spread all over. They gave Dad radiation treatments and chemotherapy, but they did not help much. He became disoriented and then had some convulsions. The doctor said the tumor had spread to the brain. They suggested irradiating the brain, but my brother and I decided it was no use. We were not surprised that the doctors and nurses agreed with us. Dad died in his sleep just a month ago today. Mom is doing really quite well.

Sincerely,

Dorothy

Author's note

This individual was in the cancer age group with a smoking history (Chapters 7 and 8) that placed him at great risk for developing cancer of the lung (Color Plates 5 through 10; Appendix A.7). This case also illustrates the propensity of lung cancer to metastasize (Chapter 2) to the brain. This occurs so commonly that if a person presents with signs and symptoms of a brain tumor, the clinician must always rule out the possibility of metastasis to the brain from a primary lung cancer.

Note an additional commonality: the cancer diagnosed as a small cell lung cancer lacks tissuelike, or "epithelial" or "glandular" differentiation as viewed in the light microscope. Lack of differentiation correlates with rapid growth and a poor prognosis.

Smoking has been shown to be the most important etiologic factor in cancer of the lung. Only fools play Russian roulette or smoke cigarettes.

Kidney cancer: Chapter 1

Dear Fred,

I need to talk to someone, and since you are my oldest friend and a urologist to boot, you have been selected. I began to pass blood in my urine so I went to the doctor. He told me that because of my age, he had to rule out bladder cancer. I was cystoscoped and the bladder was okay, but then he injected a dye in my blood and took X-rays. The dye was passed from the kidney into the urine, and there in the kidney was a tumor filling this pelvis-thing (the part that drains the urine out of the kidney). I had to wait a week because there were other tests that had to be done.

This was the worst week of my life because I had these horrible guilt feelings – I had never bought as much insurance as I should have because I wanted a nice home and things for the family. Now I am sure I have cancer and I am going to die. My wife is going to have trouble managing, and the kids won't be able to get an education. I really need some time, and I am angry besides. I didn't booze or smoke or play around, why should this happen to me? My doctor seems okay and I like him, but with something this important, I am not sure if I should let him do the surgery or go elsewhere. Please call and let me know.

Sincerely,

Charlie

Dear Charlie,

I was distressed to read your recent letter. I will be happy to come and help or do whatever is necessary. First, let me give you some advice. It sounds as though you may have a cancer of the pelvis of the kidney. This doesn't mean you will die tomorrow or the next day or at all as a result of this disease. But it does mean that unless this tumor is completely removed, it will in fact kill you someday. It may not, because about half of these tumors in the kidney pelvis, even though malignant, behave in a very benign manner, so possibly simple surgical removal of the kidney will cure you. In the event the tumor spreads, it will probably spread to the bladder. (It is a good sign that the bladder is not affected at this time.) That doesn't mean to say you are out of the woods, but it says the tumor cells have not become aggressive enough to spread through the urine to the bladder. If this spread occurs, very often it can be handled using the same type of procedure they used when you were cystoscoped. They can take out small tumors through a resectoscope inserted into the bladder through the penis. If the spread to the bladder recurs repeatedly and becomes extensive, you may lose your bladder, but this usually takes years. You'll have lots of living to do in the meantime, and it doesn't happen that often anyway. Finally, a few of these tumors spread via the lymphatics and bloodstream and go to distant organs. That is the worst possibility because the tumors do not respond to chemotherapy or irradiation and nothing can be done about them. I have my fingers crossed for you.

Yours sincerely,

Fred

Dear Charlie,

This get well card is sent with a great deal of relief. Gloria told me on the phone last night that your tumor was a grade 1, stage 1 noninvasive tumor. If you have to have a cancer, this is the best one and we are all optimistic you have been cured by surgery. Since you and I have always leveled with each other, I feel I should tell you that there is still a chance the tumor may have spread, but it is remote and with any kind of luck you're home free. I'm going to be in the Rockies this September. Let's go trout fishing!

Fred

Author's note

Fred's second letter to Charlie says it all. The degree of differentiation of tumors is graded 1 to 4 with 1 the most differentiated and least aggressive. Grade of tumor plus stage (in this case noninvasive) sets the prognosis as excellent. The combination of stage and grade together gives a more accurate prognosis than either alone (tumor grades and stages are discussed in Chapter 1).

Squamous cell cancer: Chapters 1 and 2

Dear Eileen,

I am writing to tell you about Grandpa, who is having a terrible time. Ever since Gram died four years ago he has not been quite right, if you know what I mean. He had this sore on the inside of his gum and refused to have it looked after. He claimed it was caused by his dentures. So he whittled and sanded his dentures with his knife, but the sore did not go away. We coaxed him to go to the doctor, but he refused. After about a year of tinkering with his dentures, they finally broke. The dentist told him he had cancer of the gum – a squamous cell cancer. It's strange, but that's a skin cancer. Well, he'd figured out all along he probably had cancer, but he didn't want to know. He claimed Gram was okay until they told her she had cancer, and he was damned if he would let them do it to him.

The cancer has spread to lymph glands in his neck and to the floor of his mouth and jaw. They told Grandpa they were going to do radical surgery to take out part of his tongue, jawbone on the right side, and all of the lymph glands in his neck. "The hell you say," said Grandpa. "You can't cure me, and you aren't going to cut on me."

The doctors have really been working on me to get him to submit to surgery. They say, in the first place, if he had gone and had this looked after when the sore first developed, he could have been cured with very little effort. Now they say if they do radical surgery, they can spare him a lot of discomfort and even if they take out his jaw, they can rebuild another one.

He smells pretty bad because of the infection in the tumor. Some of it is decaying, and it would appear we are in for a bad time. Grandpa just looks me in the eye and says the medical profession is after his money and he would rather give it to his grandkids than to those lousy doctors. He says he knows it is going to get bad for him, and when it gets too bad he will just cash in his chips. I don't know what to do with him, but I think it might be a good idea if you came to visit while he is still able to do the things he enjoys.

Yours sincerely,

Nelly

Author's note

In this situation the patient delayed seeing a physician because he was afraid he might have cancer. Thus he will die of a disease that, under ordinary circumstances, is curable with modern therapy. Note his strength of will, characteristic of many elderly people who have been through the school of hard knocks. It would be interesting to know the events at the terminus. Did he lose his resolve and accept therapy? Many people say they will refuse therapy, but when they are faced with dying, they often opt for treatment even though they know it offers little chance for cure. Note the commonalities: growth, invasion, and destruction of tissues, and distant metastasis (Chapter 2). Weight loss will follow, then infection and death (Chapter 1).

Testicular cancer: Chapters 1, 2, 9, and 10

Dear Dennis,

This letter is to thank you for your advice and support during our trials and tribulations with my son's testicular teratocarcinoma. Do you remember in medical school how the professors teased us when they lectured about testicular cancer and how they made light of such a grim disease? Never did I think I would come face to face with the realities in my own family. Bill is twenty-six years old. We were at the lake and he told me he had a large testicle, but it didn't hurt, so he had not done anything about it. It proved to be a teratocarcinoma that had metastasized. We did a retroperitoneal lymph node dissection and then treated him intensively with chemotherapy. He has been disease-free for two years, which means he is almost surely cured.

He had some side effects of the chemotherapy: he lost his hair – about which he was embarrassed – and although he can have sexual relations, he has no ejaculate. Luckily, he and Marge had the baby before he got sick.

I look back over this nightmare and realize in a sense how fortunate we are. He was cured with modern chemotherapy, which did not exist ten years ago. I guess temporarily losing your hair and having "dry" ejaculations is not too heavy a price to pay for life.

Your help and counsel were much appreciated,

Sam

Author's note

This individual was in the typical age group for a teratocarcinoma (twenty to thirty-three years) of the testis. He was probably embarrassed to seek medical advice, which caused delay and the development of a large tumor. Because these tumors have a tendency to metastasize (Chapter 2) via the lymphatics along the aorta, the surgeons removed all of these lymph nodes and any tumor that might have spread into them. Then the patient was given massive doses of chemotherapy with the attendant side effects, which occur because the toxic chemicals lack specificity. Even so, such cures are among the marvels of modern chemotherapy (Chapters 9 and 10), and over 70 percent of such patients are cured. The cause(s) of the differential responses of tumors to chemotherapy has not been a high priority of the medical establishment.

These tumors are extremely rare, but research on them has clarified many problems in oncology (Chapter 1).

Stomach cancer

Dear Mom,

I am enjoying my first year of residency very much, and after all the stainless steel and scientific medicine we learned it is fun to see some of the old-timers practice.

An old Scot came into clinic the other day not feeling very well with some vague upper abdominal pains that sounded like dyspepsia. I was going to give him some medicine to tide him over the weekend when the attending physician came in. The doctor is an equally dour old Scot, and he learned that after eating porridge every morning for fifty years, this patient no longer had a taste for oatmeal. He winked at me and said we better work him up. Well, we worked him up exhaustively and found he had a tiny adenocarcinoma of his stomach. People with this disease may lose their taste for a favorite food, and this old doctor knew about it. Well, we operated on him. Apparently the tumor had not spread, and we think he is going to be one of those lucky people who beats stomach cancer. When you consider that only about 10 percent of patients with stomach cancer survive, you see how lucky he really is. Old Doc MacAllister just winks and says it's all in a day's work. He got the tip-off about the oatmeal because he spends a lot of time with his patients, talks to them, teases them. I don't see how he can make a living, but he sure is a good doc.

With love,

Jennifer

Author's note

This is a clear example of good luck in life. The patient came under the care of an old-time physician who practiced the art of medicine and spent adequate amounts of time with his patients to know and understand them. The physician picked up an apparently trivial point in the clinical history which raised the possibility that the patient had stomach cancer, a disease with a cure rate of less than 10 percent. He followed through with a vigorous and thorough workup and discovered a small curable cancer of the stomach. It had not progressed to a stage where it could spread. This is clinical medicine at its best.

Melanoma: Appendix A.8; Chapters 2, 3, 7, and 8

Dear Jill,

I have the strangest story to tell you, and at the beginning I must stress how lucky I am. I went to the doctor three months ago for a Pap smear. Because I am a redhead and live in sunny California, she always checks my moles. I have about four, the largest about the size of your thumbnail. One is on my back just above my bikini top and the others are on my legs. Anyway, the one on my thigh had changed color and it itched a little. It had been a light brown and now it had a dark and light area. The doctor was worried it was becoming malignant.

She took it off, and the report from the pathologist was superficial melanoma. It had not spread at all and if you have to have a melanoma, this is the best kind. Apparently, melanomas spread widely, but mine had no evidence of invasion and the odds are better than three to one that I am cured. I prefer odds of 10 to 1 or 100 to 1, but with this kind of disease you take what you can get.

Sincerely,

Amy

Author's note

Most people have a dozen or more pigmented spots on their bodies. Because a significant number of melanomas (Appendix A.8) develop from such spots, the problem is to know which ones to treat. Redheads typically lack the pigment that protects skin cells from ultraviolet light, a carcinogen, and are thus at risk for developing melanoma (Chapters 3, 7, and 8; melanoma mortality is geographically nonrandom, see Color Plate 16). Unfortunately, too few risk factors are known for malignant disease, and more and better approaches to cancer epidemiology are needed. Changes in existing moles may be early manifestations of malignant change and warrant prompt treatment. Had this tumor infiltrated even 1.5 mm, instead of a three to one chance for cure, the odds would have been three to one in favor of death. Clearly, we need more understanding of why and how malignant cells invade and metastasize (Chapter 2). The usual noteworthy changes in pigmented lesions are itching, pain, and increase in size, all of which usually indicate the lesion is invading. The trick is to diagnose melanoma early.

Neuroblastoma: Appendix A.10; Chapters 1, 2, 9, and 10

Dear Andrea and Bob,

You did not get a Christmas card from us last year because we were overwhelmed with Gilbert's problems. But there is good news after the earlier bad news.

A year ago November, Gilbert, our bouncy nine-month-old, had a lump in his stomach. I felt it while I was bathing him. It turned out he had a malignant tumor of his adrenal gland, called a neuroblastoma. Well, it was removed early in December last year and although it seemed to be confined to the adrenal gland, the doctors, just to be sure, gave him chemotherapy. He was awfully sick for a while, but now he is healthy and his CAT scans and everything are okay. We have been assured he is one of the lucky 40 percent or so who are cured with modern treatment.

I was worried about the treatments and what they would do to him, but he seems okay.

Sincerely,

Sandra

Author's note

Neuroblastoma (Appendix A.10) is the most common tumor of children in the United States, and although it can present with fever and pain, it is more commonly found by the parent while bathing the child. These tumors are rapidly growing, invasive and can involve the adjacent organs, such as kidney and liver, by direct invasion, and they can metastasize (Chapter 2) via the bloodstream. For reasons unknown, tumors found in very young children have a better prognosis than those that develop in older ones. Prior to chemotherapy (Chapters 9 and 10) the outlook was poor, but with modern chemotherapy about 40 percent of the children can be cured.

Spontaneous regression (Chapter 1) has been reported in neuroblastoma, but its occurrence is so rare that it offers no hope for the individual.

Summary

These few letters illustrate some of the reactions of human beings to the multiplicity of diseases known as cancer. Cancer can occur in almost any tissue of the body, and because different signs and symptoms are produced by derangement of the organs, the tendency is to view cancer not as a single entity but as a series of diseases. It is no wonder that people are terrified by the diagnosis of cancer. Early diagnosis and extirpation of the tumor before it has invaded and spread offer the best hope for cure. Once cancer has spread, with only a few possible exceptions (e.g., acute lymphatic leukemia, choriocarcinoma, testicular cancer, Hodgkin's disease), chemotherapy and radiation therapy are not curative, merely palliative. Their initial positive effects, which provide freedom from pain and provide useful life, are all too soon lost as the tumor becomes resistant.

Note the commonalities of malignant tumors in various locales: a mass that grows and becomes progressively more malignant, and alterations in differentiation of the cells of the mass that correlate with rapid growth, invasion, and metastasis of the undifferentiated cells with destruction of normal cells and tissues. Weight loss (cachexia) and the inability to mount good immune and anti-inflammatory responses result in death usually by infection.

These commonalities point to the deficiencies in our knowledge of cancer. We must understand growth regulation of cells to be able to control the growth of the mass. Understanding the mechanism of progression could lead to means of preventing tumors from becoming more and more malignant with time. Understanding the mechanisms of invasion and metastasis could prevent spread. Understanding cachexia could provide useful life. Development of specific therapies could lead to cure. Because society cannot afford to treat all patients with cancer, however, there is an imperative need to identify causes of malignancy and thus minimize or prevent this dread disease. These are among the issues discussed in this book.

1

The pathology of cancer

G. Barry Pierce and Ivan Damjanov

1.1 Introduction

Our first task is to provide you with a working knowledge of the pathologic terms and concepts used throughout the text. This chapter defines terminology, compares and contrasts malignant and benign tumors, considers characteristics and behavior of malignant cells, and discusses how invading malignancies kill an individual. Tumors, with time, undergo changes that lead to autonomy. This progression of events is also examined.

An appreciation of embryology leads to a consideration of the origin of stem cells and the concepts of determination and differentiation. Both are important to understanding the origin of cancer cells, and such comprehension may lead to new modalities for treating cancers. Most textbooks of cancer biology begin with a discussion of cells. But we start with an examination of what cancer is to help you get a better grasp of the material that follows. Metastasis is difficult to understand without a prior foundation in the concepts of pathology. Similarly, carcinogenesis or chemotherapy is incomprehensible without knowledge of what a malignant cell is and how it behaves.

Much of our knowledge about tumors dates from antiquity. The streaks of hard gray tissue that extend from a tumor into the normal tissues reminded the Ancients of a crab, so they named the condition cancer (from the Greek word meaning crab). The term "tumor" denotes a mass, whether neoplastic, inflammatory, pathologic, or even physiologic. Today, tumor is used generically to describe any neoplasm, irrespective of its origin or biologic behavior. The term "cancer" is generic for any malignant tumor.

Willis (1967) defined a neoplasm as a mass, the growth of which is incoordinate with the surrounding normal tissues and that persists in the absence of the inciting stimulus. It is worthwhile considering this definition in detail. First, the mass, like any other tissue, is composed of parenchymal cells and stroma, which are the essential parts of an organ. The parenchymal cells of the mass may be well differentiated, organized as normal tissues, and proliferate slowly, or they may be poorly differentiated, rapidly proliferating, and have little or no organization. In either situation, the host is induced to supply a stroma for it. This host response is mediated by angiogenic factors (Folkman 2002; 2003), which are synthesized by the parenchyma of the tumor and stimulate proliferation of all stromal cells, including fibroblasts and vascular cells.

Accrual of mass could be the result of a decrease in cell cycle time, the period of time required for a cell to make the arrangements for cell division and to divide. More rapid cell cycles (i.e., decreased cell cycle time) would generate more cells. However, most neoplastic cells *do not* cycle faster than their normal counterparts. For example, 40 kg of gastrointestinal cells and 10 kg of white blood cells are produced annually by a human of average size (Donald Coffey, personal communication). This is a prodigious effort in cell replication when we consider that the fetus in utero requires nine months to achieve 4 kg of weight. In contrast, a patient may die harboring a tumor of less than 5 kg that took several years to develop.

Thus, tumors increase in size not because the tumor cells cycle faster than normal cells but because so many tumor cells are cycling. This fact has implications for cytotoxic chemotherapy, which is often designed to kill dividing cells. Cytototxic drugs designed to kill cells by interfering with DNA synthesis destroy tumor cells but also kill proliferating normal cells. Destruction of normal white blood cells and gastrointestinal cells may result in infection, bleeding, and diarrhea. Fortunately, normal cells recover faster from the poisoning than cancer cells do, so the clinician administers the cytotoxins in cyclic fashion, hoping to rescue the normal cells while achieving a cumulative toxic effect on the tumor.

The definition of neoplasm by Willis (1967) also states that the mass persists in the absence of the inciting stimulus. This is an important consideration and distinguishes the neoplasm from the modulations of growth that also result in changes in mass. These modulations are considered to be normal cellular responses to environmental stimuli, and they persist only as long as the environmental stimulus is present.

Some tissues are capable of renewing themselves: as normal cells become senescent and die, the cells responsible for renewal, known as stem cells, proliferate in a controlled manner to replace the precise number of cells lost. Stem cells are clearly seen in the basal layer of the normal squamous epithelium of the skin (Figure 1–1). In the presence of certain environmental stimuli, these tissues can become hyperplastic. By definition, the hyperplastic tissues or organs are larger than normal because of an increased number of normal cells, and they remain hyperplastic as long as the inciting stimulus is present. If the inciting stimulus is removed, the organ returns to normal size. As an example, the cells of the prostate gland respond to administration of testosterone by dividing, and the gland becomes enlarged (hyperplastic). Upon withdrawal of the hormone, the gland atrophies and returns to normal size. The

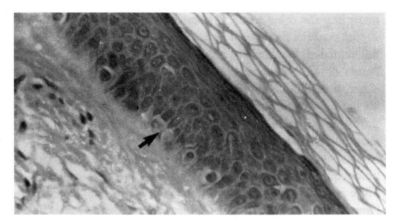

Figure 1–1. Photomicrograph of normal squamous epithelium from skin. The stem cells are located in the basal layer (*arrow*). Above the arrow, keratinocytes differentiate from their cuboidal form and mature progressively in an orderly manner to form flat surface keratinocytes lacking nuclei, which then form keratin. The basal layer of cells rests on the basement membrane, which separates epithelium and stroma, but is invisible unless special stains are employed. Aside from the layers of keratin on the surface, this is identical to squamous metaplasia.

breast undergoes hyperplasia during pregnancy, and after delivery and nursing the gland returns to normal size. Atrophy, the converse of hyperplasia, is an acquired reversible decrease in size of an organ as a result of a diminution of size and/or number of cells in the organ. In addition to withdrawal of a hormone, atrophy can be caused by a variety of factors including reduction of blood supply.

Other important modulations in cells are not necessarily associated with changes in mass, but rather with changes in differentiation. Metaplasia is a case in point: when an epithelium, for example in the respiratory tract, is chronically injured (as with use of tobacco), it may change from respiratory epithelium to squamous epithelium. The respiratory epithelium has mucous cells and ciliary cells (Figure 1–2). The latter sweep mucus containing bacteria and garbage to cough-sensitive areas in large bronchi where it is expelled by coughing. This is the house-cleaning mechanism of the airway. The metaplastic squamous epithelium, although normal in all respects but its location, is not ciliated and cannot provide this service. As a result, mucus laden with dust and bacteria accumulates behind the patches of squamous epithelium and leads to infection. The squamous metaplasia that occurs at the bifurcations of bronchi in cigarette smokers can undergo complete reversion to normal respiratory epithelium over a period of time after cessation of smoking.

Thus, metaplasia is a reversible change in phenotype in response to environmental stimuli. If the adverse influences from the environment persist the orderly arrangement of squamous cells becomes disturbed and the normal maturation of these cells is interrupted. This change is called dysplasia (Figure 1–3). Continuous exposure to carcinogenic substances in tobacco smoke will ultimately transform

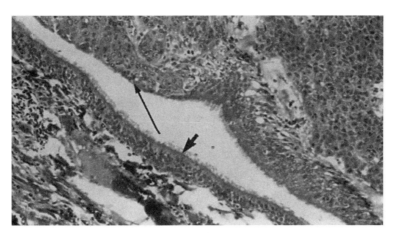

Figure 1–2. Photomicrograph of respiratory epithelium. Note ciliated cells that line the airway (*short arrow*). A large squamous cell carcinoma is growing, invading, and destroying the mucosa (*long thin arrow*). With destruction of the mucosa, the housekeeping mechanism of the airway is destroyed, predisposing to infection.

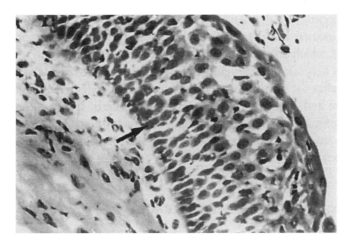

Figure 1–3. Photomicrograph of a frozen section of dysplastic squamous epithelium. The arrow points to the basal layer. Whereas in the normal situation there is orderly differentiation of basal cells to keratin (see Figure 1–1), in dysplasia the order is mixed up. Poorly differentiated cells are present near the keratin layer.

the dysplastic epithelium into malignant cells (Figure 1–4). Initially such cancers are limited to the mucosa and separated from the underlying connective tissue by a basement membrane. Intraepithelial, noninvasive carcinoma is called carcinoma in situ. Early cancers of this type are curable, but if left untreated will progress to invasive cancer, which will ultimately spread through the normal tissues and kill the host. If you look carefully, you may see such an invasive cancer in the upper part

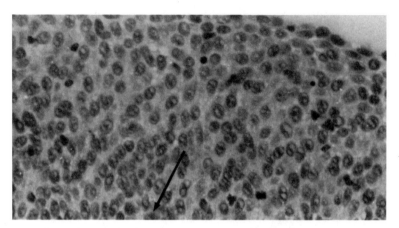

Figure 1–4. Photomicrograph of carcinoma in situ. Unlike the dysplastic specimen (Figure 1–3), no differentiation from basal (*arrow*) to superficial layer is evident. There is no evidence of invasion into the peribronchial tissue. In other words, the basement membrane has not been penetrated by the anaplastic cells.

of Figure 1–2, as it is invading the wall of the bronchus. This sequence of events, encompassing squamous metaplasia, its transition to dysplasia, and progression to squamous cell carcinoma, provides the explanation for the occurrence of squamous cell carcinomas in organs like bronchi, which under normal circumstances do not contain squamous epithelial cells.

A neoplasm is distinguished from the reversible cellular modulations by persisting after the inciting stimulus has disappeared, which makes it difficult to establish cause and relationship effects in carcinogenesis. It also means that, unlike most inflammatory or infectious diseases in which the etiologic agent can be ascertained and treated, in neoplasia, the inciting agent is long gone, so the resultant mass of cells is the target of treatment. This is an important point.

1.2   Benign versus malignant tumors

Benign tumors are slow-growing neoplasms called so to indicate that they are "of gentle disposition, or kind and innocuous," as stated in most dictionaries. Since the tumor cells divide rarely, mitoses are not usually seen in histologic sections. These tumors are composed of well-differentiated cells that closely resemble those of the normal tissue. For example, lipomas are subcutaneous tumors composed of mature fat cells. Similar tumors composed of mature, well-differentiated fat cells can occur in internal organs, as well (Color Plate 1). Hemangiomas of the skin are composed of blood vessels with normal-appearing endothelial cells. Most of these benign tumors form nodules or lumps that are of no clinical significance (Color Plates 1 through 3).

Figure 1–5. Photograph of a pheochromocytoma of the adrenal medulla. The adrenal cortex (*medium arrow*), which synthesizes cortisone among other endocrines, overlies the medulla (*long arrow*), which secretes epinephrine. Note the tumor, which lies in the medulla. It has grown, compressed, and thinned the cortex, the cells of which have progressively atrophied, died, and their stroma has formed a capsule (*short arrow*). This tumor secreted epinephrine in toxic amounts, causing acute episodic hypertension.

Many benign tumors synthesize the gene products specific to the differentiated state and also secrete proteins normally produced by fully differentiated cells (often termed luxury molecules because they are not essential for the survival of cells, but are essential for the well-being of the organism). Luxury molecules may be demonstrated in the tumor cells by special histochemical procedures, and often may be found in the circulating blood. Under normal conditions, production of molecules for specialized function is carefully controlled to meet the needs of the host, but a benign tumor produces these molecules with no regard for the needs or safety of the host. Benign tumor cells synthesize fewer of these molecules per cell than those of the normal tissues, but when the tumor becomes large, toxic amounts of the molecules can be synthesized. For example, a benign tumor of the cells of pancreatic islets of Langerhans may secrete enough insulin to cause hyperinsulinemia ("insulin overdose"), resulting in hypoglycemia and death. Similarly, a benign tumor of the adrenal medulla may produce the hormone epinephrine. The extremely high blood pressure induced by the hormone can result in death. In childhood a benign tumor of the pituitary gland may synthesize enough growth hormone to cause the child to grow into a giant.

As benign tumors grow, they often compress the adjacent normal tissues. This is an expansive type of growth that pushes the normal tissue ahead of it and compresses the thin-walled capillaries of the normal parenchyma. With insufficient blood to nourish the normal parenchymal cells, atrophy results. As the normal cells atrophy and eventually die, only the connective tissue stroma of the normal tissue remains. The stroma is compressed and forms a capsule around the tumor (Figures 1–5 and 1–6a). In most instances the well-encapsulated benign tumors pose no health

(a)

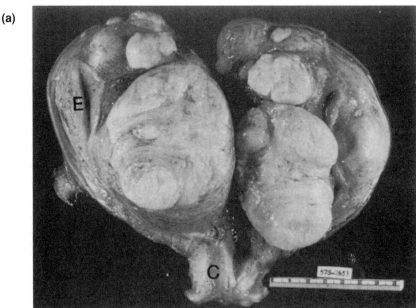

(b)   **(c)**

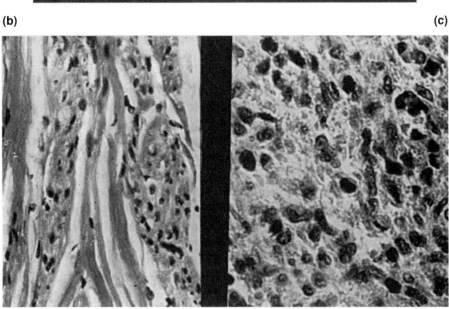

Figure 1–6. (a) Photograph of uterus opened to display multiple leiomyomas (also see Color Plate 3). Note compression of surrounding muscle and displacement of endometrial cavity by the noninfiltrative growth characteristic of benign tumors. Such tumors may cause problems during pregnancy, excessive bleeding at or between menses, but they are not life threatening to the host. Note cervix at C and endometrial cavity at E.
(b) Photomicrograph of a leiomyoma (benign tumor of smooth muscle) of the uterus. The smooth muscle cells, normal in appearance, are arranged in organized muscle bundles. The features of differentiation are contained in the abundant cytoplasm. (c) Photomicrograph of a leiomyosarcoma (malignant tumor of smooth muscle) of the uterus taken at the same magnification as the specimen illustrated in (b). Note the large and variable size of the nuclei, lack of differentiation of cell cytoplasm, and lack of organization of cells into muscle bundles. This specimen illustrates many of the features of anaplasia.

20

problems. Exceptionally, however, a benign slow growing tumor may be located close to a vital center and by compressing it endangers life. For example, meningiomas, the slow-growing tumors in the meninges (the coverings of the brain), may eventually cause death of the host by pressure atrophy and destruction of vital centers (Color Plate 2).

The diagnosis of benign tumors may be made on gross examination, because these tumors appear well circumscribed from adjacent tissue (Color Plate 3), but the final diagnosis is made histologically. Typically, the pathologist will report that the tumor is composed of uniform well differentiated cells that show no atypia and no mitotic activity.

The cells of malignant tumors have the intrinsic tendency to grow uncontrollably, invade (Color Plate 4) normal tissues, metastasize (Color Plates 10 and 11) and kill the host unless properly treated. In contrast to benign tumors (Figure 1–6b), the malignant cells divide often and accordingly mitotic figures are readily found in the histologic sections of these tumors (Figure 1–6c). The cells are pleomorphic, that is, vary in size and shape. The nuclei may be very large or small, but almost always differ from those of the normal cells in the tissue of origin. These nuclei are thus called "atypical" and "hyperchromatic" because they stain dark blue with hematoxylin in routine histologic sections. Malignant tumor cells often have prominent large nucleoli, and the chromatin is irregularly distributed within the nuclei.

Malignant tumor cells are programmed for proliferation and have a high nucleus to cytoplasm ratio. Normal liver cells which have three to four times more cyto-plasm then nuclear material (a nucleocytoplasmic ratio of 1:3 or 1:4), a malignant liver cell might have only scant cytoplasm and a nucleocytoplasmic ratio of 1:1.5. The scant cytoplasm surrounding the nucleus contains few organelles needed for the maintenanace of the differentiated state. DNA and RNA transcription and translation results in protein synthesis, which takes place in the cytoplasm and is mostly mediated by free ribosomes arranged into polysomes. In contrast to benign tumors and normal cells which contain abundant rough endoplasmic reticulum, reflecting their synthesis of luxury proteins for export, the cytoplasm of malignant undifferentiated tumor cells contains polysomes that are unattached to membranes and synthesize structural proteins required for cell division.

Malignant cells are less well differentiated than their benign counterparts, some so poorly that they defy histopathologic identification. Such tumors are said to be anaplastic ("anaplasia" literally means a condition without form, but to the pathol-ogist it represents the summation of all of the microscopic attributes by which malignant tissue is diagnosed; compare Figures 1–6b and 1–6c). The anaplastic cells invade and destroy the normal architecture of the organs, and replace it with masses of disorganized malignant cells (Figures 1–7, 1–8, and 1–9). For example, a well-differentiated adenocarcinoma may produce glandular epithelium, but the epithelium lacks the normal relationship with stroma (Figure 1–9). A less differ-entiated adenocarcinoma might make solid plug-like masses of tumor cells with-out glandular lumens. An even less differentiated one may have columns of single

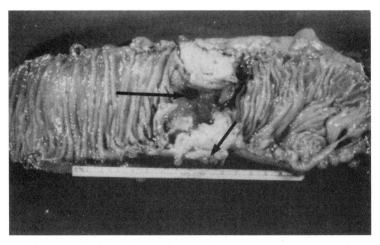

Figure 1–7. Photograph of an adenocarcinoma of the colon. The bowel has been opened along its long axis. Note the napkin ring appearance imparted by the adenocarcinoma that has invaded the mucosa, forming a malignant ulcer (*long arrow*) and through the bowel wall and into the fat (*short arrow*). This stage of malignancy carries a poor prognosis.

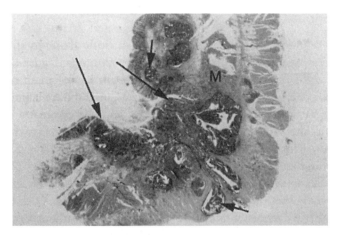

Figure 1–8. Photomicrograph taken at low power of an adenocarcinoma of the colon. Normal mucosa at the top is disrupted by a "malignant" ulcer (*between long arrows*). The base of the ulcer is composed of anaplastic cells in a fibrous stroma. Clumps of these cells (*short arrows*) have invaded beneath the epithelium at the margin of the ulcer, deep into and through the smooth muscle (M) and into the adjacent fat of the serosa. This stage of the disease has a poor prognosis.

anaplastic epithelial cells forming rows of cells that penetrate between the normal stromal cells. Finally, an adenocarcinoma may make no such organized structures, and the individual anaplastic epithelial cells would be recognized as adenocarcinoma cells only because some of them contain mucin, a marker for glandular epithelium (Figure 1–10) (see Section 1.7).

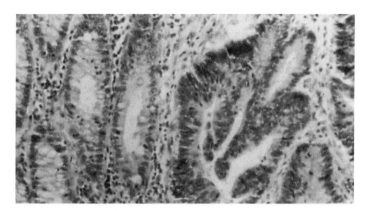

Figure 1–9. Photomicrograph taken through the margin of a low-grade adenocarcinoma of the colon. Normal glands are on the left and illustrate the orderly arrangement of the cells with basally placed nuclei and well-differentiated cytoplasm forming regular-shaped glands. Contrast this with the appearance of glands to the right where multiple nuclei are piled up, the cytoplasm is darker staining, and the glands are irregular in shape. These are evidences of anaplasia, the hallmark of malignancy.

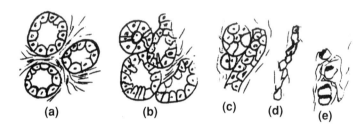

Figure 1–10. Degrees of anaplasia commonly found in adenocarcinoma cells. (a) Normal acini with regularly shaped cells and connective tissue stroma between acini.
(b) Adenocarcinoma, low grade, with slightly irregular acini with some variability in cell shape, size, and number. Note glands are back to back with no stroma between them.
(c) Adenocarcinoma high grade. Note that the cells are unable to arrange themselves in glandular acini and form plugs of anaplastic cells. (d) A further degree of anaplasia; the cells cannot form plugs, only columns of single cells. (e) The most anaplastic tumor; the cells cannot form single-cell columns faintly reminiscent of their epithelial origin. They make mucin, however, a marker for epithelium.

Contrast this appearance with that of an adenomatous polyp of the colon (Figure 1–11). These benign masses grow into the lumen of the colon, are composed of normal appearing cells, and the patient is cured by simple removal. Malignant change can occur in these lesions, but if the abnormal cells have not invaded the stalk of the polyp, again the patient is cured by removal of the polyp, its stalk, and a small cuff of normal mucosa. If not treated, the malignant cells will invade not only the stalk, but the bowel itself. A malignant ulcer results. Further invasion leads to obstruction of the bowel (see Figure 1–7).

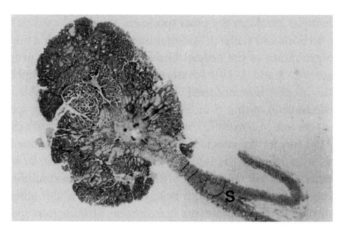

Figure 1–11. Low-magnification photomicrograph of adenomatous polyp illustrating the entire lesion. No invasion into the stalk means that cure has been achieved by local removal.

Large areas of necrosis are apparent in rapidly growing tumors. Cell death in such tumors was attributed to rapid tumor growth with inadequate blood supply. However, dead cells were often found adjacent to blood vessels. It is now known that much of the cell death is due to apoptosis (see section 1.16).

1.3   The diagnosis of benign and malignant tumors

The diagnosis of tumors is based on empirical analysis of clinical and pathologic data. Success is based solely on the knowledge, skill, and experience of the physicians assembled into a diagnostic team. Most often the team members include oncologists, radiologists, surgeons, and pathologists. The pathologist, given a specimen, is expected to render a diagnosis and define a prognosis for the patient. In addressing these tasks, every bit of information concerning the patient that can be elicited is used. The history of the patient, including age and sex, the presenting signs and symptoms, the clinical diagnosis, the organs involved, and so on: all are considered. Often a biopsy is taken while surgery is being performed because the surgeon wishes to know if the lesion is benign or malignant. The pathologist examines the gross specimen, freezes selected portions of it in a cryostat, sections and stains it, and examines the tissue with a microscope. Evidence of disorganization of the normal histologic architecture by neoplastic cells is sought (see Figures 1–8 and 1–9). Invasion into capillaries or lymphatics suggests distant metastasis may have occurred. The diagnosis, if it is possible to make one, is given to the surgeon within five to ten minutes of receipt of the specimen. The surgeon then proceeds in a manner consistent with the pathologist's diagnosis. If the pathologic diagnosis cannot be made on frozen sections, the final diagnosis must be delayed until the entire specimen is embedded and carefully sectioned. Tissue sections from formalin-fixed, paraffin-embedded tumor are stained with hematoxylin and eosin and with special stains developed to provide a better insight into the nature of the tumor than one could obtain with routine stains. Immunohistochemistry, a method based on the application of specifically labeled antibodies to cell components and molecules is often used for fine-tuning the diagnoses. Likewise, molecular biology is used more and more for assessing tumors and for choosing the appropriate therapy.

After surgery the resected tissue is preserved and examined grossly. The pathologist notes whether the tumor is confined to the organ in which it originated or whether it has penetrated the organ and invaded surrounding tissues and organs (see Figure 1–7). Regional lymph node involvement is also noted. These findings are confirmed by the microscopic examination of numerous specimens taken during the gross examination. The microscopic examination allows the tumor to be graded and confirms the stage of spread of the tumor.

1.4   Tumor grading and staging

Tumor staging is based on the clinical and pathologic assessment of the size of the tumor and its spread into the normal tissues. Staging includes data obtained during the clinical examination of the patient obtained by X-ray examination as well as during surgery. The final staging is performed on the basis of gross and microscopic data provided by the pathologist. The gross examination of the specimen aids in assessing the stage, but microscopic examination is also important because the extent of invasion can be assessed more accurately and the early involvement of small blood vessels, lymphatics, and lymph nodes can only be seen with the microscope (Figure 1–12). The patient's prognosis deteriorates progressively if the tumor has invaded the stroma of an organ but has not penetrated it. The prognosis is worse if the wall is penetrated (see Figures 1–7 and 1–8) and even worse than that if spread to lymph nodes has occurred. It is dismal if there are distant metastases (Figures 1–14 and 1–15 and Chapter 2).

The most widely used staging system is called the TNM system because it takes into consideration the size of the tumor (T), the presence or absence of metastases in the lymph nodes (N), or distant metastases (M). The TNM system is currently used worldwide, enabling clinical scientists from various institutions to standadize the staging of tumors and compare therapeutic results. A small carcinoma of the breast measuring 2 cm, that has not metastasized is designated T1N0M0. In contrast, a tumor of the same size that has already metastasized to the lymph nodes and bone will be staged as T1N1M1.

In the TNM system, the size of the tumor will determine whether the tumor will be designated as T1, T2, or T3. There is also a stage T0, which refers to

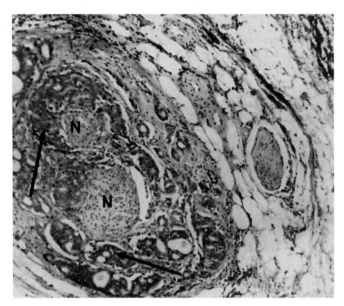

Figure 1–12. Photomicrograph illustrating invasion of cells of an adenocarcinoma of the prostate (*arrows*) along a nerve tract (N). This is a poor prognostic sign, especially when it occurs in the capsule of the gland.

carcinoma in situ, that is, preinvasive cancer that has not produced a measurable mass. Carcinoma in situ of surface epithelia, such as the cervical epithelium of the uterus or oral mucosa, is a neoplastic stage in which the tumor cells are confined to the epithelium of origin and have not transgressed the basement membrane to invade the capillaries, venules, and lymphatics of the stroma. This is the cancerous stage with the best prognosis. These lesions are cured by extirpation.

Tumor grading is based on histologic examination of tumors and is a semi-quantitative assessment of the malignancy of each tumor. In essence, the pathologist compares the tumor with its tissue of origin in an attempt to determine whether the tumor resembles it, that is, is well differentiated, or it markedly differs from the normal tissues and shows a high degree of anaplasia. In this process, the pathologist will assess the individual cells ("cytologic atypia") and the architectural organization of the tumor tissue ("architectural atypia"). In most instances, the tumors can be graded on a scale from 1 to 3, and are designated as well differentiated (grade 1), moderately well differentiated (grade 2), or poorly differentiated (grade 3).

In clinical practice, the tumor grading and tumor staging data are combined before additional therapy is recommended. The stage of the tumor at the time of diagnosis is also the most important prognostic parameter for most human tumors; it can generally predict whether a tumor can be cured or not.

Tissue/cell of origin	Benign	Malignant
Epithelial tissue	Adenoma (e.g. of colon)	Adenocarcinoma (e.g., of colon)
Mesenchymal tissue	Lipoma	Liposarcoma
	Osteoma	Osteosarcoma
Hematopoietic/lymphoid cells		Lymphoma
		Leukemia
Embryonic cells/organ primordia		Retinoblastoma
		Nephroblastoma (Wilms' tumor)
Germ cells	Teratoma	Teratocarcinoma
		Seminoma

Figure 1–13. Chart of nomenclature.

1.5 Classification and nomenclature

Tumors are named according to the tissue or organ in which they originate, and then a suffix is added to denote whether the tumor is benign or malignant (Figure 1–13). The suffix *-oma* literally means "a tumor of" and is construed to mean a benign tumor. Accordingly, a fibroma is a benign tumor of fibrous connective tissue. Lipomas (Color Plate 1), chondromas, and neuromas are benign tumors of fat, cartilage, and neural tissue, respectively. Adenoma is a benign tumor of glands, irrespective of the glandular organ in which the tumor develops. For example, an adenoma of the breast is a benign glandular epithelial tumor of the breast (see Figure A–3).

Three suffixes identify malignant tumors: *-carcinoma, -sarcoma,* and *-blastoma.* The suffix *-carcinoma* implies a malignant tumor of epithelial origin; for example, carcinoma of the lung (Color Plates 5–10) signifies a malignant tumor derived from epithelium of the lung. Adenocarcinoma of the breast signifies a malignant glandular epithelial tumor of the breast. Squamous cell carcinoma of the skin is a malignant tumor of squamous epithelium of the skin. Adenocarcinoma of the colon is a malignant tumor of the glandular epithelium of the colon, and so on. *Sarcoma* is the designation given to malignant tumors of mesenchymal tissues. For example, osteosarcoma is a malignant tumor of bone-forming cells, and chondrosarcoma (Color Plates 12 and 13), and liposarcoma, respectively, are malignant tumors of cartilage and fat cells. Finally, a group of highly malignant childhood tumors is denoted by the suffix *-blastoma.* These include neuroblastoma, originating in the neuroblasts of the adrenal medulla; retinoblastoma, originating in the retina of the eye; and nephroblastoma, originating in embryonic cells of the kidney (Figures A–16 and A–17a).

Some malignant tumors cannot be classified as carcinomas, sarcomas, or blastomas. Most important among these "other tumors" are brain tumors which originate from glia cells and are thus called gliomas. Due to their location inside the brain these tumors cannot be completely removed and thus all gliomas are in essence clinically malignant.

Malignant tumors originating from lymphocytes are called lymphomas. Although their name sounds deceptively benign, it is worth notice that there are no benign lymphomas, and all tumors in this category should be considered malignant. Likewise, germ cell tumors composed of cells resembling the seminal cells of the testis have a benign-sounding name – seminoma, but all of them are malignant. Malignant melanoma, a common malignant tumor of pigmented cells of the skin (melanocytes) is colloquially known as "melanoma," which might confuse some patients and make them believe that they do not have a malignant tumor. It is worth remembering that all melanomas are malignant, even if the ominous-sounding qualifier "malignant" was left out.

Reflecting our ignorance or out of historical considerations some tumors have eponymic names, that is, are known under the name of the physician who has described them first. Ewing's sarcoma is a poorly differentiated sarcoma of bones and soft tissue. Since the cell of origin of Ewing's sarcoma has not been identified, there are no alternative names for this neoplasm. On the other hand, Wilms' tumor is a widely used name for nephroblastoma, a childhood renal tumor; both terms are used interchangeably. In some cases, these medical eponyms are a godsend: Kaposi's sarcoma is definitely easier remembered and pronounced than the original name proposed by this famous nineteenth-century Viennese dermatologist who described and named it in Latin, "idiopathic pigmented hemorrhagic sarcoma."

1.6 Metastasis

Malignant cells have a capacity to grow along tissue spaces, especially along the nerves (see Figure 1–12). Tumor cells enter the lymphatics, venules, and capillaries and spread through the body cavities. Single cells or small clumps of tumor cells then break away from the original mass and are carried to distant organs, where they implant. This process of dissemination of malignant cells is called metastasis (see Color Plates 10 and 11). The original tumor is called the primary tumor, and the process of metastasis establishes secondary tumors in distant organs where they incite the development of a stroma and develop into new tumors, which in turn invade (see Color Plate 4) and metastasize (Figures 1–14 and 1–15). Chapter 2 presents an extended analysis of metastasis.

It should be noted that cell mobility is not a feature exclusive to malignant cells. Leukocytes and macrophages move through the tissues. Many embryonic and fetal cells migrate during intrauterine development and may invade the tissues and organs. For example, primordial germ cells of the bird invade embryonic blood vessels in the yolk sac, where they are formed and migrate (metastasize?) to the primitive gonad. How do normal migration and metastasis differ? One difference is

Figure 1–14. Photograph of gallbladder and liver. The wall of the gallbladder is normally thin and soft like a piece of velvet. In this specimen, adenocarcinoma cells of the gallbladder have grown and invaded the organ, converting the wall into a thick, cheeselike mass surrounding a gallstone (S). The tumor has invaded the liver and metastasized within the liver (*arrows*).

Figure 1–15. Photograph of the surface of a lung of a woman, which illustrates metastases of leiomyosarcoma. The pleural surface is studded with nodules of metastatic sarcoma, each capable of invading and metastasizing. Remember: when tumors metastasize, the metastases are multiple.

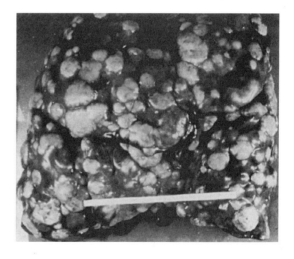

the nature of the cells involved. The germ cells are normal and under homeostatic control, whereas the malignant ones are not and form tumors where they implant. It is conceivable that the mechanisms are the same. One might ask if there is a gene for migration that is mutated, enabling malignant cells to metastasize, or if there is expression of a gene for migration in malignant cells that had been repressed since embryonic life. This issue is also discussed in Chapter 2.

1.7   Tumor markers

Tumor markers are molecules synthesized by tumors that are characteristic of the tumor, and when properly employed, aid in identification (Hammond and Taube 2002). As discussed, malignant tumor cells produce far fewer luxury molecules than normal cells, but so many tumor cells are generated that enough luxury molecules may be produced to cause clinical manifestations. For example, choriocarcinoma, the malignant tumor of placental trophoblast, can produce large amounts of human chorionic gonadotropin (HCG), which may be detected by the standard "pregnancy test." The presence of the hormone in the blood of a person with a testicular germ cell tumor indicates that the tumor contains choriocarcinoma capable of synthesizing HCG (Bagshawe 1992).

In addition to the synthesis of histiotypic (tissue-specific) molecules by malignant cells, some cancers make molecules that are biologically active but bear no apparent relationship to the histiotype of the tumor. If enough are produced for a clinical response, the syndrome is referred to as a paraneoplastic syndrome (Bunn and Ridgway 1993). For example, adrenocorticotrophic hormone (ACTH) normally is produced by the pituitary gland. However, a squamous cancer of the lung may synthesize enough ACTH to stimulate the adrenal gland to produce cortisone, causing Cushing syndrome. When the primary tumor is removed, Cushing syndrome disappears. After surgery, elevated levels of ACTH in the blood of such patients indicates that metastases have occurred, even though they may not have grown to appreciable size.

HCG and ACTH are good examples of tumor markers that are helpful clinically (Schwartz 1993), but there are others such as alpha-fetoprotein and carcinoembryonic antigen that are also useful in making diagnoses and following the course of patients. The latter are markers for malignant tumors of endodermal origin (colon, liver, stomach, etc.). Markers are sometimes elevated in non-neoplastic conditions, so care must be exercised in their use in diagnosis.

1.8   How cancer kills

1.8a   Organ failure

Tremendous reserves have been built into most of the organ systems: for example, an otherwise healthy adult can survive with one-half of a healthy kidney. After removal of one kidney, the other undergoes hypertrophy and/or hyperplasia, which compensates to a degree for the loss of tissue. A person can survive easily with a single functional lung. In the case of the liver, an animal can survive after removal of two-thirds of its liver, and, over the course of a week, regenerate the lost tissue. Thus, there is an enormous reserve of normal tissue plus a built-in mechanism to regenerate more functional tissue as required. One can see why, that

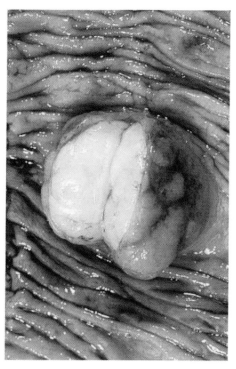

Color Plate 1. Lipoma of the intestine. The tumor is well circumscribed and it is yellow like normal fat tissue.

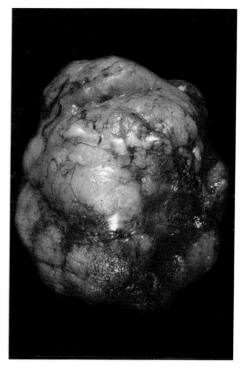

Color Plate 2. Meningioma. The tumor was removed from the intracranial cavity of a fifty-year-old man. The tumor is benign, but due to its intracranial location, it can be lethal. This typically occurs when the tumor compresses vital centers or causes increased intracranial pressure.

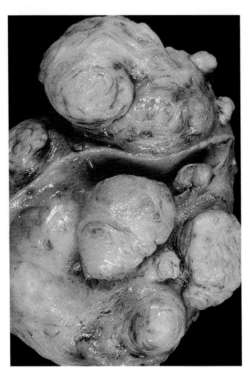

Color Plate 3. Leiomyomas of the uterus. These tumors present as well-delineated nodules in the muscular wall of the uterus (myometrium). Tumor nodules may cause enlargement of the uterus, which can compress the urinary bladder and other adjacent organs. Uterine leiomyomas are informally known as "fibroids".

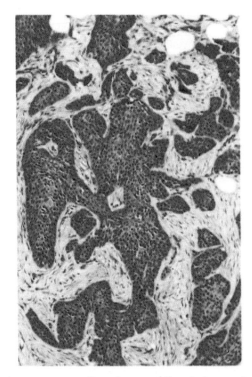

Color Plate 4. Invasive squamous cell carcinoma of the uterine cervix. The tumor is composed of irregular strands of neoplastic squamous cells.

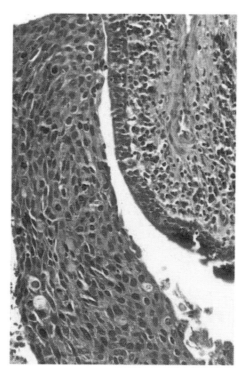

Color Plate 5. Squamous cell carcinoma of the bronchus. Compare the normal bronchial lining on the right side with the thick tumor tissue lining the bronchial cavity on the left.

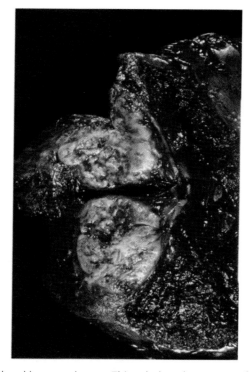

Color Plate 6. Peripheral lung carcinoma. This subpleural tumor was found in a sixty-year-old woman who was not a smoker. Histologically it was an adenocarcinoma (see Color Plate 7).

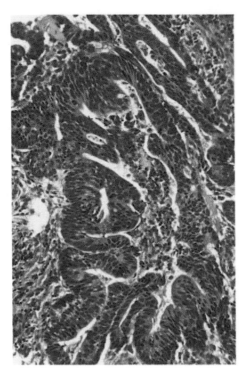

Color Plate 7. Peripheral lung adenocarcinoma. The tumor is composed of hyperchromatic cells forming irregular glandlike structures.

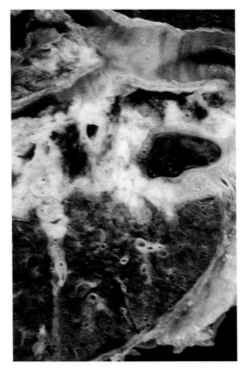

Color Plate 8. Lung carcinoma. This centrally located bronchial carcinoma was histologically diagnosed as a small cell carcinoma (see Color Plate 9).

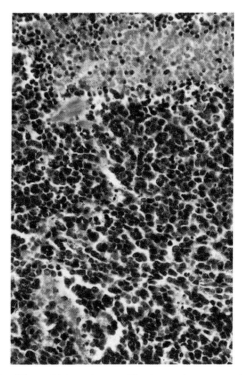

Color Plate 9. Small cell bronchial carcinoma. The tumor is composed of small blue-stained cells and shows areas of necrosis (*top*).

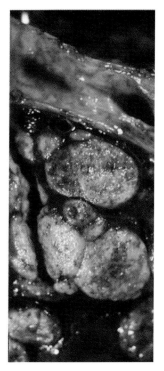

Color Plate 10. Carcinoma of the lung metastatic to the mediastinal lymph nodes. The lymph nodes appear enlarged and attached one to another. Histological examination confirmed that this "lymphadenopathy" was caused by metastatic carcinoma.

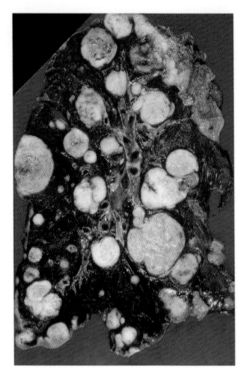

Color Plate 11. Metastases in the lung. These multiple nodules represent metastases from a primary colonic adenocarcinoma.

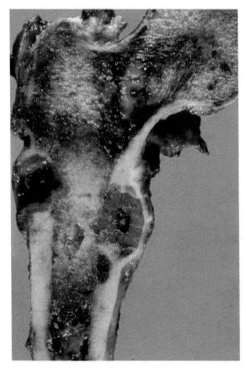

Color Plate 12. Chondrosarcoma of the femoral bone in process of destroying normal bone (for histology, see Color Plate 13).

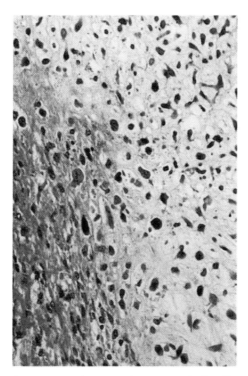

Color Plate 13. Chondrosarcoma. The tumor is composed of malignant cartilage cells. (Color Plates 1 through 13 were provided by Ivan Damjanov and Dennis Friesen)

Color Plate 14. Erythrophoroma of a gold fish, *Carassius auratus* (Photograph courtesy H.I.H. Prince Masahito and Dr. Haruo Sugano, The Cancer Institute, Japanese Foundation for Cancer Research, Tokyo).

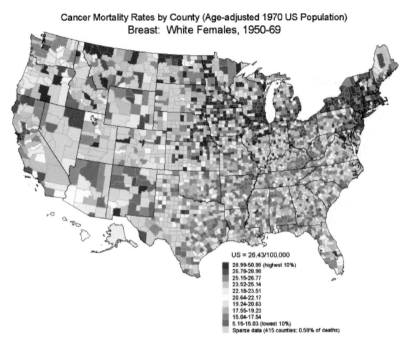

Cancer Mortality Rates by County (Age-adjusted 1970 US Population)
Breast: White Females, 1950-69

US = 26.43/100,000
28.99-50.98 (highest 10%)
26.78-28.98
25.15-26.77
23.52-25.14
22.18-23.51
20.64-22.17
19.24-20.63
17.55-19.23
15.04-17.54
5.15-15.03 (lowest 10%)
Sparse data (415 counties; 0.58% of deaths)

Color Plate 15. Breast cancer death rates among white females in the United States, 1950–1969. The geography of age-adjusted deaths is not random. Higher rates occurred in the north and northeast with lower rates in most of the south and southwest (Source: Devesa SS, DJ Grauman, WJ Blot, G Pennello, RN Hoover, JF Fraumeni Jr. 1999. Atlas of Cancer Mortality in the United States, 1950–94. Washington, DC: US Govt Print Off NIH 99-4564).

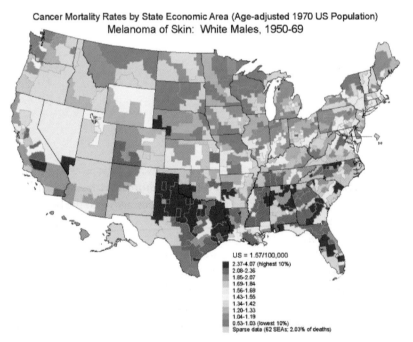

Cancer Mortality Rates by State Economic Area (Age-adjusted 1970 US Population)
Melanoma of Skin: White Males, 1950-69

US = 1.57/100,000
2.37-4.07 (highest 10%)
2.08-2.36
1.85-2.07
1.69-1.84
1.56-1.68
1.43-1.55
1.34-1.42
1.20-1.33
1.04-1.19
0.53-1.03 (lowest 10%)
Sparse data (62 SEAs; 2.03% of deaths)

Color Plate 16. Melanoma mortality in the United States is not random. Death rates among white males was higher in most of the southern states compared with northern states during the years 1950 through 1969 (Source: Devesa SS, DJ Grauman, WJ Blot, G Pennello, RN Hoover, JF Fraumeni Jr. 1999. Atlas of Cancer Mortality in the United States, 1950–94. Washington DC: US Govt Print Off NIH 99-4564).

during development of a tumor, signs or symptoms relating to organ failure are seldom seen before the tumor is far advanced. Although tumors can cause symptoms indicating an organ is affected (e.g., heaviness, and if the capsules of organs are stretched there may be pain, plus all of the nonspecific side effects of tumors such as fevers, cachexia, and weakness), there may be little or no evidence of organ failure. This explains why tumors of solid organs are silent for so long, why diagnosis is often delayed, why metastases are often present at the time of diagnosis, and why people seldom die of organ failure. Leukemia is the notable exception to this rule. Leukemic cells overgrow the marrow, causing bleeding, infection, and anemia as a result of destruction of megakaryocytes, normal leukocytes, and erythroid precursors.

1.8b Obstruction of the gastrointestinal tract, ducts, and hollow organs

The situation is somewhat different if the cancer involves the hollow organs such as the gastrointestinal tract or ducts carrying secretions from one organ to another. Carcinoma developing in such hollow organs has a propensity to invade the mucosa, ulcerate it, and cause bleeding, and thereby the possibility of anemia and weakness. The tumor also invades and grows circumferentially around the wall of the organ or duct in a shape resembling a napkin ring. This ring thickens with time, and ultimately the lumen is obstructed (see Figure 1–7). This is best seen in the descending colon where the fecal stream is solid and contributes to intestinal obstruction, a common presenting sign of cancer in that locale. Such a patient will die in about a week unless the obstruction is relieved surgically. Thus, the surgeon is often faced with a life-threatening emergency that must be relieved immediately. Even if the tumor has spread beyond the colon and the patient is incurable, relief of the obstruction affords palliation and offers useful life to the patient.

Squamous cell cancer of the esophagus results in an ulcerative lesion and causes difficulty in swallowing, which exacerbates the cachexia associated with malignancy. In addition, the tumor may penetrate the wall of the esophagus and grow into and through the wall of the adjacent trachea or bronchus causing a fistula (abnormal passageway). Aspiration of food through such passages into the bronchus and lung may result in pneumonia. Distant metastases occur, and the patient becomes cachectic and dies of infection.

Some of the cancers in the walls of hollow organs are fungating (mushroomlike in appearance) and grow into the lumen. They often occur in the right side of the colon where the fecal stream is fluid. These cause obstruction less frequently than the infiltrating napkin ring-like tumors that occur commonly in the descending colon. These fungating tumors often cause severe anemia because of massive ulceration with bleeding. This exacerbates the cachexia that occurs as the tumor grows and spreads to lymph nodes and distant organs. Again, death usually results from cachexia and infection.

Death or serious illness seldom results from the compression of ducts. For example, the ureters may be displaced by massive metastases from teratocarcinoma of the testis, but the outflow of urine is seldom obstructed sufficiently to cause signs and symptoms. A major exception to this rule is the involvement of ducts and ureters by the matted masses of lymph nodes of Hodgkin's disease. There is often a fibroblastic response (excessive growth of connective tissue) in Hodgkin's disease. The resulting scar tissue may surround and obstruct ducts. The resultant backup of secretions can give rise to symptoms and serious disease. Similarly, obstruction of the urethra by adenocarcinoma of the prostate (a portion of the urethra passes through the prostate) can cause acute and chronic retention of urine, which predisposes the individual to cystitis. A patient with invasive squamous cell carcinoma of the cervix also usually dies from the complications of urinary obstruction. In this case, the cancer cells invade around the ureters where they enter the bladder, obstruct them, and cause uremia. Invasion from an adenocarcinoma of the head of the pancreas frequently results in infiltration of and obstruction of the bile ducts, which causes jaundice.

The central nervous system is also a hollow organ with walls of various thickness. A developing tumor, whether benign or malignant, can cause disease by destroying vital centers in the brain (the thick-walled portion of the hollow nervous system). In addition, the central nervous system is contained within rigid walls, and the increase in mass of a tumor can increase intracranial pressure, causing serious effects on the brain. Moreover, tumors can grow into the canalicular system of the brain, obstructing the flow of cerebrospinal fluid with resulting damage.

If an obstruction can be removed surgically and the duct anastomosed (reconnected) successfully, then the patient is relieved of the immediate effects of the tumor. If an adenocarcinoma of the colon is close to or has involved the anus, removal of the tumor requires removal of the anus and the sphincters, and the individual is incapable of controlling expulsion of feces. An artificial anus, a colostomy, is made on the anterior abdominal wall (see letter to Uncle Harvey in the Introduction). Similarly, individuals with carcinoma of the esophagus may have a feeding tube placed directly into the stomach to bypass an obstruction. Drainage tubes may be placed in the biliary tree of the liver to relieve jaundice and to compensate for loss of bile ducts due to destruction by the cancer or the surgical procedures used. A person whose bladder has been destroyed by cancer may have an artificial bladder constructed, or the ureters, which drain the liquid waste from the kidneys, may be brought through the abdominal wall so they can drain into a bag. Similarly, shunts may be made between the canals carrying cerebrospinal fluid, draining dammed up fluid through tubes running down the neck into the thorax or abdominal cavity. Shunts relieve the increased intracranial pressure due to the backup of fluid, and the procedure eliminates stroke and other complications of the condition. These procedures are all palliative and designed to provide additional useful life.

Because of the essential role of the heart, it might be supposed that involvement of the heart would be a primary cause of death in cancer. Interestingly, primary tumors of the heart are extremely rare, and metastatic tumors involving the heart are also rare. Carcinoma of the lung by direct extension may involve the pericardium, but seldom involves the musculature or seriously interferes with the heart's function. Melanoma may metastasize to the heart, but this is an exceedingly rare cause of death.

Obstruction of ducts can lead to death, but the duct obstruction is often a presenting sign or symptom that can be relieved surgically, as described earlier. If recurrence of cancer causes reobstruction, or if obstruction occurs as a late event in the history of the illness, the surgical trauma may be too great for the debilitated patient to bear, and other palliative measures may be undertaken. In this regard, it is possible to eradicate focal tumors by irradiation, thereby relieving pain or pressure. Duct obstruction is usually not an immediate cause of death.

1.8c   Cachexia and infection

One of the syndromes most commonly associated with the late stages of malignancy is termed "cachexia" (Tisdale 2000). Cachexia results from the starvation and debilitation of the patient by the cancer (see Figure A–16). The person is wasted, weak, and incapable of mounting adequate anti-inflammatory responses. It is this wasting and the resulting incapacity that terrifies people as much as or more than the pain associated with cancer and metastasis. The immediate cause of death of such debilitated patients is infection, even with massive antibiotic therapy. The saprophytic organisms (organisms that grow on dead organic matter) that normally live in the mouth or nose, for example, invade the patient and cause pneumonia, septicemia (blood infection), abscesses, and death. The tumor represents the underlying cause, but infection is the immediate cause of death in most terminal cancer patients. Because of the efficiency of antibacterial therapy, fungi and yeast are often the agents responsible for the terminal events, and it is clear that if the cachexia could be controlled, prolongation of useful life could be achieved. Interestingly, just as the malignant neoplasm can literally starve the host, the fetus in utero also appears to have priority for the metabolic resources of the maternal organism and can lead to the death of a debilitated mother.

Cachexia is possibly a paraneoplastic syndrome in its own right (Beutler and Cerani 1986). In this regard, Japanese scientists postulated the presence of a hormone secreted by the malignant tumor that resulted in poisoning of the host. Although such a molecule has not been identified, a fragment of a protein related to an immunoglobulin molecule, called tumor necrosis factor (TNF), was found to produce many of the features of cachexia when administered to rats. TNF is synthesized by a variety of inflammatory cells that are somehow activated by malignant tumor cells (Beutler and Cerani 1986). It is likely that TNF contributes to cachexia.

As described here (and also in the letter to Uncle Harvey; see the Introduction), cachexia with superimposed infection, usually by saprophytes, is the typical cause of death in patients with cancer. The cancer is the predisposing cause of death; infection is the immediate cause.

Because certain tumors grow rapidly and are vascular, hemorrhage either from the tumor into body cavities or into the tumor itself can be a terminal event. Ascites, that is, abdominal fluid, which develops in some cancer patients, is typically blood tinged, and sometimes it can be visibly bloody. For example, hepatocellular carcinoma may cause massive bleeding into the abdominal cavity. On the one hand this tumor invades the portal vein causing an obstruction of venous backflow from the abdominal organs. Distended veins cause back-pressure into the capillaries, which in turn rupture and bleed. Hepatocellular carcinoma is, on the other hand, also associated with a bleeding tendency related to the destruction of the normal liver by the tumor. Since the liver is the main source of coagulation factors, loss of normal liver cells adversely affects blood clotting. Accordingly, these patients tend to suffer from uncontrollable bleeding; most often it occurs into the abdominal cavity.

A bleeding tendency is found also in patients with leukemia – a malignant disease of hematopoietic cells characterized by the destruction of the normal bone marrow. Leukemic patients bleed because they cannot produce platelets, which are essential for preventing excessive bleeding. Leukemic patients also suffer from repeated infections since they do not have enough leukocytes to combat bacterial infections. It bears repeating that overall, infections are the most common immediate cause of death in the terminally ill cancer patient.

1.9   Spontaneous regression

Does spontaneous regression (the complete spontaneous cure of a patient) offer hope for the cancer patient with metastasis? The answer is *no!* Although spontaneous regression is a fact of tumor biology, it is so rare that little scientific attention can be or has been directed toward it (Lewison 1976).

The classic paper on spontaneous regression involved a baby with neuroblastoma treated by Cushing and Wolback (1927). The baby was sent home to die because of the stage of the disease. Imagine the surprise when the individual, now a teenager, presented with an inflamed appendix, which was removed. While the surgeons were exploring the abdomen, they looked for evidence of the neuroblastoma and found multiple small tumors in the peritoneal cavity. These proved to be ganglioneuromas. In the embryo, neuroblasts (the normal equivalent of neuroblastoma cells) differentiate into ganglion cells. Apparently, in this particular case, all of the neuroblastoma cells (malignant) had differentiated into ganglioneuroma cells (benign). Thus, the patient had a spontaneous remission of a proved malignancy as evidenced by the differentiation of all the malignant cells.

Cases of spontaneous regression are documented in which highly malignant incurable tumors diagnosed by pathologic methods completely disappear (Lewison 1976). The mechanisms are unknown. Tumor immunologists believe they are rejected by an immune or host defensive mechanism (Sioud 2002). With the burgeoning knowledge of growth factors, it is conceivable that such tumors are responsive to negative growth factors which caused their disappearance (chalones?). Spontaneous regression probably will be explained someday on the basis of regulation of the malignant process by nonimmunologic factors. Thus, we are left with a rare phenomenon, with a paucity of experimental approaches to it, and yet with a population desperately in need of relief from the problems caused by cancers. Individuals who believe that relief will come by spontaneous regression, and therefore ignore treatment, are doomed. Their tumors will, with extraordinarily rare exceptions, kill them.

1.10 Dormancy

Do we have inactive malignant cells lurking in our bodies? Dormancy is not uncommon clinically. Two histories illustrate the phenomenon. A 60-year-old man, with only one eye, presented with an enormously enlarged liver and aspects of tumor cachexia. The history revealed that the eye was removed for a melanoma thirty years previously. Biopsy demonstrated that the liver contained melanoma of the same type removed from the eye. The accepted explanation is that the melanoma had metastasized to the liver prior to removal of the eye, and melanoma cells that had metastasized had lain dormant in the liver for many years only to be reactivated by unknown agents.

A more common example is that of a 45-year-old woman with an adenocarcinoma of the breast which was removed by radical mastectomy. In this operation the breast plus the underlying muscles and contents of the armpit are all removed. The patient was well for ten years, and then a pea-sized nodule (about 4 mm in diameter) was discovered in the mastectomy scar at an annual examination. It proved to be the same type of cancer that was present originally, and in a few months the patient succumbed with widespread metastases. Apparently, a few cells were left in the incision at the time of mastectomy and remained dormant for the ten-year period, at which time the malignant phenotype was reactivated.

The behavior of cells, normal or neoplastic, their interactions with each other and the environment, and reactions to growth factors are slowly becoming understood. It is in such mechanisms that the ultimate explanation of dormancy probably will be found. An experimental model for dormancy was developed by Fisher and Fisher (1967) in which injection of large numbers of tumor cells into the portal vein of rats caused massive liver metastases and death within a short period of time. When only fifty cells were injected in the portal tracts, the animals survived apparently tumor free, and no signs of tumor were observed when these animals were examined surgically. The animals were closed up and observed. They died within a few weeks postsurgery from the effects of massive overgrowth of tumor cells in the liver. Apparently, a few tumor cells lay dormant in the liver and were activated by the growth hormones of the repair process. This experiment is important because it demonstrates that, as in normal development, a critical number of cells is required for the expression of a phenotype, in this case the malignant one, and that environmental circumstances, in this case surgical trauma with release of growth factors, can stimulate growth of the dormant cells.

In conclusion, cancer cells are not necessarily unresponsive, autonomous cells as we have classically been led to believe. They do respond to some environmental perturbations. Intensive effort must be made to identify the positive and negative regulators of neoplastic growth.

1.11 Initiation

The development of cancer is a multistep process. The stages of carcinogenesis, initiation, latency, promotion, and progression interact sequentially in the formation of a malignancy. Initiation is the first step in which a carcinogen interacts with DNA (Friedwald and Rous 1950). It is a rapid process that permanently alters cells. Initiated cells do not develop into tumors in the absence of a suitable environment and are termed latent. If the environment contains promoting agents, proliferation of cells will be preferentially stimulated. The effects of promoting agents are reversible, and if the promoter is removed, disappearance of the expanding clones of cells will result. They will reappear if the promoter is reapplied. The second step, progression, is characterized by events that result in the autonomous state; at this time the effects are not reversible. The mechanism of progression is addressed later.

1.12 Latency

Latency is the period of time between the application of a carcinogen (Chapter 3) and the appearance of a tumor. During latency a small number of cells initiated by the carcinogen are genetically programmed for the malignant phenotype. Initiation changes the determination of stem cells from normal to neoplastic, but initiated cells do not appear different than their fellow stem cells. (Note that initiation does not alter the histiotypic determination of the stem cell, it only superimposes the malignant phenotype on it.) In this sense malignant stem cells are remarkably like determined, undifferentiated normal stem cells in the embryo, but they have acquired a change in their potential to proliferate and differentiate. As in the normal situation, latent cancer cells express the malignant phenotype only in the appropriate environments. Multiple factors are required to express embryonic phenotypes as well as malignant ones. Over a period of time, and especially with the application of promoters (Chapter 3), initiated cells begin to express the malignant phenotype, and tumors develop. To reiterate, the expression of the malignant phenotype, just like the expression of the embryonic phenotype, depends on environmental conditions. Is it possible to prevent development of tumors by gaining an understanding of the social relationships of cells in dormancy and latency?

1.13 Progression to the autonomous state

Foulds (1969) defined progression as "the gain or loss of unit characters leading to the autonomous state." A unit character could be growth rate, ability to metastasize, inability to respond to a hormone, a differentiated function, or a morphological feature. All of the events in the experiments described here illustrate the concept of progression as outlined by Foulds on the basis of studies of breast cancer in mice and man. The gains or losses of unit characters were uncoordinated and, once lost, were never regained by the tumor. Thus, tumors are dynamic and continually become more and more malignant.

Greene (1957) was interested in the early development of breast and uterine adenocarcinoma of rabbits. He transplanted rabbit tumors into the anterior chamber of the eyes of guinea pigs, an experimental form of metastasis. When the tumors were small and had not metastasized in the rabbit, they would not grow in the guinea pig's eye (an immunologically protected area). Transplants of the tumors which had grown large and had metastasized in the rabbit, grew in the eye of the guinea pig. Greene concluded that tumors undergo changes with time. In this case, early tumors could not metastasize, whereas older ones could. They had progressed, expressed the ultimate in malignancy in the host rabbit (i.e., metastasized), and were autonomous.

Charles Huggins performed a series of brilliant experiments that characterized important features of tumor development (Huggins and Hodges 1941). He found that the prostatic hyperplasia that occurs in elderly dogs was dependent upon androgens. Antiandrogen therapy reduced the hyperplasia. As a consequence, Huggins, a urologist, decided to treat human patients afflicted with adenocarcinoma of the prostate by castration to remove testosterone. Dramatic results were obtained. Bedridden men with painful metastases that were destroying their spines were relieved of pain, their spines healed, and they were able to return to work and lead productive lives. Interestingly, these individuals were never cured. The tumors eventually reappeared, were now insensitive to hormones, and the individuals died. The tumors had progressed and lost their dependence on androgens (Hanks et al. 1993).

Jacob Furth (1953) performed experiments leading to the understanding of the concept of dependency. He interfered with the regulation of thyroid function by the pituitary gland. The pituitary gland produces thyroid stimulating hormone (TSH), which stimulates the thyroid gland to produce thyroid hormone. When enough thyroid hormone is produced, it in turn causes a feedback control of TSH production by the pituitary gland, thereby maintaining status quo. Furth blocked the production of thyroid hormone in mice, which in turn caused the pituitary glands to synthesize TSH. The thyroid glands responded to the TSH by becoming hyperplastic (producing more cells to alleviate the apparent need for thyroid hormone). Eventually, metastasizing adenocarcinomas of the thyroid developed in some of them. Transplants of these thyroid metastases did not grow when transplanted in appropriate strains of animals, but if the recipients had been thyroidectomized (resulting in high levels of TSH) the tumors grew. Thus, the growth of these metastasizing adenocarcinomas of the thyroid gland depended on high levels of TSH. After repeated transplantation to thyroidectomized animals, this dependence on TSH was lost, and the tumors grew in animals with a thyroid in the presence of normal levels of TSH. In current parlance, it would be said the tumors had progressed and lost their dependence on TSH.

The development of mammary cancer in mice has also given us important insights into the events in carcinogenesis. A baby mouse may receive the mammary tumor virus when it first suckles, but tumors do not appear on the average until about forty weeks of age (latent period) (Bittner 1936; Held et al. 1994). The first tumor usually appears during pregnancy, grows rapidly (see Figure 6–7), but surprisingly regresses and may even disappear between pregnancies. The tumor reappears during the next pregnancy, and may again regress after delivery. Eventually, either during or between pregnancies, the tumor begins to grow in the unrelenting progressive manner typical of adenocarcinomas of the breast (Foulds 1969). In this case, these tumors were initially dependent on the hormones that normally control mammary development during pregnancy and lactation, but the tumors lost their dependence on those hormones as the tumors progressed to the autonomous state. Are there lessons in these facts that can be used clinically?

1.14 Selection and cellular heterogeneity

Tumors are not comprised of homogeneous populations of cells; if they were, all cells of the tumor would be expected to behave identically. For example, all cells of a tumor would have the competence to metastasize or no cells of that tumor would have that capability. Similarly, all cells of a particular tumor would be expected to respond to a single chemotherapeutic agent in a like manner. Because we know that cells within a tumor behave differently, it logically follows that tumors are composed of multiple populations of cells with different abilities to respond to environmental stimuli. Some of these cells can be selected for, as they have growth advantages over other cells. Then the phenotypic traits of these cells are expressed and the behavior of the tumor changes.

Selection has long been recognized as a mechanism of progression. Selection can occur only if there is a heterogeneous population of cells. Thus, in Huggins's adenocarcinoma of the prostate and in Furth's adenocarcinoma of the thyroid, many malignant cells were responsive to androgens and to TSH, respectively. Apparently, there were other stem cells that were independent and eventually overgrew the responsive ones, as evidenced by the eventual loss of dependence on the hormones. Cells are selected for or selected against by the environmental conditions. The net effect of progression is increased malignancy.

Selective pressures may also be generated by the tumor itself. Stevens (1967) developed a strain of teratocarcinoma called OTT6050. The teratocarcinoma cells, developed from cells of an early mouse embryo, were transplanted into the testes of adult mice. The teratocarcinomas that resulted were distributed to investigators throughout the world. Interesting cell sublines were isolated from them. One of these sublines, called F9, is an undifferentiated embryonal carcinoma, which, when appropriately treated with retinoic acid, differentiates primarily into endoderm (Strickland et al. 1980). Rizzino and Crowley (1980) demonstrated that large numbers of F9 cells grew successfully in defined media, failed to differentiate, and could be maintained indefinitely. But if small numbers of F9 cells were placed in the defined medium, they failed to grow. Rizzino and Crowley concluded that F9 cells each secreted an autocrine growth factor that reached critical concentration and stimulated cell growth, at appropriate cell density. Another subline from OTT6050, called C44, makes a toxic substance to which ECa 247 (a line of pretrophectodermal embryonal carcinoma derived from OTT6050) is especially sensitive (Gramzinski et al. 1990).

Thus, F9 and C44, components of the parent strain of OTT6050, made a growth factor or a toxic factor that affected themselves or other sublines of OTT6050. The conclusion is inescapable that autoselection by clonal lines within the tumor must play a major role in the selection that is an essential part of progression (Pierce and Parchment 1991).

If tumors are heterogeneous in their cellular composition – which indeed many, if not all, are – it logically follows that they are mosaics in exactly the same sense that the multicolored skin of the cow is a mosaic. This would mean that each subline of heterogeneous populations of cells would have a patch size. It is known from studies of early development that a threshold number of cells is required for the expression of a phenotype and its function. Thus, one could argue that in the parent OTT6050 tumor, the C44 cells would produce enough toxic material to kill ECa 247 cells only if the C44 cells had attained an adequate patch size (Pierce and Parchment 1991). Reconsider the experiments of Greene (1957) on breast and uterine adenocarcinoma of rabbits transplanted into guinea pig eye: conceivably the patch of cells capable of metastasis was not large enough to express the phenotype in the small nonmetastasizing tumors, but with time, patches of metastasizing cells became large enough to express their phenotype. Similarly, in Furth's experiments with thyroid metastases and TSH dependence, cells not dependent on TSH would eventually form a large enough patch in the metastases of TSH-dependent thyroid carcinoma cells to express their phenotype, overgrow the rest of the TSH-dependent patches, and ultimately dominate (Furth 1953). Molecular explanations for these phenomena are not known.

Heppner (1982) made fine contributions to the understanding of progression and the importance of heterogeneity of tumors. She developed subpopulations of cells from a viral-induced adenocarcinoma of the mammary gland in mice that are named 68H, 4.1, and 168. These cell lines varied in growth rates and ability to metastasize when grown subcutaneously. In addition, the growth of 68H was stimulated by media conditioned by 168, and growth of 168 was slowed by 4.1 tumor cells. Intratumor selection must play a role in progression.

What remains now is an explanation for the mechanism of producing cellular heterogeneity in tumors. The early tumor is composed of a mixture of clones representing the heterogeneity of normal cell types responding to the carcinogenic event, and, as shown earlier, selective pressures tend to reduce this heterogeneity so the tumor becomes monoclonal, homogeneous in composition, and autonomous (Pierce and Parchment 1991). Nowell (1976) has postulated the occurrence of mutations to explain heterogeneity in monoclonal tumors. Some mutations might impart a growth advantage on the mutated cell. Ultimately, a clonal patch would develop, of sufficient size, to express whatever characteristics were specified by the mutated cells. This phenomenon is called clonal expansion. It is also conceivable that chromosomal translocation could place promoters for one gene in control of another, cause a growth advantage, and/or other effects which could again lead to cellular heterogeneity and the changes in karyotype that are common in progressed cancer.

Whatever the mechanism of heterogeneity, it is clear that during progression selective events tend to make the tumor both monoclonal and more malignant. These events are made in the face of other pressures such as genetic instability that cause diversity of cell types. The result is a dynamic interaction leading to the development of tumors that are ever more anaplastic and autonomous with time. Tumors are continually changing in their biological behavior.

1.15   A developmental concept of cancer

Cancer has been described variously as aberrations of growth, differentiation, or organization of cells. These are the basic processes of development that in a coordinated and controlled manner evolve the adult organism. Prior to puberty, they are regulated for accrual of mass, but when adult stature has been achieved reregulation occurs to maintain status quo of tissues, a process known as tissue renewal. Accurate knowledge of tissue renewal dates from the use of isotopes. Prior to that, tissues of the adult were viewed as well differentiated, and little thought was given to their maintenance, much less to the presence in them of determined but undifferentiated cells from which the functional differentiated cells could be renewed or could serve as the target in carcinogenesis. In the absence of this information, an explanation was required for the mechanism by which undifferentiated malignant cells give rise to mature differentiated tissue.

It was decided erroneously that dedifferentiation was the process by which differentiated cells became anaplastic during carcinogenesis. The rationale was based on the well-known fact that dedifferentiation occurs normally in amphibians. For example, if the lens of the eye of an amphibian is removed, the cells of the iris dedifferentiate. They extrude their pigment granules and become undifferentiated in appearance, and during the recovery period, they redifferentiate not only iris but a new lens as well (Yamada and McDevitt 1974). It is the ability of dedifferentiated cells to regain potential that was repressed in the process of determination (in the case cited, the iris cells reacquired the potential to make lens) that distinguishes dedifferentiated cells from undifferentiated ones which do not regain potential. A malignant tumor has never regained embryonic potential.

As we show subsequently, the variable degrees of differentiation in carcinomas are not a reflection of varying degrees of dedifferentiation, but represent the converse, varying degrees of differentiation of malignant stem cells. Malignant stem cells are, in turn, derived from undifferentiated determined normal stem cells responsible for tissue renewal (Wylie et al. 1973). In carcinogenesis, stem cells give rise to equally undifferentiated malignant stem cells. In the adult, if they become neoplastic, they form a caricature (a gross misrepresentation) of tissue renewal, and, in the case of teratocarcinomas, of embryogenesis (Pierce and Speers 1988). To understand the developmental concept of cancer better requires a short excursion into embryology.

Embryos begin as a single cell with the genetic potential to create the adult organism, given the appropriate environment. Each of the cells at the 8-cell stage in the mouse is totipotent (Graham and Wareing 1976; Slack 1983), but speciation begins at the 8-cell stage and the potential of the involved cells is reduced by the 16-cell stage (Tarkowski and Wroblewska 1967). (Recent work with cloning has shown that *genetic totipotentiality* is retained throughout development; a brief review of cloning studies is included in Chapter 6). For example, the ball of sixteen cells is composed of an outer layer surrounding one or two inner cells. The outer layer will form extraembryonic tissues including the trophectoderm, and its descendants will attach the free-floating embryo to the mother and form the placenta. The embryo is cystic at the 64-cell stage and known as the blastocyst. The cyst wall is made up of fifty-two trophectodermal cells, and the inner cells, now twelve in number, referred to as the inner cell mass (ICM), are attached to the inner surface of the blastocyst (Figure 1–16). They will form the embryo.

The first differentiation in cells destined to become the embryo occurs in the embryo at the late blastula stage. The surface layer of cells on the ICM becomes different from the rest and forms the primitive endoderm, which will form the digestive tract of the adult. The collective noun of the ICM cells become embryonic epithelium. Mesoderm develops after collapse of the blastocyst and implantation of the embryo into the wall of the uterus. It occurs where primitive endoderm lies over embryonic epithelium (Slack 1983), and there is speculation that the endoderm induces the embryonic epithelium to form the mesodermal cells. Mesoderm has the potential for production of connective tissue, fat, bone, cartilage, and so on. The embryonic epithelium has the potential to make brain or skin.

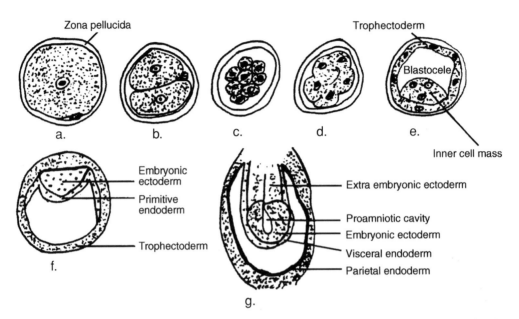

Figure 1–16. Chart of early development of the mouse. (a) Fertilized egg contained in the zona pellucida. (b) and (c) Two- and 8-cell stages. (d) Late 8-cell stage after compaction. This creates a layer of outer cells which encase one or two inner cells that will form the embryo. (e) Early blastocyst composed of thirty-two cells; the inner cell mass will form the embryo proper. (f) This is the late blastocyst, which has hatched from the zona pellucida. Primitive endoderm has differentiated from the surface of the inner cell mass and begun to line the trophectoderm. (g) Postimplantation mouse embryo. The embryonic epithelium and overlying proximal endoderm will form the embryo. Embryoid bodies resemble this part of the embryo. All of the other tissues present will form the placenta and extraembryonic membranes.

Ectoderm, mesoderm, and endoderm are referred to as the three germ layers and each of these new germ layers is said to be determined for a particular differentiation (Gehring 1968). Each has had its potential reduced in relationship to its precursor, cannot revert to the precursor, and can only express the differentiated phenotype in the appropriate environment. Note that these primitive cells are undifferentiated and do not express the gene products of the lineage for which they are determined (Figure 1–17). For example, determined skin cells have the potential to become keratinocytes and express keratin, but in the early embryo the environment allows them to proliferate only and cover the embryo. In the changed environment of the adult, these same cells express keratin.

Proof for the idea that the expression of the phenotype by determined cells depends on the appropriateness of the environment and that determination of cells is permanent stemmed from studies of drosophila (Gehring 1968). Clumps of embryonic cells named imaginal discs develop in drosophila embryos at the

Embryonic origin	Adult derivative
Ectoderm	Skin
	Brain
	Breast
	Sweat glands
	etc.
Mesoderm	Fibrous tissue (connective)
	Cartilage
	Bone
	Muscle
	etc.
Endoderm	Gut
	Liver
	Lung
	Pancreas
	etc.

Figure 1–17. Three germ layers and the tissues derived from them.

time organs form. An imaginal disc will develop into abdomen, another into wing, another into thorax, and so on. Imaginal discs determined for thorax were transplanted into the abdomens of adult flies. This foreign environment allowed for proliferation, but the cells did not express the differentiation. They were serially transplanted for several generations, and when returned to the appropriate place in the embryo, they differentiated into thorax. Two conclusions are drawn from this study: the determined state is stable and heritable, and its expression depends on environmental conditions. This leads to the interesting conclusion that the offspring of determined stem cells of the embryo which do not express the phenotype are the same as the ones in the adult that do so. The latter are regulated not for accrual of mass, as in the embryo, but for maintenance of status quo. When a cell is lost, a stem cell divides and sends another along the maturational pathway.

It was postulated that the undifferentiated determined stem cells of adult tissues respond to carcinogenesis and become stem cells of tumors (Pierce 1974). The malignant phenotype is superimposed on the histotypic determination, and the resultant tumor is a caricature of the process of tissue renewal.

As mentioned previously, after puberty when growth is complete, the rate of cell division is reregulated to maintain tissue. In a variety of tissues as exemplified by colon, intestine, skin, testis, bronchus, or bladder, tissue renewal is mediated by a population of precisely controlled stem cells. As stated, these are undifferentiated cells determined for their particular differentiation. When they divide, one cell remains behind as a stem cell; the other enters the maturational pathway, differentiates, functions, and in the intestine after about four days becomes

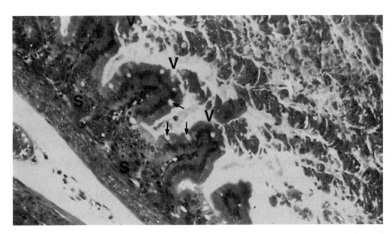

Figure 1–18. Photomicrograph of section of bowel in teratoma illustrating life cycle of interstitial epithelium. Stem cells are in the crypts (S), one remains behind as a stem cell, and the other migrates along the wall of the villus expressing the determined phenotype as it goes (mucous cell at arrows). After four to five days the cells become senescent and are sloughed from the tips of the villi (V) into the fecal stream.

senescent, dies, and is cast from the body (Figure 1–18). In the intestine, the stem cells are at the base of the villi in crypts (Snell 1978). They do not migrate along the villi; they know their place and respond to stringent homeostatic control because cell production equals cell loss. They retain the ability to respond to environmental perturbations by undergoing hyperplasia, atrophy, and so on. Stem cells are the reactive cells of a tissue; their differentiated progeny can only differentiate or die. The regulation of these processes is rapidly becoming understood with the identification of numerous regulatory molecules.

In this regard the existence of negative regulation of stem cells by chalones has been postulated. Chalones are believed to be tissue-specific, reversible inhibitors of stem cell proliferation synthesized by the differentiated cells of a tissue (Bullough 1962). If, for example, there were too many differentiated cells, enough chalone would presumably be produced by them to inhibit replication of only the stem cells of that differentiated tissue, and over time the differentiated cells would become senescent and die, restoring status quo. Conversely, if there were too few differentiated cells there would be correspondingly less chalone, and the stem cells would replicate in response to positive stimuli and restore the tissue. The need for such regulation is widely accepted, but the chalone field fell into disrepute when, despite intensive effort, molecules with attributes of chalones could not be identified (Langen 1985).

In the past it was difficult to identify the cells initially targeted in carcinogenesis because of the long latent period. Embryonal carcinoma cells, the stem cells of teratocarcinomas, develop only if primordial germ cells (the stem cell of the species) are present in the fetal testis at the time of carcinogenesis (Stevens 1967).

This established the origin of this cancer from normal stem cells that had undergone transformation. In this situation, the normal and malignant stem cells were equally undifferentiated by ultrastructural criteria. Irrespective of the mechanism of carcinogenesis, there is no need to consider dedifferentiation of the primordial germ cells as an explanation for the generation of undifferentiated cells of the embryonal carcinoma, or indeed of other malignant tumors. The determination of normal stem cells was altered in the process of carcinogenesis, however, and given the appropriate environment, the initiated cells will express the malignant phenotype that is superimposed on the normal phenotype. The malignant stem cells derived from the normal stem cells are undifferentiated from the outset and bear close resemblance to their normal counterparts (Pierce et al. 1967). The histiotypic determination of the normal stem cell is not altered during carcinogenesis because malignant stem cells, if they can differentiate, always express the appropriate determined histiotype. What appears to be altered in the malignant phenotype is the regulation of cell proliferation, differentiation, and invasiveness of the neoplastic cells. The malignant cells are no more or less differentiated than their normal precursors, have no increase in potential, and thus cannot be dedifferentiated.

Not all tissue types in the adult have stem cells. What, then, accounts for the development of tumors in these tissues? Mesenchymal tissues such as fibrous tissue, cartilage, bone, or fat do not appear to have a stem cell compartment. Under certain conditions, such as wound healing, the fibrocyte nucleus becomes vesiculated and enlarged in becoming a fibroblast. The fibroblast synthesizes DNA and divides. There is no contradiction of facts or logic in applying the concept that carcinogenesis involves undifferentiated cells capable of dividing, whether from tissues with or without a stem cell component.

But what about brain tumors? Until recently no renewal of nerve cells in the adult was ever observed. If nerve cells do not divide, are brain tumors an exception to the rule that tumors occur only in cells capable of division? Tumors of nerve cells do not occur in the adult. Neuroblastoma (the name indicates an origin in embryonic nerve cells) occurs congenitally or in very young children. The other brain tumors are derived from glial cells that are capable of cell division. Glial cells are the equivalent in the brain of connective tissue in other organs, and give rise to astrocytoma, glioblastoma multiforme, and oligodendroglioma.

Mature liver cells, given the appropriate stimulus, can similarly divide (Bucher and Malt 1971). Differentiated functional liver cells retain the capacity to divide and serve as the target in carcinogenesis. They give rise to carcinomas, and the tumor cells closely resemble normal liver cells morphologically and functionally. In addition, Sell and Leffert (1982) have reported stem cells in the liver, and it is a small leap of imagination to suppose that they too are targets of carcinogenesis.

In this regard, the stem cells of leukemias closely resemble their corresponding hematopoietic (blood-forming) stem cells (Greaves et al. 1983). Even though it has not yet been shown directly that the normal stem cell gives rise to the leukemic stem cell, in view of the experience with embryonal carcinoma, and in view of the similarity of appearance of normal and malignant stem cells, it can be stated that normal hematopoietic stem cells are the target in leukemogenesis. In a similar vein, adenocarcinoma cells of the colon resemble normal colonic stem cells and transformed fibroblasts in vitro closely resemble normal fibroblasts. Thus, it can be concluded that undifferentiated neoplastic cells originate from normal determined, undifferentiated stem cells, and the anaplastic appearance of tumors is the result of the preponderance of these cells with their inability to organize appropriately, in relationship to the number that differentiate.

We have seen in the previous discussion that malignant stem cells are no more or less differentiated than their normal stem cells, but have the malignant phenotype superimposed on the determined histiotype. What is the significance of this for cancer biology? There is no need to consider dedifferentiation as a mechanism of carcinogenesis, and the neoplastic cells, given the appropriate environment, should have capacity to differentiate. But is there evidence for this?

Numerous experiments have demonstrated differentiation of malignant stem cells into benign cells (Pierce and Speers 1988; see also Sachs 1993; and discussion in Chapter 6). The experiments have involved embryonal carcinoma cells, the stem cells of teratocarcinomas (see Appendix), squamous cell carcinoma cells (see Appendix), and leukemic cells (Gootwine et al. 1982). The malignant stem cells in each of these tumors were shown to differentiate into mature functional apparently normal cells of the appropriate lineage. It is now accepted unequivocally that some cancer cells can differentiate into terminally differentiated benign, if not normal, cells whereas others display abortive differentiation, and some display so little differentiation that it is almost impossible to identify the tissue involved and make a diagnosis. Because of the lack of differentiation in some tumors, many oncologists view cancer cells as having a block in differentiation.

Some embryonal carcinoma cells display no evidence of differentiation but can be induced to differentiate with retinoic acid or other agents (Strickland et al. 1980; McBurney 1993). This indicates that the anaplastic cells of cancer have the potential for differentiation, although they may not express it, and reinforces the idea that carcinogenesis does not alter the original histiotypic determination of the cells, it superimposes the malignant phenotype on it.

Retinoic acid is capable of inducing differentiation in a variety of other tumors. Is this the prototype experiment for a new type of therapy, differentiation therapy, for cancer with metastases?

Like the determined imaginal disc cells of the fly, cancer cells only express the normal (nonmalignant) phenotype when placed in the appropriate environment. Embryonal carcinoma cells express the malignant phenotype. But if they are placed in the blastocyst, they give rise to cells that differentiate normally. In the blastocyst they behave as normal embryonic cells in the sense that they respond to signals produced by the embryo, differentiate appropriately, and in the adult their offspring respond to homeostatic controls.

How universal is the fact that embryos can regulate their closely related cancers? It has been shown that leukemic stem cells injected into 10-day-old mouse embryos

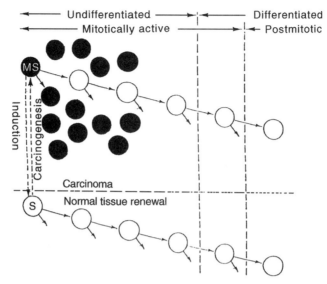

Figure 1–19. Chart of carcinoma as a caricature of the process of tissue renewal and embryogenesis. The normal process of renewal is illustrated at the bottom. Stevens demonstrated development of carcinoma from stem cells, Pierce demonstrated differentiation of malignant stem cells, Brinster demonstrated regulation of embryonal carcinoma by the blastocyst. The caricature, the gross misrepresentation, is the overproduction of malignant cells in relationship to the number that differentiate.

may be normalized to the point that the animals are chimeric in their leukocytic tissues (Gootwine et al. 1982). Neuroblastoma cells, injected into the neural crest migratory route, do not express tumors in expected numbers, but the mechanism of suppression of tumor formation is not known (Podesta et al. 1984).

Melanoma cells injected into the embryonic skin on the day normal melanoblasts arrive in the skin do not produce melanomas in expected numbers. In this situation, destruction of the melanoma cells by a mechanism responsible for programmed cell death apparently plays an important role (Parchment and Pierce 1989).

The concept that carcinoma is a caricature of the process of tissue renewal uses the term "caricature" only in the sense of a gross exaggeration of the normal (Pierce 1983). The caricature is illustrated in Figure 1–19. The malignant stem cells (MS) generate a myriad of anaplastic carcinoma cells for every cell that enters the differentiation pathway. This is the gross exaggeration.

The normal process followed by stem cells is illustrated in the lower part of Figure 1–19. The normal stem cell proliferates only in response to an unknown environmental signal that says "a senescent cell has been sloughed, and a new one is required." The malignant stem cell has the capacity for proliferation, producing many copies of itself for each one that differentiates. The mechanism for carcinogenesis is not known, but current dogma suggests mutation occurs in genes either for growth factors or receptors, or stable changes in regulatory circuits that preclude normal responses (Chapter 3). We are in the unhappy situation of knowing what cells can be made to do, but we do not know what they normally do! What is important is that an appropriate environment, such as the embryonic environment in the formation of chimeric mice, can correct the aberration. Essential to differentiation therapy is understanding differentiation mechanisms such as how the blastocyst corrects the embryonal carcinoma cell and incorporates it into the inner cell mass as a functional and apparently normal cell.

1.16 Apoptosis

Nearly a century ago it was discovered that certain cells in both adult and embryonic tissues of many organisms, from *C. elegans* to mammals, die independently of any exposure to toxic xenobiotics (Hengartner and Horvitz 1994). This naturally occurring type of cell death in healthy tissues acquired the name *apoptosis,* which described the appearance of the dying cell as subcellular structures fall away from the residual structure as "leaves fall from a tree in autumn." Apoptosis plays many important biologic roles, specifically eradicating redundant or spent cell types during development to prevent the formation of ectopic tissues (Pierce et al. 1990; Parchment 1993); removing embryonic tissues to generate the morphology of the adult (e.g., between the prospective digits in the developing hand and foot, or the tail of anura), balancing mitosis in renewing tissues to regulate tissue mass and selecting clones during lymphocyte maturation (McCarthy et al. 1992), and removing genetically damaged cells to prevent oncogenesis (Lennon et al. 1991).

Apoptosis causes the appearance of several distinctive cytologic features. The cytoplasm and chromatin condense (*pyknosis*), and the plasma membrane contracts, causing the dying cell to withdraw from its neighbors except at points of adhesive plaques; yet most, if not all, of the intracellular organelles, including mitochondria and lysosomes, remain active. Then the nucleus fragments into vesicles called apoptotic bodies that contain condensed chromatin fragments surrounded by condensed cytoplasm. Near the end of the process, the apoptotic cells and bodies are recognized and phagocytosed by adjacent cells, to be destroyed subsequently in lysosomes. Although molecular markers in some model systems can be used to monitor apoptosis, at this time there appears to be no universal molecular marker. Degradation of genomic DNA into nucleosome-sized fragments, activation of a Ca^{2+}, Mg^{2+}-dependent endonuclease, flip-flop of phosphatidyl serine from the inner to the outer leaflet of the plasma membrane, and up-regulation of cell surface receptors is a complex and tightly regulated sequence of events important for the normal biology of the cells (Adams 2003).

These cytologic features provide a straightforward distinction between apoptosis and a second process of cell death called *necrosis,* in which the dying cell shows swollen cytoplasm and mitochondria, clumped chromatin, and nuclear membrane and plasma membrane rupture, although late-stage apoptosis in the absence of phagocytosis and lysosomal degradation can resemble necrosis (Kerr et al. 1972).

Perhaps the most important distinction between apoptosis and necrosis is a bio-logical one: apoptosis is a controlled process that keeps intracellular substances contained within membranes at all times and therefore does not culminate with an inflammatory reaction; necrosis results in cytolysis and an inflammatory reaction which can often cause serious tissue damage (Kerr et al. 1972). Especially in the renewing epithelial tissues of the adult, apoptotic cells can be found completely surrounded by differentiated or proliferating cells.

This precise control of apoptosis, that is, its regularity, predictability, and localized nature, led to the notion of an intracellular, genetically programmed pathway that kills a cell from within and protects surrounding cells from the presumed lethal hit inside the cell, and hence the term "programmed cell death." The presence of isolated dead cells in the midst of a milieu of viable cells did not seem explainable unless there was a mechanism for such cellular suicide, and suicide is a reasonable explanation for the lack of adverse effects on neighboring cells. However, extracellular substances, such as tumor necrosis factor and monoclonal antibodies against epitopes on T- and B-cells, trigger apoptosis, and in the murine blastocyst, extracellular H_2O_2 triggers apoptosis of a few cells, expressing a developmental program that is no longer needed but spares the majority (Pierce et al. 1989; Parchment 1993). These data prove that death from within ("suicide") is not required to obtain an isolated apoptotic cell in the midst of living cells. Highly selective apoptosis can also be caused by extracellular factors to which all cells in the vicinity are exposed but to which only a handful respond (Parchment 1993). These more recent data indicate that apoptosis can be "murder" as well as suicide.

Recently, oncogenes and suppressor genes have been implicated in the regulation of apoptosis and the role of cell cycle progression in the commitment to the cell death program, both transcription factors and signal transduction mechanisms. Many of the functions of proto-oncogenes are required for apoptosis, and dysfunction of oncogenic forms can inhibit apoptosis and lead to an accumulation of cells (hyperplasia) that can contribute to malignancy (see Chapter 5).

1.17   Summary

This is the background on which the medical profession currently works: once cancer cells have progressed and metastasized, they cannot be extirpated or killed with X-rays; the only hope is treatment with cytotoxic agents. These lack specificity and destroy rapidly growing normal cells as a complication of treatment. The oncologist has little range for error in using these extremely toxic poisons, but this is the best hope for palliation and, rarely, a cure for these patients.

The need for an understanding of the pathology of cancer is underscored by the fact that carcinoma of the pancreas, for example, has a cure rate of about 5 percent. Does not that tell us that current treatments are inadequate? Should we persist in administering cytotoxic therapy with its deleterious side effects in such

tumors, hoping that a particular patient will be one of the lucky few? Should we
admit we really don't have a therapy for such malignancies and expedite the search
for new cytotoxins, or even better, for alternatives to cytotoxic treatment for such
carcinomas? The key to this approach may lie in the observation that embryonic
fields in every situation tested are able to reregulate malignant cells. In order to
develop new and effective modalities for treating cancer, we must first understand
the mechanisms of determination (Plasterk 1992) and the controls of proliferation
and differentiation in the embryo.

2

Invasion and metastasis

Robert G. McKinnell

The significance of metastases is that they "form an ineluctable hindrance to successful therapy."

Alton Ochsner and M. DeBakey 1942

The detection of metastases constitutes decisive evidence for categorizing a proliferating primary lesion, previously of uncertain potential, as neoplastic and "malignant," and the phenomenon is a topic of unparalleled importance in cancer medicine and biology.

D. Tarin 1992

2.1 Introduction

Ochsner and DeBakey's use of the terms "ineluctable hindrance" in describing the effects of metastasis on cancer therapy, written well over a half century ago, is essentially valid today. Metastasis is feared by both patient and physician and that fear is not without merit. Metastases remain the principal cause of death by cancer. But, changes in the understanding of invasion and metastasis may soon necessitate reconsideration of this dismal prospect for cancer patients. The phenomena of cancer invasion and metastasis, with its many steps forming a "cascade," have remained unchanged. However, the molecular events of metastasis, in particular metastasis suppressor genes and their expression, are now recognized and their understanding is evolving. Implicit with that understanding is the possibility, indeed the likelihood, of chemotherapeutic agents that specifically target metastasis suppressor genes. Understanding invasion and metastasis is critically important to the understanding of the pathogenesis of cancer.

It has been said that cancer is infrequently a localized disease. This is because of the propensity of malignant neoplasms to disseminate early in the disease and to grow as secondary tumors in the body of the host. This often occurs before the primary tumor is discovered. The establishment of secondary tumors, no longer contiguous to the primary tumor, is known as metastasis (Figures 1–15 and 2–1, Color Plates 10 and 11). Normal cells are strictly confined to their proper anatomic domains; the proclivity of malignant cells for growth at sites remote from their origin is the cardinal attribute of malignancy. Although cancer has been known at least since the time of Hippocrates, who named this group of diseases (Ewing 1928) (Chapter 1), it was not until Recamier (1829) described a brain tumor that had developed from a primary carcinoma of the breast that such secondary tumors were named. He coined the term "metastasis" to describe the local infiltration, invasion, and growth of secondary tumors. Later, Stephen Paget observed that metastatic tumors are not capriciously distributed in the body; rather, their distribution follows a pattern specific for each cancer. He likened cancer cells disseminated randomly by the vasculature to seeds that grow best in hospitable soil, that is, the tissue microenvironment of specific organs is receptive to certain cancer cells while other organs are less welcoming (Paget 1889, see box, this chapter). The "hospitable soil" requirement, of course, varies with the particular needs of the cancer cells and because of this, distinct patterns of metastasis occur with specific cancers.

James Ewing, in the first issue of the *Journal of Cancer Research*, compiled characteristics of cancer cells; he listed hyperplasia,[1] loss of polarity, promotion of connective tissue growth, anaplasia, infiltrative growth,[2] and metastasis. Of these six characteristics of cancer, metastasis was asserted to be the "most convincing evidence of lawless growth" (Ewing 1916). Although metastasis remains "the most convincing evidence" of cancer, it is clearly not entirely lawless. This is because the dissemination of malignant cells from a primary tumor with subsequent growth as multiple secondary colonies is known to follow a number of predictable steps.

Lawful (nonrandom) aspects of metastasis include the stepwise process of cell activities that leads to successful dissemination of malignant cells. Steps in the "metastatic cascade" are found among the cellular activities of normal cells. It has been postulated that knowledge of the mechanisms of normal cell translocation and subsequent growth at a new site may lead to enhanced understanding of the comparable phenomena in neoplastic cells. The predictability, in general, of metastatic tumor sites is another aspect of the lawful behavior of translocated cancer cells. Clearly, metastasis is not the result of nonspecific, random, or haphazard events – were it so, there could be no comprehension of this cardinal aspect of malignant cell behavior.

It is perhaps useful to consider the significance of therapeutic approaches to cancer in the context of the present chapter. Cancer treatment (Chapters 9 and 10) includes surgery, radiation, and chemotherapy. If a tumor is localized, then surgery is

---

[1] Ewing's article was written in 1916 and his use of the term "hyperplasia" is now antiquated. Contemporary pathologists use the term "increased cellularity" or "cellular hyperproliferation" (Kern and Vogelstein 1991) and retain hyperplasia only for benign conditions of increased mitotic activity.

[2] "Invasion" is a more generally recognized term.

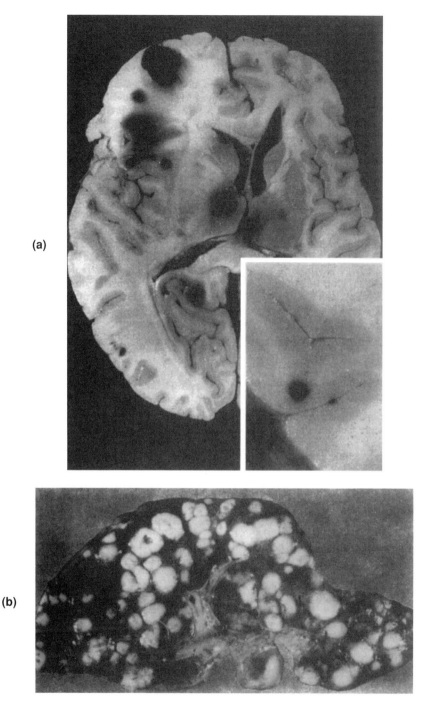

Figure 2–1. The devastation of metastatic cancer to human tissue: (a) A gastric adenocarcinoma metastatic to the brain (inset shows a metastatic lesion in the cortex-subcortical white matter interface). (From Mørk, S., Laerum, O. D., and deRidder, L. 1984. Invasiveness of tumours of the central nervous system. In *Invasion, experimental and clinical implications*, ed. M. M. Mareel and K. C. Calman, pp. 79–125. Oxford: Oxford University Press. By permission of Oxford University Press.) (b) Multiple metastatic lesions to the liver. (From Warren, J. R., Scarpelli, D. G., Reddy, J. K., and Kanwar, Y. S. *Essentials of General Pathology*. New York: Macmillan Publishing Company. Copyright © 1987 The McGraw-Hill Companies. Reproduced with the permission of the McGraw-Hill Companies.)

sufficient treatment for the obvious reason that a localized cancer ceases to exist with surgical ablation. Unfortunately, as stated above, cancer is infrequently a localized disease. Radiation therapy is also directed at the limited area thought to be the site of the primary tumor, and, with few exceptions, is palliative. Chemotherapy, however, is designed to eliminate disseminated cancer cells at *all* sites in the body, which is of course not possible with either surgery or radiation. Thus, chemotherapy is therapy of metastasis. However, until chemotherapeutic agents specifically target by repression those genes permissive for metastasis or enhance the expression of metastasis-inhibiting genes, chemotherapy will remain imperfect and have limited successes.

Cancers are not composed of homogeneous populations of cells. Cancer cells exhibit heterogeneity with respect to growth potential, chromosome constitution, cell surface receptors, sensitivity to chemotherapeutic agents, enzymes, angiogenesis-stimulating capability, morphology, and related to all of the preceding, competence for invasion and metastasis. Tumors are "societies" of diverse cell types, not all of which have the ability to form metastases and, thus, not all of which need be the target of exceptionally toxic chemotherapeutic drugs. The minority of cells that are invasive and which do form metastases must have the competence to perform a series of steps known as the metastatic cascade.

2.2   The metastatic cascade

One definition of cascade is a series of small waterfalls over steep rocks. Diversion or disruption of any one of the component falls stops water flowing at the lower parts of the cascade and at the mouth of the cascade. So it is with the metastatic cascade. A series of interrelated events, a multistep process, leads to metastasis. Disruption of any of the events in the cascade, at least in theory, restrains the establishment of disseminated cancer.

The interrelated events of the metastatic cascade are listed in Table 2–1 and illustrated diagrammatically in Figure 2–2. They are considered individually in the text that follows.

Table 2–1. *Events in the metastatic cascade*

- Disruption of the basement membrane
- Cell detachment (separation)
- Cell motility
- Invasion
- Penetration of the vascular system
- Circulating cancer cells
- Arrest (stasis)
- Extravasation and proliferation

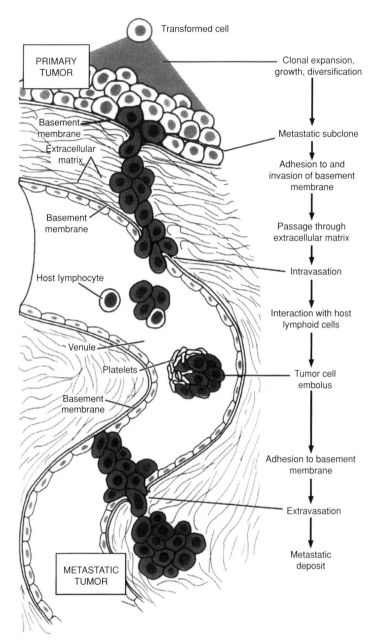

Figure 2–2. Cascade of events that lead to metastasis. Reading from the top of the diagram to the bottom, a transformed cell gives rise to mitotic progeny that form a primary tumor. The tumor is a mosaic of cells, some of which have the competence to give rise to metastatic colonies. The metastatic subclones (indicated with darkened cytoplasms and nuclei) bind to and disrupt the basement membrane on which they rest. They invade the extracellular matrix. Subsequently, they bind to and disrupt the basement membrane of a vessel as a prelude to intravasation. Thus, invasion requires, among other traits, cell motility and ultimately cell detachment from the primary tumor. The cells must also have the competence to penetrate the vascular or lymphatic system (intravasation) and circulate (either alone or as a component of an embolus) until they arrest by adhesion to the basement membrane of a vessel at a distant site. The nascent metastatic cells then exit the vessel (extravasation) at a site where they form a metastasis. (From Cotran, R. S., Kumar, V., and Robbins, S. L. 1994. *Pathological Basis of Cancer*, 5th ed. Philadelphia, PA: W. B. Saunders Company.)

2.2a Disruption of the basement membrane and lytic activity in the extracellular matrix

Most malignancies are carcinomas, which are tumors of epithelial cells (Chapter 1). Epithelial cells are supported by a basement membrane, a thin sheet of extracellular material, which separates epithelial cells from the vascularized extracellular matrix (ECM). Blood capillaries and lymphatic vessels are not found in normal epithelia, and epithelial cells must derive their sustenance and oxygen from subjacent ECM. Transformed epithelial cells become more complex with time and give rise to a diversity of cell types, the "heterogeneity" referred to previously. Subsets of malignant cells acquire the competence to degrade and invade the basement membrane and subsequently the subjacent connective tissue. Basement membranes are again breached when carcinoma cells gain access to vessels for dissemination. The controlled enzymatic digestion of ECM components is a major step in invasion and metastasis (Wilson and Sinha 2004; Egeblad and Werb 2002; Bogenrieder and Herlyn 2003).

Northern Leopard Frogs, *Rana pipiens*, are vulnerable to the Lucké renal adenocarcinoma (Chapter 6). The tumor is spontaneously metastatic. Basement membranes (BM) contain type IV collagen[3] as well as entactin, laminin, fibronectin, and heparan sulfate. The elaboration of an enzyme that degrades type IV collagen was sought and found in the Lucké renal adenocarcinoma (Shields et al. 1984). Why is this important? Disruption of the BM is absolutely a prerequisite for passage of cancer cells into the extracellular matrix. Spontaneously occurring mammary cancers of mice and human tumors similarly produce type IV collagen-degrading enzymes, which facilitate migration of cancer cells through the membrane and into the stroma.

Non-BM extracellular matrix proteins include collagen and other non-collagenous proteins such as fibronectin, elastin, and proteoglycans. Digestion of one or more of these proteins is essential for passage of tumor cells through the stroma. The frog renal tumor is aggressively invasive in vitro and metastatic in vivo at temperatures of 18° C and above, but fails to metastasize at cooler temperatures. Culture of frog tumor fragments at metastasis-restrictive and metastasis-permissive temperatures permits identification of enzymes that correlate with, and may be important to, metastatic behavior (McKinnell and Tarin 1984). Frog renal adenocarcinoma maintained at 30° C elaborates high levels of an enzyme that degrades type I collagen. Minimal enzyme activity was detected in tumor maintained at reduced, nonmetastatic temperatures. Normal kidney released negligible levels of collagen-degrading activity at both high and low temperatures. One interpretation

---

[3] Different types of collagen are formed of different but related collagen molecules. Further, specific amino acids (proline, lysine, and cysteine) of the collagens are modified so that several types of collagen are recognized, namely types I through V. The collagen types are found in particular anatomic locations; for example, type I is found in skin, bone, tendon, and connective tissue; type IV is found in, among other places, the basement membranes of epithelial cells.

of these experimental data is that normal frog kidney is metabolically less active (or moribund) at 30° C compared with the Lucké renal adenocarcinoma, which would account for the minimal collagenolytic activity of the normal kidney. This notion was shown to be incorrect by measuring glucose utilization of normal and malignant kidney. Normal renal tissue was obviously alive and functioning at both the initial low and subsequent elevated temperature (Figure 2–7). What was demonstrated here is a metastasis-permissive lytic response of the neoplastic tissue (collagenase release) that does not occur with viable kidney held under comparable conditions (Ogilvie, McKinnell, and Tarin 1984). The elaboration of collagenase correlates with invasion and metastasis in the frog renal tumor and thus strongly supports the role for this enzymatic activity in malignancy.

Another experiment supporting the metastatic role of proteolytic enzymes is that of a transformed rat sarcoma cell line. The rat tumor is metastatic in a lung colonization assay and secretes a matrix metalloproteinase. If the expression of that enzyme is inhibited, the competence to form experimental metastatic colonies is lost (Hua and Muschel 1996).

The matrix metalloproteinases (MMP) consist of at least twenty-five enzymes and each is known by number. MMPs include collagenases, gelatinases, stromelysins and matrilysins and these enzymes have the competence to degrade essentially any protein of the ECM (Wilson and Sinha 2004). Elevated MMP activity is found in all malignancies. While our focus on MMPs here concerns the enzymatic degradation of components of the basement membrane and the stromal extracellular matrix, MMPs have been implicated in a number of cellular events that are critical for cancer progression. These include, in addition to invasion, control of cell migration, growth, apoptosis and angiogenesis (Wilson and Sinha 2004; Egeblad and Werb 2002).

A marine cartilage extract known as AE-941 is under investigation for its antimetastatic properties. AE-941 is reported to markedly inhibit several matrix metalloproteinases (Gingras et al. 2001). Similarly, a reduction in the potential for tumor cell colonization of the lung (Figure 2–3) by a rat cell line resulted from the use of a tissue inhibitor of metalloproteinase (TIMP) (Alvarez, Carmichael, and Declerck 1990). Because expression of MMPs favors invasion and metastasis, one would expect that all TIMPs would similarly inhibit invasion and metastasis. Such is not the case. Some TIMPs enhance cancer progression, growth, and angiogenesis and also inhibit tumor cell apoptosis (Egeblad and Werb 2002). The role in cancer of TIMPs is complex and still unfolding.

There is abundant evidence of MMPs' role in metastasis from animal and human studies, especially where enhanced expression of the enzymes results in tumors becoming more aggressively malignant, and the converse, that is, when aggressive tumors become more benign with downregulation. Enhanced expression of MMPs in human cancer is related to tumor stage, intensified invasion and metastasis, and perhaps most grimly, reduced survival time (Egeblad and Werb 2002).

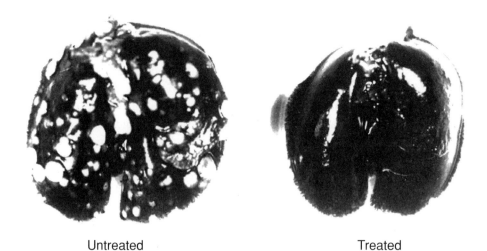

Untreated   Treated

Figure 2–3. Lungs of animals injected with a highly metastatic rat embryo cell line. The lungs on the left were of untreated hosts that received tumor cells plus sterile buffer intraperitoneally. The lungs on the right were of an animal that received tumor cells and was treated with intraperitoneal injections of a recombinant human inhibitor of metalloproteinases (rTIMP). Lung tumor colonies were inhibited by 83% in the rTIMP-treated animals. (From Alvarez, O. A., Carmichael, D. F., and Declerck, Y. A. 1990. Inhibition of collagenolytic activity and metastasis by a recombinant human tissue inhibitor of metalloproteinases. *J Natl Cancer Inst* 82:589–95.)

Cathepsins consist of a number of enzymes that catalyze the hydrolysis of peptide bonds. Laminin, fibronectin, collagen, and proteoglycan are extracellular matrix and basement membrane components. Increased levels of cathepsin B (CB) are associated with degradation of these substances in many solid tumors. Cathepsin B's localization in invasive cells and capillaries of human prostate cancer strongly supports its role in metastasis. Can the information about the presence of a particular enzyme active in a tumor, in this case CB, predict aggressiveness and outcome of a common malignancy? That rhetorical question may have an answer with human prostate cancer, especially if considered in relation to an enzyme inhibitor. Stefin A is an inhibitor of cysteine proteases, including cathepsin B, and may serve as protection against proteolytic activity during invasion. Levels of CB and stefin A vary in prostate cancer even within the same histological and morphological grade (Sinha et al. 1999, 2001). What is of interest here is that within a particular prostate cancer, heterogeneity (see above and Chapter One) is found in the ratio of cathepsin B and stefin A. Cathepsin B may be less than, equal to, or greater than stefin A. When the level of cathepsin B exceeds that of stefin A, there is a more dismal prognosis. In this particular case, the ratios identify aggressiveness within a given grade (Chapter 1) of cancer and uphold the notion of the critical importance to invasion of an enzyme and its inhibitor (Sinha et al. 2002). It may be anticipated that identification of aggressive cancers will aid in selecting effective treatments.

Plasminogen activators have the ability to cleave plasminogen to give plasmin. Plasmin, the active form, hydrolyzes fibronectin, laminin, and other glycoproteins found in the basement membrane and the extracellular matrix. Plasmin also has the competence to activate other proteases. As with enzymes important for invasion and metastasis (described above), inhibitors of both plasminogen activators and plasmin are found in tissues. These complexities of enzymatic activities associated with invasion and metastasis are reviewed by Wilson and Sinha (2004).

Although transgression of the basement membrane with access to the stromal extracellular matrix is an important early event for carcinoma cells, those cells must be competent to disseminate and survive in the circulation if metastasis is to occur. In other words, cells must detach and move away from the primary tumor.

2.2b   Cell detachment

As soon as it was recognized that metastasis results from the translocation of cancer cells, it became obvious that a fundamental property of those cells which detach must be reduced cohesion. Ordinarily, normal cells do not separate from their cohorts in a tissue; they stay in place. Cancer cells, however, exhibit a deficiency in cohesiveness, and this reduction or loss is essential in the metastatic cascade.

A number of years ago, Coman studied both normal and malignant lip and cervical cells. He reported that it required more force to separate normal cells from each other than it did to separate their malignant counterparts. The cell pairs were impaled on microneedles, and the force necessary to separate them was calculated by the bend of the needles as the cells were pulled apart. Normal pairs clung to each other tenaciously resulting in considerable cell distortion during separation, in marked contrast to the cancer cells, which separated easily (Figure 2–4) (Coman 1944). An alternate method to demonstrate loss in cohesion is to shake tissue fragments contained in a culture medium and count the cells that detach. More cells were detached from cancers of the stomach, colon, rectum, thyroid, and breast than from their normal tissue controls. These simple observations led to an analysis of factors affecting cohesion of cancer cells.

Normal frog kidney cells do not separate from each other when agitated in a buffered calcium- and magnesium-free electrolyte solution at any physiologic temperature. However, with gentle agitation frog kidney cells that are malignant (the Lucké renal adenocarcinoma) readily detach from a tumor fragment at elevated physiologic temperatures (18° to 28° C), but not at temperatures cooler than 18° C (Seppanen et al. 1984). In this study, conditions that permit decreased cell cohesion are similar to conditions that permit the elaboration of type I collagenase. Next, it was observed that ethylene diamine tetraacetate (EDTA) inhibited tumor cell dissociation. This is in contrast to what was expected; EDTA has been used for years in embryologic studies to enhance dissociation. However EDTA, in addition to its role in dissociation, also inhibits collagenase. Excess cysteine inhibits both cell detachment and the enzymatic activity of type I collagenase. These observations are

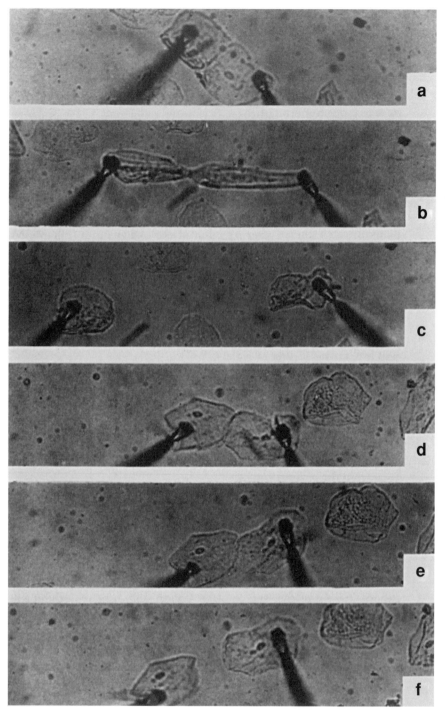

Figure 2–4. Separation of living squamous epithelium cells from a normal lip and their malignant counterparts from a carcinoma of the lip. Cell pairs are impaled on microneedles attached to a micromanipulator. One of the needles is moved while the other remains stationary. Note that the normal cells (a) are tenaciously attached to each other, and the microneedles stretch and distort the cells (b) prior to their being separated (c). In contrast, cancer cells (d) readily detach with minimal distortion and little movement of the microneedles (e and f). (From Coman, D. R. 1944. Decreased mutual adhesiveness, a property of cells from squamous cell carcinomas. *Cancer Res* 4:625–9.)

consistent with the hypothesis that cell detachment in the Lucké renal adenocarcinoma is mediated, at least in part, by the activity of an enzyme that is, or is similar to, a collagenase.

The Lucké renal adenocarcinoma cultured under noninvasive conditions (18 degrees or less) is highly differentiated, whereas the tumor appears anaplastic when maintained under conditions that are permissive for invasion (temperatures above 18 degrees) (McKinnell et al. 1986). In other words, poorly differentiated frog tumors metastasize more frequently than highly differentiated ones. The relationship of differentiation state and invasion competence of the frog malignancy holds for other tumors as well. The expression of the calcium-dependent cell adhesion molecule E-cadherin is related to retention of the epithelial state and normal morphology with a noninvasive phenotype. Carcinoma cells that do not express E-cadherin lose their characteristic morphology and become invasive (Frixen et al. 1991). Suppression of invasion is correlated in vitro with activity of E-cadherin (or P-cadherin) (Foty and Steinberg 1997). As stated earlier in this chapter, interruption of any step in the metastatic cascade should abrogate the malignant process, and this applies to the cell–cell adhesion molecule E-cadherin (Fujimoto et al. 1997). Transfection of invasive cells with E-cadherin cDNA, with the resumed production of this molecule important in cell adhesion, blocks invasion. Clearly, detachment of cancer cells from the primary tumor mass is of critical importance for metastasis.

Detached cells depend on motility for passage through the extracellular matrix and for ultimate contact with capillaries or lymph vessels. Factors related to cell motility are described next.

2.2c   Cell migration and motility

Cells, or small cell clumps that appear to have become detached from a primary tumor, are regularly observed by pathologists. The observation of detached cells was interpreted to mean they were released from the restraints of the tumor mass and had locomotive competence (Strauli and Haemmerli 1984). Nearly a century and a half ago, the pioneer German pathologist Rudolf Virchow reported morphologic observations consistent with the concept that cancer cells possess "amoeboid" motility (Virchow 1863). Confirmation was made of cancer cell motility in vitro (Figure 2–5) (Lucké 1939; Enterline and Coman 1950) and in vivo (Farina et al. 1998).

The phenomenon of cancer cell migration leading to metastasis (in contrast to the molecular and biochemical events that take place during migration) can be described in several steps. As has been stated before concerning any of the cascade of events leading to metastasis, disruption or termination of cell motility and migration would hinder or restrain metastasis.

The solid malignant lump or mass cannot migrate. Individual cancer cells must break away from the primary lump. Since most cancers are carcinomas, the detached motile cells must breach the basement membrane and migrate across that structure as individual cells. Once in the stromal extracellular matrix, the cancer cells must

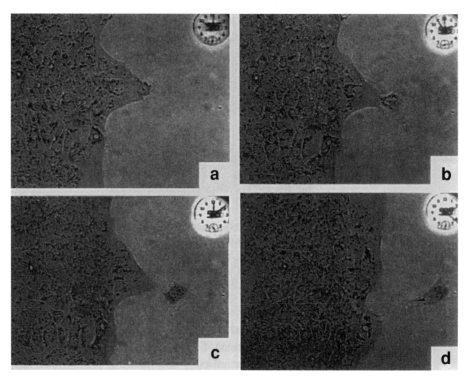

Figure 2–5. A culture of the Lucké renal adenocarcinoma showing detachment and outmigration of a tumor cell. The initially attached cell (a) begins its detachment (b) and wanders away from the tissue mass (c and d). Motility of this cell was recorded during a period of 4.5 hours. (From Lucké, B. 1939. Characteristics of frog carcinoma in tissue culture. *J Exp Med* 70:269–76. By permission of The Rockefeller University Press.)

move to access blood vessels and migrate through the vessel wall. This is called intravasation. Migration in the vascular system is passive but rapid (translocation perhaps is a better term for this passive movement). Most cancer cells do not survive the vascular rapid transit system but those that do must now stop and begin their migration out of the vasculature (termed extravasation). For continued cell division and growth to occur, the extracellular matrix must be accessed at a hospitable and appropriate site. Migratory activity of cancer cells results from an interaction of the extracellular matrix with cells. Interaction of cell surface receptors with ligands in the ECM profoundly affects cell detachment and migration.

Cancer cell motility can be quantified by an in vitro assay that counts cells that squeeze through microscopic pores in a membrane. The porous membrane separates the cells from a substance which elicits a migratory response. A frog tumor cell line will migrate across eight micrometer pores in response to the chemoattractant fibronectin. As might be expected with cells from an ectothermic animal, the number of cells that migrate through the pores with any given time is directly proportional to temperature (within physiological limits). The migratory response is not unique to the tumor cell line and is shared with cells from primary frog tumors and normal embryos (Hunter, Tweedell, and McKinnell 1990). While this is an expected observation, it emphasizes that cancer cell behavior is frequently similar to or a variant of normal cell behavior.

If malignant cells with a high potential for metastasis differ in cell motility from malignant cells which are less aggressive, then that difference may become useful in identifying patients for intensive early treatment. Moreover, malignancies with high cell motility and high metastatic competence suggest a need to focus drug research on agents that affect motility directly. It has been shown that cell lines obtained from a rat prostate cancer that exhibit high cell motility also have high metastatic competence. The expression of glyceraldehyde-3-phosphate dehydrogenase, an enzyme involved in the increased glycolysis seen in many tumors, correlates with cell motility and metastatic potential in the rat cells (Epner et al. 1993). If human prostatic cancer cells behave similarly to the rat prostate cell lines, then it might be feasible to predict which malignancies have a high potential for metastasis on the basis of the expression of the glycolytic enzyme. Whether or not this ever proves to be the case, the study is useful to illustrate the thinking of biologists who seek to exploit differences between normal and malignant cells for the treatment of cancer.

Motility is a characteristic of cancer cells freed of the restraining organization of the ECM. Chemoattractants that are important in motility have been identified (McCarthy et al. 1985), but what initiates locomotion? The autocrine cytokine known as autocrine motility factor (AMF) and activation of its receptor (AMF-R) stimulates cell motility. Higher gene expression of AMF-R in some lung cancer cell lines is related to increased cell motility compared with cell lines with lower expression. Patients with a high expression of the receptor gene have a more dismal outlook than those with a lower expression (Takanami and Takeuchi 2003).

Discussed previously were enzymes important in lytic degradation of basement membranes and the extracellular matrix. One of the matrix metalloproteinases (MMP2) partially digests basement membrane laminin, thereby opening a formerly hidden ("cryptic") site which induces cancer cell motility (Pirila et al. 2003). Of course, response of cells to motility clues in the microenvironment is a normal part of development (Quaranta 2002) and commonalities of tumor cell behavior and normal cell behavior is expected and observed.

The cytoplasmic microtubule complex (CMTC) (Figure 2–6) is structured of tubulin and implicated in directional migration. The chemotherapeutic vinca alkaloids vinblastine, vincristine, vindesine, and the synthetic compound nocodazole are agents that depolymerize microtubules and also inhibit invasion. Chemotherapeutic agents, including 5-fluorouracil and mitomycin C, which do not depolymerize microtubules, do not inhibit invasion in vitro (Mareel et al. 1982; Mareel, DeBaetselier, Van Roy 1991). Adriamycin, another type of chemotherapeutic agent, inhibits tumor cell motility and invasion in vitro at a concentration that does not have an effect on tumor cell growth (Repesh et al. 1993).

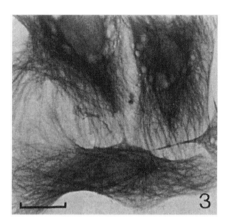

Figure 2–6. Light microscope photograph of frog tumor cells cultured on glass and stained with an antiserum against tubulin. Microtubules are revealed as fine lines in the cytoplasm of the three cells in this photograph. (From McKinnell, R. G., DeBruyne, G. K., Mareel, M. M., Tarin, D., and Tweedell, K. S. 1984. Cytoplasmic microtubules of normal and tumor cells of the leopard frog. Temperature effects. *Differentiation* 26:231–4.)

It can be seen from this discussion that regulation of cancer cell motility is probably accomplished via a number of molecular mechanisms. Which of these, or of others yet to be discovered, that can be exploited for the control of metastasis remains to be seen.

2.2d   Invasion

Carcinoma cells that have detached from the primary tumor mass, penetrated the basement membrane, and exhibited motility may now come in contact with the extracellular matrix (ECM) or connective tissue (Figures 2–2 and 2–7). Capillaries in the ECM are a biologically significant destination for motile tumor cells. In a sense, the problem of gaining access to capillaries is similar to that of breaching the basement membrane; that is, the cells need to digest a pathway to nearby small vessels.

Invasion assays of various kinds have been developed. The assays are designed to discriminate between normal noninvasive cells and their malignant counterparts. Obviously, carcinoma cells, which rest on a basement membrane, must transgress that membrane to gain access to connective tissue and capillaries. Various basement membranes, both real and artificial, have been used in invasion studies. Recently, sea urchin embryo basement membranes have been used for that purpose. The sea urchin basement membrane is biochemically similar to its mammalian counterpart; for example, it contains laminin, fibronectin, collagen type IV, and heparan and chondroitin sulfate proteoglycans. It is of interest to note that malignant cells of mammals recognize and invade the sea urchin membrane, which not only indicates the usefulness of the invertebrate membrane in experimental studies but "suggests that molecules participating in basement membrane recognition and invasion have been functionally conserved during the time separating vertebrates from invertebrates" (Livant et al. 1995).

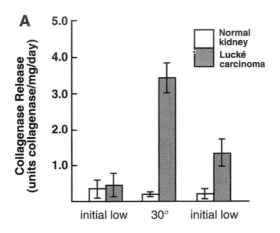

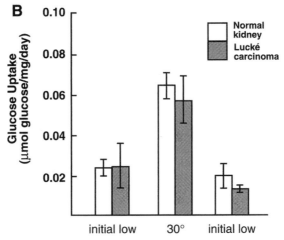

Figure 2–7. (a) Collagenase release by normal and neoplastic renal tissue at two temperatures. The initial low temperature, 7° C, is metastasis inhibiting; the second temperature, 30° C, is metastasis permissive. The Lucké renal adenocarcinoma releases significantly greater quantities of type I collagenase than does normal renal tissue cultured under identical conditions. (b) Glucose uptake by normal renal tissue at both temperatures attests to the viability and functioning of the normal tissue. Thus, the type I collagenase response illustrated in (a) is tumor specific and probably related to invasion. (From Ogilvie, D., McKinnell, R. G., and Tarin, D. 1984. Temperature-dependent elaboration of collagenase by the renal adenocarcinoma of the leopard frog, *Rana pipiens*. *Cancer Res* 44:3438–41.)

2.2e Penetration of the vascular system

Cancer cells disseminate via capillary and lymphatic vessels. The thick wall of an artery is rarely penetrated. It is believed that cancer cells can enter newly formed capillaries more easily than preexisting capillaries because of defects in the new blood vessels, such as gaps between endothelial cells and a discontinuous or absent

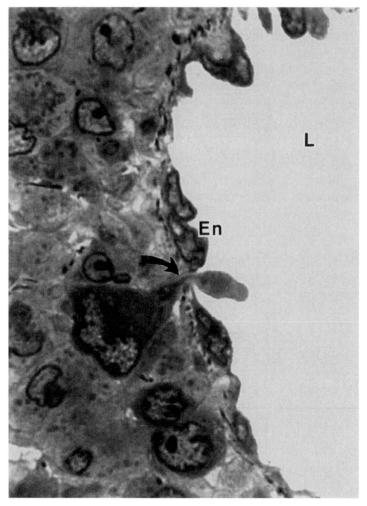

Figure 2–8. Cancer cells gain access to blood circulation by diapedesis. The cancer cell traverses a small opening (*arrow*) between endothelial cells (En) of the vessel to gain access to the lumen (L). (From Song, M. J., Reilly, A. A., Parsons, D. F., and Hussain, M. 1986. Patterns of blood-vessel invasion by mammary tumor cells. *Tissue Cell* 18:817–25. By permission of Churchill Livingstone.)

basement membrane (Folkman 2003). Even if defective capillaries are not present, the entrance of tumor cells into vessels may be facilitated by the production of type IV collagenase (Liotta et al. 1979; Shields et al. 1984). Neoplastic cells gain access to small vessels by digesting an opening in the outside wall of the vessel through which they migrate to the lumen (Figure 2–8). The process of slipping through the opening in the vessel is known as diapedesis (Engell 1955). Fortunately, most of these cells are damaged or destroyed upon entrance to the circulation, and only a minority of cells live to express their metastatic potential.

2.2f Cancer cells in the circulation

Cells thought to have originated in cancers were detected in the blood of a cancer patient well over a century ago (Ashworth 1869). With remarkable insight, the author wrote, "The fact of cells identical with those of the cancer itself being seen in the blood may tend to throw some light upon the mode of origin of multiple tumours existing in the same person." As many as three million cancer cells per gram of rat mammary carcinoma enter the circulation per day (Butler and Gullino 1975). What is the implication of cancer cells in the circulation? Quite simply, malignant cells in blood vessels are obviously a requisite for hematogenous dissemination. It is worth noting here that the density of capillaries in a tumor is positively correlated with metastatic behavior (Bosari et al. 1992). Obviously, capillaries provide an avenue for entrance into the vascular system; the more avenues, the greater the traffic flow. With increased vascular traffic, enhanced metastasis is expected.

The fate of the cancer patient cannot be foretold by the presence of circulating cancer cells (Snellwood et al. 1969; Griffiths et al. 1973; Salsbury 1975). Indeed, eight of fourteen inoperable cancer patients with shunts to recirculate peritoneal fluid containing viable cancer cells to the general circulation had no evidence of metastatic cancer (Tarin 1992; Tarin et al. 1984). Even when comparing cancers of the same histological origin, some patients with the shunts were found to be metastasis-free in contrast with other patients who revealed metastases at autopsy. All patients with the shunts were exposed to a vast number of circulating cells but metastases, as indicated, appeared in fewer than half of the patients.

While patient fate cannot be predicted by the presence of blood-borne cancer cells, it must be noted that hematogenous dissemination and subsequent metastasis obviously cannot occur without blood-borne neoplastic cells. Consequently, surgeons attempt to minimize the release of cancer cells during operations.

2.2g Arrest of circulating cancer cells (stasis)

Cancer cells in the circulation adhere to each other and to lymphocytes and platelets, forming emboli that may adhere to the inner surface of capillaries. A fibrin-containing thrombus (blood clot) forms, which stabilizes the embolus. Growth of the tumor cells and possible occlusion of the capillary follows. The stabilized cancer cells interact with the capillary endothelial cells, causing endothelial cell retraction with the exposure of the subjacent basement membrane. Metastasis is not an efficient process; some cancer cells may die within the thrombus, and the fibrin may dissolve, resulting in the subsequent loss of the stabilized tumor cell mass to the general circulation.

Treatment with warfarin, an anticoagulant, enhanced survival time of terminally ill, small cell lung carcinoma (Color Plate 9) patients but had little effect on survival of patients with other tumors (Zacharski 1984). Presumably the efficacy of warfarin, and other anticoagulants, derives from the dissolution of the clot, resulting in exposure of the released cancer cells to the hazards of the blood circulation (Hejna, Raderer, and Zielinski 1999). Salsbury (1975) was probably correct when he speculated that cancer cells "may do less harm to the host in the circulation than elsewhere." Warfarin used as a rat poison results in death to the rodent by internal bleeding. The anticoagulant activity that kills rats is the same property that investigators sought to exploit with the small cell lung carcinoma patients.

Warfarin has been suggested for use with other problems in cancer treatment. Venous thromboembolism may occur with long term use of central venous catheters to administer cancer chemotherapy. Anticoagulant prophylaxis is being studied as a mode of prevention of the very serious health risks from unwanted coagulation due to the catheters. The use of warfarin in medicine illustrates the potential of the prudent use of a toxic substance for the benefit of patients.

The role of a fibrin enclosure of a tumor cell nidus (cluster) and lysis of that enclosure is complex. The fibrin enclosure may act as a barrier to undefined and unknown noxious factors and stabilize the trapped tumor cells, resulting in enhanced invasion. Accordingly, digestion of the enclosure (fibrinolysis) may decrease invasion. Alternatively, fibrinolytic activity may enhance invasion. Digestion of the fibrin barrier may increase cell detachment and active invasion. Thus, fibrinolysis may either inhibit or enhance metastatic behavior depending on the particular tumor.

2.2h   Extravasation, growth of metastases, and metastasis of metastases

Extravasation of cancer cells requires cell motility and transgression of the basement membrane of the capillary. By active migration, the cancer cells move between the retracted endothelial cells, through the breached basement membrane, and into the intercellular matrix of the connective tissue (Figure 2–9). Resumption of neoplastic growth with the proliferation of tumor cells may either be immediate or after a variable period of dormancy (Chapter 1). The nidus of cancer cells is restricted in growth to a lump 0.5 to 1 mm in diameter due to the limited nutrient and oxygen supply available by diffusion from blood. Without angiogenesis, the center of the tumor cell colony becomes necrotic. However, new capillaries are recruited into the metastases that will enhance their survival. This occurs via activation of an angiogenesis switch. Vascularization of the new tumor colony allows blood to deliver essential oxygen and nutrients and permits the removal of toxic metabolites from the metastatic mass. Growth follows. With growth, the whole process repeats, and the metastatic colony will give rise to additional metastatic colonies; thus, the metastasis of metastases.

2.3 A multiplicity of genes are associated with metastasis

Some cancers are highly metastatic – others exhibit little tendency to disseminate through the bloodstream. Is it possible to discriminate between these extremes in neoplastic cell behavior by the expression of particular genes? Identification of such genes would be useful in prognosticating which cancer patient is unlikely to suffer metastasis at diagnosis and which patient is vulnerable to early spread of the neoplasm with an unfortunate outcome. Knowing which neoplasms metastasize exuberantly, seemingly without restraint, presents the possibility of identifying cellular processes that govern metastasis and offers the hope of exploiting that knowledge for the control or cure of cancer. It is not surprising, therefore, that genes that control metastasis have been sought. Although literally hundreds of individual genes important for the process of metastasis have been identified, most of the genes show only an association with metastasis and a clearly cause/effect relationship between the gene and metastasis has not been demonstrated. Metastasis is a complex series of events, and it is likely that the phenomenon is under the control of many genes, which when activated or inhibited, in a coordinated manner, result in the inappropriate cell behavior of invasion and metastasis. It is not growth per se of the primary tumor that results in metastasis; rather, it is the inactivation of tumor suppressors and the activation of genes that enhance metastasis working together that result in this dismal event.

Induction of metastatic behavior was demonstrated by transfection of cultured rat fibroblast cells with human DNA containing *ras* oncogenes. These rat cells in turn produced lung metastases in 100 percent of nude mice injected with the cells. Increased production of type IV collagenase was correlated with metastatic competence (Thorgeirsson et al. 1985). This is an elegant study that associates competence of tumor cells to grow as metastatic colonies with the production of type IV collagenase – it should not be interpreted that metastatic behavior is dependent upon the presence and expression of a single gene.

*Nm*23 is a gene with a long evolutionary history and it may provide insight into the metastatic process. Reduced expression of this gene is correlated with metastasis in some, but not all, malignancies. The gene, *nm*23, is generously expressed in normal mice and in melanoma clones with a low metastatic potential but is poorly expressed in clones with a high metastatic potential (Steeg et al. 1988; Freije, MacDonald, and Steeg 1996). The gene is found in human breast cancer where relatively high *nm*23 mRNA levels (compared to other metastatic cancers) are reported to be correlated with well-differentiated cancer, lack of lymph node metastasis, and longer disease-free survival (Hennessy et al. 1991).

*Nm*23 is highly conserved in bacteria (Mundoz-Dorado, Inouye, and Inouye 1990), slime molds (Wallet et al. 1990), *Drosophila* (Biggs et al. 1990), frogs (Kim et al. 2000), and rats (Kimura et al. 1990) and its appears to be ubiquitously found in all organisms. Certainly, the retention of molecular similarity across such phylogenetically diverse and ancient groups argues strongly for an important role for *nm23*. The gene produces nucleoside diphosphate (NDP) kinase. NDP kinase (and thus *nm23*) catalyze transphosphorylation of a phosphate from a nucleoside triphosphate to a nucleoside diphosphate. However, the biochemical mechanisms that specifically inhibit metastasis (where such inhibition is recognized) are not known. Much remains to be learned about the regulatory roles of the metastasis suppressor gene *nm23* in cancer biology (Salerno et al. 2003).

Other genes have been identified as potentially important in metastasis. One family of genes that encodes for a glycoprotein believed to be involved in cell adhesion is designated CD44. The gene is located on the short arm of chromosome 11. The CD44 exons are transcribed in different combinations (a phenomenon known as alternative splicing), giving rise to variant isoforms depending on cell activity. Matsumura and Tarin (1992) reported that they could distinguish between metastatic tumors of the breast or colon versus those with no metastasis on the basis of the overproduction of specific variants of the CD44 genes. They suggested that overexpression of the differently spliced genes could be a useful tumor cell marker (Chapter 1) and assist in the identification of tumors with metastatic potential using surgical biopsy (Tarin 1996).

The suppression of metastasis in prostate cancer is associated with a gene located on the short arm of human chromosome 11. The gene is designated KAI1 (*kang ai* is Chinese for "anticancer"), also known as CD82, and its expression may interfere not only with prostate cancer metastasis (Dong et al. 1995) but also metastasis in breast, bladder, pancreatic, ovarian, colorectal and other cancers. The KAI1 gene expression may be related, in some cases, to the expression of p53.

Because metastasis remains the principal cause of death in cancer, there is much interest in understanding metastasis suppressor mechanisms. The suppressors discussed here and others that are not have been reviewed recently (Shevde and Welch 2003).

2.4   Soil and seed hypothesis of Paget

Cancer cells that gain access to the vascular system are disseminated throughout the body. If tumor cells were to arrest, escape from the vasculature, and grow as secondary colonies at *random* anatomic sites, then there would be no pattern of specific target organs associated with particular metastatic neoplasms. This is not what happens. It has long been recognized that prostatic carcinoma metastases are commonly found in bone, small cell carcinoma of the lung (Color Plate 9) has a predilection for the brain, and neuroblastoma metastases frequently occur in the liver. The hypothesis that certain cancer cells grow at specific secondary sites is known as the "soil and seed" theory of cancer cell metastasis (Paget, 1889; see box this chapter).

An alternative view to this hypothesis, which does not necessarily exclude the concept that certain tissues or organs are particularly favorable for the growth of some cancers, has been suggested. It simply states that organ preference is a function of entrapment of cancer cells in the first capillary bed encountered. Tumor cells released into capillaries, venules, or lymphatics and transported to the heart may be caught in the sieve of the pulmonary capillaries. Tumors that release cells to the hepatic portal vein may colonize the liver. Subsequently, the hepatic or pulmonary colonies may release cells that express their affinity for a particular organ or tissue (Sugerbaker 1981; Tarin 1992). Therefore, the location of a metastasis may in some cases be related to blood-borne cells being trapped in a capillary bed.

The pattern of organ involvement in metastasis probably depends on both the mechanical sieving effect of capillary beds *and* the presence of specific microenvironmental conditions in certain organs that favor the adhesion of tumor cells in a capillary at a particular site, the escape from that capillary, and the subsequent growth of a metastatic colony at the nonrandom site.

STEPHEN PAGET, M.A., F.R.C.S.
(Founder of the Research Defence Society).

Stephen Paget: No "ploughman" was he!

Were the distribution of metastatic cancer colonies truly haphazard or fortuitous, there would be little possibility of understanding the subtleties of cancer spread. That of course is not the case. It has been known for over a century that many cancer cells have a propensity to colonize certain organs in preference to others. Stephen Paget (1889) sought to demonstrate that "the distribution of the secondary growths is not a matter of chance" and he contemplated the question "What is it that decides what organs shall suffer in a case of disseminated cancer?"

An example: Stephen Paget considered metastasis in 735 cases of breast cancer in which autopsies were performed. He reported 241 instances of metastasis to the liver in contrast to only 17 metastatic colonies to the spleen. Paget noted that the artery to the spleen is larger than the artery to the liver and consequently the spleen must be exposed to more blood-borne cancer cells than the liver. However, the liver's lower exposure to cancer cells resulted in a metastatic rate more than 14 times greater than that of the spleen. Clearly the distribution of metastatic colony growth was neither passively determined nor random. Paget presented other evidence of metastatic "predisposition." In his discussion of metastasis, he analogously ventured "(W)hen a plant goes to seed, its seeds are carried in all directions; but they can only live and grow if they fall on congenial soil." That sentence is the origin of the "soil and seed" concept of cancer cell metastasis.

Paget, born 17 July 1855, was the youngest child of the distinguished Sir James Paget. The father provided a rich environment for Stephen and his siblings. Sir James counted among his friends Tennyson, Browning, Huxley, and Pasteur. He had a special relationship with his much admired compatriot Darwin. The excitement of meeting these of his father's acquaintances must have stimulated the desire to learn in young Stephen. The elder Paget was a student of botany and perhaps some of his love of plants rubbed off on Stephen, who as described above, likened metastatic cells to the seeds of plants. Stephen became a student at Christ Church, Oxford. He studied medicine at Saint Bartholomew's Hospital Medical School and later served as a senior surgeon at several London hospitals. Paget died in 1926 (Poste and Paruch 1989).

Stephen Paget's love of learning resulted in his being the author of many books, including discourses on surgery, experiments on animals, John Hunter, and Louis Pasteur.

Paget spurned those who only enumerated cancer records as mere "ploughmen"; in contrast, those who sought *insight* into the nature of the "seed" (and of course "soil" too) were likened to "scientific botanists." Stephen Paget's work on metastasis has withstood the scrutiny of scientific evaluation for well over a century and in this sense, he was incontrovertibly a "scientific botanist."

2.5 Is metastasis limited to malignant cells?

If one assumes that the malignant tumor cell is inappropriately expressing a preexisting normal cell program of physiological invasion, then the fundamental difference between normal and malignant cells is regulation.

Kohn and Liotta 1995

Although Ewing (1916) considered metastasis the most convincing evidence of cancer (and few would disagree with him), many if not all aspects of metastatic behavior can be observed and studied in normal cells. Frost and Levin (1992) correctly stated, "There is nothing that a metastatic cell can do that is not a routine task for normal cells." The reason for listing attributes of normal cells that (seem to) mimic malignant behavior is twofold. First, the inappropriate behavior of cancer cells seems not to derive from cell properties that develop de novo but is related to the expression of otherwise normal cell attributes at an inappropriate time or place. Second, in at least some instances, the normal cell analogue may be more convenient to study than the same characteristic in malignant cells.

Motility is a characteristic of many normal cells as it is of malignant cells. Embryonic and adult cell types that exhibit cell motility are neutrophilic, eosinophilic, and basophilic granulocytes, lymphocytes, monocytes, and macrophages of the blood (Armstrong 1984b), neural crest cells (Halloran and Berndt 2003), myogenic stem cells of the somites involved in the formation of the wings and legs of chick embryos (Rees et al. 2003), and primordial germ cells (Raz 2003). While most developmental biologists subscribe to the notion of motile cells in development, the "active" migration of *some* normal cell types has recently been questioned (Freeman 2003).

Primordial germ cells (PGCs), early progenitors of sex cells found in the embryo in chordates, do not have their embryologic origin in the gonad; rather, because of cell motility and other activities, they translocate from their site of origin to the gonad. In mammalian embryos, the PGCs develop first in the posterior yolk sac near the allantois. The PGCs subsequently translocate via the hindgut to hindgut mesentery, from there to the dorsal body wall, and then to the germinal epithelium adjacent to the developing mesonephros (Witschi 1948). One could use Paget's venerable soil and seed aphorism to describe the PGCs as seed and the germinal epithelium as soil. PGCs that reach the hospitable germinal epithelium flourish and ultimately give rise to functional gametes after their metastasis-like trek.

Another instance of the translocation of yolk sac cells involves hematopoietic stem cells. These embryonic cells are liberated into the circulation where they are subsequently "seeded" to embryonic blood-forming organs (e.g., bone marrow, spleen, thymus) (Moore and Owen 1967). This is an example of the blood-borne dissemination of cells that flourish and grow after arrival in hospitable soil.

Related to natural developmental translocation of blood stem cells is bone marrow transplantation in adult humans. The bone marrow, containing hematopoietic stem cells, is transplanted directly to the recipient's venous circulation where the inserted blood-forming cells have access to all of the organs of the body. Despite access to all anatomic sites, the hematopoietic stem cells localize and repopulate only the host's bone marrow, where sustained hematopoiesis occurs. The "homing" of the hematopoietic progenitor cells to bone marrow precludes the need to transplant the cells surgically into the marrow.

Neural crest cells have their origin at the dorsal surface of the neural tube in the embryo. Several migratory pathways permit mitotic descendants of neural crest cells access to the skin where they become pigment cells, to the adrenal gland where they become medullary cells, laterally to the spinal cord where they give rise to spinal ganglia, and to other sites where they give rise to other normal tissues (Perris and Bronner-Fraser 1989). Similarly, the parathyroid gland cells translocate from the dorsal parts of the third and fourth pharyngeal pouches in the human embryo to the thyroid gland (Moore and Persaud 1993).

Although the examples given thus far are from normal biology, there are pathologic but nonmalignant examples of conditions that mimic metastasis. One such condition is endometriosis, in which the lining of the uterus, the endometrium, grows in inappropriate locations in the body. The misplaced uterine tissue may be found in the ovaries, uterine tubes, uterine ligaments, and other places (Fox and Buckley 1992). Just as the structure and function of a malignant metastasis are true to the tissue of origin (e.g., metastatic malignant thyroid nodules accumulate iodine not unlike normal thyroid), endometrial tissue that has migrated to a distant site retains its original behavior, and cyclic bleeding occurs at menstruation. Of course, a significant difference of metastasis, when compared to endometriosis, is the capacity for continued growth and expression of other aspects of the malignant phenotype.

Perhaps it would be instructive to consider several specific steps in the metastatic cascade that occur in normal embryos (Figure 2–9). During early development, the chicken (also mouse and human) embryo consists of two discoidal layers. The upper layer, the epiblast, is epithelial and rests on a basement membrane. Mesoderm (the "middle layer") formation occurs when epiblast cells migrate toward the center of the blastodisc (Spratt 1946) and become less differentiated (it would be appropriate in the present context to designate these cells as "anaplastic"). Loss of differentiation occurs with a down-regulation of the cell adhesion molecule E-cadherin. Earlier in this chapter, E-cadherin was described in the context of the loss of cell differentiation, cell detachment, and metastasis (Mareel et al. 1993). The embryonic premesoderm cells detach and invade through the primitive streak after loss of the basement membrane barrier, a process analogous to carcinoma cell transgression of the digested basement membrane. One might wish to study chick embryos because of the strict chronologic predictability of the appearance of migratory cells, cells which detach from each other and pass through the basement membrane. Primitive streak and mesoderm formation highlight what has been stated before, namely, that activities of malignant cells have their counterparts in normal cells. In this instance, those activities in the normal cells (i.e., in the chicken embryos) are easily studied and the embryos are inexpensive and widely available.

Cancer cell and normal cell transgression across the basement membrane in vivo have an in vitro analogue. Migration through a reconstituted basement membrane matrix was proposed as a method to discriminate between normal and malignant cell types – it was postulated that the former could not but the latter could migrate through the reconstituted basement membrane. However, it has been shown that a number of malignant cell types failed this bioassay, whereas certain other normal cells readily invaded cells (Noel et al. 1991).

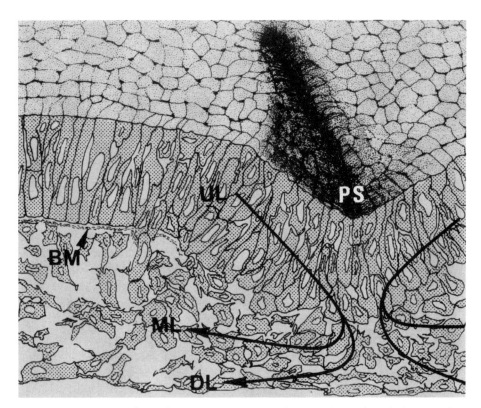

Figure 2–9. Diagram of morphogenetic movements that give rise to the primitive streak, mesoderm (middle layer), and some cells of the endoderm (deep layer) of the chick embryo. Note that cells lateral to the primitive streak (PS) in the upper layer (UL) are highly differentiated and rest on an intact basement membrane (BM). Cells become progressively more anaplastic with loss of the basement membrane at and near the site of the primitive streak (PS). The cells that detach and migrate through the primitive streak and form the middle layer (ML), as well as those which contribute to the deep layer (DL), have a reduced expression of the cell adhesion molecule E-cadherin. This example from normal embryology parallels steps in the metastatic cascade in which cancer cells digest the basement membrane, become anaplastic, and detach and migrate from the primary tumor ultimately to form metastatic colonies. A down-regulation of expression of E-cadherin takes place during the steps in the malignant process. Primitive streak formation illustrates that normal cells share many activities of cancer cells. (Reprinted from Harrisson, F., Andries, L., and Vakaet, L. 1988. The chick blastoderm: Current views on cell biological events guiding intercellular communication. *Cell Diff* 22:83–106, with kind permission from Elsevier Science Ireland Ltd., Bay 15K, Shannon Industrial Estate, Co. Clare, Ireland.)

Fertilization in humans occurs outside of the uterus. The developing embryo does not reach the uterus until it has cells competent for implantation into the endometrial lining. Implantation is a process in which the endometrium is invaded by aggressive embryonic trophoblast cells. These cells have the capacity to digest the endometrial lining and its connective tissue matrix as well as the walls of maternal arterioles. Were this aggressive invasive activity not to occur, implantation would fail and pregnancy would not ensue. The trophoblast releases proteolytic enzymes, which are implicated in this normal but highly invasive activity. Again, compare this normal phenomenon of proteolytic enzyme activity with its malignant counterpart discussed earlier in this chapter.

Dissemination of malignant cells often occurs via a vascular route. There are examples of vascular dissemination of normal cells. Just described is the trophoblast of pregnancy, which has the competence to digest maternal arterioles, thereby providing access to the maternal circulation. As a result of this, trophoblast cells intravasate (enter the vasculature) and are found not uncommonly in the blood of pregnant women.

Primordial germ cells of the chick embryo (and other bird embryos) intravasate, disseminate, and extravasate in the process of moving from the germ crescent to the gonad primordia via the vascular system (Swift 1914). Although these cells do not have the malignant phenotype and do not form tumors, they are nevertheless behaving much as many metastatic tumor cells do.

Many more examples could be cited to support the contention that metastasis of malignant cells has similarities to phenomena observed in normal cell biology. Perhaps these examples are adequate to remind the student that tumor cells do not have extraordinary capabilities – they are cells with ordinary capabilities expressed in an inappropriately extraordinary manner.

2.6 How do we know a metastasis to the liver is not a primary neoplasm of the liver?

A skeptic may remark that a malignancy in the brain is a brain tumor and a malignancy in the liver is a liver tumor. How does one distinguish between a primary brain tumor and a metastatic growth in the brain that had its origin elsewhere? Or, how does one distinguish between a liver cancer and a metastasis from some other place to the liver? Because the intrinsic differentiated state is generally not lost during acquisition of the malignant phenotype, the metastatic tumor faithfully preserves the cell and tissue morphology of the primary tumor regardless of its anatomic location (Chapter 1). Thus, we can distinguish a metastatic tumor cell colony in a particular organ from a primary tumor in that organ because the metastatic tumor generally retains histologic characteristics of its tissue of origin – the primary tumor does this too and it thus may be a relatively simple matter to distinguish one from the other (on the basis of microscopic anatomy, i.e., histopathology). This is not to say that the origin of all malignancies can be ascertained. Highly anaplastic cancers occur, making it difficult, if not impossible, to determine from what tissue the tumor originated. Not knowing what kind of cancer a patient has makes treatment a more complicated matter. Fortunately however, in less anaplastic malignancies, metastatic tumors are recognized because of their morphologic similarity to the primary tumor.

2.7   Why study metastasis?

If about two-thirds of cancer patients already have cancer that has metastasized at the time of diagnosis, what good is there in knowing the biology of metastasis? In the words of Frost and Levin (1992), "knowledge of how the metastases developed offers little insight about control." Many believe the issue to be addressed is effective treatment, not more biology. The academic scholar who loves knowledge for knowledge's sake would find this attitude puzzling. An analogy from another discipline concerns those who would seek to understand why an archeologist studies ancient culture when contemporary populations of those same people may be hungry. The archeologist studies previous cultures because they are interesting per se and solutions to today's problems may lie in an examination of the past. Similarly, a cell biologist may find metastasis interesting and this interest may have salubrious results. It is rarely possible to prognosticate the value of basic research prior to its being done. Understanding the subtleties of invasion and metastasis may lead to clinical applications not possible to anticipate with present knowledge.

Malignant neoplasms are comprised of a mosaic of cells with differing characteristics (tumor cell heterogeneity). One of the differences in tumor cells is the competence to produce metastatic colonies. Only a very small minority of cancer cells have this competence. Treatment of that subpopulation of tumor cells clearly should be a target of therapy. It may become possible to treat the *minority* of cells with lethal metastatic potentialities without killing the mass of less destructive non-metastatic cells, thereby leaving a patient with less damage to his/her normal cells than with conventional cytotoxic chemotherapy.

What are some other practical applications of the study of metastases? Markers (discussed in Chapter 1) may be found to be associated with particularly aggressive tumors with a high potential for metastasis (Tarin 1996, Sinha et al. 2002). However, all such tumors within that category may not be similarly aggressive. Would it not be useful to advise a patient who has a less aggressive cancer of the great likelihood that he or she probably will have many years of symptom-free good health and to go about living as usual? In contrast, sure knowledge obtained from a marker for metastasis that the patient has a poor prognosis allows him or her time to make rational decisions for family and self. That information, although not related to a cure, is useful and may well develop from the study of genes associated with metastasis.

Although two-thirds (some say up to 75 percent) of cancer patients have metastases at the time of presentation, one can take heart that about one-third *appear* not to. Metastasis can be abrogated, in theory, at any point in the cascade. With greater knowledge, it may become possible to repress *potential* metastasis in those patients who appear to be free of that dismal complication. Temporal (or permanent) inhibition of metastasis in a patient free of secondary tumors would have benefits for both the patient and physician.

Millions of dollars have been spent looking for a cure for cancer. Most of the "cures" (Chapters 9 and 10) kill cancer cells, and, in the process, kill a significant number of normal host cells, thus proving to be highly toxic. It would be a boon to cancer patients if a method could be devised to contain a primary tumor – to cage it – so it could not spread. Further, for those unfortunate two-thirds who already have metastatic cancer, containing their metastases while undergoing therapy is not a worthless goal. Perhaps restraint of metastasis of metastases will provide a somewhat larger window of time for effective therapy of existing tumor burden.

2.8 Summary

Metastasis accounts for most of the lethality of cancer and is feared for good reason. However, this most malignant aspect of cancer has been shown to depend on a number of interrelated steps known as the metastatic cascade. Interruption or significant perturbation of any of the steps has the potential for abrogating malignant spread. The steps of the cascade seem to be understandable and ultimately controllable because the pathways to malignant dissemination follow the routes of normal cell behavior, normal but with the trek sidetracked by untimely and disadvantageous cell function. The business of the cancer cell biologist is to set the pathway straight (normal), and, by doing so, more effective therapy will unquestionably emerge.

An epilogue concerning natural selection and metastasis: Bad and good news for the cancerpatient

The astute student must have noted while reading the previous sections of this chapter that an individual cancer cell has a perilous life indeed. Perhaps the best example of survival hazard is the incredible loss of cancer cells in the vasculature. What kinds of cells are lost? The answer to this rhetorical question is those cells are lost that are poorly suited for growth as metastatic colonies. It has been said that in politics many are called but few are chosen. This applies to cancer cell metastasis. The millions upon millions of cancer cells face lethality at each step in the metastatic cascade. Only the finest cells (in terms of their ability to grow at secondary sites) survive. The surviving cells are far from wimpy. That is the bad news. It is the task of the oncologist to eradicate or kill these powerful survivors and obviously, the task is enormous and extraordinarily complicated. Is there any good news? Of course there is. Cancer cells are the cells of the victim – they are not invaders from outside the patient (a nice polysyllabic term for this is "autochthonous"). They are self, not foreign. If they required significantly different growth factors, nutrition, pH, temperature, etc., they would simply not survive. The body of the host selects cancer cells that retain commonalities with normal cells. The good aspect of this is that everything that molecular and cell biologists learn about normal cells is applicable to cancer cells. The twenty-first century is a period of incredible new insight into the workings of living cells. Oncology will certainly share in the benefits of new knowledge and hopefully, novel methods for the control of this loathsome malady will emerge.

3

Carcinogenesis

Alan O. Perantoni

For the first time in the history of the world, every human being is now subjected to contact with dangerous chemicals, from the moment of conception until death.

Rachel Carson, *Silent Spring*, 1962

3.1 Introduction

Our environment has been described as a "sea of carcinogens" awash with a variety of chemicals and, to a lesser extent, with oncogenic viruses and high-energy radiations, all of which may contribute significantly to cancer incidence in humans. Although much of our attention has been focused on the proliferation of synthetic chemicals and their waste byproducts, an examination of these substances has revealed that, in fact, most chemicals are not carcinogenic. Testing by the U.S. National Toxicology Program (NTP), an agency within the Department of Health and Human Services charged with the responsibility of defining human carcinogens in the United States (Huff et al. 1988), and the International Agency for Research in Cancer (IARC), a part of the World Health Organization (WHO) with a similar but broader mission (Tomatis 1988), has shown that only one-fourth to one-third of those substances even suspected of being carcinogenic on the basis of their chemistries actually are cancer-causing agents. In fact, of the more than 80,000 chemicals listed for commercial use with the NTP, thus far only 228 have been categorized as being "known" or "reasonably anticipated" as human carcinogens according to the tenth *Report on Carcinogens* (released December, 2002). Thus, despite the initial bleak portrayal, these findings lead us optimistically to the possibility that cancer might be reduced or even eliminated through the identification and removal of carcinogens from the environment, assuming the numbers of carcinogenic substances are indeed limited and that we are primarily responsible for their introduction into the environment.

Although industrial contributions have captured considerable attention and may, in fact, account for a small share of cancers (perhaps 1 percent to 4 percent of cancers in the United States) (Tomatis 1988), studies now underscore the potential of naturally occurring, or endogenous carcinogens (Gupta and Lutz 1999): substances which, by their nature, cannot be eliminated. Furthermore, we have come to recognize that cancer patients harbor heritable predispositions for neoplasia, which in studies of twins seems to account for some 26 percent of all cancers (Hemminki and Mutanen 2001). Thus, not all smokers develop lung cancer, despite comparable exposure levels to the more than forty chemical carcinogens detected in cigarette smoke. This chapter focuses on several aspects of carcinogens and the carcinogenic process, including the classes of carcinogens, the kinetics and dynamics of carcinogen–organism interaction, theories of carcinogenesis, government regulation of carcinogenic substances, and the continuing argument of risk versus benefit in regulation.

3.2   What is a carcinogen?

Simply stated, a carcinogen is any substance or agent that significantly increases tumor incidence. Operationally, that is, in a regulatory sense, the definition is considerably more complex. The NTP includes epidemiological evidence of a causal relationship between exposure to a substance and cancer, carcinogenicity studies in experimental animals, and structural or mechanistic relationships with known carcinogens in its determinations. However, chemical testing in rodents still provides some of the most compelling evidence of carcinogenic activity since human data are infrequently available and rodents are susceptible to a variety of chemical agents. Rodents, though, tend to exhibit greater sensitivities to carcinogens than humans, which can lead to overestimates of carcinogenic risk for humans, as will become clear in later discussions. Conversely, testing procedures may overlook other incomplete carcinogens, that is, agents, like promoters, which are incapable of inducing tumors by themselves but can accelerate the passage of altered cells through individual stages of the carcinogenic process. Because virtually all substances that have been identified as human carcinogens by epidemiologists are also carcinogenic in bioassay animals (Wilbourn et al. 1986), animal studies remain one of the most conservative and only definitive means for testing. However, to be categorized by the NTP as "known to be a human carcinogen," there must be epidemiological evidence such as studies of occupational risks or the inadvertent induction of tumors in humans receiving drug therapy for a medical condition. On the other hand, those "reasonably anticipated to be human carcinogens" are implicated primarily through carcinogenicity studies in experimental animals or through structural relationships with known well-defined classes of substances. Due to space limitations, only the major classes of carcinogens are discussed here. For more information on individual or groups of chemicals, the IARC Monograph series or the NTP website provides continuous comprehensive updates on the various classes of carcinogens.

Although discussion here of the various types of carcinogens focuses on those described as "complete," to better understand what this means in terms of tumor induction, a description of the putative stages of the carcinogenic process is necessary.

3.3   Carcinogenesis as a multistage process

The consensus among cancer researchers is that carcinogenesis is a multistage genetic and epigenetic process, requiring some two to seven mutations on average. This results in a shift in homeostasis from a regulated population of cells destined to differentiate and eventually become apoptotic to one exhibiting unregulated pro-liferative control, autonomous growth, reduced apoptosis, angiogenesis, tumor cell invasion, and metastasis (Cohen and Ellwein 1991; Herzig and Christofori 2002). This theory is based on observations in both experimental animals and human populations: the latency of tumor formation following carcinogen exposure, the apparent sequence of pathologic events described in the development of several types of tumors, age-specific tumor incidence, the existence of cancer-prone fami-lies, and the many oncogenes or tumor suppressor genes associated with the various pathologic events of tumorigenesis.

As mentioned in Chapter 1, carcinogenesis is often characterized by four sequential stages: initiation, promotion, progression, and malignant conversion (Figure 3–1). In this process, initiation occurs when a carcinogen interacts with DNA, producing a strand break or, more often, an altered nucleotide called an adduct. Then, if the genome is replicated before repair enzymes can correct the damage, a DNA polymerase may misrepair the damaged sequence and permanently fix a heritable error in the genome with replication. Since only cells with replicative ability are therefore at risk, it is believed that the small population of progenitor or stem cells involved in normal tissue renewal are the actual targets of carcinogenesis.

The presence of a DNA adduct then causes the insertion of an incorrect nucleotide in the DNA strand opposite the damage, resulting in a mutation of the affected gene. For some agents, for example, ultraviolet light or the mycotoxin aflatoxin-B1, the characteristic genetic alteration is sufficiently unique to implicate these carcinogens in causation. Thus, some carcinogens provide evidence (a 'smok-ing gun') of their involvement based upon the detection of a specific adduct or mutation in an oncogene or tumor suppressor gene sequence, yielding a biomarker for exposure to that carcinogen. Indeed, agencies such as the IARC constantly eval-uate the potential use of putative biomarkers for assessing carcinogen exposures in human populations.

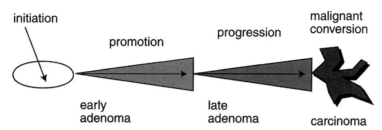

Figure 3–1. Carcinogenesis, a multistage process, begins with the genetic event of initiation followed by selective expansion of altered cells during promotion to form benign early adenomas. In the absence of continued promotion, the adenomas regress and disappear. With a second genetic event, a small number of promoted adenomas progress to form late adenomas, some of which may then undergo malignant conversion.

As for biological consequences, the vast majority of these random misincorporations are either neutral to the cell because they fail to involve a cancer susceptibility gene or occur in cells that have differentiated and are not at risk. However, if the mutation occurs in a sequence that encodes a growth regulatory protein, for example, such an error may, under certain circumstances, provide that cell with a selective growth advantage. Those circumstances include exposure to a class of compounds known as promoters. Promoters are believed to preferentially select for or stimulate proliferation of initiated cells to form multiple benign tumors or hyperplastic lesions, and they represent the second stage of carcinogenesis. The effect of promotion is completely reversible, and withdrawal of the promoter results in the disappearance of the expanding clones of cells or hyperplastic nodules. Promoters, therefore, are not considered genotoxic or mutagenic, that is, able to induce heritable alterations or mutations but instead function as mitogenic agents, which increase cell growth through an epigenetic mechanism, involving gene regulation at the transcriptional or posttranscriptional level. They are also poor inducers of malignant conversion (discussed later).

Finally, a second genetic event or series of events is proposed that allows some permanent selective growth advantage to initiated cells or somehow increases the probability a cell will become neoplastic. This stage, called progression, provides the impetus for conversion from benign adenomas to infiltrative and finally metastasizing neoplasms and represents the longest period in tumorigenesis. The alterations during progression could arise from additional exposures to a carcinogen, spontaneous mutations due to the natural infidelity of enzymes involved in replication, or genomic instabilities induced by initiating mutations. Regardless of the mechanism, the outcome is an irreversible change in the cell that allows expression of the neoplastic phenotype. In light of this theory, a "complete" carcinogen is an agent capable, through single or multiple exposures, of providing an initiating insult, promoting a selective growth advantage for altered cells, and generating lesions during progression that result in malignant conversion.

3.4   Chemical carcinogenesis

The English surgeon Percivall Pott is generally regarded as the father of carcinogenesis studies for his astute recognition that an environmental agent was responsible for tumor induction in chimney sweeps in London (Pott 1775). In a series of papers, he characterized a high incidence of scrotal cancer associated with this occupation and suggested that deposits of soot in the scrotal area were causative. As a consequence, other European nations legislated bathing and clothing requirements for chimney sweeps and, thus, virtually eliminated scrotal cancer as an occupational hazard (Butlin 1892). This clearly demonstrated not only that environmental substances could induce tumor formation but also that by identifying and eliminating the etiologic agent from the immediate environment, tumor incidence could be significantly reduced.

Others later established associations between exposures to soot and cancer incidence for a variety of professions, for example, gardening (Earle 1808) and tar manufacturing (Bell 1794), but more than a century passed before an animal model was developed to directly test the idea that chemicals cause tumors and to permit purification of the etiologic agents. Subsequently, Yamagawa and Ichikawa (1918) demonstrated that chronic application of coal tar to the ears of rabbits could induce skin carcinomas. Thus, chemists could examine the carcinogenicity of purified fractions from tar products. Their detailed study established a clear multistage process consisting of early irritation and proliferation of cells, papilloma formation at the site of exposure in treated animals, neoplasia, and even metastases in several animals. Finally, Cook, Hewett, and Hieger (1933) described the successful extraction of a crystalline substance responsible for the carcinogenicity of coal tar on rabbit skin. The compound benzo[a]pyrene (BP), a member of a class of carcinogens called polycyclic aromatic hydrocarbons (PAH), has since been studied extensively as a metabolism-dependent chemical carcinogen and has proven invaluable in elucidating the mechanisms for carcinogen activation. Since the discovery of BP, hundreds of other occupational or environmental chemical carcinogens have been identified (Table 3–1), including a variety of organic substances (e.g., PAH and aromatic amines), as well as a number of environmentally important inorganic metals and minerals, such as nickel and asbestos. Despite the diversity of chemistries, more than 95 percent of the various carcinogenic chemicals fall into one of three major categories: alkylating agents, aralkylating agents, or arylhydroxylamines (Figure 3–2). Although these agents differ in the type of residue transferred to create a DNA adduct, they share certain characteristics. Most notably, they are either intrinsically reactive with DNA or can be metabolically activated to a stable DNA-reactive form. These so-called electrophiles bond with the electron-sharing atoms of the DNA nucleotides, such as ring nitrogen or exocyclic oxygen atoms (Figure 3–3), to form stable altered nucleotides or adducts (Hemminki 1994).

Because of their ability to react with DNA and cause misincorporation of nucleotides in the genome, many, but not all, carcinogens are also mutagenic (Shelby

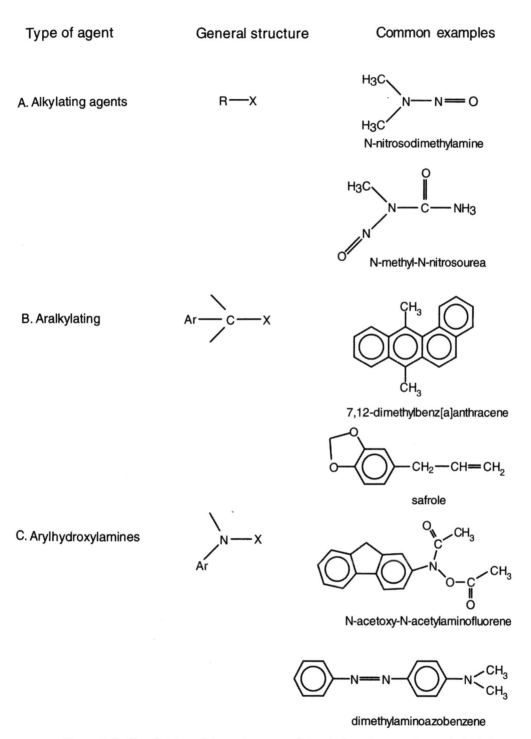

Type of agent	General structure	Common examples

A. Alkylating agents — R—X — N-nitrosodimethylamine, N-methyl-N-nitrosourea

B. Aralkylating — Ar—C—X — 7,12-dimethylbenz[a]anthracene, safrole

C. Arylhydroxylamines — N—X — N-acetoxy-N-acetylaminofluorene, dimethylaminoazobenzene

Figure 3–2. Classification of the major types of chemical carcinogens. The majority fall into three categories based on their donor groups in adduct formation: alkylating, aralkylating, and arylhydroxylamines.

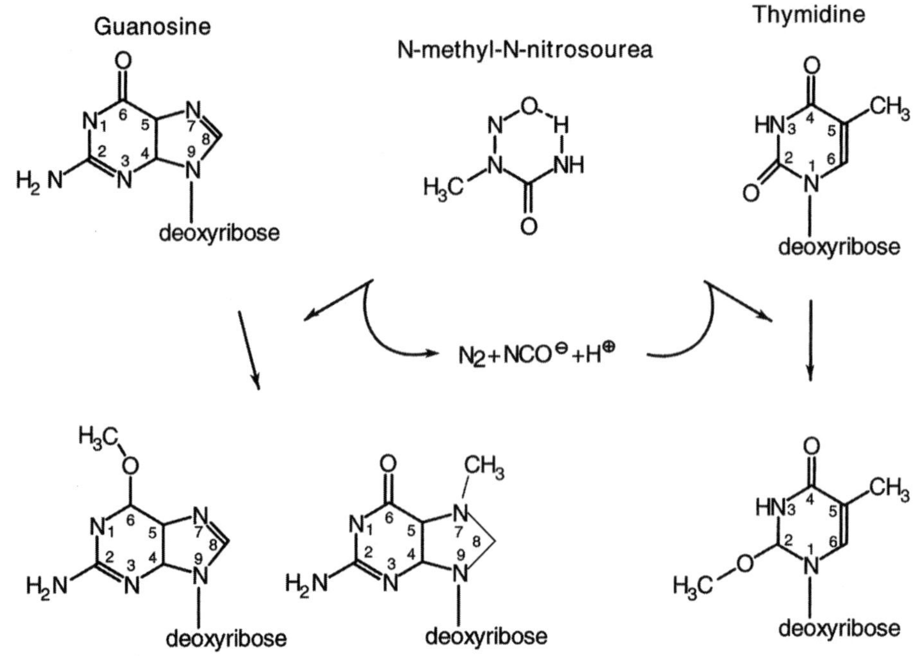

Figure 3–3. Principal DNA adducts formed with the direct-acting alkylating agent N-methyl-N-nitrosourea (MNU). The carcinogen MNU donates a methyl group primarily to the ring N-7 and the exocyclic O-6 in guanosine and the exocyclic O-2 of thymidine to cause mutations in the genome.

and Zeiger 1990). One of the most sensitive and rapid assays for mutagenicity called the Ames test for its developer employs strains of *Salmonella typhimurium*, which are readily mutable to growth independence from histidine supplementation of growth media. In the presence of a mutagen (such as a mutagenic carcinogen), an inactive enzyme involved in bacterial histidine synthesis can be mutated to an active form, allowing the bacterium to grow in a medium that lacks histidine. The problem historically with this assay was that bacteria do not contain the drug or carcinogen-metabolizing enzymes found in mammalian cells, so carcinogens requiring metabolic activation depend on the extracellular addition of mammalian enzymes for mutagenicity. More recently, these bacterial strains have been genetically engineered to express human drug-metabolizing enzymes, allowing chemicals to be activated within the cell to increase sensitivity in the assay (Josephy 2002).

Many of the drug/carcinogen-metabolizing enzymes exist in the endoplasmic reticulum, so inclusion of microsomal preparations from rodent livers in bacterial assays is often sufficient to permit activation. Other carcinogens, however, appear to be nongenotoxic by current means of testing, yet increase tumor incidence. This may be due in part to the shortcomings of testing procedures. Asbestos, for example, induces large chromosomal alterations and not point mutations, the endpoints for

Table 3–1. *Known and suspected carcinogens identified by the National Toxicology Program*

Part A. Known to be a Human Carcinogen
Aflatoxins
Alcoholic Beverage Consumption
4-Aminobiphenyl
Analgesic Mixtures Containing Phenacetin (See Phenacetin and Analgesic Mixtures Containing Phenacetin)
Arsenic Compounds, Inorganic
Asbestos
Azathioprine
Benzene
Benzidine (See Benzidine and Dyes Metabolized to Benzidine)
Beryllium and Beryllium Compounds
1,3-Butadiene
1,4-Butanediol Dimethylsulfonate (Myleran®)
Cadmium and Cadmium Compounds
Chlorambucil
1-(2-Chloroethyl)-3-(4-methylcyclohexyl)-1-nitrosourea (MeCCNU)
bis(Chloromethyl) Ether and Technical-Grade Chloromethyl Methyl Ether
Chromium Hexavalent Compounds
Coal Tar Pitches (See Coal Tars and Coal Tar Pitches)
Coal Tars (See Coal Tars and Coal Tar Pitches)
Coke Oven Emissions
Cyclophosphamide
Cyclosporin A (Ciclosporin)
Diethylstilbestrol
Dyes Metabolized to Benzidine (See Benzidine and Dyes Metabolized to Benzidine)
Environmental Tobacco Smoke (See Tobacco-Related Exposures)
Erionite
Estrogens, Steroidal
Ethylene Oxide
Melphalan
Methoxsalen with Ultraviolet A Therapy (PUVA)
Mineral Oils (Untreated and Mildly Treated)
Mustard Gas
2-Naphthylamine
Nickel Compounds (See Metallic Nickel and Nickel Compounds)
Radon
Silica, Crystalline (Respirable Size)
Smokeless Tobacco (See Tobacco-Related Exposures)
Solar Radiation (See Ultraviolet Radiation Related Exposures)
Soots
Strong Inorganic Acid Mists Containing Sulfuric Acid
Sunlamps or Sunbeds, Exposure to (See Ultraviolet Radiation-Related Exposures)
Tamoxifen
2,3,7,8-Tetrachlorodibenzo-*p*-dioxin (TCDD); "Dioxin"
Thiotepa
Thorium Dioxide
Tobacco Smoking (See Tobacco-Related Exposures)
Ultraviolet Radiation, Broad Spectrum UV Radiation (See Ultraviolet Radiation Related Exposures)
Vinyl Chloride
Wood Dust

(*continued*)

Table 3–1. (*continued*)

Part B. Reasonably Anticipated to be a Human Carcinogen

Acetaldehyde
2-Acetylaminofluorene
Acrylamide
Acrylonitrile
Adriamycin® (Doxorubicin Hydrochloride)
2-Aminoanthraquinone
o-Aminoazotoluene
1-Amino-2-methylanthraquinone
2-Amino-3-methylimidazo[4,5-*f*]quinoline (IQ)
Amitrole
o-Anisidine Hydrochloride
Azacitidine (5-Azacytidine®, 5-AzaC)
Benz[*a*]anthracene (See Polycyclic Aromatic Hydrocarbons)
Benzo[*b*]fluoranthene (See Polycyclic Aromatic Hydrocarbons)
Benzo[*j*]fluoranthene (See Polycyclic Aromatic Hydrocarbons)
Benzo[*k*]fluoranthene (See Polycyclic Aromatic Hydrocarbons)
Benzo[*a*]pyrene (See Polycyclic Aromatic Hydrocarbons)
Benzotrichloride
Bromodichloromethane
2,2-bis-(Bromoethyl)-1,3-propanediol (Technical Grade)
Butylated Hydroxyanisole (BHA)
Carbon Tetrachloride
Ceramic Fibers (Respirable Size)
Chloramphenicol
Chlorendic Acid
Chlorinated Paraffins (C12, 60% Chlorine)
1-(2-Chloroethyl)-3-cyclohexyl-1-nitrosourea
bis(Chloroethyl) nitrosourea
Chloroform
3-Chloro-2-methylpropene
4-Chloro-o-phenylenediamine
Chloroprene
p-Chloro-o-toluidine and p-Chloro-o-toluidine Hydrochloride (See p-Chloro-o-toluidine and
    p-Chloro-o-toluidine Hydrochloride)
Chlorozotocin
C.I. Basic Red 9 Monohydrochloride
Cisplatin
p-Cresidine
Cupferron
Dacarbazine
Danthron (1,8-Dihydroxyanthraquinone)
2,4-Diaminoanisole Sulfate
2,4-Diaminotoluene
Dibenz[*a,h*]acridine (See Polycyclic Aromatic Hydrocarbons)
Dibenz[*a,j*]acridine (See Polycyclic Aromatic Hydrocarbons)
Dibenz[*a,h*]anthracene (See Polycyclic Aromatic Hydrocarbons)
7*H*-Dibenzo[*c,g*]carbazole (See Polycyclic Aromatic Hydrocarbons)
Dibenzo[*a,e*]pyrene (See Polycyclic Aromatic Hydrocarbons)
Dibenzo[*a,h*]pyrene (See Polycyclic Aromatic Hydrocarbons)
Dibenzo[*a,i*]pyrene (See Polycyclic Aromatic Hydrocarbons)
Dibenzo[*a,l*]pyrene (See Polycyclic Aromatic Hydrocarbons)

1,2-Dibromo-3-chloropropane

1,2-Dibromoethane (Ethylene Dibromide)

2,3-Dibromo-1-propanol

tris(2,3-Dibromopropyl) Phosphate

1,4-Dichlorobenzene

3,3'-Dichlorobenzidine and 3,3'-Dichlorobenzidine Dihydrochloride (See 3,3'-Dichlorobenzidine
    and 3,3'-Dichlorobenzidine Dihydrochloride)

Dichlorodiphenyltrichloroethane (DDT)

1,2-Dichloroethane (Ethylene Dichloride)

Dichloromethane (Methylene Chloride)

1,3-Dichloropropene (Technical Grade)

Diepoxybutane

Diesel Exhaust Particulates

Diethyl Sulfate

Diglycidyl Resorcinol Ether

3,3'-Dimethoxybenzidine (See 3,3'-Dimethoxybenzidine and Dyes Metabolized to
    3,3'-Dimethoxybenzidine)

4-Dimethylaminoazobenzene

3,3'-Dimethylbenzidine (See 3,3'-Dimethylbenzidine and Dyes Metabolized to
    3,3'-Dimethylbenzidine)

Dimethylcarbamoyl Chloride

1,1-Dimethylhydrazine

Dimethyl Sulfate

Dimethylvinyl Chloride

1,6-Dinitropyrene (See Nitroarenes [selected])

1,8-Dinitropyrene (See Nitroarenes [selected])

1,4-Dioxane

Disperse Blue 1

Dyes Metabolized to 3,3'-Dimethoxybenzidine (See 3,3'-Dimethoxybenzidine and Dyes
    Metabolized to 3,3'-Dimethoxybenzidine)

Dyes Metabolized to 3,3'-Dimethylbenzidine (See 3,3'-Dimethylbenzidine and Dyes Metabolized
    to 3,3'-Dimethylbenzidine)

Epichlorohydrin

Ethylene Thiourea

di(2-Ethylhexyl) Phthalate

Ethyl Methanesulfonate

Formaldehyde (Gas)

Furan

Glasswool (Respirable Size)

Glycidol

Hexachlorobenzene

Hexachlorocyclohexane Isomers

Hexachloroethane

Hexamethylphosphoramide

Hydrazine and Hydrazine Sulfate (See Hydrazine and Hydrazine Sulfate)

Hydrazobenzene

Indeno[1,2,3-*cd*]pyrene (See Polycyclic Aromatic Hydrocarbons)

Iron Dextran Complex

Isoprene

Kepone® (Chlordecone)

Lead Acetate (See Lead Acetate and Lead Phosphate)

Lead Phosphate (See Lead Acetate and Lead Phosphate)

Lindane and Other Hexachlorocyclohexane Isomers

2-Methylaziridine (Propylenimine)

(*continued*)

Table 3–1. (*continued*)

5-Methylchrysene (See Polycyclic Aromatic Hydrocarbons)
4,4'-Methylenebis(2-chloroaniline)
4-4'-Methylenebis(*N,N*-dimethyl)benzenamine
4,4'-Methylenedianiline and 4,4'-Methylenedianiline Dihydrochloride (See
    4,4'-Methylenedianiline and its Dihydrochloride Salt)
Methyleugenol
Methyl Methanesulfonate
N-Methyl-*N*'-nitro-*N*-nitrosoguanidine
Metronidazole
Michler's Ketone [4,4'-(Dimethylamino)benzophenone]
Mirex
Nickel (Metallic) (See Nickel and Nickel Compounds)
Nitrilotriacetic Acid
o-Nitroanisole
6-Nitrochrysene (See Nitroarenes (selected))
Nitrofen (2,4-Dichlorophenyl-*p*-nitrophenyl ether)
Nitrogen Mustard Hydrochloride
2-Nitropropane
1-Nitropyrene (See Nitroarenes [selected])
4-Nitropyrene (See Nitroarenes [selected])
N-Nitrosodi-*n*-butylamine
N-Nitrosodiethanolamine
N-Nitrosodiethylamine
N-Nitrosodimethylamine
N-Nitrosodi-*n*-propylamine
N-Nitroso-*N*-ethylurea
4-(*N*-Nitrosomethylamino)-1-(3-pyridyl)-1-butanone
N-Nitroso-*N*-methylurea
N-Nitrosomethylvinylamine
N-Nitrosomorpholine
N-Nitrosonornicotine
N-Nitrosopiperidine
N-Nitrosopyrrolidine
N-Nitrososarcosine
Norethisterone
Ochratoxin A
4,4'-Oxydianiline
Oxymetholone
Phenacetin (See Phenacetin and Analgesic Mixtures Containing Phenacetin)
Phenazopyridine Hydrochloride
Phenolphthalein
Phenoxybenzamine Hydrochloride
Phenytoin
Polybrominated Biphenyls (PBBs)
Polychlorinated Biphenyls (PCBs)
Polycyclic Aromatic Hydrocarbons (PAHs)
Procarbazine Hydrochloride
Progesterone
1,3-Propane Sultone
?-Propiolactone
Propylene Oxide
Propylthiouracil
Reserpine
Safrole

Selenium Sulfide
Streptozotocin
Styrene-7,8-oxide
Sulfallate
Tetrachloroethylene (Perchloroethylene)
Tetrafluoroethylene
Tetranitromethane
Thioacetamide
Thiourea
Toluene Diisocyanate
o-Toluidine and o-Toluidine Hydrochloride(See o-Toluidine and o-Toluidine Hydrochloride)
Toxaphene
Trichloroethylene
2,4,6-Trichlorophenol
1,2,3-Trichloropropane
Ultraviolet A Radiation (See Ultraviolet Radiation Related Exposure)
Ultraviolet B Radiation (See Ultraviolet Radiation Related Exposure)
Ultraviolet C Radiation (See Ultraviolet Radiation Related Exposure)
Urethane
Vinyl Bromide
4-Vinyl-1-cyclohexene Diepoxide
Vinyl Fluoride

Source: This list was derived from the 10th Annual Report of the NTP, released December 17, 2002, and evaluations of individual chemicals can be obtained on the Internet at ntp-server.niehs.nih.gov

the bacterial mutation assays (Jaurand 1989). Pure promoters like phenobarbital are nontransforming and nonmutagenic by themselves but can enhance tumorigenicity severalfold in combination with weakly transforming doses of a carcinogen. For others, like the herbicide amitrole, which induces thyroid tumors in experimental rodents, current assay systems may lack the enzymes for metabolizing carcinogens to reactive electrophilic forms (Kraus et al. 1986) or nongenotoxic mechanisms, such as the observed disregulation of thyroid hormone secretion, may be responsible for its tumorigenic activity (Hurley 1998).

Although many carcinogens exhibit mutagenic activity, as just described, not all carcinogens have been shown to be mutagenic, nor are all mutagens carcinogenic. Perhaps the best example involves certain nucleotide analogues that serve as potent mutagens in bacterial assays but are noncarcinogenic in animal carcinogenicity tests. Also, certain organophosphate pesticides are highly mutagenic in the Ames test but fail to induce tumors in two-year rodent bioassay. In this case, it has been proposed that these chemicals fail to cause cell proliferation, while structurally related carcinogens induce proliferation (Cunningham and Matthews 1995).

The most frequently studied carcinogens include the N-nitroso compounds and aflatoxins (alkylating agents), the polycyclic aromatic hydrocarbons (aralkylating agents), and the aromatic amines and aminoazo dyes (arylhydroxylamines). With the exception of certain N-nitroso carcinogens, these substances are all metabolized by the P-450 mixed-function oxidase enzymes (P-450 cytochromes or CYPs), a super-family of drug-metabolizing microsomal enzymes, to electrophilic intermediates that react with DNA bases. Although structure/activity studies of these compounds have resolved details of cellular mechanisms for metabolic activation, they have been less useful in defining events of carcinogenesis. For this, studies of direct-acting agents, such as the N-nitrosoureas, have clarified the chemistry of carcinogen-target interactions.

3.4a   Organic compounds

Alkylating agents – chemicals that transfer alkyl groups, often methyl or ethyl groups, to nucleotides to form DNA adducts (see Figures 3–2 and 3–3).

The N-nitroso compounds, especially the nitrosamines, are perhaps the most insidious and therefore potentially most hazardous of the various carcinogens. Epidemiologic studies have provided no definitive link of members of this class of carcinogens to human cancer, with the exception of the use of the chloroethylnitrosureas, chemotherapeutic agents for gastrointestinal cancers (Boice et al. 1983). However, the potent carcinogenicity of several of these compounds in every species tested (including nonhuman primates; Magee and Barnes 1956; Kelly et al. 1966), the hepatotoxicity of certain nitrosamines in workers exposed occupationally (Kimbrough 1983), and the ubiquitous nature and volatility of these chemicals strongly suggest that they pose a significant health hazard to humans (Bartsch et al. 1992).

Activation of these compounds often requires biotransformation either enzymatically by oxidation (as for dimethylnitrosamine) or directly by alkali-mediated hydrolysis (as for the direct-acting carcinogen methylnitrosourea). In either case, a methyl group (CH_3-) or an ethyl group ($CH_3 CH_2-$), depending on the chemical, is available for the modification of a DNA base. Methylation occurs predominantly at exocyclic oxygen moieties or ring nitrogens (see Figure 3–3).

The problem with these compounds is not only their ubiquitous distribution in the environment but also the ease with which they are generated endogenously. Formation of N-nitroso compounds by the nitrosation of secondary amines occurs fairly readily under acidic conditions, for example, in the stomach. It can also occur under nonacidic conditions, for example, bacterial metabolism in the gut, curing of tobacco leaves, or preservative treatment of food products with sodium nitrite (Lijinsky 1999). Thus, compounds like dimethylnitrosamine can be readily produced in the body or in consumed products, but whether or not this represents a significant health risk has not been established.

Other alkylating agents of human consequence include members of the numerous naturally occurring plant or fungal toxins, many of which function as natural pesticides or antibiotics in the environment and also exhibit carcinogenic activity in mammals (Ames et al. 1987). Over three hundred mycotoxins have been identified, some of which are strong mutagens and carcinogens. The potent rodent liver carcinogen aflatoxin B_1, produced by the common mold *Aspergillus flavus,* can heavily contaminate grains, vegetables, and nuts on which the mold thrives. Aflatoxin is especially a problem in areas of the world where methods of food preservation are deficient. Its general distribution and potency has been considered serious enough in the United States to warrant limitations on levels in such food products as peanut butter.

Aflatoxin was first identified as the hepatotoxic substance in contaminated feed in England that devastated the domesticated turkey population in 1960 (Blount 1961) and was subsequently shown to induce liver tumors in several species (IARC 1992). Its carcinogenicity for humans is suggested in epidemiological evaluations of African (Alpert, Hutt, and Davidson 1968) or Asian (Lunn 1997) populations with a high incidence of liver tumors and in the mutational spectrum of genetic lesions found in the p53 suppressor gene from hepatocellular cancers (Hsu et al. 1991). However, the multiplicity of factors identified in these populations (notably, chronic infections with hepatitis B virus) obscures the role of aflatoxin B_1 in human liver tumorigenesis.

Metabolic activation of aflatoxin B_1 involves oxidation of the molecule by the P-450 mixed-function oxidases/CYPs to generate an intermediate which reacts with a cyclic nitrogen in guanine, forming the adduct 8,9-dihydro-8-(N7-guanyl)-9-hydroxyaflatoxin B_1 in the DNA strand (Essigmann et al. 1982). Because the N7-guanine-AfB$_1$ adduct is excreted in urine following aflatoxin exposure in both rodents and humans, it has been possible to develop rapid bioassays that allow its quantitation. Such studies have demonstrated the efficacy of using measurements of adduct formation as a means of human cancer risk assessment in certain exposed populations and has provided an effective biomarker for clinicians intent on reducing aflatoxin exposure in these populations (Groopman and Kensler 1999).

Aralkylating agents – chemicals that transfer aromatic or multiringed compounds to a nucleotide to form an adduct (see Figure 3–2).

Polycyclic aromatic hydrocarbons (PAH), the principal group of aralkylating agents, remain an occupational problem in several industries. As previously mentioned, exposure to the hydrocarbons in soot increased the incidence of scrotal cancer in British chimney sweeps. More recently, the crude oils used in cotton spinning (Heller 1930) or tool setting (Waldron Waterhouse, and Tessema 1984) were reported to cause scrotal cancer in workers whose pants became saturated with them. Structurally similar compounds have been found in combustion products (Grimmer and Misfeld 1983) and as such are added continuously to the environment. This includes products such as benzo[a]pyrene and the potent 7,12-dimethylbenz[a]anthracene, which are generated in cigarette smoke and on charcoal-grilled meats (Lijinsky and Shubik 1964). These compounds readily induce tumors in laboratory animals, causing rapid tumorigenesis in rat mammary tissue following ingestion (Huggins, Grand, and Brillantes 1961), the major route of exposure for humans.

Metabolism of PAH has been investigated extensively, and certain generalizations can be made from these studies. The hydrocarbons are not reactive with DNA unless

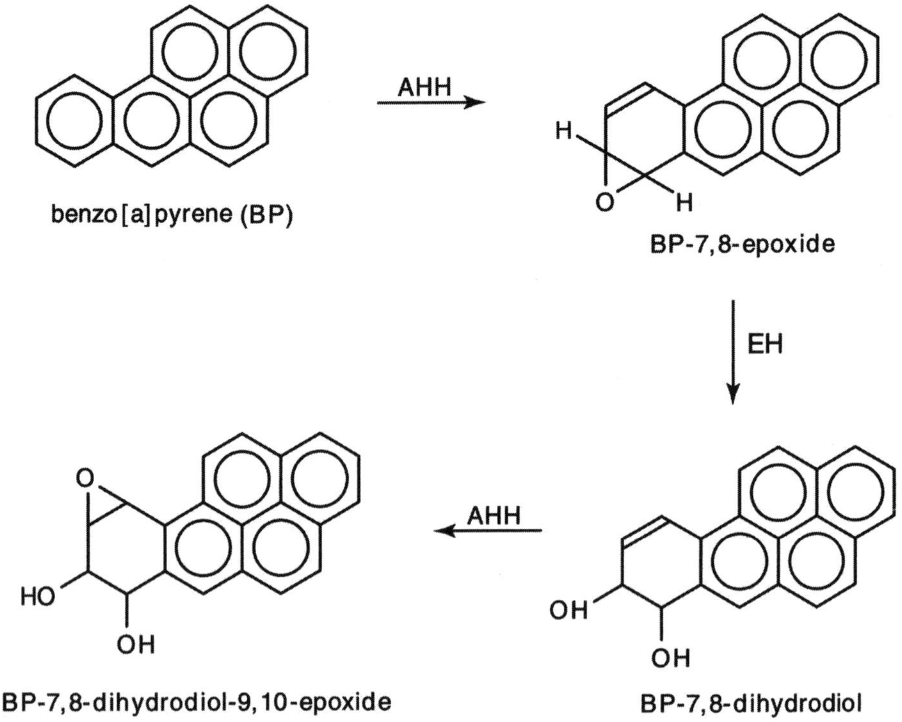

Figure 3–4. Metabolic activation of the aralkylating agent benzo[a]pyrene (BP). Activation is mediated by the P-450 microsomal enzymes and specifically arylhydrocarbon hydroxylase (AHH), which oxidizes the carbon-carbon double bond to form an epoxide. Another enzyme, epoxide hydratase (EH), destroys the epoxide, and AHH further metabolizes the BP to form the presumed reactive intermediate.

first metabolized. The reactive metabolites appear to be dihydrodiol epoxides, especially those formed in the bay region, as shown, which are generated by two consecutive cycles through the P-450 mixed-function oxidase system/CYPs (Figure 3–4). The intermediates are sufficiently stable to allow their passage to the cell nucleus but also highly reactive with DNA to cause mutations and tumor formation (Dipple et al. 1999).

Arylhydroxylamines – chemicals that transfer aromatic amines to nucleotides to form adducts (see Figure 3–2).

   The carcinogenic activity of the aromatic amines, a major group of arylhydroxylamines, was established from epidemiological studies of workers in the dyestuff industry. Occupational exposures to aniline dyes caused a high incidence of bladder cancer (Rehn 1895), and the etiologic agents most responsible were eventually identified as 2-naphthylamine and benzidine (Case et al. 1954). In a study of workers involved in the distillation of 2-naphthylamine, nearly all heavily exposed individuals subsequently developed bladder cancer (Case 1969). Similarly, 2-naphthylamine was implicated in the high incidence of bladder tumors in the manufacturing of rubber (Case and Hosker 1954). Bioassays for carcinogenic activity in rats and dogs showed these compounds to be active primarily in the bladder following dietary exposures (Radonski 1979). While industrial exposures have now been largely controlled, aromatic amines, such as 2-naphthylamine or 4-aminobiphenyl, also exist in tobacco smoke. Furthermore, recent focus has been given to the dietary heterocyclic aromatic amines (HAAs), which have been documented in cooked meats, fish, and poultry (Turesky 2002). The more than twenty chemicals in this family of compounds are potent mutagens and carcinogens in experimental animals, and specific members, for example, 2-amino-3-methylimidazo[4,5-f]quinoline (IQ), are designated as "reasonably anticipated to be a human carcinogen" based upon these observations. While conclusive evidence of human carcinogenicity is currently lacking due to the complexity of exposure for compounds that are generated by the cooking process, the occurrence of urinary metabolites or HAA-DNA adducts in tissues as biomarkers of exposure has been documented in human subjects.

The arylhydroxylamines, like PAH, are metabolized by the P-450 mixed-function oxidase/CYP enzymes, and commonly undergo N-oxidation to generate reactive intermediates. In the case of 2-naphthylamine, the chemical is first enzymatically oxidized to generate an N-hydroxyl intermediate, which may be further metabolized in liver or kidney to form a stable glucuronide conjugate that is passed to the bladder (Figure 3–5) (Kadlubar, Miller, and Miller 1977). In the bladder, the final activated electrophile, presumed to be a nitrenium cation, is formed, which can react with DNA in bladder epithelia (Orzechowski, Schrenk, and Bock 1992). Some HAAs can also be conjugated to form N-acetoxy or N-sulfonyloxy esters, which in turn generate mutagenic and carcinogenic nitrenium ions (Turesky 2002).

3.4b   Inorganic compounds and asbestos

Certain inorganic metals and minerals exhibit carcinogenic activities or are associated with elevated risks for cancer in humans. These include arsenic, cadmium, chromium, lead, nickel, and asbestos. The unequivocal effect of asbestos on tumorigenesis both in experimental animals and humans and its extensive presence in the environment, such as in cement construction materials, insulation, and fireproofing, have established it as a significant health hazard.

The term "asbestos" encompasses a variety of silica fiber types. Although one form of asbestos, the fibrous serpentine magnesium-containing chrysotile $(Mg_6Si_4O_{10}(OH)_8)$, represents more than 90 percent of the mined asbestos in the United States, carcinogenic activity is generally greater with the fibrous amphiboles (actinolite, amosite, anthophyllite, crocidolite, and tremolite) and especially iron-bound crocidolite $(Na_2(Fe3+)_2(Fe2+)_3Si_8O_{22}(OH)_2)$ or calcium-containing tremolite $(Ca_2Mg_5Si_8O_{22}(OH)_2)$. The various mineral forms of asbestos generally reflect differences in fiber structure, differences that affect the ability of the fibers to be retained in the lungs upon inhalation. For example, chrysotile fibers often occur as clusters of curly fibers that penetrate the lungs inefficiently and are

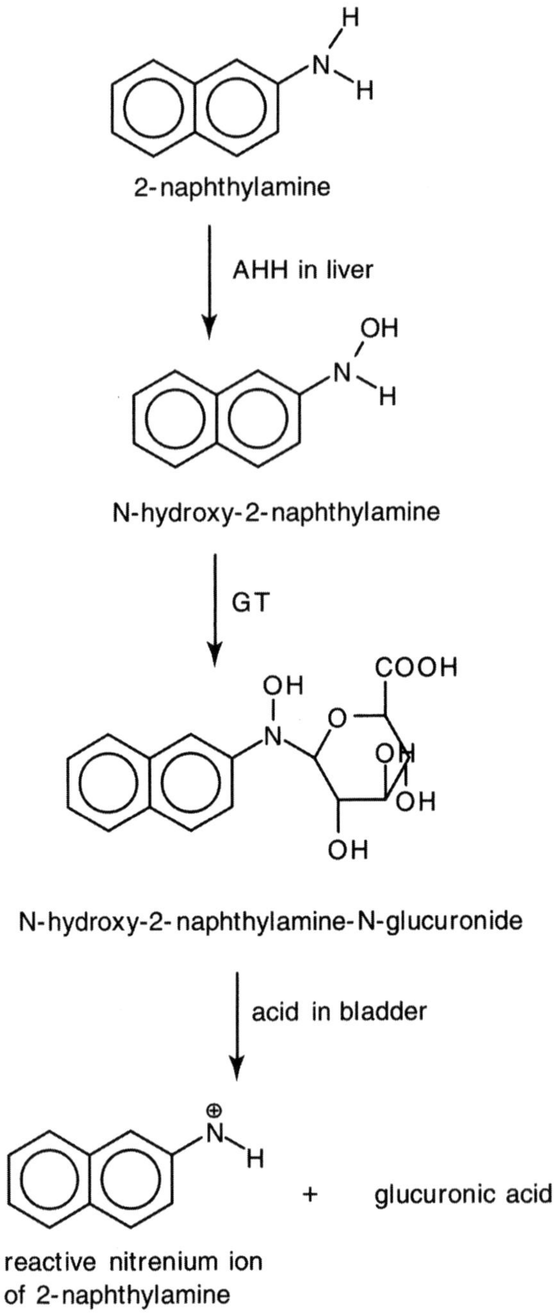

Figure 3–5. Metabolism of the arylhydroxylamine 2-naphthylamine. AHH activity in the liver oxidizes the nitrogen to generate the N-hydroxy derivative. The conjugating enzyme glucuronide transferase (GT) adds the sugar residue to the carcinogen. This stabilizes the compound until it reaches the bladder, where acid conditions cause the formation of the carcinogenic nitrenium ion.

cleared far more rapidly than the amphiboles. Conversely, the amphiboles (a large group of structurally similar silicate minerals) such as crocidolite exist as individual rods that readily penetrate deep into the lungs and may remain there for several months to years (Mossman et al. 1990).

From epidemiological studies, it is apparent that all individuals carry a significant number of asbestos fibers in their lungs and that includes infants as well (Haque and Kanz 1988). However, workers exposed to asbestos in the mining, milling, or manufacture of insulation carry a much greater lung fiber burden than the general population and have nearly a tenfold greater risk for lung cancer. In addition, cigarette smoking acts synergistically with asbestos to enhance lung cancer risk ten times above incidences for asbestos exposure alone and five times for smoking alone (Saracci 1987).

Asbestos workers present with either lung carcinomas or mesotheliomas of the pleura or peritoneum. As many as 20 to 25 percent of heavily exposed workers develop lung cancer (Lemen, Dement, and Wagoner 1980). Because the occurrence of mesotheliomas is extremely rare and almost totally associated with asbestos exposure (Wagner, Sleggs, and Marchand 1960), epidemiologists were readily able to establish asbestos as the etiologic agent. Although all types of asbestos fibers can cause chromosomal aberrations (Sincock, Delhanty, and Casey 1982; Hei et al. 1992) as well as lung tumors in rats (Davis et al. 1978), studies of fiber biodistribution in asbestos workers with mesothelioma suggest that the lung burden of amphiboles is significantly greater than chrysotile fibers (MacDonald and MacDonald 1987; Wagner et al. 1988), as would be predicted on the basis of the fiber structures described here. These studies therefore are consistent with the idea that the amphiboles present a significantly greater health risk than the chrysotiles. Although epidemiologic evidence seems to indicate that the crocidolite fibers are more potent inducers of mesothelioma than chrysotile fibers, evidence for a similar circumstance in lung cancer induction is far less compelling. The fact that fiber size, shape, and composition varies in different asbestos deposits causes problems interpreting the epidemiologic studies. Sufficient evidence, however, suggests that both crocidolites and chrysotiles may be relevant to lung cancer induction in asbestos workers (Stayner, Dankovic, and Lemen 1996).

While the mechanism responsible for its carcinogenic activity is not known, asbestos fibers are mutagenic in assay systems that detect large DNA deletions but not in the Ames test, so this carcinogen behaves quite differently from other chemicals. In vitro, the fibers can disrupt the mitotic spindle to cause chromosomal damage (Ault 1995), and in vivo, the inflammation resulting from the presence of the fibers may cause the release of reactive oxygen free radicals, which initiate DNA damage, or cytokines which promote cell proliferation (Mossman, Kamp, and Weitzman 1996).

The absolute stability of the asbestos fibers and their ability to travel long distances in the air as well as epidemiological and toxicological evidence support the tight controls currently placed on the manufacture and use or removal of amphibole-containing products. Furthermore, the growing literature implicating the chrysotiles in tumor induction reinforces the concept that both classes of asbestos should be treated with equal caution, and in fact the Occupational Safety and Health Administration (OSHA) has revised standards of acceptable occupational exposure levels to include all forms of asbestos.

What of nonoccupational exposures to asbestos? Low levels of fibers persist in outdoor air, which probably accounts for the fact that all individuals possess a small but significant lung burden. Interestingly, levels of airborne fibers are no greater in buildings with damaged asbestos materials than are found outdoors. Consequently, medical experts have suggested that the elevated levels of fibers, which can occur with removal, may actually increase the health risks to the asbestos (Mossman et al. 1996). Because the cancer risk from those materials already in place could conceivably be negligible, the costly issue of asbestos removal from buildings warrants further investigation before significant resources are committed to its elimination.

3.4c Naturally occurring chemicals

Finally, and perhaps of greatest importance, is the category of generally unexplored naturally occurring chemical carcinogens, which have only recently received attention. Although the U.S. government regulates synthetic compounds and food additives, there is currently no control over natural substances such as pesticides or antibiotics produced by plants to protect themselves from microbial or animal predators. It has been estimated that humans are exposed to 10,000 times more of these natural pesticides than synthetic ones, and, in fact, these natural pesticides may constitute as much as 5 percent to 10 percent of a plant's dry weight. It has been estimated that the average adult in the United States ingests some 1,500 mg of natural pesticides per day. Although most of these compounds remain untested, preliminary studies indicate that they are not without risk. Of seventy-one ubiquitously distributed substances evaluated thus far, thirty-seven are carcinogenic in rodents (Ames and Gold 2000). In addition to the multiplicity of chemicals, some of these compounds may occur naturally in volumes as high as parts per thousand, whereas exposure to synthetic pesticides is generally in parts per million or less. Thus, as Ames et al. (1987) have asserted, "nature is not benign." Granted, this area requires further investigation to assess the risks from these substances, but they deserve greater attention in light of public pressure to eliminate beneficial synthetic compounds, which may in fact contribute little to our total carcinogen burden.

3.5   Radiation

3.5a Ultraviolet radiation

Although most agree that, of the exogenous agents, chemicals are responsible for the greatest share of cancer, chronic exposure to ultraviolet light as a consequence of increased leisure time or to radon because of improved standards of household insulation has refocused attention on the various forms of radiation. Ultraviolet light is continuously bathing our environment. Although predominantly from sun, artificial lighting ensures, to a lesser but significant extent, that this exposure will be continuous because both fluorescent and the increasingly popular tungsten-halogen fixtures emit significant amounts of UV radiation. As the ozone is depleted from the stratosphere, prospects are that exposure levels to ultraviolet irradiation will also increase.

Ultraviolet radiation includes wavelengths between 200 and 400 nm (visible light ranges from 400 to 700 nm) and is often subdivided into three regions: UV-A, 320 to 400 nm; UV-B, 280 to 320 nm; and UV-C, 200 to 280 nm. Biologic effects are elicited primarily with UV-B radiation, which is directly absorbed in DNA and induces the acute symptoms of sunburn and the adaptive responses to exposure, hyperpigmentation and skin thickening. Because stratospheric ozone and layers of dead skin effectively absorb UV-C radiation, exposure in this region appears to be rather limited. Little is known regarding the biologic activity of UV-A radiation, although levels do not vary with time of day or season, suggesting a lack of association with factors that play a role in skin cancer induction. Furthermore, UV-A wavelengths are not absorbed directly in DNA but rather through reactive intermediates such as oxygen radicals, which can damage DNA indirectly. There are, however, indications that it can effect tissue injury in conjunction with UV-B radiation or certain photosensitizing chemicals (Urbach 1993).

Circumstantial experimental and epidemiological evidence has implicated UV radiation in the induction of both basal and squamous cell carcinomas, the most common but generally curable forms of cancer (Urbach 1993; Marks 1996b). As discussed in Chapter 7, although nearly 100 percent curable with early detection and treatment, these skin cancers have been observed with increasing frequency as leisure time has expanded. Epidemiological studies have shown that tumor incidence correlates positively with circumstances that elevate cumulative skin exposures to UV radiation. Thus, tumor incidence increases with decreasing latitude (Elwood et al. 1974). Furthermore, tumors arise predominantly in weakly pigmented individuals or ethnic groups (Chuang et al. 1990) and generally appear on those body surfaces that receive the greatest exposure, such as the head and neck (Haenszel 1963). Also, individuals in occupations that require greater outdoor exposure clearly have a higher tumor incidence (Vitaliano 1978).

Ultraviolet radiation catalyzes the formation of cyclobutane pyrimidine dimers (Beukers and Berends 1960) and [6–4] pyrimidine-pyrimidinone photoproducts (Figure 3–6), both of which result in C-to-T or CC-to-TT transition mutations in DNA if not repaired. The involvement of dimer formation in carcinogenesis is strongly supported by studies of the genetic defect xeroderma pigmentosum (XP), a complex of disorders characterized by deficient excision repair of ultraviolet-induced pyrimidine dimers and a high skin cancer incidence (Kraemer, Lee, and Scotto 1987). Furthermore, the 'signature' tandem transition mutations serve as biomarkers of UV exposure and are demonstrable in tumor susceptibility genes, such as *ras*,

cyclobutane dimer   6-4' photoproduct

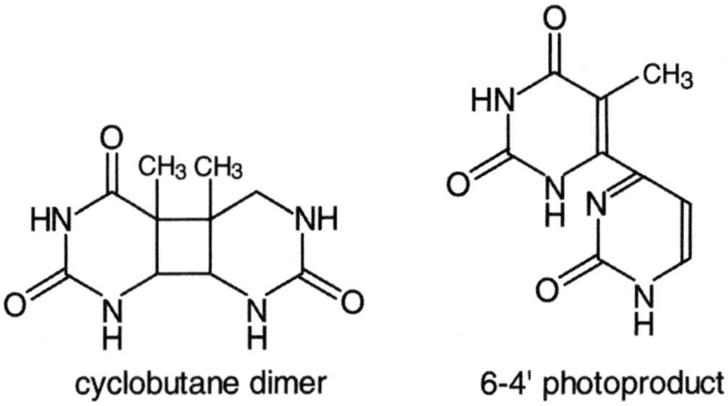

Figure 3–6. Ultraviolet light causes the formation of two principal adducts in DNA: the cyclobutane dimers and 6–4 photoproducts of thymidine. Both products require excision repair processes to correct.

p53, or *patched*, which have been analyzed in skin cancers (Sarasin 1999). Finally, in vitro studies have shown that the UV action spectrum for transformation of hamster embryo cells (Doniger et al. 1981) or human embryonic fibroblasts (Sutherland et al. 1981) is consistent with that for UV induction of dimer formation.

Malignant melanoma, the highly aggressive cutaneous cancer of neuroectodermal melanocytes located in the skin, is of growing concern because it affects primarily young adults, is increasing in incidence faster than any noncutaneous cancer, and is often lethal. Although melanoma incidence is similarly associated with exposure to sun (IARC 1992) and is increased in XP patients (Kraemer et al. 1994), the relationship is more complex than for other skin tumors and seems to involve intense, intermittent exposures. For example, melanoma incidence in individuals with outdoor occupations is actually lower than for those receiving intermittent exposures (Lee and Strickland 1980; Garland et al. 1990). However, the incidence does increase with decreasing latitude (Armstrong 1984) and predominates in white-skinned populations (Muir et al. 1987) as do other skin cancers. Other risk factors include the extent of intermittent exposure to sun, the numbers of dysplastic nevi (possible precursors of melanoma), and susceptibility to sunburn; frequency of sunburn shows no correlation (Armstrong 1988). For risk factors associated with skin cancer, see Chapters 7 and 8.

3.5b   Ionizing radiation

Ionizing radiation is a well-established human carcinogen first noted by experimentalists and clinicians during the development of Roentgen's cathode tube, the basis for the X-ray machine. Not recognizing its health consequences, early radiologists often used their own hands to focus the electron beam in primitive X-ray machines, resulting in the frequent induction of skin cancer (Frieben 1902). Later, luminescent dial painters in watch factories suffered a high incidence of osteosarcomas, a tragedy attributed to radium ingestion when painters orally formed pointed tips on paintbrushes contaminated with radium-bearing paint (Martland 1931). The radium was localized to and retained by cells in the bone due to its ability to mimic calcium, thus making bone the immediate target of radium's high-energy alpha emissions.

Ionizing radiation has been clearly linked to the excess cancer cases in populations exposed to nuclear detonations. An intensive epidemiological investigation of some 86,572 atomic bomb survivors who were exposed to neutron and gamma radiation in Hiroshima and Nagasaki has revealed an increased risk of cancer mortality from leukemias especially (about 20 percent of cancers), and to a lesser but significant extent from tumors of the digestive tract (especially esophagus, stomach, and colon), lungs, liver, bladder, female breast, and ovaries (Shimizu et al. 1989; Shimizu, Kato, and Schull 1990; Pierce et al. 1996). Interestingly, the increases occurred in direct proportion to normal increases in cancer with aging; that is, tumor latency was apparently not abbreviated, but the number of observed versus expected cases increased. It is also noteworthy that all measures of germline genetic damage in this population have thus far demonstrated no increases attributable to parental exposures, and, therefore, the offspring of survivors of ionizing radiation are apparently not at additional risk to any adverse genetic effects (Neel et al. 1990). Study group estimates now project that fewer than a thousand deaths due to cancer in the more than 86,000 survivors will be attributable to atomic bomb irradiation (Lenihan 1993). These observations demonstrate the relative carcinogenic impotence of ionizing radiation when not tissue localized and suggest that current low-level environmental or health-related exposures may be of little consequence.

On the other hand, tissue localization of an ionizing radiation source can have adverse health consequences. In an inadvertent human exposure during an above-ground U.S. nuclear test, an unexpected wind shift carried high levels of the radioactive isotope iodine 131 over the Marshall Islands. Subsequent epidemiologic analysis of the indigenous population on one of the heavily exposed islands revealed thyroid tumors in more than 50 percent of children under the age of ten at exposure (Conrad, Dobyns, and Sutow 1970). Because thyroid tumors were extremely rare in this population, the high tumor incidence could be clearly attributed to localization of this moderate-energy beta and gamma emitter in the thyroids of exposed children.

The single exposure due to the Chernobyl accident has resulted in a similar pattern of tumor induction. In April, 1986, the nuclear reactor sustained a steam explosion, blowing the cap off the reactor, and causing a meltdown of the graphite core. This released the volatile radioisotopes in the reactor, including predominantly iodine 131, and exposed 10 to 20 million people in Belarus, northern Ukraine, and Russia to high levels of radiation. Follow-up studies of those populations have demonstrated a highly significant age-dependent increase in the induction of papillary thyroid cancers (Williams 2002). More than 900 pediatric cases were reported by 1998 in Belarus and the Ukraine, yielding a ten-fold increase in incidence in these areas. When age was factored in, the relative risk for thyroid cancer rose to over 200 for children exposed at ages 0 to 1 versus 6 for children exposed at age 10. The high incidence in children is at least partially due to the mode of exposure, since isotope-contaminated milk is the most common route.

The mechanism of carcinogenesis from ionizing radiation is believed to involve either direct ionization of DNA or indirect formation of mutagenic oxygen free radicals. Due to the tissue penetrance of certain types of ionizing radiation, oxygen free radicals can be generated at the DNA level by ionizing the shell of hydration surrounding the DNA, thus making it a readily available target for these highly reactive and extremely short-lived molecules. Since scavengers for these molecules can reduce the amount of DNA damage by as much as 70 percent, a mechanism involving free radical formation is believed to play a major role in mutation induction by ionizing radiation. Once formed, the reactive oxygen species (.OH [hydroxyl radicals], H_2O_2 [hydrogen peroxide], $1O_2$ [singlet oxygen], or O_2- [superoxide radicals]) can induce more than thirty different DNA adducts as well as DNA-protein cross-links (Feig, Reid, and Loeb 1994). The free radical–generated mutations may result directly from the specific adducts formed, or they may be caused indirectly by free radical–induced alterations in the DNA polymerases. Although adduct formation does occur, it is generally agreed that DNA double-strand breaks (DSBs) are responsible for most ionizing radiation–induced mutations. These dangerous breaks are repaired by nonhomologous end joining (NHEJ) or homologous recombinational repair (HR), two complementary mechanisms that contribute to genomic stability. Misrepair of these difficult lesions may lead to chromosomal abnormalities such as rearrangements, and loss of either mechanism results in hypersensitivity to ionizing radiation, chromosomal instability, and tumorigenesis (Khanna and Jackson 2001).

The various forms of ionizing radiation described are carcinogenic when presented at unusually high doses, but what about the chronic low levels to which most are exposed routinely, for example, chest X-rays, dental exam X-rays, endogenous isotopes, and cosmic or terrestrial irradiation? It is estimated that natural sources of external radiation, that is, cosmic rays and isotopes in the soil and air, and of internal radiation, that is, potassium-40, deliver in combination only about 1–2 mSieverts (mSv) per year of whole body dose, and artificial sources, such as medical X-rays add only another 0.5 mSv (National Research Council 1998). While these exposures seem limited, ionizing radiation characteristically generates clusters of free radicals which, in turn, produce clusters of DNA damage. The multiplicity of damage makes the repair of such lesions inherently more difficult. Thus, even limited exposures could yield significant genetic damage if the random event occurs in a critical regulatory element. Although many efforts have been made to quantify tumor incidence from low-dose exposures, the weakness of ionizing radiation as a carcinogen and the qualitative similarity between spontaneous and radiation-induced classes of tumors have hampered such attempts. The sizes of experimental groups needed to reach significance are enormous and beyond the capabilities of most institutions. To circumvent these problems, epidemiologists have extrapolated risks based on data from occupational or public exposures and made estimates by inference. The problem with this approach is knowing the shape of the curve to apply when extrapolating from higher dosage portions of exposure/incidence curves to low-dose exposure. Nonthreshold linear dose-response curves may lead to an overestimate of cancers but are a conservative approach to risk estimates because one assumes that any amount of exposure has an effect. This model appears to be appropriate for nonleukemic cancers in atomic bomb survivors (Shimizu et al. 1990). A nonthreshold linear-quadratic or curvilinear representation, however, is accepted as the most accurate predictor of risk for leukemias due to ionizing radiation, again based on results from studies of atomic bomb survivors (Shimizu et al. 1990). Under either scenario, low-dose exposures are predicted to increase tumor incidence, and, as a result, federal regulations mandate that exposures to radiation must be kept "as low as reasonably achievable."

3.5c   Endogenous ionizing radiation

Finally, although impossible to evaluate the hazard, we are continuously exposed to endogenous levels of ionizing radiation estimated to constitute 11 percent of the annual average total radiation dose to a person in the United States (NCRP 1987). Such isotopes as 40K, 14C, 3H, and 226Ra are assimilated internally and may even incorporate into DNA. One might argue that due to their proximity and continuous impact on cellular targets, these endogenous isotopes are of greater consequence to cellular transformation than is exposure to exogenous sources.

3.6   Radon

More recently the volatile isotope radon, a decomposition product of uranium-238, has been recognized as a significant environmental carcinogen. Its carcinogenic potential was first suggested from studies of uranium miners, who showed a nearly 50 percent incidence of lung cancer in a single region of Czechoslovakia (Holaday 1969). Since then, more than twenty epidemiological evaluations, including pooled studies, of radon as a risk factor in lung cancer demonstrate a consistent association between exposure levels and excess risk in underground hard rock miners, for example, those who mine uranium, iron-ore, or fluorspar (Samet and Eradze 2000). In unusual circumstances, radon exposures to miners have approached the relatively low levels commonly reported for homeowners (Tirmarche et al. 1993), or conversely, high levels in homes have even exceeded those of uranium miners (Ennemoser et al. 1993, 1994). In each case, lung cancer incidence was increased significantly, suggesting that exposure is not without risk in populations receiving significant levels.

The health consequences are not directly due to the radon itself, which is a ubiquitously distributed noble gas and is concentrated in enclosed structures such as houses, schools, or commercial buildings. Instead the offending agents are radon's

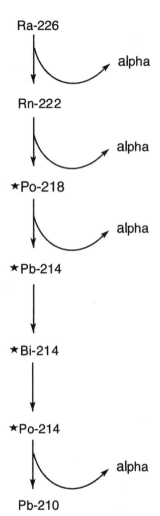

Ra-226

alpha

Rn-222

alpha

★Po-218

alpha

★Pb-214

★Bi-214

★Po-214

alpha

Pb-210

Figure 3–7. Radioactive decay of radon. The inert gas radon decomposes to form several particulate daughters, including the two polonium isotopes.

particulate daughters (e.g., polonium-218 and polonium-214), which can be deposited and accumulated in the lungs (Figure 3–7). These isotopes are high-energy alpha emitters with considerable potential for ionizing damage, but, due to their short effective range and inability to penetrate the outer layer of skin, they are inconsequential outside the body. Thus, they are harmful only when inhaled, and their effects are localized to the lungs or bronchi.

Because radon exposure represents more than 50 percent of the total ionizing radiation to which individuals in the United States are subjected, its health risks cannot be ignored (Samet and Eradze 2000). However, considerable uncertainty in risk assessment does exist due to numerous biological variables and difficulties in obtaining accurate estimates of exposures. Newer, more sensitive techniques may eventually resolve some of these issues. For example, glass can be used as a long-term index for radon exposure, since the long-lived particulates become embedded in it and can be quantified. To develop estimates of attributable risk for lung cancer in the general population, currently extrapolations are made from pooled miner data or direct estimates are made in those populations with case-controlled studies. Either approach is subject to problems of uncontrollable errors in dosimetry, which significantly reduces the sensitivity of such studies. Furthermore, some studies ignore important variables such as occupation, smoking habits, family cancer history, diet, and residential history. Based on the National Research Council's predicted relative lung cancer risk for women of 1.2 to 1.5 from household radon exposure though, most studies may not be of sufficient size to test the hypothesis that radon is a residential risk factor (Neuberger 1992). Even efforts to critically evaluate and pool epidemiologic studies of residential exposures have been unable to establish an association between radon and lung cancer risk in non-smokers (Neuberger and Gesell 2002).

Animal studies also support the epidemiologic conclusion that radon/radon daughters are carcinogenic. Rats and hamsters develop adenomas, adenocarcinomas, and squamous cell carcinomas of bronchiolar/alveolar origin in a dose-dependent manner following inhalation of radon (IARC Working Group 2001). Furthermore, cumulative exposures in rats at levels comparable to lifetime exposures in high-radon houses results in a lung tumor incidence that is directly proportional to the exposure rate of alpha particles and that the maximal incidence is dependent upon both the exposure rate and the cumulative dose (Monchaux and Morlier 2002). Thus, radon has been declared a human carcinogen by the NTP and IARC on the basis of both epidemiologic studies of occupational exposures and compelling animal studies; however, the issue of its role in lung cancer for the general population remains too complex for a clear understanding. Although the EPA asserts that radon causes a significant share of lung cancers in the United States, and therefore levels of exposure should be controlled in the home, the jury is still out on the health consequences of these low level exposures.

3.7   Viral carcinogenesis

No area of study has been more productive in terms of our basic understanding of the molecular mechanisms of tumorigenesis than that of viral carcinogenesis. Dissection of the genome from the various chicken and mouse tumor viruses has revealed more than a hundred sequences capable of transforming mammalian cells. Furthermore, it has led to major advances in our understanding of the control of signaling and proliferation in normal cells. The RNA tumor viruses or retroviruses selectively incorporate transforming or oncogenic genes from a host cell or insert into the host genome and reregulate host sequences using viral promoters. On the other hand, DNA tumor viruses contain viral sequences that sequester and suppress the activity of cellular proteins that normally inhibit cell growth.

While clearly implicated in the etiology of chicken and mouse tumors, evidence of viral involvement in human tumorigenesis is somewhat less compelling. The literature is replete with associations of viral sequences in certain tumors, for example, Epstein-Barr virus (EBV) and Burkitt's lymphoma or hepatitis B virus (HBV) and liver cancer, but direct proof of a role in carcinogenesis has remained somewhat elusive. The reason is that human tumor viruses are generally not complete carcinogens but rather only assist in tumor development through the establishment of long-term persistent infections. Perhaps the best case has been demonstrated for the human papilloma DNA viruses (HPV) and anogenital cancers, especially cervical cancer. Epidemiologic studies (discussed in Chapter 7) have long implicated a venereal component in cervical carcinoma (Kessler 1976), but initial attempts to correlate the now reputed etiologic agent HPV with these neoplasms were unsuccessful due primarily to the multiplicity of HPV viral types. Some seventy-five different HPVs have since been reported (DeVilliers 1989), of which ten are categorized as "high-risk" types due to their involvement in cases of severe dysplasia and neoplasia (Werness, Münger, and Howley 1991). Recent examination of cervical carcinomas now has shown that more than 90 percent of these cancers contain "high-risk" HPV sequences of which half contain HPV-16, and women infected with a high risk form have a one in thirty chance of developing cervical cancer (Howley and Münger 1999). Molecular fragmentation of the "high-risk" forms has revealed two viral oncoproteins E6 and E7 that interact respectively with cellular growth suppressor gene proteins p53 or pRb (Münger et al. 1989; Scheffner et al. 1990), important regulatory factors for cell cycle progression that are discussed in Chapter 5. In the majority of cervical cancers, HPV is incorporated in the human genome as opposed to remaining episomal. This causes enhanced expression of the viral transforming proteins E6 and E7. These HPV oncoproteins are capable, either independently or in cooperation with other transforming genes, of immortalizing (Hawley-Nelson et al. 1989) or transforming (Phelps et al. 1988) a variety of cell types in vitro. Although these observations in no way prove HPVs function in human cervical cancers, the epidemiological reports combined with the molecular findings provide compelling evidence that they participate in the neoplastic process and, on the basis of their late detectability, probably do so during the tumor progression stage.

Other viruses with substantial epidemiological evidence of human tumor causation include EBV as an etiologic agent in Burkitt's lymphoma; HBV and hepatitis C virus (HCV) in hepatocellular carcinoma, nasopharyngeal carcinoma, and Hodgkin's lymphoma; and RNA human T-cell leukemia virus type 1 (HTLV-1) in adult T-cell leukemia. It has been estimated that worldwide, 15 percent of all tumors have viral association and that the incidence is three times greater in developing countries, presumably due to cooperating factors such as aflatoxin contamination of foods in the induction of hepatocellular carcinoma or malarial infections in the development of Burkitt's lymphoma (Parkin et al. 1999). As more information becomes available, other viruses may be implicated in human cancers. Already there are suspects, for example, the polyoma virus SV40, which contains a transforming gene and can induce tumors in animal models (Butel and Lednicky 1999). Others will undoubtably follow.

3.8 Endogenous carcinogenesis

Investigations of etiologic agents in carcinogenesis have focused primarily on exogenous effectors, but what about endogenous processes? Might they account for events in carcinogenesis? Despite the evolution of highly accurate proofreading functions in the nucleus, the cellular processes involved in DNA synthesis and repair are inherently mutagenic and therefore may present a risk for carcinogenesis. Additionally, the multistep nature of carcinogenesis and the aneuploidy common in tumor cells during the latter stages of tumorigenesis have led to speculation of a mutator phenotype. This is a genetic sequence, which when mutated could increase the rate of spontaneous mutations throughout the genome. Such a mutation would thus destabilize the genome and theoretically would allow the fixation of the multiple genetic alterations frequently observed in neoplasms, especially the more aggressive forms.

Loeb, Loeb, and Anderson (2003) describe several mechanisms whereby the genomic instability associated with tumor progression could occur without participation of an exogenous carcinogen. DNA is in a dynamic equilibrium due to its inherent instability in aqueous environments and the processes that maintain normal nucleotide sequences. On the one hand, depurination of DNA, that is, purine release from its deoxyribose moiety (3×10^{-11} events/base/s) due to hydrolysis of glycosylic bonds in the genome is a fairly common event (Lindahl and Nyberg 1972). The estimated 10,000 depurinations generated each day may increase the likelihood that misincorporation of a base during repair or replication will produce a spontaneous mutation (Nakamura and Swenberg 1999). Apurinic sites are created by a repair enzyme that recognizes an altered base or adduct and eliminates it from the DNA strand. This produces a gap in the sequence prior to replication, which a repairing polymerase preferentially fills with deoxyadenosine, causing G-to-T or A-to-T transversions. Alternatively, hydrolytic deamination of 5-methylcytidine to thymidine can also occur, albeit with lower efficiency, yielding C-to-T transitions (Lindahl and Nyberg 1974). Indeed, these transition mutations at CpG sites are commonly observed in some cancer susceptibility genes (Rideout et al. 1990).

Another potential source of endogenous transforming mutations may be from certain biochemical processes, for example, respiration, inflammation, and phagocytosis, which generate significant quantities of oxygen free radicals, especially hydroxyl radicals. These processes are estimated to yield 20,000 altered bases in DNA/human cell/day (Beckman and Ames 1997). This frequency far exceeds rates of spontaneous mutations, but it is based on measurements of free radical–modified nucleotides in urine. The estimates probably exaggerate the frequency of spontaneous mutation because the modified nucleotides are presumably released during phagocytosis in a non-nuclear environment laden with free radicals. Even so, cells have evolved several mechanisms for eliminating these reactive oxygen species or damage resulting from their activity, suggesting that they may play a significant role in producing mutations and therefore influence the carcinogenic process. This is consistent with recent findings that have associated malignancy with chronic inflammatory processes capable of generating high levels of oxygen free radicals. For example, chronic ulcerative colitis predisposes to adenocarcinomas of the colon, which arise from inflamed epithelia with multiple chromosomal aberrations (Brentnall et al. 1996), and, in addition to the gastric mucosa, infections of *Helicobacter pylori* are associated with development of mucosa-associated lymphoid tissue lymphomas and a high frequency of a specific chromosomal translocation (Ye et al. 2003). Thus, endogenous production of reactive oxygen species normally or with inflammation may participate in carcinogenesis.

Transforming mutations may also result from the infidelity of the enzymes responsible for DNA replication and repair. Although base complementarity, base selection by DNA polymerases, 3' to 5' proofreading activity, and postreplicative mismatch correction together ensure the high level of accuracy of replication, the latter three processes offer potential targets for the mutator phenotype. Replication is a highly accurate process, yielding mutation rates as low as one alteration/billion nucleotides polymerized, and is dependent principally upon the activity of DNA polymerase-δ. If its proofreading ability is abrogated in transgenic mice, the animals develop a high incidence of lymphomas and carcinomas (Goldsby et al. 2002), supporting the concept of the mutator phenotype. In other contexts, a single base substitution in the proofreading function of *E. coli* DNA polymerase III increases mutation rates by 10,000-fold (Fowler, Schaaper, and Glickman 1986). Similarly, mutations in mismatch repair genes enhance mutation rates as much as 100-fold (Cox 1976) and predispose to hereditary nonpolyposis colorectal cancer (Ionov et al. 1993). Somatic or germline mutations in these genes then might predispose individuals or entire families to neoplasia. It is noteworthy that mutations derived from the lack of DNA polymerase fidelity in bacteria are predominantly GC-to-AT transition mutations (Schaaper and Dunn 1987). Similarly, spontaneous mutations observed in cultured mammalian cells are predominantly GC-to-AT transitions (De Jong, Gorsovsky, and Glickman 1988) as are the mutations identified most frequently in human tumor studies of activated *ras* oncogenes (Burmer and Loeb 1989).

As our understanding of molecular mechanisms in the cell expands, so does the list of potential targets of a mutator phenotype (Loeb et al. 2003). In addition to replicative and repair enzymes, mutations in checkpoint pathways involving proteins such as p53, which normally blocks cell division if there is substantial DNA damage and promotes apoptosis if the damage is irreparable, clearly allow for greater genetic instability and are associated with cancer. Other possible sites include the mitotic spindle checkpoint genes (Masuda and Takahashi 2002), such as MAD1 and MAD2, telomere maintenance genes (Chang, Khoo, and DePinho 2001), which affect chromosomal stability, and also genes that regulate apoptosis. Alteration of a regulatory component within any of these pathways could conceivably enhance genetic instability and promote tumorigenesis, thus yielding a mutator phenotype.

Thus carcinogenesis may be caused by processes inherent in cells and not necessarily from exposures to exogenous agents. That is not to say that occupational exposures to carcinogens do not accelerate this tumorigenesis. There is no question that exogenous chemicals can significantly increase the risk of cancer; however, the major mechanism of carcinogenesis might be completely internal, especially for certain embryonal tumors that arise in the pediatric population and, in all likelihood, develop during gestation when exposures to exogenous chemicals should be limited.

3.9   Metabolism of xenobiotics

The kinetics of organismal processing of carcinogens and the dynamics of carcinogen–target interaction, that is, the mutagenic events involved in carcinogenesis, until the discovery of oncogenes, received the greatest attention in cancer research. This was due, at least in part, to the number of toxicologists–pharmacologists involved in the cancer effort. In general, the principles of toxicology and xenobiotic processing are similarly applicable to studies of carcinogens. The dose of a carcinogen that reaches a transforming target depends on the kinetics of its processing by the organism: absorption, distribution, metabolic activation, detoxification, and excretion of the carcinogen. All of these processes in combination modulate the carcinogen's effectiveness. The dynamics of interaction of an "ultimate" carcinogen, that is, the active electrophile, with DNA targets determine the formation and permanent fixation of transforming lesions.

3.9a   Host defenses

Despite the formidable task of devising a protective scheme against the variety and quantity of xenobiotics to which we are exposed, evolution has provided humans with a fairly effective shield. Regardless of form or method of presentation of foreign substances, cellular mechanisms at several levels modify and often eliminate the risks of chemical injury, including carcinogenesis. These mechanisms may affect organismal uptake, distribution, metabolism/activation, and elimination of these chemical substances/carcinogens.

To begin, chemical uptake is fairly well restricted in most cases and regulated at the level of inhalation, ingestion, or surface contact. The skin effectively blocks the entry or absorption of most carcinogens, although intake is more dependent upon the properties of the chemical, for example, its charge and hydrophobicity, which determine its ability to penetrate tissue. However, the layer of keratinized dead skin cells can trap and eliminate, upon sloughing, many external electrophiles. It is also thick enough to prevent certain types of ionizing radiation such as alpha particles from reaching living tissue, except through inhalation. Ingested carcinogens experience the intense acidity of the stomach, which hydrolyzes and inactivates certain chemicals, such as those with diazo groups. Stomach acids, however, may actually aid in the production of other carcinogens like the nitrosamines, as described previously in this chapter. For inhaled substances, particulates are restricted by size and trapped and eliminated by the mucous and cilia lining the respiratory passages. Once in the alveoli, the uptake of a gaseous substance is also inherent in its chemistry, but because of the nature of the the breathing process, exposures may be quite brief.

Those substances that survive the environment of the stomach may find targets in the gut during absorption, but cell turnover is so rapid as to limit the retention of any initiated cells. Additionally, selective absorption further restricts carcinogen internalization. Most chemicals enter cells by passive diffusion, and as such their lipid solubility determines their rate of absorption. It also influences the ability of a cell to store these carcinogens because the lipophilic substances can be sequestered for long periods in fatty tissues. Thus, the lipid solubility of the polycyclic hydrocarbon 7,12-dimethylbenz[a]anthracene (DMBA) may in part explain its potency as a mammary carcinogen because it may be stored and concentrated in the fatty tissue of the mammary gland. There are also carcinogens that structurally mimic important biologic molecules and are therefore similarly transported and concentrated. Azaserine, a diazo amino acid that structurally resembles the amino acid glutamine, is actively concentrated by amino acid transport pathways in the pancreas and kidney where it exerts its carcinogenicity (Longnecker and Curphey 1975).

If inhaled and taken up, a chemical quickly enters the blood stream and is distributed in tissues throughout the body. However, following gut absorption, a chemical first passes through the liver where a superfamily of hemoproteins has evolved for the metabolism and elimination of ringed chemicals, that is, monooxygenation of lipophilic substances to form more hydrophilic compounds. In mammals, the cytochrome P-450 mixed-function oxidases (CYPs) are Phase I enzymes that convert chemicals, usually through oxidation, to reactive electrophiles, allowing Phase II enyzmes to add a hydrophilic group, for example, glutathione, to produce more water-soluble metabolites for distribution and excretion. The CYP superfamily may be divided into two major functional classes: (1) steroid- and cholesterol-synthesizing (metabolism of endogenous compounds), and (2) foreign substance or xenobiotic-metabolizing (metabolism of exogenous chemicals). Both classes are localized in the cellular microsomal membranes, that is, the endoplasmic reticulum, and are expressed at the highest levels with the greatest complexity of function in the liver. Their expression is not restricted to the liver; however, and various isoforms can be detected in several tissues. Certain forms are expressed only in specific nonhepatic tissues. Because of their broad but overlapping substrate specificities, the xenobiotic-metabolizing CYPs represent a formidable barrier to a remarkable range of absorbed lipophilic substances.

Over fifty CYP gene families have been identified in upper eukaryotes, of which at least eighteen occur in mammals but only 3 (CYP1,2, and 3) appear to play a significant role in carcinogen metabolism (Nelson et al. 1993). This multiplicity of CYPs is thought to have resulted directly from selective pressures imposed on evolving animals by diversification in the plant community, which elaborated a variety of novel toxic natural pesticides (Nelson and Strobel 1987). Indeed, the majority of CYP isoforms function specifically in the metabolism and elimination of xenobiotics present in the air or in our diet. Unfortunately, as a result of cellular efforts to eliminate some of these toxins, the cell may in some cases metabolically activate them to highly mutagenic and/or carcinogenic forms. Thus, while protecting an organism from the accumulation of large numbers and quantities of toxic environmental chemicals, some may be inadvertently metabolized by the CYPs to harmful forms.

3.9b   Inducibility of xenobiotic metabolism

The discovery of the CYP (P-450) proteins resulted from studies of an hepatic microsomal pigment, which produced a characteristic absorbance maximum at 450 nm when reduced and bound by carbon monoxide (Garfinkel 1958; Klingenberg 1958). Then, in seemingly unrelated studies, chronic administration of the barbiturate phenobarbital to rats was shown to reduce the sedative effect of phenobarbital by stimulating its own metabolism (Remmer 1962). These separate observations were subsequently unified by the demonstration that CYP proteins play a role in drug and carcinogen metabolism (Cooper et al. 1965). Critical to these findings was the fact that the CYPs are inducible. In this case, induction of the CYPs is similar to the classic process observed in bacteria for the *lac* operon. Essentially, enzyme activity is increased by elevating the levels of transcription and translation from the gene that encodes the enzyme. Furthermore, chemicals generally induce levels of the same CYPs that metabolize them. Thus, phenobarbital stimulates its own metabolism in liver by increasing the expression of a specific CYP isoform. Although each isoform is induced by and metabolizes a characteristic broad spectrum of substrates, general categories of chemical induction have been identified: the polycyclic aromatic hydrocarbons, peroxisome proliferators, glucocorticoids, and phenobarbital. Through the use of cell culture systems and mouse transgenic models, understanding of the basis for the transcriptional regulation of the major CYP isoforms has been achieved (Sonoda et al. 2003).

Ah receptor and induction. The chemical specificity of inducer compounds reflects the ability of the chemical to interact with a specific cytoplasmic receptor. This has been effectively demonstrated for the Ah (aromatic hydrocarbon) receptor, a cytosolic protein with high affinity for PAH. The Ah receptor (AhR) was discovered using the potent CYP1 gene family inducer 2,3,7,8-tetrachlorodibenzo-p-dioxin (TCDD), a chemical that enhances the metabolism of PAH. With a radiolabeled form of TCDD, Poland, Glover, and Kende (1976) found a receptor protein in mouse liver cytosol with an extremely high affinity for TCDD. Since then several PAHs have been shown to interact with the AhR, and subsequent studies have indicated that this receptor belongs to the basic helix-loop-helix PER-ARNT-SIM (PAS) domain family of DNA-binding transcription factors that function as biosensors. Normally, the AhR is stabilized in the nucleus through interaction with three other proteins: hsp90, hsp90 co-chaperone p23, and X-associated protein 2 (Carlson and Perdew 2002). Upon complexing with its ligand, the AhR is translocated to the nucleus where the complex heterodimerizes with AhR nuclear translocator protein (ARNT), enhancing transcription of CYP1 isoforms by binding specifically to a dioxin-responsive element (DRE; 5'-TNGCGTG-3') in the gene's promoter region. Induction through this pathway is not required for normal development or survival, since knockout mice that are null for members of the CYP1 family generally live a normal lifespan (Gonzalez and Kimura 2003), suggesting that these enzymes have evolved solely for purposes of xenobiotic metabolism. Loss of the AhR itself, however, resulted in abnormal immune and liver functions, indicating that the receptor plays a greater developmental role than simply processing of xenobiotics.

The AhR is not the only receptor for activating CYPs. Inducers of peroxisome proliferation (for oxidation of long-chain fatty acids) activate the associated CYP4 through binding with peroxisome proliferator-activated receptors (PPAR; Johnson et al. 2002). This nuclear receptor then heterodimerizes with the retinoid X receptor (RXR) and binds to a peroxisome proliferator response element, a consensus sequence in the CYP4A DNA promoter region. While some drugs that induce peroxisome proliferation increase liver tumor incidence, the primary function of CYP4 appears to be the regulation of carbohydrate and fatty acid homeostasis, as noted in studies of null mice that lack PPAR (Lee, Olson, and Evans 2003). For tumor promoter phenobarbital, two different CYPs, CYP2B and CYP3A, can be induced following exposure. Transcriptional activation of CYP2B requires the androstane receptor and, for CYP3A, the pregnane X receptor (Corcos and Lagadic-Gossmann 2001). Both nuclear receptors also heterodimerize with RXR for CYP promoter binding and induction.

CYP (P-450) induction and carcinogenesis. How does CYP induction relate to carcinogenesis? The effects of induction can be profoundly beneficial or detrimental. Amino azo dye-induced carcinogenesis is markedly reduced in rodents first treated with the weak PAH carcinogen methylcholanthrene, which apparently stimulates CYP-mediated azo dye metabolism (Conney 1982). Conversely, treatment of mice null for the AhR with benzo[a]pyrene failed to induce liver or skin tumors as in wild-type control animals (Shimizu et al. 2000), and loss of inducible CYP1B1 drastically reduced the incidence of DMBA-induced lymphomas in mice (Buters et al. 1999), suggesting that induction is required for carcinogenesis. Furthermore, differences in levels of carcinogen metabolism occur in strains of mice with dissimilar abilities to induce CYP activity. Studies of Ah responsive (inducible) and nonresponsive (noninducible) mice have shown that nonresponsive mice have lower levels of the AhR and a weaker affinity of the inducing chemical for the AhR. Tumor incidence increases in responsive mice when the carcinogen is administered locally and in nonresponsive mice when administered systemically (Okey et al. 1986). The

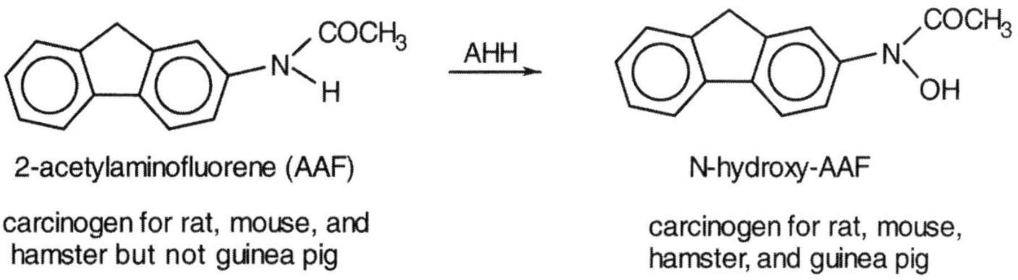

Figure 3–8. Species-specific metabolism of 2-acetylaminofluorene (AAF). AAF is a carcinogen in rat, mouse, and hamster but not guinea pig because it lacks the ability to oxidize AAF. Studies of these differences provided the first evidence that some carcinogens require metabolic activation.

explanation for this may be that responders with inducible liver CYPs can clear carcinogens from their system during the first pass of the chemical through the liver, but nonresponders cannot (Okey 1990). Conversely, high levels of inducible tissue CYPs in the responders may only result in greater levels of carcinogen activation in locally exposed tissues where inactivating pathways may be deficient. For humans, the situation is even more complicated. Polymorphisms have been identified in specific CYP isoforms, which yield enzymes with substantially decreased abilities to metabolize carcinogens; however, there is as yet no compelling epidemiological evidence indicating that these CYPs function as cancer risk factors.

3.9c   Metabolic activation of chemical carcinogens

Most chemical carcinogens require metabolic activation to elicit their carcinogenicity. This may be due to the fact that metabolism-independent chemicals can react with a multitude of nucleophilic molecules outside of the cell or cell nucleus and therefore cannot reach the DNA unless present in very high concentrations, although recent evidence suggests that most reactive intermediates are sufficiently stable to penetrate the cell and nuclear membranes. Alternatively, metabolism-dependent chemicals as procarcinogens can be distributed systemically as nonreactive substances, enter cells in a variety of tissues, and become activated to their "ultimate" reactive carcinogenic forms by the cellular CYP proteins. The classic studies of the Millers (1970) demonstrated for the first time that unreactive chemical carcinogens or procarcinogens can be metabolized to more reactive intermediates or ultimate carcinogens. Studies of 2-acetylaminofluorene (AAF) established a species-specific pattern of carcinogenesis in which rats, mice, hamsters, and rabbits were susceptible to tumor formation but guinea pigs were not (Figure 3–8). However, an N-hydroxy derivative of AAF induced tumors even in the guinea pig, showing that it lacked the ability to metabolize AAF to its N-hydroxy form. This intermediate did not react directly with DNA, however, suggesting further metabolism was necessary for activation and that guinea pig cells retained the metabolic capability to process the N-hydroxy intermediate to its ultimate carcinogenic form.

Although most environmentally important carcinogens require metabolic activation, some occupational substances are direct-acting or metabolism-independent compounds that can induce tumors systemically. Although highly reactive substances often decompose too rapidly to be distributed beyond the site of administration, other compounds like N-ethyl-N-nitrosourea are sufficiently stable to penetrate transplacental barriers and induce, at high incidence, tumors of the central and peripheral nervous systems in fetal rats (Druckrey, Ivankovic, and Preussmann 1966). Thus, the direct-acting carcinogens cannot be ignored.

3.9d   Inactivation of chemical carcinogens

In addition to the role of activation in susceptibility to carcinogenesis, inactivation plays a similarly important function. During CYP induction, several conjugating or detoxifying enzymes called Phase II conjugating enzymes are simultaneously induced: liver cytosolic glutathione S-transferase, liver microsomal epoxide hydrase, acetyltransferase, and glucuronide transferase. These enzymes collectively add polar groups to CYP/P-450 reaction products, increasing their water solubility or hydrophilicity and allowing their elimination by excretion. In liver, these enzymes are present at high levels, like the CYPs; however, tissues other than liver generally do not maintain these levels. Regardless, DMBA/TPA-induced skin tumorigenesis was significantly increased in mice lacking the inactivating enzymes glutathione transferase P1 and P2 (Henderson et al. 1998), indicating that the Phase II enzymes do impact carcinogen metabolism in tissues other than the liver.

Although the desired effect of xenobiotic conjugation is the detoxification and removal of hazardous compounds, this mechanism is also not foolproof. For example, liver-mediated glucuronidation of the bladder carcinogen 2-naphthylamine actually stabilizes the intermediate until it reaches the bladder where it is further metabolized to its ultimate carcinogenic form (see Figure 3–4).

3.9e   Systemic distribution of chemical carcinogens

Once a carcinogen has passed through the liver, it is available for distribution throughout the organism and metabolic activation by CYP enzymes in individual tissues. For transport, a variety of serum proteins are available. Lipophilic compounds can be sequestered by proteins like albumin and shunted to adipose tissue where they can be temporarily stored unmodified. Certain metals such as cadmium may be bound by either transferrin or metallothionein and concentrated in tissues with a capacity for iron storage. As already mentioned, it seems remarkable that even highly reactive substances like N-ethyl-N-nitrosourea are disseminated uniformly throughout the tissues not only in the host but also in fetuses exposed transplacentally. It appears then that distribution may not be a limiting factor in carcinogen exposure, especially in circumstances where the liver is unable to clear the first pass of carcinogen from circulation. This would allow unreacted chemicals to reach tissues without the liver's detoxification mechanisms.

3.9f Mechanisms for carcinogen suppression/chemoprevention

Since the process of tumor formation is lengthy, it may be possible to block neoplasia at multiple stages from uptake through metastasis. The fact that most environmental agents require metabolic activation provides opportunities for manipulating the initiating process to suppress carcinogenic activity. To this end, a group of chemopreventative blocking agents has been identified that prevents carcinogens from reaching or reacting with a target gene (Wattenberg 1996). These function as inhibitors of metabolic activation, enhancers of detoxification pathways (conjugating enzymes, discussed previously), and scavengers of electrophilic intermediates (the ultimate carcinogens). Once a genetic lesion has been sustained, however, another class of chemicals called suppressors can inhibit or prevent an initiated cell from becoming neoplastic. These include chemicals which induce cells to terminally differentiate, block oncogene activation, or selectively inhibit the proliferation of initiated cells.

As described in the chapter on epidemiology (Chapter 8) and in Appendix A, dietary constituents may provide significant protection against certain forms of cancer. Dietary factors seem to modulate the risks of colorectal, lung, cervical, and stomach cancers, for rates are lowest in those groups that consume the highest levels of fruits and vegetables (Hirayama 1979; Boyle, Zaridze, and Smans 1985). Examining these foods has led to the identification of several natural substances that block carcinogen-target interactions by interfering with processes described earlier. Diallyl sulfide, a member of the organosulfur compounds found in garlic and onions, the grape-derived polyphenol resveratrol, the plant phenol and curry spice curcumin, and the widely distributed and predominant dietary flavonoid quercetin, all decrease tumor incidence in carcinogen-treated rodents and function at the level of carcinogen activation/detoxification (Kris-Etherton et al. 2002; Tamimi et al. 2002).

Chlorophyllin, a water-soluble form of chlorophyll, has been shown to interact with aflatoxin (possibly reducing its bioavailability) and to decrease aflatoxin-initiated hepatocarcinogenesis (Breinholt et al. 1995). Currently it is under test as a cancer chemopreventive agent on individuals at high risk for hepatocellular carcinoma due to exposures of aflatoxin and hepatitis virus in Qidong, People's Republic of China. Studies have revealed significantly lower levels of excreted aflatoxin-adducts, although it is too early to know whether this will result in a lower tumor incidence, as occurred in animal studies (Egner et al. 2001). Such studies suggest that simple dietary modifications may significantly impact tumor development, especially in high-risk populations.

The scavengers of reactive carcinogens, notably the antioxidants, have drawn considerable attention since the tocopherols (vitamin E) in wheat germ oil were found to inhibit carcinogenesis from coal tar (Davidson 1934). Several naturally occurring antioxidants, such as vitamins C and E and β-carotene, have been shown in animal experiments to inhibit the formation of chemically induced tumors (Steinmetz and Potter 1991). Vitamins C and E also restrict the endogenous formation of nitrosamines. The plant flavonoid ellagic acid reduced benzo[a]pyrene-diol-epoxide-initiated lung tumor formation (Chang et al. 1985) and 3-methylcholanthrene- and DMBA-induced skin tumor incidence when applied orally to rodents at the time of carcinogen exposure. Similarly, β-carotene inhibited colon tumor formation induced with dimethylhydrazine. On the basis of experimentation with the ultimate carcinogenic metabolite of benzo[a]pyrene, it appears that these antioxidants elicit their effects at the level of the ultimate carcinogen. While epidemiologi studies suggest a role for these antioxidants in carcinogenesis, clinical studies of vitamins C, E, or β-carotene have failed to decrease lung (Alpha Tocopherol, Beta Carotene Cancer Prevention Study Group 1994) or colorectal (Greenberg et al. 1994) tumor incidence with long-term intervention.

In addition to natural products, efforts have been made to identify non-dietary chemicals that can suppress tumorigenesis in individuals predisposed to specific neoplasms. The selective estrogen-receptor modulator (SERM) raloxifene reduced the risk for invasive breast cancer in postmenopausal women by a remarkable 76 percent over a three-year period (Cummings et al. 1999). Non-steroidal anti-inflammatory drugs (NSAIDs) such as the cyclo-oxygenase-2 (COX2) inhibitor celecoxib significantly decreased the number of colorectal polyps (nearly 30 percent) after only six months of treatment in patients with familial adenomatous polyposis, a genetic condition which predisposes to colorectal cancer (Steinbach et al. 2000). As we improve our understanding of the various molecular mechanisms contributing to tumorigenesis, we will rapidly develop therapies for cancer prevention that target these pathways.

Elizabeth Cavert Miller with husband James

Born in Minneapolis, MN on May 2, 1920, Elizabeth Cavert received a bachelor's degree with highest honors in Biochemistry from the University of Minnesota in 1941. Moving then to the University of Wisconsin to pursue a Ph.D., she was directed toward a double major in Biochemistry and Home Economics, despite her impressive academic credentials, due to the limited academic opportunities for women at the time. While attending a graduate level biochemistry class at the University, she met and subsequently married her teaching assistant James A. Miller, also a highly regarded graduate student in Biochemistry. This began a devoted personal and highly productive professional relationship that lasted forty-five years. Upon receiving her Ph.D. in 1945, she joined her husband at the renowned McArdle Laboratory for Cancer Research also at the University of Wisconsin where they teamed up to make several seminal discoveries in the area of chemical carcinogenesis. These included the first demonstration that many chemical carcinogens require metabolic activation, that once activated the 'ultimate' carcinogen functions as a reactive electrophile to interact with DNA and protein, and that the enzymes responsible for carcinogen activation are inducible microsomal oxidases. These contributions were recognized with numerous honors, including the prestigious Bristol-Myers, Bertner, General Motors, Papanicolaou, and Rosenstiel awards, membership in the National Academy of Sciences (USA), and presidency of the American Association for Cancer Research. Elizabeth C. Miller died of renal cell carcinoma on October 14, 1987. Having devoted her life and career to cancer research, it might seem ironic that cancer would eventually claim it but instead simply reflects its prevalence and apparent indiscriminant nature.

3.10   Modulation of carcinogenesis

After a carcinogen has been taken up, processed, and transported to its target (presumably DNA), the effect of a carcinogenic lesion, such as a DNA adduct, may be modified by a number of inherent cellular and organismal processes. DNA repair or replication, cellular proliferation, immune competence, cell differentiation, apoptosis, or hormonal responsiveness may enhance or suppress carcinogenicity by affecting the fixation and maintenance of an initiating lesion.

The interaction of a carcinogen with DNA results in a variety of genetic lesions, which in some cases reflect the type of adduct formed, thus providing a biomarker for the carcinogen in human tumors (Hemminki et al. 2000). A carcinogen may generate a hydroxyl, alkyl-, aralkyl-, or arylamine-modified nucleotide, a frameshift mutation due to the intercalation of a ringed compound into the DNA strand, or strand breaks that may cause sequence additions or deletions. Oxygen free radicals commonly cause the formation of 8-oxo-deoxyguanosine, thymine glycol, and 5′-hydroxyuracil – all of which are mutagenic (Marnett 2000). The 8-oxo-dG adducts can trigger the misincorporation of dATP (Shibutani, Takeshita, and Grollman

1991) in the complementary DNA strand opposite the adduct during replication, producing a G to T transversion, which is commonly observed in cancer susceptibility genes. The adduct also destabilizes neighboring bases so multiple mutations in adjacent bases can occur (Kuchino et al. 1987). For the metabolism-independent alkylating agent N-ethyl-N-nitrosourea (ENU), O^2-ethyldeoxythymidine persists in DNA following carcinogen exposure, causing misincorporation of dTTP opposite the modified base (Jansen et al. 1994) and generating the T to A mutations commonly observed in tumors from ENU-exposed tissues (Perantoni et al. 1987, 1994). Therefore, the ability of a chemical to behave as a carcinogen may reflect the persistence of the adducts it generates in DNA.

Examination of adduct formation by the other classes of carcinogens has established preferred sites for carcinogen-base interactions (see Figure 3–3). The alkylating agents, in general, attack exocyclic oxygens and ring nitrogens, especially N-7 in the purines. The aralkylating agents react preferentially with exocyclic nitrogens, and the arylamines form adducts primarily with either ring or exocyclic nitrogens in the purines. The qualitative distribution of adducts formed therefore is inherent in and restricted by the chemistry of the class of carcinogen.

The fixation of mutations in the genome depends on a dynamic equilibrium between cell replication and DNA repair processes. Prior to repair, normal DNA replication can produce mismatches where DNA adducts occur, generating point mutations. Thus, the most sensitive cell cycle phase to mutagenesis is the S phase or DNA replicative phase. Perhaps the clearest example of replicative sensitivity to carcinogenesis involves the transplacental induction of neurogenic tumors in rats. ENU is 50 times more effective at inducing these tumors in the fetus than in the adult animal, presumably because there is substantially more cell replication in fetal neuroectodermal tissue than in the adult (Druckrey et al. 1966).

Transcriptional activity also profoundly influences the extent of mutagenesis by a carcinogen. Excision repair accounts for virtually all adduct repair in DNA and is accomplished by two different mechanisms: a slow inefficient general genome repair mechanism primarily for untranscribed DNA and a rapid transcription-coupled repair process involving endonuclease activity (Huang et al. 1992). The general repair process depends on the recognition of DNA damage by some type of DNA surveillance complex that is still poorly understood. For transcription-coupled repair, the damage may be identified by a stalled RNA polymerase. In studies of UV repair in cultured mammalian cells, excision of UV-induced dimers in a transcriptionally active gene occurred preferentially in the transcribed or coding sequence (Mellon, Spivak, and Hanawalt 1987). Similarly, repair of methylated adducts induced by N-methyl-N-nitrosourea (MNU) occurred significantly faster in transcriptionally active sequences (LeDoux et al. 1991). Through the study of genetic complementation groups in patients with deficiencies in DNA repair processes, that is, xeroderma pigmentosum (XP), a series of genes has been cloned that function both in DNA repair and transcription (Friedberg 2001). More recently the role of these genes has been examined in the context of the mouse genome (knockout or gene targeting technology), which has clarified their involvement in transcription, repair, and carcinogenesis.

XP is a genetic disorder characterized by a cutaneous hypersensitivity to ultraviolet radiation and a predisposition to all forms of skin cancer. The genetic lesions are predominantly associated with excision repair mechanisms, especially a process called nucleotide excision repair (NER). NER identifies and corrects damage caused by ultraviolet light and aryllating chemical carcinogens such as benzo[a]pyrene through removal and resynthesis of about twenty-five to thirty nucleotides of DNA surrounding the adduct. The current model proposes that when an RNA polymerase II complex encounters an adduct in the DNA, transcription is arrested. This in turn causes the recruitment of several proteins to the damaged site. The generated multiprotein complex determines the appropriate repair mechanism and assembles the proper factors to deal with it. If the damage involves formation of an oxidative adduct, then base excision repair (BER) will correct the lesion by removing the altered nucleotide base with a base-specific DNA glycosylase. If the damage instead requires NER, then a multiprotein complex that includes transcription factor IIH, which contains two DNA unwinding proteins XPB and XPD, and two endonucleases (XPG and XPF for cutting DNA 3' and 5' to the damage, respectively) is assembled (Friedberg 2001). NER is eventually completed with a DNA polymerase, its accessory proteins, and a ligase, which restores DNA strand integrity.

Several other factors have been implicated in pathways of carcinogenesis and are mentioned briefly here. Considerable effort has been focused on immune surveillance, a postulated process mediated by the immune system for the selective elimination of aberrant preneoplastic or neoplastic cells. It was hoped that an understanding of this process would lead to anticancer therapies that could enhance surveillance for these cells and perhaps even lead to the development of antitumor vaccines. After all, tumor cells are genetically unstable and develop thousands of mutations, which result in the expression of many altered proteins, any of which might provide a basis for an immune response. In this regard, vaccines of tumor-specific antigens can protect mice from a subsequent lethal challenge of injected tumor cells but are ineffective when administered following tumor cell injections (Lollini and Forni 2002). Since no common human tumor-associated antigen has been discovered and since we are unable, in most cases, to predict what type of neoplasm an individual will develop, preimmunization is currently not an option. Furthermore, in the absence of neoplastic disease, preventative therapies run the risk of precipitating an autoimmune crisis that could be equally debilitating. In the case of patients with neoplastic disease, clinical trials of tumor vaccines have thus far proven disappointing in that tumor regression is infrequent and unpredictable. This may be due in part to the type of antigen used in vaccine development as well as the heterogeneous nature of tumors. It may be possible, however, to address the reasons for these failures and develop more effective therapies. In the past, tumor antigens have not included proteins required for tumor cell survival or proliferation. More recently, immunotherapy has targeted the oncogene receptor HER-2/neu, which is required for tumor cell proliferation and progression, and vaccination significantly decreased tumor incidence in mice transgenic for this oncogene (Rovero 2000). In the future, it may be possible to combine this type of therapy with other forms to elicit a more effective treatment and reduce problems of tumor cell resistance.

The stage of tissue development modifies carcinogenesis in a variety of ways, but perhaps most significantly by production of a high proportion of cycling stem cells in undifferentiated tissues and the frequent absence of carcinogen-metabolizing enzymes in these undifferentiated tissues. As already discussed, rat neurogenic tissues are considerably more sensitive to carcinogenesis by direct-acting alkylating agents (require no metabolism) in fetal development than in adult tissues, probably because of the high proportion of replicating stem cells. Similarly, the less differentiated proliferative cells in the lower layers of skin are more susceptible to carcinogenesis than the highly keratinized differentiated nonproliferative upper layers. In the case of metabolism-dependent carcinogens, the undifferentiated proliferating tissues generally express lower levels of P-450 proteins. This reduces tissue sensitivity to carcinogenesis unless the carcinogen is metabolized elsewhere and transported to the target tissue as can occur in transplacental exposures.

Factors involved in growth suppression are discussed in Chapter 5 but probably markedly influence carcinogenesis at multiple levels. The suppressive factors are thought to affect primarily initiated cells by inhibiting the formation and progression of autonomous colonies. It is clear that initiated cells can remain quiescent in the suppressive environment of normal cells. The classic example of this is the suppression of the malignant phenotype of mouse embryonal carcinoma cells upon implantation into a normal mouse blastocyst. These tumor cells, when placed in an embryo, can be distributed throughout the animal, yet their tumorigenic behavior is suppressed and they function normally in their respective tissues (Mintz and Illmensee 1975).

In summary, several factors modulate carcinogenesis once the carcinogen has reached its target and initiated the process of tumor formation. These factors include properties inherent in the chemicals themselves, the tissues, the repair processes, and the cellular environment surrounding the altered cells. The combined influences of these various factors determine whether or not a cell will begin its journey on the road to independence.

3.11 Tumor promotion

The carcinogens described thus far all have genotoxic effects on cells; that is, they can directly cause mutations through interactions with DNA. As already mentioned, at least one class of compounds involved in carcinogenesis shows little if any genotoxicity and is therefore believed to act through 'epigenetic' mechanisms, involving changes in gene expression at the transcriptional or post-transcriptional levels. These are the so-called promoters, which are believed to stimulate growth or block differentiation/apoptosis preferentially in initiated cells. The existence of this group of substances was originally suggested in early mouse skin painting experiments, when it was observed that chemical irritation enhanced the rate of skin tumor formation (Berenblum and Shubik 1947). In these experiments, mouse skin was first exposed to a PAH carcinogen that induced few skin tumors and then to repetitive skin paintings of croton oil, a noncarcinogenic plant extract that markedly increased tumor incidence. The active component of the oil was identified as 12-O-tetradecanoylphorbol 13-acetate (TPA) (Hecker 1967), a substance now used routinely in skin tumor promotion studies because of its potent promoting activity. Following a single "nontumor-inducing" dose of an initiating carcinogen, repeated exposures to TPA induced the appearance of multiple proliferative lesions or adenomas of the skin. Maintenance of the early lesions required the continuous presence of the TPA, and withdrawal of TPA resulted in their complete regression and disappearance.

The phenomenon of promotion has also been observed in other organ systems. The classic rat liver carcinogenesis studies of Peraino, Fry, and Staffeldt (1971) identified phenobarbital as an effective promoter of hyperplastic nodule formation and eventually of hepatocellular carcinoma. More recently, rat renal cortical and transitional cell tumors were found to be promotable with barbital (Diwan, Ohshima, and Rice 1989).

The chemicals used in promotion protocols are generally of little environmental consequence to humans. However, certain chemicals that function as promoters in bioassay systems have been used therapeutically on humans. For example, the skin promoter benzoyl peroxide is frequently employed in the treatment of certain skin ailments. Perhaps of greater importance are possible endogenous promoters. No intensive study of these substances has been undertaken, but certainly growth factors and hormones are likely candidates. TPA mimics the mitogenic activity of epidermal growth factor (EGF), which induces proliferation in a wide variety of cell types. It is possible therefore that EGF or its family members (e.g., transforming growth factor-α) function as promoters for certain initiated cells. Hormonal status or responsiveness also influences tumorigenesis by providing selective growth-promoting stimuli. For example, prostate tumor cells are often initially androgen dependent, and tumor growth can be slowed by orchidectomy, which reduces testosterone levels by 90 percent. Because of their role as inducers of proliferation, these substances may be important endogenous promoters of tumorigenesis. Elucidation of endogenous promoters therefore may lead to novel therapies for tumor management.

Although the epigenetic mechanisms responsible for promotion remain poorly understood, examination of various promoting agents has established that hyperplasia is an absolute prerequisite for promotion. Based on studies of specific target tissues for promoters or genetic susceptibility for promotion (Sisskin, Gray, and Barrett 1982; Loury, Goldsworthy, and Butterworth 1987), all promoters induce cell proliferation. Induction of proliferation, however, is not sufficient for promotion; for example, acetic acid, a potent inducer of hyperplasia in skin, is an inefficient promoter (Slaga, Bowden, and Boutwell 1975). Exactly what the additional requirement might be is unclear. For promotion in mouse skin, two events have been described: one for conversion of an initiated cell to one that can be selectively propagated and the second for expansion or propagation of the converted population (Drinkwater 1990).

If an initiating mutation in a gene enhances responsiveness to growth stimulation by a growth factor or inhibits the ability of a stem cell to terminally differentiate or become apoptotic, then a promoter such as TPA might induce proliferation selectively in those initiated cells. Alternatively, TPA, like other promoters, is cytotoxic. If the normal cells surrounding the initiated cells are killed during exposure, the initiated cells might be released from regulation and additionally stimulated by the various growth factors secreted during tissue repair. Finally, promotion correlates well with a chemical's ability to inhibit gap-junctional intercellular communication (Trosko 2001), providing another mechanism whereby initiated cells can escape regulation by surrounding normal tissues.

DNA methylation can play an important role at both initiation and promotion stages. The extent of methylation (5-methylcytosine residues) in gene promoter regions correlates inversely with the levels of transcription allowed from a gene. Thus, a gene promoter that is hypomethylated is often strongly expressed and one that is hypermethylated is poorly expressed. In tumor cells, one can readily envision transcriptional activation of an oncogene by hypomethylation to increase cell proliferation or hypermethylation of a tumor suppressor gene to enhance tumor cell survival. Indeed both situations have been observed. In addition, methyl-deficient diets, treatment with the liver promoter phenobarbital, or genetic loss of DNA methyltransferase I all cause generalized tissue hypomethylation and facilitate carcinogenesis (Counts et al. 1996; Gaudet et al. 2003). Such events may complement any carcinogen-induced genetic lesion to permit the selective survival and growth of an initiated cell.

Although any of the mechanisms just described could account for the selective growth or loss of normal growth regulation of initiated cells, promoters like TPA can have genotoxic effects as well and are weakly carcinogenic, perhaps as a result of oxygen free radical formation during exposure (Cerutti 1985). Also, promoters can enhance carcinogenesis, albeit weakly, even when applied prior to the initiating carcinogen. This would seem to negate the concept of clonal expansion of initiated cells during the promotion stage. Thus, unanswered questions remain on the mechanisms of promotion and its role in carcinogenesis.

3.12   Tumor progression

Once a cell has been initiated and that population of cells specifically expanded by promoters, the altered cells can either regress following promoter removal or experience another genetic event to facilitate tumor cell progression, a process which enhances the biological and clinical aggressiveness of a tumor cell (Nowell 2002). The specific events involved in the progression phase are not completely understood primarily because of the multiplicity of changes that occur, many of which do not participate in malignant conversion. Regardless, the process includes a defect in apoptosis, an increase in the proliferative fraction of cells, a decrease in tendency to terminally differentiate, a shift toward autonomous growth, genetic instability, and invasive and metastatic behaviors. With the exception of certain types of leukemia, tumor cells at this stage generally show considerable heterogeneity and become aneuploid. The heterogeneity results from pressures on cells to differentiate and from the selective growth pressures just described. One might predict even greater heterogeneity during progression, but instead actually observe a decrease in heterogeneity as populations of tumor cells undergo selection. Additionally, the genome becomes unstable due to an alteration in a genome maintenance function, for example, DNA mismatch repair, causing chromosomal alterations (mutator phenotype) with increasing frequency during tumor progression. These lesions do not accrue randomly, and tumors often show remarkable uniformity in their chromosomal aberrations. In nearly every category of tumor, specific primary and secondary chromosomal alterations are consistently selected and may contribute to the progression process.

As the progression phase ends, tumor cells have converted to the neoplastic phenotype. They are invasive, sometimes metastatic, and highly autonomous. They can now erode their surrounding normal tissue barriers, penetrating cellular layers as well as the basal laminar matrices. They can escape both physical and growth regulatory restraints imposed by surrounding normal tissues.

3.13   Alternative pathways for carcinogenesis?

Most cancer researchers support the initiation/promotion/progression/conversion concept of carcinogenesis, and, indeed, certain molecular events in tumor formation associate with specific phases of the process at least in promotable tissues such as skin (Yuspa 1994). Vogelstein proposed that tumors result from the accumulation of genetic events and has documented a series of genetic alterations, both dominant and recessive, in tumorigenesis of the colon (Fearon and Vogelstein 1990; Kinzler and Vogelstein 1996). Although the sequence of events may not be absolute, clearly certain events must precede others in tumor formation, a finding consistent with this multistage model.

3.14   Federal regulations

Considering the variety and quantity of synthetic chemicals added to the environment annually, what type of protection are we provided? Do governments do anything to regulate exposures to potentially hazardous substances? Outside of the United States, the WHO, as mentioned earlier, has created the IARC for monitoring global cancer incidence, identifying carcinogens, defining carcinogen mechanisms, and developing programs for cancer prevention worldwide. However, it functions primarily in an information-gathering and advisory capacity, and therefore implementation of cancer control measures throughout the world is not part of its mission. Thus, individual states are responsible for providing protection for their populations.

In the United States, various Executive Branch agencies have been charged by Congress with the responsibility of defining substances of concern and setting allowable levels of exposure in various environments. For example, pesticide exposures are regulated by the Food Quality Protection Act, landmark legislation passed in 1996 that allowed the Environmental Protection Agency (EPA) to use current science in determining safe levels of pesticide residues (reasonable certainty of no harm) in all foods. Under the Federal Insecticide, Fungicide, and Rodenticide Act (FIFRA), the EPA registers pesticides and regulates their use in order to prevent adverse health or environmental consequences. Under the Federal Food, Drug, and Cosmetic Act (FFDCA), the EPA sets legally permissible levels of pesticides in all foods (raw or processed), regulates on the basis of any deleterious health consequences, not just carcinogenicity, provides additional protections for children due to issues of greater chemical sensitivity, considers the benefits of pesticide use in determining tolerances (e.g., when a pesticide prevents greater health consequences), calls for a complete reevaluation of existing chemical-specific tolerance levels within ten years, provides for periodic review of all pesticides, mandates a process for the accelerated consideration of "safer" pesticides, and prevents states from establishing tolerances that differ from those of the EPA. The tolerance limits are established by the EPA based upon scientific studies from a variety of sources, including the manufacturer. Enforcement of these limits is the responsibility of the Food and Drug Administration (FDA) for most plant-based foods and the Department of Agriculture/Food Safety and Inspection Service for most animal products.

For exposures outside of foods, some of the same agencies bear responsibility through the same or similar legislation. For example, asbestos exposures in any consumer products, including foods, adhesives, or pharmaceuticals are regulated and monitored by the FDA through the FFDCA. The EPA controls asbestos use through several pieces of legislation: the Clean Air Act; Clean Water Act; Comprehensive Environmental Response, Compensation, and Liability Act; Resource Conservation and Recovery Act; Safe Drinking Water Act; Superfund Amendments and Reauthorization Act; and the Toxic Substances Control Act. Furthermore, work place limits have been established and enforced by the Occupational Safety and Health Administration (OSHA).

As part of this overall effort to characterize potential environmental hazards, the National Toxicology Program (NTP), as described earlier, has been charged with the responsibility of verifying the safety of various nominated chemicals. The NTP is an interagency program that coordinates toxicology efforts within the National Institutes of Health's National Institute of Environmental Health Sciences (NIEHS), the FDA's National Center for Toxicological Research (NCTR), and the Centers for Disease Control and Prevention's (CDC's) National Institute for Occupational Safety and Health (NIOSH). Information on the regulation of the various chemicals evaluated by the NTP is available through their website (http://ntp-server.niehs.nih.gov) in the Report on Carcinogens.

Guidelines for protection in the workplace were established in the Occupational Safety and Health Act of 1970. In this document, the Occupational Safety and Health Administration (OSHA) was charged with defining health and safety standards for the workplace and enforcing those standards. Its mission to reduce exposures to hazardous substances has in some cases virtually eliminated certain occupational cancers. Vinyl chloride, a polymerizing agent in plastics manufacturing that induced a high incidence of the rare cancer angiosarcoma, is no longer a health problem in industry despite its continued use because of OSHA standards. Establishing appropriate industry standards has been a difficult problem for this organization, but efforts are being made by the current administration to devise risk-assessment guidelines for all federal agencies.

3.15   Summary

Carcinogenesis is a multistage process initiated and/or promoted by any of a variety of external chemicals, radiations, or endogenous processes. The ability of a carcinogen to induce tumor formation is intrinsic to the nature of the carcinogen and its ability to interact with a host and modify target DNA. Until the 1980s, cancer investigators focused on a characterization of the various carcinogens and their gross interactions with the host, for example, the metabolism of chemicals and DNA-adduct formation. This period also resulted in the development of model systems for studying tumor formation in most tissues. The rapid development of molecular technologies in the 1980s ushered in a new era in carcinogenesis studies, allowing researchers to focus on the specific targets of carcinogens, that is, the transforming lesions. The discovery of cancer-related genes has many scientists convinced that we are rapidly approaching an understanding of the neoplastic process. Furthermore, this knowledge has provided significant insight into the fundamental mechanisms of cell growth regulation and differentiation, as well. Chapter 5 addresses this aspect of carcinogenesis: the putative targets and how they function in carcinogenesis.

4

Genetics and heredity

Robert G. McKinnell

[A]ll cancers exist in both hereditary and nonhereditary forms . . . the heritable cancers are all uncommon.

A. G. Knudson 1985

For decades there has been no doubt that cancer is genetic, in the sense that transformation of a normal cell to invasive and malignant growth is due to changes in the DNA. But most cancer is genetic only at the level of the transformed cell, not in the germline of the patients.

M-C. King, S. Rowell, and S. M. Love 1993

4.1 Introduction

The heredity of cancer deals with familial aspects of neoplasia. "Familial aspects of neoplasia"means simply that more members of a family suffer from cancer than would be expected. Human pedigrees are examined to detect if a cancer is indeed familial, that is, is the cancer found in greater abundance than expected and thus perhaps inherited from previous generations? Understanding the heredity of cancer requires an investigation not only of the pedigree of families, that is, genealogy, but also of chromosomes, genes, and gene products. In this sense, this chapter concerns both the genetics and heredity of cancer. The term "genomics"refers to the structure and function of the entire genome of a species, including the complete DNA sequence, the regulation of its expression and how its genes and gene products function in a species. A related field is "oncogenomics"and that term pertains to the genomics of cancer. Oncogenomics is a rapidly evolving discipline resulting from new technology. One aspect of the new technology, DNA microarray analysis, will be considered in this chapter. In contrast to microarray analysis, the study of human

126

pedigrees is an ancient endeavor. For example, familial clustering of breast cancer was known to the Romans of the first century (Lynch 1985a).

It is not the intention of this chapter to provide information to diagnose self or others who are at high risk for, or indeed have, a cancer that is hereditary. That evaluation is made only after a thorough analysis of a family's cancer history compiled with the assistance of a genetic counselor or medical geneticist. That history includes data on all types of cancer in the family, age(s) of onset of the cancer(s), the occurrence of multiple primary cancers, and a number of other factors that aid in the evaluation (Lynch and de la Chapelle 2003). A list of cancer genetic resources, with web site addresses and telephone numbers, is found in the paper by Sifri, Gangadharappa, and Acheson (2004).

Tumors that differentiate inappropriately and which, when properly treated, differentiate normally and sometimes give rise to benign and postmitotic cells were discussed in Chapter 1. One might argue that a cancer, which has the potential to give rise to normal cell progeny, must have an intact genome (i.e., the cancer cell DNA does not differ from normal cell DNA) or only minor genetic defects that can be bypassed with appropriate treatment (Sachs 1987). However, it is proposed here that tumors with a normal or near-normal genome have an aberrant genetic component which most likely consists of abnormal gene regulation and expression. Some genes are inappropriately expressed in these and all cancer cells, and whether the inappropriate gene expression is due to genetic mutation or a non-genetic "epigenetic" phenomenon, the fact remains that the genome is not properly programmed. Because of this, we can say that all cancer has a genetic basis. While this statement is true, most cancer is *not* hereditary (see quotation of Knudson in the epigraph that heads this section). How does one escape from the conundrum that all cancers are genetic but hereditary cancers are rare? There is no escape. Read on.

4.2 Chromosomes and cancer

4.2a Aneuploidy

Chromosomes are comprised of DNA (and protein) and they transmit genetic information. For those reasons, this chapter starts with chromosomes. Normal cells are said to be euploid. Euploidy simply refers to a normal complement of chromosomes and can be described as a chromosome number that is either diploid or an exact multiple of the diploid number (this definition permits the polyploid cells of a normal liver and some other organs to be characterized as euploid even though they do not have the diploid number). "Aneuploidy," on the other hand, designates an abnormal number of chromosomes. Aneuploid cells have either too many or too few chromosomes. Missing chromosomes may result in incomplete genomic DNA and the loss of critically important genes. Extra chromosomes may lead to unbalanced gene expression. The terminology used to characterize chromosome abnormalities is illustrated and described in Gardner and Sutherland (2004).

An interest in aneuploidy as it relates to normal development and cancer has a long history. Theodor Boveri (1862–1915) studied the effect of aneuploidy in sea urchins. He observed that when sea urchin eggs were fertilized with two sperm instead of one, an additional cleavage center was formed. As a result, the sea urchin chromosomes were distributed to the daughter cells unequally, resulting in aneuploidy. The "ruinous" developmental aberrations of the sea urchin larvae were attributed to the aneuploidy (Boveri 1907). If the fate of a population of *normal* dividing cells is profoundly affected by aneuploidy, could a population of cells become *malignant* because of chromosomal imbalance? Boveri (1914) speculated that malignancy could indeed occur as a result of a "certain abnormal condition of the chromosomes." While this notion is completely reasonable (especially with the enhanced vision of cytogenetics in the new millennium), he abandoned this concept, at least temporarily. Why? Because of the skepticism and hostility of his colleagues and other "experts." What a pity! Certainly, Boveri's early thoughts on the role of aneuploidy in embryonic development have been amply sustained in plants (for example, see studies of the genus *Datura,* Blakeslee 1934) and in animals (Fankhauser 1945). The abnormalities of some cloned embryos (DiBerardino, 1997) and the morphologically aberrant progeny of animals exposed to mutagens (McKinnell et al. 1980) was shown to have a chromosomal basis. Boveri was clearly correct, and abnormal chromosomes are now known to be associated with abnormal development *and* cancer.

Boveri would be ecstatic to learn of the two-volume compendium that catalogs 22,076 cases of human neoplasms with known chromosomal aberrations (Mitelman 1994). This publication has 4,252 pages devoted to neoplasms of hematologic origin (73 percent) with the remainder being solid tumors. More is known of cancer cytogenetics in patients from the western world; thus it is not surprising to learn that the patient data is biased toward Europe and the United States with very few cases from Africa, South America, and Australasia. Boveri has been vindicated – few would argue with his hypothesis in the early years of the twenty-first century. And, the work of chromosomal characterization of human cancer continues (e.g., Nilsson et al. 2003; Hoglund et al. 2003).

The development of cancer cytogenetics had a slow start. One reason for the sluggishness in scientific advance was the historic difficulty of ascertaining the correct number of chromosomes in humans. How was one to recognize an abnormal number of chromosomes in a cancer cell if the proper and correct number was not known in normal cells? Boveri and many other investigators believed the diploid number of chromosomes in humans was forty-eight. Other less precise estimates put the diploid number variously at eight to more than fifty (Kottler 1974; Trask 2002). It was not until 1956 that the *correct* human diploid number of forty-six was published. Chromosomes were counted in 265 dividing human embryo cells, and all but four showed the number forty-six (Tjio and Levan 1956). Since that time, no deviation from the number forty-six in normal diploid cells of humans has been reported by any competent observer.

4.2b   Euploidy does not preclude genetic change

Aneuploidy is unequivocal evidence of an altered genome. As already discussed, there was early interest in establishing the relationship between chromosomes and cancer. The real issue, however, is not an aberrant chromosome number as a cause of cancer but whether a change in the genome can be causally associated with cancer. Obviously, aneuploidy is one kind of genomic alteration, yet genomic changes, which do not result in a changed chromosome number, can occur. It is not apparent, when examining an euploid set of chromosomes obtained from a malignancy, whether mutations, which may have caused that malignancy, have occurred. A neoplasm of the Northern Leopard Frog, *Rana pipiens,* serves to illuminate several aspects of chromosomes and neoplasia. The diploid number of chromosomes in normal cells of the frog is twenty-six (Figure 4–1). The chromosome complement of the frog neoplasm, the Lucké renal adenocarcinoma, is also twenty-six (Figure 4–2) and thus the frog cancer is euploid (DiBerardino, King, and McKinnell 1963). An extra haploid set of chromosomes, resulting in triploid cells with thirty-nine chromosomes, is not incompatible with either normal development or neoplasia (Figure 4–3). Consistent differences in chromosome morphology (shape) between normal and malignant cells have not been detected (nuclear transplantation studies of this renal adenocarcinoma are discussed in Chapter 6).

The cytogenetics of diploid frog tumors may lead one to suggest that the frog neoplasm is an exception to the notion that cancers arise as a change in the complement (number) or morphology of chromosomes. Other exceptions exist. For example, there has been a failure to detect karyotypic aberrations in the chromosomes of Li-Fraumeni syndrome patients (Malkin et al. 1990; see discussion in section 4.5b). Other cancers where normal karyotypes have been observed (among abnormal karyotypes) include astrocytomas (Orr et al. 2002), chondrosarcomas (Mandahl et al. 2002), pediatric brain tumors (Chumas et al. 2001), and acute myeloid leukemia (Döhner et al. 2002; Baldus et al. 2003).

Modern cytogenetics is concerned with the structural rearrangements of chromosomes as well as numerical changes. There may be translocations, deletions, and inversions, but none of these structural anomalies necessarily involves a change in the chromosome number. How are the structural rearrangements detected? The primary mode of detection is by banding of stained chromosomes. Banding is the longitudinal cross-staining that varies between chromosomes but is constant for each chromosome pair (Figure 4–4). Conventional banding permits the detection of chromosomal structural change. More subtle chromosomal aberrations may be

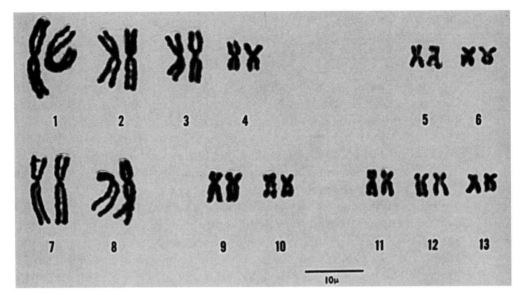

Figure 4–1. Chromosomes derived from normal adult tissue of *Rana pipiens*. The diploid number is 26. (From DiBerardino, M. A., King, T. J., and McKinnell, R. G. 1963. Chromosome studies of a frog renal adenocarcinoma line carried by intraocular transplantation. *J Natl Cancer Inst* 31:769–89.)

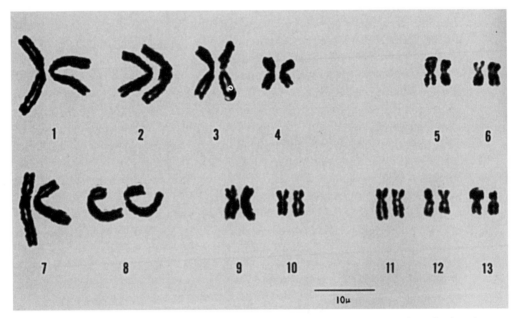

Figure 4–2. Chromosomes of a spontaneous renal adenocarcinoma allografted to the anterior eye chamber of a normal adult *Rana pipiens*. The chromosome number of this neoplasm is 26. (From DiBerardino, M. A., King, T. J., and McKinnell, R. G. 1963. Chromosome studies of a frog renal adenocarcinoma line carried by intraocular transplantation. *J Natl Cancer Inst* 31:769–89.)

130

Figure 4–3. Chromosomes of a virus-induced triploid renal adenocarcinoma with 39 chromosomes. These chromosomes are of a metaphase plate and have not been arranged in the form of a karyotype as illustrated in Figures 4–1 and 4–2. (From Williams III, J. W. W., Carlson, D. L., Gadson, R. G., Rollins-Smith, L. A., Williams, C. S., and McKinnell, R. G. 1993. Cytogenetic analysis of triploid renal carcinoma in *Rana pipiens. Cytogenet Cell Genet* 64:18–22. By permission of Karger, Basel.)

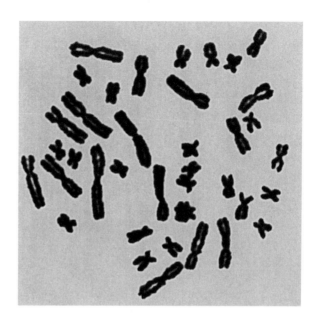

detected with the use of multiple fluorescent dyes that label each chromosome with its own distinct color. This method, known as FISH for "fluorescence in-situ hybridization", permits detection of "extremely subtle" changes in chromosome composition (Trask 2002).

Even with chromosome banding, genetic alterations are not easy to detect. The average chromosome band is composed of about 5×10^6 nucleotide pairs. Structural aberrations (i.e., deletions or duplications) of 2×10^6 nucleotide pairs would probably go unnoticed (Rowley 1977). Over a thousand genes could be lost or duplicated with no cytogenetic stigmata. *For this reason, failure to detect chromosomal changes does not prove the lack of genetic modification.*

As indicated, *failure to detect* altered chromosomes in a particular neoplasm does not rule out genetic change. Detection of a consistent chromosomal aberration associated with a particular malignancy, however, is irrefutable evidence of a genetic event that is either causal to the malignancy or a consequence of malignancy progression.

4.2c   Cancers with chromosomal aberrations

Chronic granulocytic leukemia: The first malignancy associated with a specific chromosomal aberration. As noted above, improved cytogenetic techniques led to the identification of the correct chromosome number for normal humans (Tjio and Levan 1956) and the search was on for consistent (i.e., nonrandom) chromosomal anomalies associated with specific neoplasms. In a remarkably short time, a consistent chromosomal aberration was found in the cells of patients with chronic granulocytic leukemia (also known as chronic myelogenous leukemia, chronic myelocytic

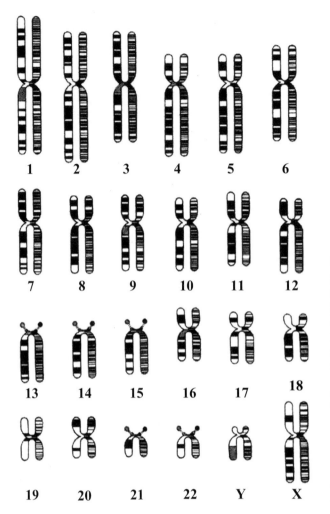

Figure 4–4. Banding pattern of human chromosomes. G-banding patterns of mid-metaphase are represented in the left chromatid; G-banding patterns of late prophase chromosomes are represented in the right chromatid. (Adapted from Yunis, J. J. 1976. High resolution of human chromosomes. *Science* 191:1268–70. Copyright © 1976 American Association for the Advancement of Science.)

leukemia, chronic myeloid leukemia, or CML). The leukemia is characterized by excessive and abnormal proliferation of malignant hematopoietic stem cells of the bone marrow.

A brief history of the discovery of this chromosomal aberration is worthwhile because this particular leukemia and its unique chromosomal marker have become the model for understanding cytogenetics and the molecular basis of cancer. A minute (a tiny chromosome fragment) was found to "replace" one of the small chromosomes in some cells of seven patients with chronic granulocytic leukemia (Nowell and Hungerford 1960; Wang et al. 2001). Because both Nowell and Hungerford were affiliated with institutions in Philadelphia (Nowell with the University of Pennsylvania and Hungerford with what is now the Fox Chase Cancer Center), the tiny chromosome fragment became known as the Philadelphia chromosome and was designated Ph. More than 90 percent of patients with chronic granulocytic

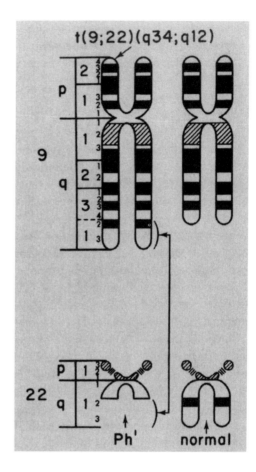

Figure 4–5. Diagrammatic portrayal of the reciprocal translocation in cases of chronic myelogenous leukemia. A portion of chromosome number 22 moves to chromosome number 9. The diminished chromosome number 22, known as a minute, is designated the Philadelphia (Ph) chromosome. (From Whang-Peng, J., Lee, E. C., and Knutsen, T. A. 1974. Genesis of the Ph chromosome. *J Nat Cancer Inst* 52:1035–6.)

leukemia have the Ph chromosome in their leukemic cells, and an additional 5 percent can be shown to have the chimeric gene known to be associated with the chromosome.

The seminal discovery of the Ph chromosome was made at a time when it was possible to judge the loss or gain of genomic material. The missing material of the small chromosome was estimated by the Fox Chase laboratory to be about 2×10^7 nucleotide pairs, or about 0.5 percent of the entire diploid genome (Rudkin, Hungerford, and Nowell 1964). The genetic material was either lost or translocated to another chromosome. Translocation proved to be the case as was shown in banded chromosomes a decade later (Rowley 1973). The missing material from Ph (now shown to be an abnormal chromosome 22) was discovered to be translocated to the long arm of chromosome 9 (Rowley 1973; Figure 4–5).

The translocation results in a "chimeric" or hybrid gene. The hybrid gene BCR-ABL is formed from the fusion of 5′ genetic material from the breakpoint cluster region (BCR) located on chromosome 22 with the 3′ portion of the Abelson murine viral proto-oncogene (ABL) of chromosome 9 (see Chapter 5 for a discussion of what oncogenes are). The new gene is transcribed and translated into a BCR-ABL fusion protein that is intimately associated with the genesis of chronic granulocytic leukemia. The understanding of the Ph chromosome follows closely the evolution of advances in chromosome technology and molecular biology. In this case, it has permitted the identification of a specific hybrid gene that is involved in the malignant process.

Some patients, although lacking the Philadelphia chromosome, nevertheless express the ABL gene on chromosome 22. The gene has moved without the usual Philadelphia reciprocal translocation (Morris et al. 1986). This has been referred to as a "masked Ph chromosome". There are two important lessons to be learned here. First, the really important aspect of genetic change in chronic granulocytic leukemia is the juxaposition of ABL with BCR. The second lesson relates to a seemingly normal karyotype. As indicated in the discussion on exceptions to Boveri's hypothesis, chromosomes may appear normal in a stained metaphase preparation but in fact contain altered genomic material.

Chronic granulocytic leukemia is obviously a genetic disease. In this case, an altered genome results in cancer. But is it hereditary? The answer to that query is probably "no." That chronic granulocytic leukemia is an acquired genetic condition is supported by studies of identical twins, when only one of the twins has a Ph chromosome and manifests the disease.

Other chromosomal changes and variant translocations occur in chronic granulocytic leukemia in addition to the common reciprocal translocation between chromosomes 9 and 22. Furthermore, the 9 to 22 translocation is not uniquely associated with CML. A minority of patients with acute lymphoblastic leukemia (ALL) and adult acute myelogenous leukemia (AML) also manifest the translocation.

Burkitt lymphoma. Burkitt lymphoma, the most common malignancy of African children, is named for the English surgeon Denis P. Burkitt (see essay of Mr. Burkitt and his interest in diet and cancer, Chapter 8), who first described the neoplasm for the medical world. This B-lymphocyte malignancy commonly afflicts young individuals, who frequently present with massive jaw neoplasms that grossly distort the face (Figure 4–6). The initial report of the cancer originated from Kampala, Uganda, but other patients were reported from a diversity of tropical African locations. The neoplasm is also found in tropical South America and New Guinea. Furthermore, the adrenal glands, kidneys, liver, and other organs may also be involved (Burkitt 1958). The common (endemic) form of this tumor is African and is almost always associated with the Epstein-Barr herpesvirus (EBV, Chapter 3), but a less common (sporadic) form of the lymphoma occurs in North America and it sometimes associated with EBV infection. Burkitt lymphoma occurs mostly in areas where malaria is common, and that disease may have an effect on the role of EBV in the disease.

Cytogenetic analysis of cells obtained from Burkitt lymphoma patients revealed a loss of genetic material in the distal region of the long arm of chromosome 8 translocated as an extra band to the long arm of chromosome 14 (Zech et al. 1976).

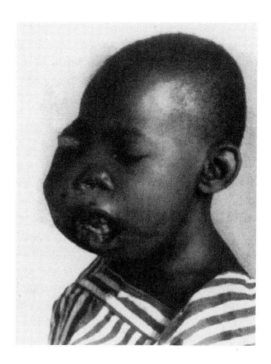

Figure 4–6. Child manifesting the severe facial distortion characteristic of some Burkitt lymphoma patients. (From Burkitt, D. 1958. A sarcoma involving the jaws of African children. *Br J Surg* 46:218–25. By permission of Blackwell Science Ltd.)

As a result, the c-MYC proto-oncogene of chromosome 8 is juxtapositioned to the locus that codes for the immunoglobulin heavy chain on chromosome 14. The 8:14 translocation is most common (about 80 percent), but 8:22 and 2:8 variant translocations occur also. The variant translocations result in c-MYC juxtapositioned to immunoglobulin light chain gene loci. The translocations cause a deregulation and abnormal expression of c-MYC, an event which seems to be a critical step in the pathogenesis of this neoplasm.

As in chronic granulocytic leukemia, Burkitt lymphoma is clearly a genetic disease; that is, it results from a change in the genome of the patient. Is the malignancy hereditary? Most patients are afflicted with EBV infection and some may have been exposed to malaria. The EBV infection, exposure to malaria, and genetic vulnerability may be involved in a causal way with endemic Burkitt and with the chromosome translocation associated with Burkitt lymphoma. The translocation deregulates the c-MYC gene, which in turn is involved in the development of the cancer. Ordinarily, EBV infection and exposure to malaria are not hereditary, and thus endemic Burkitt lymphoma is not considered a hereditary disease.

4.3 Chromosome damage, mutation, and vulnerability to cancer

Genoclastic agents (agents that cause chromosome damage) have been identified as substances that can cause cancer (see Chapter 3), and many, perhaps most, carcinogens are mutagens. Chromosome damage occurs with ultraviolet radiation (Chapter 7 and 8), X-rays, and other forms of radiation. Data derived from the survivors of the atomic bomb dropped at Hiroshima serve to estimate the cancer risk induced by radiation (Straume et al. 2003). Further, the survivors manifest chromosomal anomalies (Miller 1966). Uranium miners, a group that suffers from increased risk for pulmonary cancer, are exposed to radiation, and the chromosomal aberrations they carry are a sensitive biologic indicator of their exposure to that radiation (Brandom et al. 1978). Of course, observations of chromosome damage do not establish that the cancers under consideration are genetic; that is, the observed chromosome damage may be independent from the cancer, which may arise for other reasons. Nevertheless, the observations of genomic damage associated with increased cancer risk is a relationship that continues to elicit the interest of cell biologists.

Research into cancer prevention has been thought to be potentially the most beneficial area for cancer investigation (Chapter 8). Why discuss this issue in a chapter on cancer heredity and genetics? The reason relates to "spontaneous" muta-tion. A natural rate of mutation (due to natural causes) seems to be unrelated to factors in the environment that can be controlled by reasonable human effort. Although 80 percent to 90 percent of cancer may be produced by human-generated or human-controllable environmental insult, the remaining basal rate of 10 per-cent to 20 percent may not be preventable (Knudson 1985). Some humans, whose remains persist as fossils, clearly had cancer. These ancient individuals were not exposed to smokestack emissions, man-made environmental radiation, automobile exhausts, preserved foods, and ubiquitous pesticides. Nevertheless, some became afflicted with cancer. Although knowledge concerning the genesis of cancer gleaned from the study of mutagenic agents may be useful in understanding the cell alter-ations leading to malignancy, and perhaps ultimately to the prevention of much environmentally induced cancer, it is unlikely this knowledge will ever eliminate a basal low rate of cancer due to "spontaneous" mutation.

4.4 Hereditary cancers

Virtually every cancer exists in both a genetic ("heritable predisposition") form and a noninherited (sporadic) form (see epigraph of Knudson at the head of this chapter). Obviously not all cancers can be considered in this introductory textbook. Only a few examples have been chosen here, either because of their prevalence or because they are well known examples of genetically transmitted risk factors for cancer.

4.4a Retinoblastoma

Although retinoblastoma (Rb) is rare, it is the most common eye neoplasm of children. The cell type that evidences malignancy is an embryonic stem cell, as indicated by "blastoma" in the name. Because immature, proliferating retinal stem cells become progressively less common as an individual ages and because terminally differentiated retinal cells cannot become malignant, the cancer rarely afflicts adults. The malignancy, if untreated, moves from the eye along the optic nerve to the brain, metastasizes to other organs, and eventually causes death. Even with successful treatment of the retinoblastoma, the patient has an increased vulnerability to other cancers later in life.

A specific deletion (loss of a small region), or a mutation in the long arm of chromosome 13 is related to the pathogenesis of Rb. Every cell in the body of a child with inherited Rb has the deletion or mutation in one homologue of chromosome 13. This obviously includes all cells of both retinas. These children are at risk for developing Rb, but only with the occurrence of another mutation or deletion in the other copy of the gene. The second event led to the "two-hit" concept of Knudson, who suggested that in the inherited form one mutation is present prezygotically and found in all cells of the body. A second mutation occurs at the same locus in the other homologue postzygotically. Tumor initiation occurs with the second mutation, hence the name of the "two hit" hypothesis (Knudson 2002). Of course, two hits are also required in nonhereditary retinoblastoma.

The conditions that lead to the second mutation or deletion in hereditary Rb (that is, the conditions that yield homozygosity in the cells of one eye) are such that there is significant risk of developing the malignancy in the other eye. This is why both eyes may become afflicted in hereditary Rb. Both alleles in the same retinal cell must be mutant in nonhereditary retinoblastoma but, in contrast with hereditary Rb, the likelihood of this occurring in both eyes is so low that for all practical purposes it does not occur. Bilateral Rb is thus diagnostic for the hereditary type.

More than one-third of Rb cases are inherited. The remaining cases are sporadic (not hereditary). About one half of the children of an affected parent will carry the altered gene, and of these almost all will manifest retinoblastoma. Significant differences are noted between inherited and sporadic Rb. The age of onset of the inherited type is approximately ten months, which is on average eight months earlier than the sporadic type.

The specific chromosomal alteration in Rb is the loss of genetic material that limits growth. The growth limiting genes, deleted in Rb, were originally designated as "anti-oncogenes" but are now better known as "tumor suppressor genes". Tumor suppressor genes were recognized for the first time in retinoblastoma (see the discussion of suppressor genes and the Rb locus in Chapter 5 and a brief description of the pathology of Rb in the appendix).

4.4b   Wilms tumor

Wilms tumor (WT), also known as nephroblastoma, is a rare pediatric malignancy derived from metanephrogenic stem cells. The prefix "nephro" refers to the kidney and the root "blastoma" indicates that the tumor arises in embryonic stem cells. Cells in the tumor may differentiate into imperfect kidney tubules and glomeruli. While nephroblastoma is rare, it is the most common solid tumor of childhood. The tumor grows rapidly and often becomes metastatic to the lungs (see description of the tumor in the Appendix).

The bilateral early onset form of Wilms tumor is uncommon. The vast majority of WT is unilateral and sporadic (Feinberg and Williams 2003). Bilateral WT afflicts children at younger ages than unilateral Wilms. Knudson and Strong (1972) suggested that like Rb, Wilms tumors also arise after two separate genetic "hits." With inherited bilateral WT, the first "hit" occurs prezygotically with one mutated allele found in all of the cells of the body. The cancer occurs after a separate and subsequent mutation at the same locus in cells of the target tissue resulting in both alleles being affected. Accordingly, two "hits" are required, one prezygotically and one postzygotically. In sporadic WT, each "hit" occurs postzygotically. In either early bilateral or later unilateral sporadic WT, the cancer is thought to ensue only after two mutations.

WT may occur with several other clinical conditions. Wilms tumor that is found with the eye condition aniridia (absence of the iris), genital and urinary malformations, and mental retardation is known as the WAGR syndrome (W for Wilms, A for aniridia, G for genital, and R for retardation). Most patients with bilateral Wilms tumors are afflicted with the WAGR syndrome, which in turn is associated with deletions in the short arm of chromosome 11. The tumor suppressor gene WT1 is deleted or altered with the WAGR chromosomal aberration. Wilms tumor is also associated with Denys-Drash and with Beckwith-Wiedemann syndromes. Point mutations in WT1 occur with Denys-Drash. Mutations in a neighboring locus, also on the short arm of chromosome 11, are associated with Beckwith-Wiedemann syndrome. Furthermore, homozygous deletions of WT1 occur in a small number of sporadic WT cases. While WT1 is associated with a specific locus on chromosome 11, in fact, chromosomal aberrations of great complexity occur with WT and the aberrations appear not to be random (Kullendorff et al. 2003). Clearly, Wilms tumor biology is complex, and the childhood malignancy can arise via several developmental routes (Feinberg and Williams 2003).

4.4c Hereditary conditions that increase cancer risk

One may ask whether there are diseases related to chromosome damage that manifest an increased risk for cancer. The answer to that question is yes. For example, individuals with xeroderma pigmentosum do not repair radiation-induced damage to DNA. This genetic disorder, an autosomal recessive disease that occurs infrequently in the United States (about 1 in 250,000), does not result in the inheritance of cancer, but rather the individuals inherit a deficiency in DNA repair that means they are extraordinarily vulnerable to skin cancer (Cleaver 1968, Cleaver and Crowley

2002). Similarly, Bloom syndrome (a condition with growth retardation, sensitivity of skin to light, elf-like features, and a vulnerability to repeated infections), Fanconi anemia (congenital anemia with dwarfism, mental retardation, and other defects), and ataxia telangiectasia (abnormal dilation of small blood vessels on exposed skin with cerebellar atrophy and confinement to a wheelchair before the end of the second decade of life) are autosomal recessive genetic disorders characterized by chromosomal fragility with resultant chromosomal breakage and rearrangements. Not surprisingly, patients with these disorders experience an increased vulnerability to selected cancers (Meyn 1995).

4.5   Familial cancer syndromes

Families with an unusual clustering of cancer incidence, for example, colon cancer in males and ovarian and breast cancer in females, are known from contemporary reports (Lynch et al. 1981) and from antiquity. Familial colon and breast cancer are discussed here. Because of the high prevalence of prostate cancer and the increased longevity of males in the United States and much of western Europe and because studies of twins (monozygotic vs. dizygotic) consistently reveal a stronger hereditary component in the risk of prostate cancer than the risk of any other human cancer, it is also discussed in section 4.5c.

4.5a   Colon cancer

New cases of colorectal cancer in 2004 will affect an estimated 146,940 individuals with a death toll of about 56,730 in the United States (Jemal et al. 2004). About 10 percent of cancer deaths are caused by this neoplasm, and it is the third most common cause of death by cancer in both men and women. Most cancers of the colon, like most other cancers, are thought to be sporadic, random, and not due to heredity. The remaining cases, perhaps as many as a third (Lichtenstein et al. 2000) are thought to be heritable. A family history of colorectal cancer may ensue from environmental similarities, coincidence, or heredity. Two principal hereditary conditions lead to familial colon cancers: familial adenomatous polyposis (FAP) and hereditary nonpolyposis colon cancer (HNPCC).

Patients with familial adenomatous polyposis, an autosomal dominant trait, comprise less than 1 percent of the total of individuals with colorectal cancer. They are afflicted with many (hundreds to thousands) benign adenomatous polyps of the mucosa of the large intestine that, although not present at birth, appear as early as the first decade of life. The adenomatous polyps of FAP are precursors of colorectal cancer. Patients with FAP are almost certain to develop cancer by age forty and, if untreated, will probably die of colorectal cancer before the age of sixty.

FAP is related to a mutation of a cancer suppressor gene (Chapter 5), known as the adenomatous polyposis coli (APC) gene, located on the long arm of chromosome 5

(Bodmer et al. 1987). The APC gene specifies for a cytoplasmic protein, the expression of which is related to proliferation and maturation of colonic epithelial cells. Failure to produce proper APC protein results in enhanced proliferation of cells and failure of the maturation process. Mutation of the APC gene is associated with essentially all cases of FAP. Curiously, many individuals (40 percent to 50 percent) who are not afflicted with FAP but who have benign colonic polyps or the "sporadic" form of colon cancer also have a mutation or deletion of the APC gene. Thus, the APC gene obviously plays an important role in most cases of colorectal cancer. Although mutation of the APC gene is necessary for colorectal cancer, it is insufficient for the expression of the malignant phenotype. Carcinogenesis ordinarily results from a multistep process. Hence, other genetic changes are required for colorectal cancer. These changes involve the early loss of methyl groups in the DNA of adenomas, perhaps leading to aneuploidy, mutation of the k-*ras* oncogene, deletions of the *p53* tumor suppressor gene, and deletions of the DCC (*d*eleted in *c*olon *c*ancer) antioncogene located on the long arm of chromosome 18. It is believed that malignancy ensues from FAP due to the accumulation of genetic changes, some of which are described here.

Intestinal polyposis occurs in individuals with Gardner's syndrome, another autosomal dominant condition. These individuals are also afflicted with abnormal growths, extra teeth, cancers at other sites, and bone abnormalities. Gardner's syndrome occurs at half the frequency of the relatively rare FAP. An autosomal recessive condition leading to adenomatous polyposis and tumors of the central nervous system is known as Turcot syndrome. Gardner's syndrome, Turcot syndrome, and FAP may occur simultaneously in a family.

Hereditary nonpolyposis colorectal cancer (HNPCC), hereditary colon cancer without adenomatous polyposis of the large intestine, accounts for about 5 percent of all colon cancer, but both HNPCC as well as FAP, are far less common than sporadic colon cancer of unknown etiology (Lynch 1985b). HNPCC, inherited as an autosomal dominant trait, is characterized by multiple colon cancers at an early age.

Two genes that account for the vast majority of HNPCC have been isolated. Both are associated with DNA repair. Various kinds of errors occur in DNA replication in dividing cells. One kind of error is the mismatching of nucleotides in the complementary DNA strands. Failure of repair permits the retention and accumulation of thousands of DNA errors (mutations) located throughout the entire genome, and some of these mutations lead to cancer. A mutated gene designated MLH1, located on the short arm of chromosome 3 (Papadopoulos et al. 1994), is involved in about 30 percent of HNPCC tumors, and another gene, MSH2, located on the short arm of chromosome 2, is related to approximately 60 percent of HNPCC cases (Peltomäki et al. 1993). The identification of two genes, both involved in mismatched DNA repair, which are damaged in HNPCC, provides striking evidence of the relationship of the loss of normal gene repair and resultant mutation with the genesis of cancer (Cleaver 1994, Bale and Brown 2001).

4.5b Breast cancer

An estimated 217,440 new cases of breast cancer in U.S. women are expected in 2004 with death occurring to 40,110 women (Jemal et al. 2004). Prior to 1985, breast cancer claimed more lives of women than any other cancer. Since that time, lung cancer has surpassed breast cancer as a cause of death among women, but breast cancer prevalence and mortality remain formidable. In the United States, women now have a lifetime risk of developing breast cancer of one in eight (Miki et al. 1994). New breast cancer cases are expected to afflict about 1,450 men with about 470 deaths due to that neoplasm in 2004.

As would be expected from so common a malignancy, many risk factors, both environmental and genetic, have been identified. Environmental factors (see Chapter 8) are thought to cause most breast cancer. Heritable factors are estimated to account for 10 to 25 percent of breast cancer (Lichtenstein et al. 2000; Wooster and Weber 2003). However, because breast cancer afflicts so many women, the minority of women who have a heritable form of the disease comprise a rather large number.

Two important genes associated with inherited susceptibility to breast cancer are BRCA1 (BR for breast, CA for cancer) and BRCA2. Mutations of these genes convey a lifetime risk for breast cancer of 60 to 85 percent. Individuals with mutations to BRCA1 also have an increased risk for ovarian cancer. Mutations to BRCA1 and 2 probably account for no more than 2 or 3 percent of all breast cancers (Wooster and Weber 2003).

BRCA1 is physically mapped to the long arm of chromosome 17 (Hall et al. 1990), and the gene and its protein product of 1,863 amino acids have been characterized (Miki et al. 1994). BRCA2, located on the long arm of chromosome 13, accounts for almost as much inherited vulnerability to early-onset breast cancer as does BRCA1 but appears to be less involved in ovarian cancer risk.

Another mutant gene related to breast cancer occurs at an even lower prevalence than those already discussed. It accounts for less than 1 percent of breast cancers that arise in women prior to age forty. These rare breast cancers are the most common malignancy of a group of familial cancers, known as Li-Fraumeni syndrome (LFS), that includes, in addition to breast cancer, soft tissue sarcomas, osteosarcomas, brain tumors, acute leukemia, and adrenocortical carcinoma. The familial predisposition to cancer in LFS patients led to the notion that the condition was inherited (Li and Fraumeni 1969). It has been postulated that LFS follows from mutations (or deletions) of the tumor suppressor gene p53 (Chapter 5) located on the short arm of chromosome 17 (Isobe et al. 1986).

Identification of these several susceptibility genes cannot cure breast cancer, but they can help identify individuals who are at high risk for this malignancy. Other susceptibility genes are being studied. Added to the earlier studies are contemporary molecular procedures with the potency for identification of whole families of genes being expressed in a cancer and during malignant progression, and there has emerged the potential for detection of familial risk and early treatment, evaluation of the effectiveness of that treatment, and development of novel treatments for breast cancer with, hopefully, greater efficacy.

As the quotations at the beginning of this chapter suggest, breast cancer due to heredity is only a small proportion of breast cancer cases. Women with inherited susceptibility generally become afflicted with breast cancer at an earlier age than those without inherited susceptibility. Breast cancer associated with BRCA1 mutations has an incidence that peaks in women prior to fifty years of age. Breast cancer incidence peaks one or two decades later among women who do not have detectable BRCA1 mutations (i.e., women with sporadic breast cancer). About 25 percent to 35 percent of women who are diagnosed with breast cancer between the ages of twenty and twenty-nine have inherited susceptibility. There is no age group at which most breast cancer is of individuals with inherited susceptibility. However, given the prevalence of this malignancy (with over 200,000 new cases annually), it is worth repeating that the minority of women who have inherited susceptibility have become a relatively large group when compared with other genetic diseases.

Why are women with inherited susceptibility generally diagnosed with breast cancer at a younger age? It is believed that breast cancer, as so many other cancers, ensues from not one mutation but from several; that is, cancer evolves from a multistep process (King et al. 1993). Inasmuch as every cell in the body of a person carrying an allele for breast cancer susceptibility already has one of these mutations, it follows that fewer added genetic changes are necessary for malignancy, resulting in the inherited form occurring at a younger age.

4.5c   Prostate cancer

With an estimated 230,110 new cases and about 29,900 deaths in the United States in 2004 (Jemal et al. 2004), prostate cancer is the most common internal cancer and the second most common cause of cancer deaths in males. Of obvious concern is the estimate that about two-thirds of men between age sixty and seventy have small prostatic tumors. Environmental factors are thought to be responsible for somewhat more than half of prostate cancer cases. The relatively large remainder are thought to be caused by heredity. Prostate cancer that occurs early (prior to age fifty-five) has a greater proportion of hereditary disease than that of elderly prostate patients. While nonhereditary "sporadic" cancer is always more common than that caused by hereditary factors, the large role for hereditary factors in prostate cancer is unusual. Indeed, concordance of prostate cancer in monozygotic (identical) twins compared with dizygotic (fraternal) twins reveals a greater role for genetic factors in this malignancy than in any other human cancer (Lichtenstein et al. 2000). It is estimated that men with first-degree relatives with prostate cancer have more than a two-fold increased risk compared with men who have no family history. With two or more first-degree relatives, the risk increases to more than five-fold (Zeegers,

Jellema, and Ostrer 2003). Clearly, there is a real familial aspect of prostate cancer, but the precise role of several individual genes is not known and is still under investigation (Nelson et al. 2003).

4.5d Microarray technology as a way of examining many genes simultaneously

The identification of BRCA1 and BRCA2, as well as other specific genes associated with cancer risk, were heralded in their day as singularly important advances in the understanding of cancer. This was entirely appropriate and those genes are no less important now. However, cancer is an incredibly complex disease. Humans are not an inbred species, and neither are their cancers. If we were as highly inbred as some mice used in cancer research, we might expect somewhat greater uniformity in human cancers. However, gross and microscopic anatomy differ from individual to individual. Even within a tumor, the histopathological appearance varies from one site to another, and of course, it varies with time. Similar tumors in different individuals vary with regard to age of onset and their response to therapy. Just as every individual has a characteristic thumb print and a face that is recognized as unique, so too are human cancers distinct. Those genes, which are essential for survival and are held in common by all living cells, are expressed (i.e., are transcriptionally active) in a tumor. In addition, a multiplicity of genes peculiar to a particular malignancy in a genetically unique individual, are also transcriptionally busy. Comparison of gene expression in the normal tissue of origin to gene expression in tumors is a method of identification of marker genes associated with malignancy. Such comparisons will assist in evaluating the efficacy of therapeutic procedures.

As suggested above, traditional genetic research involves the study of one gene at a time. How then is it possible to study hundreds, or even thousands, of genes simultaneously? DNA microarray technology permits this kind of analysis. The technology exploits the long-known ability of mRNA to hybridize to appropriate sequences of DNA. In a slide or "chip" with many DNA fragments of known sequence, the expression of many genes can be estimated by measuring the amount of mRNA bound to each DNA site. An expression profile (an expression "signature") characteristic of a particular malignancy emerges with identification of expressed genes. Profiles of tumors may reveal patterns of gene activation and expression that may be predictive of future behavior of the cancer. For example, the ability to recognize breast cancer with a high metastatic potential (Chapter 2), compared with breast cancer with a lower metastatic potential, is emerging from the study of gene expression profiling (Chapter 5, see also Van't Veer and Weigelt 2003; Wooster and Weber 2003). The studies have the potential of predicting not only metastatic potential but also the target organ(s) of metastasis. Perhaps most exciting of all is the potential for development of therapeutics that will prevent metastasis.

4.6   Summary

This chapter began with the premise that essentially all cancer has a genetic component. However, when specific cancers known to be hereditary are considered, it is estimated that they comprise a minority of malignancies. What is the difference? Cancer is a disorder of cells, and cells cannot function in the absence of a genome and its controlling factors. Therefore, all cancer involves the genome and the regulation of that genome, and thus all cancer has a genetic component. But when we consider specific cancers with known hereditary transmission, we find they are all uncommon. With present knowledge, it seems most malignancies are of the sporadic, nonhereditary type. Nevertheless, an enhanced understanding of cancer will surely ensue from a greater understanding of the specific genomic changes and inappropriate gene regulation that are the hallmark of cancer cells.

5

Cancer-associated genes

Alan O. Perantoni

5.1 Introduction

Over the past twenty years, more than a hundred genes have been identified that can convert nontumorigenic tissue-cultured test cells from rodent cell lines to a transformed phenotype, that is, foci of piled-up and crisscrossed cells that grow in soft agar and form tumors when explanted into immunocompromised rodents. These dominant transforming genes, or oncogenes, encode proteins involved in signal transduction or cell cycle regulation and affect a variety of normal cellular functions, including apoptosis, proliferation, cell differentiation, and adhesion. When overexpressed or structurally altered by point mutation, deletion, or fusion, most oncogenic proteins selectively induce the proliferation of cells that express them. More recently, studies of families predisposed to specific types of cancers have yielded a growing number of recessive tumor suppressor genes. Loss of suppressor function by deletion or point mutation of both gene copies allows cells to proliferate unregulated or with reduced restraints. The discovery of both oncogenes and suppressor genes has proven pivotal in our understanding of the mechanisms of carcinogenesis by allowing scientists to focus on the genetic targets of carcinogens and thus finally link the details of carcinogen activation, adduct formation and repair, and neoplastic conversion and metastasis with definable molecular events. Furthermore, targeting of these genes with drugs that modulate their function has already led to effective therapies for human cancers.

5.2 What is an oncogene?

An oncogene is an altered form of a normal cellular gene called a *proto-oncogene*. It encodes a regulatory protein with dominant transforming properties; that is, a single copy of the altered sequence can transform a cell and expression of the normal sequence cannot block its transforming ability. The term "oncogene" is really a misnomer because no one believes these genes evolved specifically for the induction of cancer but rather that the normal proto-oncogenes have been altered by one of several mechanisms to cause transformation. It is not surprising, then, that proto-oncogenes function in the regulation of normal cell growth and/or differentiation.

Initially, the principal method for identification of oncogenes was through the isolation of the transforming genes in oncogenic *retroviruses*, although certain oncogenes have been discovered only in preparations of mammalian tumor DNAs. More recently, newly discovered genes, even as normal sequences, have been routinely assessed for transforming ability when overexpressed in cultured cells. Regardless of source, the end point of detection is the ability of DNA preparations to transform certain rodent cell lines when those sequences are incorporated by transfection, a process involving cellular uptake of purified precipitated DNA. Alternatively, genetic experiments in which genes are overexpressed globally or in a tissue-specific manner in mice have identified additional transforming sequences.

The first oncogenic virus, an RNA tumor virus or retrovirus, was described in 1911 by Peyton Rous as a filterable agent (capable of passing through a small pore filter) from a homogenate of a chicken tumor. The filtrate could induce sarcomas in chickens upon reintroduction (Rous 1911). The etiologic agent was subsequently shown to be a tumor virus (Rous sarcoma virus, or RSV) and belonged to a family of avian sarcoma viruses, the study of which eventually led to the discovery of oncogenes.

The retroviral genome exists as single-stranded RNA that must be reverse transcribed to generate the double-stranded sequences required for incorporation into a host's DNA. Characteristically, it contains three genes necessary for the production of viral particles: *gag*, *pol*, and *env* (Figure 5–1). The *gag* gene encodes the synthesis of a core protein that packs the viral RNA in the viral particle; the *pol* gene contains the sequence for production of a reverse transcriptase or RNA-dependent DNA polymerase; and *env* provides the glycoproteins of the viral envelope. The retroviral genome terminates at both ends in sequences called long terminal repeats (LTRs), which provide the sites for integration into host DNA and promote the transcription or expression of viral sequences in the host genome. Viruses in this configuration are weakly or chronically transforming in host cells; however, when isolated from transformed cells and reintroduced into normal cells, they become highly transforming (acute transforming) retroviruses. For RSV, a comparison of viral recombinants with acute transforming ability and viral mutants that had lost their transforming ability revealed a specific region of the genome associated with tumor induction. This region was distinct from the three defined viral sequences (Kawai and Hanafusa 1971; Bader 1972). The novel sequence was designated v-*src*, for viral sarcoma gene. It was subsequently isolated and surprisingly shown to be homologous to a host-derived sequence c-*src* [cellular *src* gene] (Stehelin et al. 1976). The virus had actually

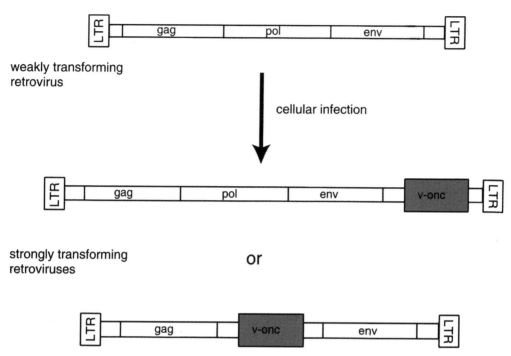

Figure 5–1. Structure of retroviruses before and after host integration. Prior to infection, a retrovirus contains three essential genes: *gag, pol,* and *env.* With these genes the viruses are only weakly transforming. By integrating into a host genome, the virus occasionally removes a copy of a host sequence, which if oncogenic, can convert the retrovirus to a highly transforming virus.

incorporated a gene from the host chicken cell that it had infected. Presumably, the acute transforming virus was formed by fortuitous recombination between the chronic transforming virus and the c-*src* sequence in the host genome. The v-*src* gene encoded the synthesis of a single phosphoprotein p60 (Purchio et al. 1978), which Hunter and Sefton (1980) subsequently demonstrated to be a tyrosine kinase, an enzyme that specifically phosphorylates tyrosine residues in substrate proteins to regulate their function. It was propitious that researchers began with RSV because this virus, unlike most other oncogenic retroviruses, incorporated the oncogene sequence downstream of the *env* sequence, so the additional gene did not interfere with viral replication. Most other retroviruses lose their ability to replicate during host integration because the oncogene sequence is inserted within the *pol* gene. Helper viruses with normal *pol* function are then required to rescue the transforming virus with its associated oncogene.

Using the same approach as for RSV, several other v-*oncs* or viral oncogenes that have a normal counterpart in the eukaryotic cell (c-*onc* or proto-oncogene) have been isolated from chicken or mouse tumors. Studies of these virally transmitted cellular sequences, however, indicate they are not accurately reproduced by the

virus during viral incorporation and that the sequence differences between viral
(v-*onc*) and cellular (c-*onc*) forms actually account for the oncogenicity of the v-
onc gene (Bister and Jansen 1986). Mutations, including base pair substitutions,
deletions, and insertions, arise at high frequency during viral transcription and
splicing. Because transforming mutations in the v-*oncs* are necessary for tumor
formation, the changes observed in viral oncogene sequences give indication of what
mutations in proto-oncogenes might be transforming under other circumstances
(e.g., in chemically induced rodent tumors or in human tumors).

   The significance of retroviral oncogenic sequences as dominant-transforming
genes in normal cells remained largely unappreciated until Weinberg's (Shih et al.
1981) and Cooper's (Krontiris and Cooper 1981) laboratories independently
reported that DNA from a human tumor could transform mouse cells in culture. Subsequently, the labs of Barbacid (Pulciani et al. 1982), Weinberg (Shih and
Weinberg 1982), and Wigler (Goldfarb et al. 1982) separately cloned the transforming sequence and established its homology with members of the *ras* oncogene
family, which had previously been isolated from an avian sarcoma virus. Since
then, several other cellular oncogenes isolated from retroviruses have been shown
to cause cultured mouse cells to assume a transformed phenotype in transfection
assays.

5.3   Proto-oncogenes function in signal transduction, cell cycle regulation, differentiation, or programmed cell death (apoptosis)

Proto-oncogenes encode proteins that participate in the regulation of growth, differentiation, or death of normal cells, and all are involved at various levels in signaling
from the membrane to the nucleus. In normal signaling, or signal transduction, the
interactions of the cell with its extracellular environment are translated to the nucleus
to elicit a nuclear or transcriptional response. Signaling is often initiated by the binding of a secreted growth factor or cytokine (a ligand) to a ligand-specific receptor
anchored in the cell's plasma membrane. The resulting conformational change in
the receptor then triggers a cascade of events generally involving the phosphorylation of specific tyrosine, serine, or threonine residues in an intermediary protein
messenger, which participates in the transmission of a signal to the nucleus. The
end result is transcriptional activation (expression) or repression of a set of genes
with appropriate response elements in their promoter/enhancer regions. Signaling
may be initiated by any one of several proto-oncogene–encoded growth factors that
interact with membrane-bound receptors (e.g., *int-2/fgf3* or *hst-1/fgf4*, members
of the fibroblast growth factor family, or *sis,* platelet-derived growth factor). Other
proto-oncogenes encode the receptors themselves, such as *erb* B-1/EGFR, the epidermal growth factor receptor. The *ras* family members are localized to the inner side
of the cell membrane and participate in signal transduction from the membrane.

The products of others, like *jun*, which encodes a transcriptional-activating factor, or *cyclin D*, which regulates cell progression through the cell cycle, are localized to the nucleus.

In addition to their positioning and functions, the regulatory importance of the proto-oncogenes is also inferred from their conservation in evolution. Homologous *ras* sequences have been identified from yeast and fruit flies to humans (Santos and Nebreda 1989), and the multiplicity of family members that have been identified suggest that they regulate a wide variety of cellular functions. In fact the Ras super-family consists of over 150 members that are subdivided into six separate families and that function as small GTPases (Ehrhardt et al. 2002), which will be described later. There is redundancy of function found in the subfamilies, which provides some assurance that their regulatory capabilities will be maintained within a cell. For example, the classical p21 *ras* gene family members, H-*ras*, the K-*ras* splice variants 4A and 4B, and N-*ras*, differ significantly in their introns, or noncoding regions, but generate reasonably homologous proteins, which can substitute for one another biochemically. That is not to say that they behave identically but rather that they may overlap in function in vivo.

Proto-oncogene involvement in differentiation has been well documented through the elucidation of differential expression patterns during tissue develop-ment as well as through gene targeting studies. In any tissue, genes are transcribed in a temporal and tissue-specific manner, such that the profile of gene expression differs substantially during embryonic development relative to adult expression pat-terns. During embryonic and fetal development of the mouse, for example, the *ret* proto-oncogene, a receptor tyrosine kinase for the secreted ligand glial-derived neu-rotrophic factor (GDNF), is expressed at high levels in the tips of the ureteric bud (progenitor for the collecting duct system in the kidney) during kidney development but later is down-regulated in adult kidneys (Schuchardt et al. 1994). However, if expression is lost in the embryo through gene targeted deletion, the ureteric bud does not branch, severely inhibiting the formation of normal kidney structures. Thus, appropriate expression of a proto-oncogene during development can be critical to normal tissue differentiation.

As one might predict, those genes involved in cell proliferation and inhibition of apoptosis or programmed cell death are prominently expressed during differ-entiation, since there is a regulated but massive expansion of cells in embryonic tissues. Thus, the growth-associated proto-oncogene and transcription factor *N-myc* is required for early development, and targeted homozygous deletion of this gene results in embryonic lethality (Charron et al. 1992). Through partial or tissue-specific deletion, N-*myc* has been implicated in the growth and differentiation of such diverse tissues as the cerebellum and the kidney (Knoepfler, Cheng, and Eisen-man 2002; Bates et al. 2000). Conversely, transgenic dysregulated overexpression of N-*myc* in adult animals induces a variety of neoplasms that depend upon the speci-ficity of its misexpression (Rosenbaum et al. 1989; Norris et al. 2000). In general, the evidence suggests that the reported amplification of N-*myc* in neuroblastomas may provide more than just a prognostic indicator for these tumors but rather also function in tumorigenesis. These observations would seem to indicate that tumor development is at least in part context dependent.

Similarly, the cell survival protein Bcl-2 (B-cell lymphoma gene-2), a pivotal anti-apoptotic factor, may play a role in both differentiation and tumorigenesis. It was first implicated as an oncogene due to its constitutive expression in human follicular lymphomas with chromosome 14:18 translocations, which link the immunoglobu-lin heavy chain gene with *bcl-2* (Tsujimoto et al. 1985). It has since been associated with tumorigenesis in other lymphomas (Hillion et al. 1991) and chronic lym-phocytic leukemias (Adachi et al. 1989) when fused to the immunoglobulin light chain gene. In normal development, its absence yields neonates that are signifi-cantly smaller than normal and that become deficient in lymphocytes with matura-tion (Nakayama et al. 1994). Transgenic mice overexpressing *bcl-2* following birth have lymphocytes with an increased lifespan and a higher incidence of lymphomas (Strasser, Harris, and Cory 1993). In addition, when c-*myc* is simultaneously over-expressed, tumor incidence, latency, and aggressiveness are all increased (Strasser et al. 1990). Such observations are consistent with the involvement of *bcl-2* in both differentiation and neoplasia, but may indicate that *bcl-2* plays a rather passive role in tumor development by regulating cell survival and thus increasing the probability for fixation of a transforming mutation (Coultas and Strasser, 2003).

5.4   Genetic approaches to delineate proto-oncogene function

Studies that correlate proto-oncogene expression with specific developmental pro-cesses, although helping to identify significant macromolecules, by themselves pro-vide little insight into the actual cellular functions of these genes. For this purpose, several powerful technical innovations have created a window into the cell, allowing regulation of any given gene at an investigator's discretion. As described above, one approach involves the use of gene targeting by so-called knockout mutations for specific genes. In initial applications of this technology, genomic DNA was modified to eliminate a gene function in virtually every cell in a mouse, which resulted in the loss of its function in all cells that expressed it. Since many genes are normally turned on and off in various tissues during development, their premature loss or aberrant expression may yield little knowledge of their true tissue-specific functions. This is especially problematic if embryonic lethality precludes an examination of tissue or organ maturation. More recently, it has become possible to circumvent this problem and selectively eliminate expression of a gene in defined populations of cells at any stage of development using binary transgenic systems (reviewed by Lewandoski, 2001). One popular approach involves an effector site-specific DNA recombinase, for example, bacteriophage P1's Cre, which cleaves recognized sites (loxP) in a target gene to modulate its function (Figure 5–2A). If the effector DNA recombinase is then placed under the regulation of a tissue-specific promoter, the target gene will only be modified in cells that activate this promoter. Thus, the given gene function will be artificially deleted only in those cells. Conversely, by inserting a 'floxed' (flanked by loxP sites) stop codon in a gene with a constitutively expressed promoter, it is also possible to selectively induce the overexpression of a gene, creating a tissue-specific transgenic mouse. The advantage to this binary system is that once a targeted DNA sequence has been 'floxed', it can be incorporated into established mouse genetic lines with Cre under the regulation of a variety of different promoters, thus allowing regulation of target gene expression in many different tissues. As a variation on this theme, inducible promoters in the Cre effector lines are available which allow temporal regulation of a 'floxed' target gene. One such system involves the use of the tetracycline analog doxycycline. In this system, doxycycline interacts with a transactivator protein that either turns on the expression of Cre recombinase, which in turn modifies a 'floxed' target gene to either activate or inactivate its expression. Target gene expression can then be regulated in a temporal manner.

Depending upon the gene of interest, there are a variety of strategies for introducing mutations that will result in the cellular loss of normal function of a particular gene product. In the case of proto-oncogenes with receptor function as shown (Figure 5–2B), the genetic sequences can be engineered so they maintain their ability to bind ligands but fail to convey a signal upon binding, a so-called dominant negative. Therefore, when incorporated into embryonic stem cells or a cultured cell line, they sequester ligands from the normal receptors and block specific signaling pathways. Although genetic redundancy often prevents loss of proto-oncogene function despite the loss of a normal proto-oncogene product, occasionally germline targeting does allow a glimpse at the cellular function of these genes in the developing embryo. Thus, an examination of *ret* function through targeted deletion, as described above, has implicated this proto-oncogene in the development of both the enteric nervous system and kidney (Schuchardt et al. 1994). For transcription factors such as N-*myc*, generally the DNA-binding domain of the protein is altered or deleted to prevent the factor from interacting with DNA, thus blocking its ability to activate N-*myc*-mediated transcription. Germline loss of the N-*myc* transcription factor activity (Stanton et al. 1992; Charron et al. 1992) results in severe malformation of certain neural tissues, while overexpression specifically in neuroectodermal cells induces neuroblastoma (Weiss et al. 1997), demonstrating a role in tumorigenesis.

As already described, it is also possible to evaluate the activity of a gene product by overexpressing or up-regulating its function. In this regard, when normal transforming growth factor-α, a ligand for the EGFR, is overexpressed in a transgenic mouse, it induces hepatocellular carcinomas and mammary and pancreatic abnormalities in the animals (Jhappan et al. 1990). Conversely, by overexpressing inhibitory proteins of particular oncogenic pathways, such as the inhibitor of Wnt signaling Dickkopf-1 (DKK-1), tumorigenicity can be suppressed (Mikheev et al. 2004).

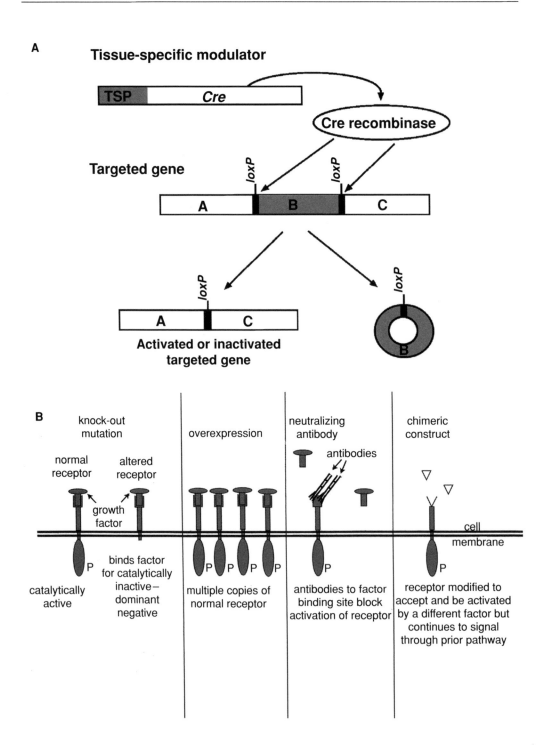

As signaling pathways become defined, the function of specific proto-oncogenes in development has been clarified. A case in point is the participation of *ras* in the differentiation of neuroectodermal cells (the embryonic tissue that gives rise to nerve cells). A rodent cell line of neuroectodermal origin (PC12 cells) can be induced in culture to express a neuronal phenotype when stimulated with nerve growth factor (NGF). Microinjection of anti-*ras* antibody blocks this differentiation (Hagag, Halegua, and Viola 1986), and a *ras* protein in a constantly signaling or activated form induces neuronal differentiation (Bar-Sagi and Feramisco 1985). Thus, *ras* may participate in the NGF-mediated signaling pathway for the development of neuroectoderm. This also demonstrates another methodology for functional evaluation of gene products: the use of specific neutralizing antibodies to remove the activity of a proto-oncogene product. A similar approach has been taken in studies of the erbB-2/HER-2, which has resulted in the development of a potential palliative therapy for breast cancer. An antibody preparation called Heceptin or TrastuzuMab, which targets HER-2, has been applied with some success against advanced stage HER-2$^+$ breast cancers in clinical trials (Baxevanis et al. 2003).

The application of chemical or natural biological inhibitors has also become quite popular in the analysis of specific signaling pathways. Interest in targeting a particular regulatory protein such as a proto-oncogene stems from the possibility of not only understanding the biology and biochemistry of that protein, but also of discovering a potential chemotherapeutic agent that could inhibit the activity of its oncogenic form. Thus, the chemical inhibitor azatyrosine blocks activation of the MAPK pathway downstream of *ras*, but it acts preferentially on PC12 cells transformed with oncogenic *ras* to induce their differentiation (Monden et al. 1999). It is now also clear that most critical signaling pathways are negatively regulated by one or more naturally occuring peptides or proteins. For example, secreted Wnt proteins normally interact with members of the membrane-bound Frizzled-receptor family for activation of the Wnt signaling pathway; however, a member of the Secreted Frizzled-Related Protein family (sFRP) can bind to the Wnt proteins, preventing

---

Figure 5–2. Common methodologies used for investigating gene function. **A.** Binary system for gene targeting in a tissue-specific manner. Cre recombinase expression is regulated by a tissue-specific promoter (TSP). When expressed, Cre recognizes loxP sites in a target gene and eliminates the intervening sequence with recombination. If the eliminated sequence contains domains important to protein function, the activity of the protein will be lost. However, if it instead contains a stop codon that prevents expression, the protein will be activated. **B.** One form of knock-out mutation involves creation of an inactive dominant-negative protein, a truncated nonsignaling form that sequesters its natural ligand, preventing signal transduction. Overexpression enhances signaling through specific pathways and is used to identify downstream targets in a pathway. Small molecules, neutralizing antibodies, or naturally occurring protein inhibitors are commonly used to block signal transduction. Finally, where the ligands for a receptor have not been defined or purified, chimeric constructs circumvent these problems by fusing a known ligand-binding domain with a receptor of interest.

their interaction with the receptor and thus inhibiting Wnt signaling (Yoshino et al. 2001). While naturally produced during embryonic development, the sFRPs may eventually prove useful in targeting cells dependent upon Wnt signaling for survival and growth.

Occasionally, a signaling component is identified without its effector, for example, a membrane receptor with an undefined activating ligand. In this case, it is possible to study its signaling function by splicing known ligand-binding elements with sequences of the catalytic or functional domain of the proto-oncogene. For example, studies of the membrane-bound proto-oncogene *neu* were initially hampered by the fact that its ligand had not been identified. Because *neu* shares considerable homology with *erbB*, the gene that encodes the EGFR, chimeric receptor molecules composed of the ligand-binding portion of the EGFR and the intracellular signaling domain of the *neu* protein could be generated (Lehvaslaiho et al. 1989). When introduced into mouse cells, the chimeric receptor localized to the membrane and was activated by treatment with epidermal growth factor, a ligand for the EGFR. This allowed events mediated by the functional domain of *neu* to be monitored within the cells.

5.4a   DNA microarray analysis – global gene expression or genomic profiling

The importance of the most recent breakthroughs in sequencing of the human and mouse genomes and their associated expressed genes cannot be overstated in facilitating our understanding of gene regulation. While still in its infancy, microarray analysis is yielding volumes of information on transcriptional regulation based upon complete gene expression databases (for review, Guo 2003). The technology for microarray chips was actually developed by and adapted from semiconductor manufacturing of microchips. It generally involves the application of photolithography and solid-phase chemistry to the production of microarray chips that contain as many as 50,000 sets of densely packed oligonucleotide probes (from Affymetrix) – each representing an antisense fragment of a potentially expressed sequence. Since a typical mammalian cell is estimated to express some 20,000 different genes at one time, it is possible to identify a complete gene expression profile in a single experiment. In such expression studies, RNA isolated from as few as a single cell can be amplified by reverse transcription-polymerase chain reaction (RT-PCR; for animation, see http://www.bio.davidson.edu/courses/Immunology/Flash/RT_PCR.html).

This process incorporates labeled nucleotides into the generated cDNA molecules, which are then hybridized directly to a gene array chip. Detection of amplified signals is usually performed using fluorescence, which is both highly sensitive and quantifiable. Data analysis software then allows comparisons of gene expression profiles from a variety of sources. An examination of *c-myc*-induced gene expression in cultures of primary human cells has identified over two hundred genes that are upregulated and more than two hundred that are downregulated compared to controls and include genes critical to cell cycle regulation, DNA repair, and protein synthesis (Menssen and Hermeking 2002). Furthermore, comparative studies of in vitro expression patterns in cases of proto-oncogene overexpression, for example, *myc* or *ras*, with tumor tissues associated with deregulation of these genes indicate that the in vitro models are quite predictive of profiles that arise in vivo (Huang et al. 2003) and may help characterize patterns found in other neoplasms. Similarly, the loss of proto-oncogene function can be assessed by microarray analysis as another approach for elucidation of their signaling targets. Finally, this technique has been applied now to tumors at various stages of neoplasia and, at least for breast cancer, the expression profiles in primary tumors compared to lymph node metastases have been deemed to be excellent predictors of cancer recurrence (Huang et al. 2003). Of course there are significant limitations to this approach, especially with regard to the current data mining/analysis tools and the frequently overwhelming lists of transcriptionally regulated genes and the fact that translational and posttranslational changes are ignored by this methodology. Despite these limitations, DNA microarray is poised to become one of the most valuable technologies in biomedical research as well as in clinical diagnostic labs for identifying so-called 'signatures' of disease, that is, the microarray gene expression profile specifically linked to a given pathology.

Although DNA microarray or genomic analysis is technically further advanced at the moment, its application is of course limited to studies of DNA and mRNA. However, translational and posttranslational modifications of proteins are also critical to growth regulation and therefore tumorigenesis. For example, transcription factors such as the Signal Transducer and Activator of Transcription (STAT) family members may be phosphorylated on specific tyrosine or serine residues, and the ability of these factors to translocate to the nucleus and activate STAT-dependent transcription is determined by their pattern of phosphorylation. These patterns then are regulated by tyrosine or serine/threonine kinases that function upstream of STAT phosphorylation and not at the level of transcription. Proteomic techniques for evaluating complex mixtures of proteins with their associated modifications are just becoming available. The methods generally involve two-dimensional gel electrophoresis, high-performance liquid chromatography, capillary electrophoresis, or antibody-based protein chips and often include ultrasensitive mass spectrometry for protein identification. The advantage to these methods is that differentially modified proteins can be distinguished, but, unlike DNA or RNA, proteins cannot be amplified to enhance signals from components represented by the small copy numbers typical of regulatory factors. Methodologic development of approaches in proteomics is currently an area of intense focus and innovation.

5.5 Classification of proto-oncogenes/oncogenes

Proto-oncogenes or oncogenes integrate into critical areas of growth control at a number of different levels. Activities are not limited to cell cycle regulation, but include signal pathways, especially at points of crosstalk, where signaling in one pathway may be picked up and transmitted along another; DNA repair factors such as telomerase that manages DNA loss from chromosomal telomeres; adhesion molecules that regulate tissue mobility and involve both cell-to-cell and cell-to-matrix proteins; and angiogenic factors that induce capillary formation to provide a blood supply for tumor growth.

Examination of the human genome suggests that more than 20 percent of the 32,000 encoded proteins are involved in signal transduction at some level. These include more than 500 protein kinases (proteins that phosphorylate protein substrates) and 100 protein phosphatases (proteins that dephosphorylate protein substrates). These enzymes reversibly modify tyrosine or serine/threonine residues. Currently, from the more than 100 characterized oncogenes, protein kinases, and particularly the protein tyrosine kinases (PTKs), comprise the largest group. The PTKs catalyze the transfer of the γ-phosphate from an ATP molecule to the phenolic hydroxyl group of an available tyrosine in a target protein. Of the PTKs, some sixty encode receptor protein tyrosine kinases (RTKs), that is, membrane-bound signal initiating factors (Blume-Jensen and Hunter 2001).

Some of the better characterized proto-oncogenes are classified according to their functional role and position in pathways of signal transduction and subcategorized as growth factors/cytokines (section 5.5a), receptor (5.5a) or nonreceptor (5.5b) tyrosine kinases, GTP-binding proteins (5.5c), serine/threonine kinases (5.5d, 5.5e), or nuclear proteins (5.5f)/transcription factors (5.5g) (Table 5–1). In this order, they reflect the sequence in which the normal proteins transmit an extracellular signal to the nucleus, resulting in transcriptional activation of specific responding genes. The common feature among these signal-transducing proteins is their ability to interact directly with proteins that immediately precede or follow them in the sequence and to regulate or be regulated by phosphorylation and/or dephosphorylation reactions as a result of these interactions. Each category then represents a different level in the cascade of phosphorylation events. As described later, aberrant continuous or constitutive signaling from any of these proteins is often associated with transformation. These categories, however, exclude one critical group of regulatory proto-oncogenes: the cell cycle genes. In many cases, the outcome of signal transduction is modulation of cell cycle proteins to either induce or inhibit cell proliferation. It is not surprising, then, that one or more of these proteins may be affected in carcinogenesis. Their involvement in tumor formation is described at length later in this chapter.

5.5a Growth factors and their receptors

Beginning with known growth factors encoded by proto-oncogenes, for example, platelet-derived growth factor (*sis*), fibroblast growth factor (Fgf) family members *int-2*/*fgf3* and *hst*/*fgf4*, and transforming growth factor-α (*tgf-α*), all act through

Table 5–1. *Classifications of proto-oncogenes/oncogenes with examples*

Type	Location	Function	Human tumor association
I. Growth factors	Secreted		
sis		PDGF family	sarcoma
hst		FGF family	gastric cancer
int-2		FGF family	mammary cancer
II. Receptor TK*	Membrane		
erbB		EGF receptor	glioblastoma
erbB-2/HER2/neu		EGF receptor family	colon/mammary/gastric/ ovarian/lung
met		HGF receptor	renal/thyroid/liver/sarcoma
ret		GDNF receptor	thyroid
III. Nonreceptor TK	membrane/cytoplasm/nucleus		
src		membrane-coupled signaling through integrins/RTKs/ GPCR#	colon/mammary/ pancreatic
abl		activates signaling through ras/rac/ Stat/ PI3K/BclX$_L$	leukemias – CML & ALL
IV. GTP-binding	membrane		
K-*ras*		MAPK signaling	colon/pancreas/leukemia
V. S/T¶ kinases	membrane/cytoplasm		
raf		phosphorylates MAPKK	colon/melanoma
protein kinase C		Ca^{2+} release	Lung/breast/brain/thyroid
VI. Transcription factors		nucleus	
N-*myc*		DNA binding	neuroblastoma
jun		DNA binding/transcription	osteosarcoma/skin cancer
β-catenin		TCF activation	colon/kidney/breast
VII. Cell cycle	nucleus		
cyclin D1		activates CDK4/6	breast/lung/endometrium/ bladder
Mdm2		inactivates p53 and pRb	sarcomas
VIII. Apoptosis	cytoplasm/mitochondria		
Bcl-2		inactivates Bax and Bak	lymphomas

TK* = tyrosine kinase; GPCR# = G protein coupled receptor; S/T¶ = serine/threonine

cell membrane-bound receptors with tyrosine kinase activities and therefore initiate cascades of phosphorylation by activating these receptors. The proto-oncogene receptors are grouped according to sequence homologies and are therefore classified as members of receptor superfamilies for specific growth factor ligands: for example, the receptor tyrosine kinase epidermal growth factor receptor (EGFR) family, which includes EGFR or *erbB-1*, *erbB-2/neu/HER-2*, and *erbB-4*; platelet-derived growth factor receptor (PDGFR) family of *pdgfr-α/β*, *csf-1r*, *kit*, and *flk2/3*; Fgf receptor family of *fgfr-1-4;* nerve growth factor receptor (NGFR) family *trkA-C;* hepatocyte growth factor receptor (HGFR) family of *met* and *ron;* insulin receptor *ros;* and GDNF receptor *ret*.

In general, the membrane receptors consist of three principal domains: an extracellular amino-terminal ligand-binding domain, a hydrophobic transmembrane region, and an intracellular carboxy-terminal tyrosine kinase domain. In the absence of a ligand, the receptor exists as a monomer in the membrane, and activation is characterized by a series of events that alter its configuration, allowing its intracellular domain to interact with other membrane-bound and cytoplasmic proteins (reviewed in Schlessinger 2000). For example, when TGF-α or EGF binds the EGF receptor, a conformational change in the receptor allows it to dimerize with another EGF receptor molecule (Yarden and Schlessinger 1987). The dimer promotes phosphorylation (autophosphorylation) on specific intracellular tyrosine residues of the receptor itself (Honegger et al. 1990). These phosphorylation events generate binding sites on the receptor for and recruitment of a group of proteins with SH2 (*src* homology 2) or PTB (phosphotyrosine binding) domains (Figure 5–3). The SH2 domain in a protein recognizes a 1 to 6 amino acid motif on the C-terminal side of the phosphorylated tyrosine residue in the receptor (Songyang et al. 1993), while the PTB domain requires 3 to 5 amino acids on the N-terminal side of the phosphotyrosine for binding (Margolis 1999). It is the phosphotyrosine residue in each case that allows proteins to interact with the receptor. The adapter protein, growth factor receptor–bound protein 2 (Grb2), contains an SH2 domain that can bind the tyrosine-phosphorylated EGF receptor (Lowenstein et al. 1992, Kizaka-Kondoh, Matsuda, and Okayama 1996). Grb2 also forms a protein complex through an SH3 domain, which binds a proline-rich motif PXXP in a cytosolic protein called Son of Sevenless (Sos). This complex is then recruited to the phosphotyrosine residue in the EGF receptor through the Grb2 SH2 domain (Buday and Downward 1993). The end result is that Sos is brought to the membrane, where it functions as a guanine nucleotide exchange factor (i.e., it catalyzes replacement of a bound nonsignaling GDP molecule with a signaling GTP molecule) to activate the membrane-linked p21*ras* protein, the next step in the cascade. The Grb2/Sos complex can be recruited through interaction with the Shc protein as well. When tyrosine phosphorylated, the Shc protein forms a complex with Grb2/Sos through the Grb2 SH2 domain. This ternary complex then associates with the EGFR through a Shc PTB domain (Batzer et al. 1995). Since signal activation from the receptor can occur through two independent sites, namely SH2 and PTB recognition sites, the loss of a Grb2-binding site in the receptor does not block signaling. In both cases, a Grb2/Sos complex is formed and p21*ras* becomes activated or capable of signaling (Margolis 1999). Furthermore, even if the EGFR has lost its autophosphorylation sites, the Shc/Gbr2/Sos complex can interact instead with the erbB2/neu/HER2 receptor following EGF ligand binding to the EGFR and activate p21*ras* and mitogenesis, demonstrating the redundancy of signaling mechanisms (Sasaoka et al. 1996). The role for these adapter proteins in the neoplastic process is suggested in transgenic studies in which unregulated Shc or Grb2 expression dramatically accelerated mammary tumor development when coexpressed with a viral oncogene polyomavirus middle T antigen (Rauh et al. 1999).

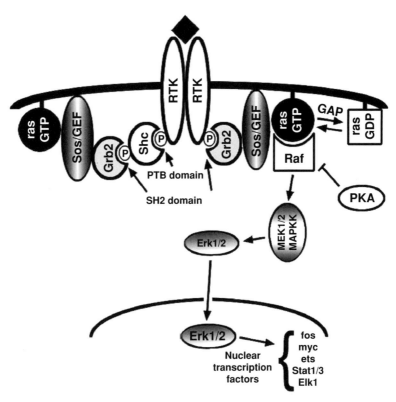

Figure 5–3. MAP Kinase signaling. While multiple MAPK pathways have been described, the one shown is commonly involved in growth regulation. It is initiated when a ligand binds a receptor tyrosine kinase (RTK), which allows autophosphorylation of the receptor, creating SH2 and PTB binding sites. A variety of proteins can then interact with the receptor, but most notably the Grb2 protein, which transports the guanine nucleotide exchange factor (GEF) Sos to the membrane to phosphorylate and activate ras. When bound to GTP in its active configuration, ras then interacts with the MAP kinase kinase kinase Raf. Raf signals through the MAP kinase kinases MEK1 + 2, which phosphorylate the MAP kinases ERK1 + 2. The activated ERKs are then transported to the nucleus where they regulate multiple transcription factors involved in growth. This pathway is negatively regulated at ras by a GTPase-activating protein (GAP) and at Raf by Protein Kinase A (PKA).

The phosphorylated tyrosine residues in the RTKs, including the EGFR, provide docking sites for other proteins with SH2 domains to activate additional signaling pathways. These include phospholipase C-γ (PLCγ), Src, GTPase-activating protein (GAP), or phosphatidylinositol 3-kinase (PI3K) (Fantl et al. 1992; Kashishian, Kazlauskas, and Cooper 1992). In each case, autophosphorylation of the receptor is responsible for the interaction, but the resulting proximity then allows the tyrosine kinase activity of the receptor to phosphorylate tyrosine residues in the target proteins.

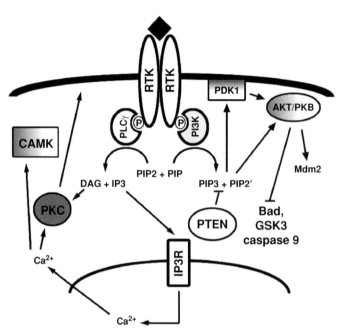

Figure 5–4. Other major signaling pathways involving the RTKs. Both phospholipase Cγ (PLC) and phosphatidylinositol 3-kinase (PI3K) contain SH2 domains, which facilitate their interaction with an RTK. Once activated, both mediate signaling through specific phosphoinositides. PLC metabolizes phosphatidylinositol 4,5-biphosphate (PIP2) to inositol 1,4,5-triphosphate (IP3), which stimulates the release of sequestered Ca2+ from various cellular organelles, and diacylglycerol, which activates Protein Kinase C (PKC). PI3K, on the other hand, phosphorylates PIP2 and phosphatidylinositol 4-phosphate (PIP) to generate phosphatidylinositol 3,4,5-triphosphate (PIP3) and phosphatidylinositol 3,4-diphosphate (PIP2′). These metabolites activate AKT, a critical antiapoptotic factor. The tumor suppressor tyrosine phosphatase PTEN blocks AKT signaling by dephosphorylating PIP3.

Signaling from PLCγ is mediated by the metabolic products of PLCγ activity and involves members of the protein kinase C (PKC) superfamily in a very complex series of reactions (Rameh and Cantley 1999). PLCγ is catalytically activated by an RTK such as the EGFR through phosphorylation at the membrane (Figure 5–4). Once active, it hydrolyzes phosphatidylinositol 4,5-biphosphate (PtdIns(4,5)P2) to inositol 1,4,5-triphosphate (Ins(1,4,5)P3), which stimulates the release of sequestered intracellular calcium, and diacylglycerol, which in turn activates PKC, a serine/threonine kinase. The released Ca^{2+} binds to calmodulin to induce the activity of calcium/calmodulin-dependent target kinases. Interestingly, diacylglycerol and the tumor promoter TPA bind the same regulatory domain of PKC. This interaction increases the enzyme's hydrophobicity and allows it to be recruited to the intracellular side of the membrane (Newton 1995). Both also increase the efficiency of PKC's associated serine/threonine kinase activity. PKC then increases the calcium flux from outside the cell (Nishizuka 1992), which further enhances PKC kinase activity. The various phospholipid molecules generated by PLCγ all tend to prolong the activation of PKC, presumably sustaining its kinase function in certain signal transduction pathways involved in growth control regulation.

PI3K is activated when it complexes directly or indirectly with a phosphotyrosine residue in an RTK, such as the EGFR, or an adapter protein (see Figure 5–4). Activated PI3K phosphorylates PtdIns(4)P and PtdIns(4,5)P2, forming PtdIns(3,4)P2 and PtdIns(3,4,5)P3, which in turn activate the serine/threonine kinase Akt/PKB, a retroviral oncogene (Bellacosa et al. 1991) and a major cell cycle regulator (Sen et al. 2003). The phosphotidylinositol lipids are believed to mediate the recruitment of Akt to the membrane through its pleckstrin homology (PH) domain. Once at the membrane, Akt is phosphorylated on threonine 308 by the serine/threonine kinase PDK1, causing partial activation of the Akt kinase, while the additional phosphorylation at serine 473 induces maximal kinase activity from Akt (Luo, Manning, and Cantley 2003). PDK1 (Balendran et al. 1999), Akt itself (Toker and Newton 2000), and Integrin-linked Kinase (ILK) (Lynch et al. 1999) can all interact with and phosphorylate serine 473. Major targets of Akt include the pro-apoptotic BAD and caspase-9 proteins, which are inactivated with phosphorylation, glycogen synthase kinase (GSK3β), which, when phosphorylated, allows the accumulation of the growth-stimulating proto-oncogene β-catenin, and Mdm-2, which promotes cell cycle progression when phosphorylated (Steelman et al. 2004). In addition, it inactivates cell cycle inhibitors that will be discussed later, that is, Cip/Kip family members p21 (Li, Dowbenko, and Lasky 2002) and p27 (Liang et al. 2002). More recently, the mammalian Target of Rapamycin (mTOR) protein, itself a serine/threonine kinase, has been shown to positively regulate cell growth and stimulate protein translation by activating both the ribosomal S6 kinase (S6K1) and the oncogenic protein elongation initiation factor 4E (eIF4E) (Mamane et al. 2004; Ruggero et al. 2004). The end result of Akt signaling is inhibition of apoptosis and activation of the cell cycle and cell growth, obvious conditions for cell transformation. It is not surprising then that Akt has been implicated as a significant prognostic factor in a variety of tumors (for example, Schlieman et al. 2003) and is overexpressed early in tumor development (Roy et al. 2002). The hypothetical value of targeting the PI3K-Akt pathway in tumor therapy has accelerated the identification and testing of several potential inhibitors of this signaling pathway, some of which are currently in clinical trials (Luo et al. 2003).

5.5b Nonreceptor tyrosine kinases

An additional group of proto-oncogene products with tyrosine kinase activity is localized to the cytoplasmic side of the plasma membrane. The *src* gene family (*fgr, fyn, hck, lck, lyn, src,* and *yes1*) comprises the majority of the known nonreceptor kinases. The src protein exemplifies the concept of allosteric regulation in kinase activation. In its inactive state, the c-src protein is repressed by intramolecular interactions. A C-terminal phosphotyrosine residue (Tyr530 in humans) creates a binding site for the central SH2 domain within the protein itself. This tyrosine residue is absent from the oncogenic retroviral form of the protein, suggesting that loss of this site derepresses the kinase domain of src, allowing it to be constitutively activated. Normally, c-src may be derepressed by a protein tyrosine phosphatase that removes the phosphate group from Tyr530, eliminating the SH2-binding site (Bjorge, Jakymiw, and Fujita 2000). However, a subset of human colon tumors are characterized by truncation of the C-terminal aspect of the src protein ending at Tyr530 (Irby et al. 1999). The mutant protein is transforming and contains a constitutively active kinase domain. In addition, the normal protein has an SH3 domain that interacts with a binding site near the kinase domain. This interaction also represses kinase activity and therefore must be relieved by associating with other proteins to elicit full tyrosine kinase activity. The downstream targets of src signaling associated with transformation have not been clarified, but the transcription factor STAT-3 and cell cycle activator c-myc both seem to be involved (Bowman et al. 2001).

5.5c   GTP-binding proteins: *ras* activation

A major group of intracellular membrane-bound proteins includes the *ras* proto-oncogene family members, notably, K-*ras*, H-*ras*, and N-*ras*, which are frequently mutated in human tumors. These proteins function as molecular switches, coordinating signals from RTKs with a variety of different downstream pathways that mediate proliferation, apoptosis, cell motility, angiogenesis, and differentiation. Thus, they are positioned in critical processes that are commandeered in tumorigenesis and are therefore potential targets for chemotherapeutic intervention (Downward 2002). For activation, the *ras* proteins (p21*ras*) must first insert into the inner cell membrane through a series of post-translational processing events. These involve terminal cysteine farnesylation (addition of a 15-carbon isoprenoid chain) and carboxy terminal methylation, and in some cases palmitoylation, to convert the carboxy terminus to a hydrophobic region capable of insertion into the cell membrane. Without these modifications, the protein remains cytosolic and inactive. Once incorporated into the membrane, it remains in a state of dynamic flux: active and signaling when the protein binds GTP and inactive upon GTP hydrolysis to GDP by the p21*ras* (see Figure 5–3). The state of signaling depends upon its interactive partners. As described above, a Grb2/Sos complex forms on a ligand-bound RTK. As you will recall, Sos is a guanine nucleotide exchange factor (GEF), which facilitates the exchange of inactive bound GDP for an active GTP molecule in a target protein. Thus, Sos mediates the conversion of p21ras protein from an inactive to an active and signaling form. The GTP alters the conformation of the ras protein, allowing it then to interact downstream with proteins involved in a variety of signaling pathways (Chardin et al. 1993; Gale et al. 1993). These downstream effectors include some proteins already mentioned, namely, PI3K, PDK/Akt, and PLCγ as well as the serine/threonine kinase RAF (Cox and Der 2003; Downward 2002; Macaluso et al. 2002).

Signaling is then sustained until p21 *ras* activity is modified by a group of GTPase-activating proteins (GAPs). GAPs are directly phosphorylated and activated by the RTKs. They inhibit p21 *ras* activity by accelerating the hydrolysis of GTP to GDP. Because p21 *ras*-mediated signals are generated by the conformational change associated with the presence of GTP, GAPs negatively regulate the activity of the normal p21 *ras* by limiting the duration of the signal. In tumor cells that contain a mutation in *ras,* signaling through p21 *ras* occurs continuously when the GTPase activity is lost, since p21 ras is no longer able to interact with a GAP. Alternatively, the loss of GAP function results in the same outcome (Downward 2002). In both circumstances, the p21 *ras* remains in its active signaling conformation.

5.5d   Cytoplasmic serine/threonine kinases

The proto-oncogene serine/threonine kinase Raf proteins (also called mitogen-activated protein kinase kinase kinase or MAPKKKs) were the first characterized downstream targets of p21 *ras* signaling in the cascade of events leading to transcriptional activation and growth regulation. Using purified p21 *ras* and Raf proteins, direct binding of p21 *ras* with Raf was observed (Zhang et al. 1993; Warne, Viciana, and Downward 1993), and Raf selectively bound the GTP-charged or activated form of p21 *ras* (see Figure 5–3). It is now apparent that p21 *ras* translocates Raf to the cell membrane where it is activated (Leevers, Paterson, and Marshall 1994). Of the three forms identified in mammals, Raf-1/c-Raf, A-Raf, and B-Raf, only B-Raf is directly activated by interaction with p21 *ras,* while the other two require further modifications (Hagemann and Rapp 1999). In human malignancy, constitutive Raf activation is associated with transforming mutations in the *ras* genes; however, mutations also occur directly in the various forms of Raf. In particular, mutations in B-Raf have been observed in 66 percent of human malignant melanomas (Davies et al. 2002). Mutations were all located in the kinase domain of the Raf protein and produced a constitutively active kinase that was transforming in NIH/3T3 cells independent of Ras activity.

Once its kinase function is activated by interaction with p21 *ras,* Raf then associates with a family of proteins called mitogen-activated protein kinase kinases (MAPKK 1 and 2 or MEK1 and 2) (see Figure 5–3). Raf directly phosphorylates these serine/threonine kinases on two separate serine residues in generating their active or signaling forms in the cascade (Schaeffer and Weber 1999). The MEKs then, in turn, activate members of the serine/threonine MAPK family, Extracellular Signal-Regulated Kinase (ERK)1 or ERK2 (Seger et al. 1992), by phosphorylation of both threonine and tyrosine residues in a Thr-Glu-Tyr motif in the kinase activation loop (Payne et al. 1991). This increases the kinase activity of the MAPKs one thousand-fold.

To establish an association between Raf signaling and MAPKK or MAPK activity, several methods were devised to modulate Raf and monitor its effects downstream. By fusing a steroid-binding domain from a steroid receptor to the catalytic domain of the proto-oncogene *raf*-1, the activity of the Raf-1 protein can be directly regulated using the steroid, because the Raf protein is activated when the steroid interacts with the steroid-binding domain. Using this approach, Samuels et al. (1993) reported that two classes of cellular kinases, MEKs and ERKs, both become phosphorylated in cells treated with the steroid. In this case, the steroid activates the kinase activity of the Raf-1 protein. Raf, in turn, induces phosphorylation of the MEKs and ERKs, demonstrating that these two kinases receive phosphorylation signals either directly or indirectly from Raf and are therefore downstream of Raf signaling in Raf-mediated signal transduction.

When activated, the ERKs phosphorylate one or more of the fifty ERK targets identified in the cytoplasm and nucleus, showing that the activated ERKs are also translocated through the nuclear membrane. Among the nuclear targets are the members of the ets family of transcription factors, which includes Elk1 (Yordy and Muise-Helmericks 2000). This induces expression of another transcription factor, Fos, which heterodimerizes with another ERK target c-Jun to form the active transcriptional complex AP-1. AP-1 then stimulates expression of cell cycle regulatory proteins such as the D cyclins (Pruitt and Der 2001) (see Figure 5–3).

Although the description of this mitogen-stimulated signaling pathway provides the basic details for one of the better defined signal transduction pathways, the actual mechanisms are considerably more complex than presented. It is known that multiple forms of kinases exist at each level, so there are, for example, three proto-oncogenic forms each of the p21ras and Raf proteins (Moodie and Wolfman 1994). Furthermore, as already mentioned, there are multiple signaling pathways initiated potentially by the same signal, but which yield quite different outcomes, depending upon the available players. Thus, in addition to Raf, ras may activate PI3K, which, in turn, activates the antiapoptotic factor Akt. It also stimulates Rac/Rho activity, which regulates the actin cytoskeleton, altering cell shape and motility. Furthermore, Ras can directly activate Tiam1 (Lambert et al. 2002) to regulate Rac activity rather than indirectly activating it through PI3K, so there are multiple routes to the same endpoint. Finally, ras-induced phosphorylation of PLC activates PKC and causes Ca^{2+} release, which modulates cell differentiation. Thus, signaling is context dependent. One cell type may express components for Akt stimulation, while another expresses factors for Ca^{2+} regulation. Depending upon the context, the ras pathways can also provide opposing biological effects. In tumor cells, for example, ras activates PI3K/Akt, inhibiting apoptosis; however, ras/Raf signaling can cause either a pro-apoptotic or an antiapoptotic effect, depending upon the signaling balance from the various pathways (Cox and Der 2003).

5.5e Suppression of *ras* signaling

Whereas p21*ras* activation stimulates signaling through the MAPK pathway, there are at least two proteins in the signaling pathway that negatively regulate or suppress signaling. As already mentioned, GAP proteins inhibit p21*ras* signaling by facilitating the metabolism of p21ras-bound GTP to GDP, thus converting the conformation of p21ras to a nonsignaling form. GAPs may then function as tumor suppressors, and indeed the Ras-GAP protein called NF1 behaves in this manner. Neurofibromatosis type I is an autosomal dominant syndrome associated with a high incidence of neurofibromas in the peripheral nerve sheath. It is characterized by loss of heterozygosity for the *NF1* gene through germline deletion and/or somatic point mutation (Serra et al. 2001).

cAMP also down-regulates or inhibits signaling through the Ras/MAPK pathway. It accomplishes this by activating a serine/threonine kinase, protein kinase A (PKA), which blocks the transmission of signaling between Raf and p21*ras* and thus prevents activation of the MAP kinase cascade (Wu et al. 1993; Cook and McCormick 1993) (see Figure 5–3). The mechanism involves PKA-mediated phosphorylation of Raf on serines 43, 233, and 259, creating binding sites for the 14-3-3 proteins, which interact with phosphorylated serines within so-called 14-3-3 motifs. The 14-3-3 proteins sequester Raf protein to prevent its interaction with p21*ras* and thus block its recruitment to the cell membrane where it is normally activated (Dumaz and Marais 2003). This may explain the long-established effects of cAMP as a negative growth regulatory and differentiation-inducing molecule. Because signal transduction through p21ras has been implicated in tumorigenesis, it may be possible to preempt signaling through p21ras and inhibit abnormal cell growth using molecules that activate PKA. Accordingly, overexpression or cAMP stimulation of PKA has been shown to suppress the transformed phenotype in ras-transformed cells (Budillon et al. 1995; Bjorkoy et al. 1995).

5.5f Nuclear signaling

Several links between the phosphorylation cascade in the cytoplasm and signal transduction in the nucleus have been established. As already mentioned, interactions between nuclear transcriptional activation factors and the MAPKs have been reported in vitro, and it appears that an intervening nuclear membrane does not prevent such interactions because both serine/threonine kinases ERK1 and ERK2 are translocated to the nucleus in growth factor–stimulated cells (Lenormand et al. 1993). Additionally, tyrosine kinases have been localized to the nucleus. The proto-oncogene c-*abl* encodes a tyrosine kinase, which contains a nuclear localization signal and a DNA-binding function (Wang 1994). When translocated to the nucleus, it elevates transcription levels in conjunction with other transcription factors or by direct phosphorylation of RNA polymerase II (Baskaran, Dahmus, and Wang 1993). Even *src* family members appear in the nucleus under certain circumstances.

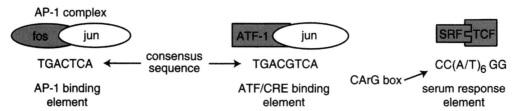

Figure 5–5. Transcription factor recognition sites. Transcription factors generally polymerize to form homo- or heterodimers of factor proteins that recognize specific consensus sequences in the DNA based on the particular complex generated.

The tyrosine-phosphorylated STAT proteins, which mediate proliferation-inducing cytokine signaling, are tyrosine phosphorylated at the cell membrane. The phosphotyrosine residue creates an SH2 binding site, allowing STAT dimerization and subsequent nuclear translocation (Silvennoinen et al. 1993; Bromberg and Darnell 2000). Thus, numerous nuclear-translocated proto-oncogene proteins are capable of mediating signals to the genome, but the details of their involvement in the various pathways, especially with regard to target specificity, are not completely understood.

5.5g   Transcriptional activation

Several transcription factors are encoded by proto-oncogenes (Figure 5–5), for example, *erb*A, *ets, fos, jun, myb,* and *myc.* These nuclear proteins regulate the expression of genes involved prominently in growth regulation or apoptosis. Regulation is achieved through the ability of the transcription factor to recognize and interact with specific DNA sequences found in the promoter/enhancer regions of those genes to be expressed. The AP-1 complex and its components have been extensively studied because of the pivotal role they play in transmitting signals from various growth factors, cytokines, or stress factors. The complex is a homodimer or heterodimer of members of the Jun, Fos, Atf, or Maf (musculoaponeurotic fibrosarcoma) families. Of these, c-*jun* and c-*fos* were among the first identified and were discovered as retroviral oncogenes (v-onc). Interestingly, other members of these families, for example, *junB* or *junD,* may function as tumor suppressors because of their abilities to stimulate apoptosis. The behaviors of these genes have been extensively examined through the use of knockout and transgenic technologies (Eferl and Wagner 2003).

The ability of the AP-1 complex to regulate transcription is mediated at the transcriptional level by controlling *jun* or *fos* expression (to increase levels of the complex) and at the post-translational level by phosphorylation at multiple sites in the Jun or Fos proteins (to enhance AP-1 complex activity, that is, increase its ability to promote transcription). In fact, EGF induces the phosphorylation of c-Jun on serine 63 and 73 through either the MAPK family member Jun N-terminal kinase (JNK1/JNK2; Dérijard et al. 1994) or Erk1/Erk2. Phosphorylation at these sites enhances AP-1 activity. On the other hand, glycogen synthase kinase-3β (GSK3β) phosphorylates c-Jun at threonine 239 (Deng and Karin 1994), which inhibits DNA binding, and activating stimuli such as EGF or TPA treatments causes this site to be dephosphorylated (Morton et al. 2003). The viral oncogenic form V-Jun contains a deletion near the N-terminus of the protein that ablates the requirement for JNK phosphorylation of Ser63 and 73 to activate the protein. The c-Fos protein, on the other hand, is phosphorylated by Erks at the carboxy-terminal end. V-Fos is characterized by a C-terminal deletion that eliminates a serine normally phosphorylated by an Erk. Thus, viral activation results from a loss of regulatory sites in these proteins.

The c-Fos protein interacts with the c-Jun protein to generate a heterodimer, which recognizes the consensus sequence TGACTCA (Chiu et al. 1988). This sequence is referred to as the AP-1/TPA-response element (TRE) because the AP-1 complex mediates transcriptional activation by TPA through binding to this sequence. AP-1/TRE and TRE-like sequences are located in promoter/enhancer regions of several genes, most of which stimulate growth through the upregulation of proliferation inducers (e.g., cyclin D1, EGFR, and HB-EGF) or repression of proliferation inhibitors (e.g., INK4A/p16 or p53) or of apoptotic factors (e.g., the Fas ligand) (Shaulian and Karin 2002; Eferl and Wagner 2003). C-Jun or c-Fos can also dimerize with an Activating Transcription Factor (ATF) to form a heterodimer that preferentially recognizes a cAMP Response Element (CRE) of sequence TGACGTCA (Hai and Curran 1991). This occurs following cell stimulation with neurotransmitters or polypeptide hormones that activate either protein kinase A (PKA) with cAMP or calmodulin-dependent kinases (Misra et al. 1994) with calcium. This mechanism allows cross-talk between signaling pathways because C-Jun synthesis can modulate expression from both the AP-1/TRE and the ATF/CRE elements.

The *jun* promoter/enhancer contains a modified AP-1/TRE, which is recognized by Jun/ATF heterodimers, and a Smad3/Smad4-binding element for enhanced activation by transforming growth factor-β (TGFβ) (Wong et al. 1999). The c-*fos* promoter also consists of multiple elements. In addition to the ATF/CRE element, it includes another growth factor–stimulated binding site called the Serum Response Element (SRE) (Treisman 1990). Transcriptional activation is enhanced from the SRE by interaction of the constitutively bound protein Serum Response Factor (SRF) with a DNA sequence called the CArG box (CC(A/T)$_6$GG) in the SRE. An ets family member such as the Elk1 protein, a ternary complex factor (TCF), is also required to maximize the response to growth factor stimulation of c-*fos* transcription (Shaw, Schroter, and Nardheim 1989). Following its mitogen-induced phosphorylation, Elk1 is recruited to the SRE by the SRF. Transcription is also enhanced through a Sis-Inducible Element (SIE), which binds STAT-3 to upregulate c-fos transcription (Yang et al. 2003).

The prototypical promoter/enhancer region therefore incorporates multiple recognition elements. The *fos* enhancer/promoter actually contains binding sites for an ets family member, an SRF, an AP-1-like element which allows it to self-regulate, and STAT-3 (Treisman 1994; Yang et al. 2003). For c-*met*, which encodes a receptor tyrosine kinase for hepatocyte growth factor, the promoter/enhancer region includes a consensus sequence for the transcription factor PEA3 (AGGAA(G/A)), which is induced by growth factors and TPA, and two AP2 sites (CCC(A/C)N(G/C)$_3$) (Gambarotta et al. 1994). Because these transcriptional elements in general can act additively or synergistically in transcriptional activation or repression, signaling from multiple pathways may elicit dramatically different responses in the nucleus. (For an on-line database and search tool, see http://cmgm.stanford.edu/help/manual/databases/tfd.html).

Interaction of a transcription factor with its consensus site, phosphorylation of the factor itself, or multiple transcription factor-specific binding elements are not the only regulatory mechanisms for gene expression. Transcription factors may require transcriptional coactivating or corepressing proteins to modulate gene expression and regulate biologic function. The transcription factor Tcf4, which targets Wnt signaling, serves as a repressor in the absence of secondary messenger β-catenin. Without Wnt signaling, Tcf4 associates with the corepressor protein Groucho; however, with Wnt activation, β-catenin is stabilized in the cytoplasm and translocated to the nucleus. It then associates with Tcf4, overcoming its repressive effects and inducing transcription of Wnt-dependent target genes (Barker, Morin, and Clevers 2000). Thus, coactivators, such as β-catenin, or corepressors, such as Groucho, may participate in determining transcriptional activity for a given gene.

5.6   Regulation of DNA synthesis and the cell cycle

Once the signals for growth stimulation have reached the nucleus, they are translated into cellular actions by a complex of proteins that mediate progression of the cell through the sequential phases of the cell cycle (Figure 5–6): G_0, a quiescent noncycling state which accounts for the vast majority of nonproliferating cells in an adult; G_1, the preparation for DNA replication; S, DNA synthesis to duplicate the entire genome; G_2, preparation for nuclear and cell division; and M, mitosis. This is a highly regulated process during which the genome is replicated completely and only once for each cycle. The DNA is allocated such that one complete copy resides in each cell progeny following cell division. It is controlled by the stage-specific activation of at least five members of the cyclin-dependent kinase (CDK) family, all serine/threonine kinases: G_1 – CDK2, CDK4, and CDK6; S – CDK2; G_2 and M – CDK1; and CDK7, which associates with cyclin H to form a CDK-activating kinase (Vermeulen, Van Bockstaele, and Berneman 2003). The CDKs are expressed constitutively throughout the cell cycle and are activated by interaction with members of the cyclin family, which are expressed in a stage-specific manner.

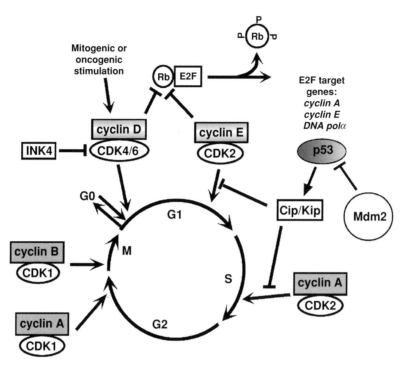

Figure 5–6. Cell cycle regulation. Members of the Cyclin-Dependent Kinase (CDK) family associate with stage-specific members of the cyclin family to initiate and sustain cell cycle progression. The G1 complexes hyperphosphorylate the Rb tumor suppressor protein, causing its inactivation and releasing the E2F transcription factor to induce expression of G1 and S phase cyclins. Negative regulation of cycle progression is provided at multiple levels: INK4 family proteins on the G1 CDKs, Cip/Kip family proteins on the G1 CDK-cyclin complexes, and p53 by upregulating expression of the Cip/Kip family members.

Cyclins D1, D2, and D3 activate CDK4 and CDK6 to mediate cell entry into G_1 and are present as long as growth factor stimulation continues (Assoian and Zhu 1997). The retinoblastoma tumor suppressor protein pRB figures prominently in regulation of the early stages of cell cycle progression by sequestering the transcription factor E2F, and this will be discussed in greater detail later in the chapter. Cyclin E binds to CDK2 to mediate cell exit from G_1 and into S phase. During S phase, cyclin E is removed by degradation (Won and Reed 1996), releasing CDK2 for interaction with cyclin A, which regulates progression through S phase (Girard et al. 1991). Furthermore, the CDK2/cyclin A complex inactivates E2F and thus downregulates the transcription of E2F-responsive genes, including cyclin E (Krek, Xu, and Livingston 1995). Finally, cyclins A and B in combination with CDK1 control entry into mitosis, after which cyclin levels plummet; however, the cycle can be repeated with continued mitogenic stimulation and upregulation of cyclin D. Each of the cyclins contains sequences that allow its immediate and rapid proteolysis when it is no longer required, thus also inactivating the CDK to which it is bound (Rechsteiner and Rogers 1996).

Factors that positively regulate cell cycle progression have been implicated in carcinogenesis either through overexpression or mutation. For example, elevated expression of cyclin D1 has been observed in human esophageal tumors, lung, breast, bladder, and squamous cell carcinomas (Hall and Peters 1996), leading to the suggestion that the normal cyclins represent a class of proto-oncogenes. In B-cell malignancies, a t(11;14) chromosomal translocation results in the juxtaposition of cyclin D1 and the gene for immunoglobulin heavy chain, which causes the misexpression of this cyclin (Weisenburger et al. 1987). For the CDKs, overexpression of CDK4 has been reported for some gliomas, sarcomas, and melanomas (Wolfel et al. 1995), while mutations in CDK4 and CDK6 that interfere with CDK-inhibitor protein binding have been observed in neuroblastomas (Easton et al. 1998). Thus, both cyclins and CDKs may function as oncogenes.

In addition to loss of cyclin proteins, the cell cycle is negatively regulated by a family of suppressors called CDK inhibitor (CKI) proteins (see Figure 5–6), which block CDK activity through direct interaction either with the CDK or the CDK/cyclin complex (Peter and Herskowitz 1994). Two families of proteins have been identified: the INK4 family members (p15/INK4b, p16/INK4a, p18/INK4c, and p19/INK4d), which inactivate the G_1 CDKs, and the Cip/Kip family members (p21/Waf1/Cip1, p27/Cip2, and p57/Kip2), which block activity of the G_1 CDK/cyclin complexes, causing a G_1 arrest. The INK4 proteins bind CDK4 and CDK6 to prevent their association with cyclin D (Serrano, Hannon, and Beach 1993). The p21 and p27 proteins are thought to play an especially significant role in inhibiting the activity of CDK2/cyclin E and are considerably less effective at blocking the activity of complexes that include cyclin D (Blain, Montalvo, and Massague 1997); however, CDK4/cyclin D binds and removes both Cip proteins (Alt, Gladden, and Diehl 2002; Muraoka et al. 2002). This suggests that, following mitogenic stimulation, the accumulation of CDK4/cyclin D complexes sequesters the Cip proteins, allowing the CDK2/cyclin E complexes to become activated (Sherr and McCormick 2002). Once active, the complex phosphorylates p27, triggering its degradation (Vlach, Hennecke, and Amati 1997). The process is reversed when mitogenic stimulation stops and cyclin D transcription ceases. In the absence of CDK4/cyclin D complexes, the Cip proteins accumulate and inhibit activity of the CDK2/cyclin E complex, causing G_1 arrest. While this explanation assimilates current thinking on cell cycle regulation, it is clear that the model is continuing to evolve. It has now been demonstrated that CDK2 null homozygotes are viable and show no proliferative disorder, suggesting that it is not essential for cell cycle activity (Ortega et al. 2003). However, it is possible that other CDKs may substitute for CDK2 activity. In addition, loss of cyclins E1 and E2 manifests as disorders in proliferation, but animals remain viable (Geng et al. 2003). Furthermore, loss of cyclin A2 does result in embryonic lethality (Murphy et al. 1997), suggesting that cyclin E is not essential for cell cycle progression and that possibly CDK1 and cyclins A2 and B are the only components necessary to allow the cell cycle to proceed (Roberts and Sherr 2003), but the jury is still out on this.

As one might predict based upon their biological activities, the CKIs function as tumor suppressors, which has been aptly demonstrated in genetic studies. For example, mice lacking INK4 proteins p16 and p19 develop spontaneous tumors with an abbreviated latency and are highly sensitive to tumorigenesis with carcinogen treatment (Serrano et al. 1996). Similarly, the loss of p21/Cip1 (Adnane et al. 2000) or p27/Cip2 (Nakayama et al. 1996) results in increases in spontaneous tumor development, consistent with a role as a suppressor. Transforming growth factor-β (TGFβ) is a potent antimitogenic factor that arrests cells in late G_1 apparently by stimulating transcription of both p15/INK4b (Hannon and Beach 1994) and p27/Cip2 (Reynisdottir et al. 1995).

Of particular interest with regard to the CKI proteins is that p21/Cip1 expression is induced by the tumor suppressor protein p53 (el-Deiry et al. 1994). P53 is activated following cellular DNA damage to block cell proliferation in order to give the cell sufficient time to repair the damage or redirect the cell toward apoptosis. Cell cycle inhibition is apparently elicited through p21 activity, but this will be discussed at length later in this chapter.

The function of the active CDK/cyclin complexes is to phosphorylate target proteins at specific motifs as well as sequester the CKIs. One target of considerable study has been the tumor suppressor pRB (Pines 1994), which was discovered initially in pediatric patients suffering from a genetic predisposition to the rare cancer retinoblastoma. When pRB (or RB family members p107 and p130) becomes hyperphosphorylated and thus inactivated, it releases the transcription factors E2F-1 and DP-1, which induce transcription of target genes such as cyclins A and E through consensus sequences (TTTCGCGC) for E2F, allowing cells to enter S phase (Ewen 1994). In turn, the formed CDK2/cyclin E complex phosphorylates pRB to sustain it in its hyperphosphorylated and inactive state throughout the cell cycle (Neuman et al. 1994, Weinberg 1995).

5.7 Other mechanisms for the regulation of signaling

As details of the various proto-oncogenes have become available, increasing evidence shows unified and interactive pathways of molecules that generate, transmit, and interpret signals from extracellular stimuli to the genome. Thus, families of membrane receptors, many of which contain considerable sequence homology in the intracellular domains of the proteins, initiate common pathways of secondary messengers through a number of regulatory proteins, such as phospholipase C, protein kinase C, or p21ras, or through direct interaction with structural proteins anchored in the cell membrane. The signaling appears to involve a flow or cascade of phosphorylation, initally on tyrosine residues of RTKs and subsequently on serine or threonine residues of cytoplasmic and nuclear kinases. It has long been known that regulatory enzymes in metabolic pathways are activated and/or inhibited by phosphorylation reactions, which alter the conformation of the modified protein to modulate its interactions with substrates, cofactors, or other proteins – so-called allosteric regulation. In metabolism, phosphatases are invariably involved in reversing the effects of phosphorylation reactions. Now both tyrosine- and serine/threonine-specific phosphatases have been identified as major components of growth signaling pathways and have been shown under some circumstances to act as both positive and negative regulators of signaling (Hooft van Huijsduijnen 1998; Li and Dixon 2000). Although estimates place the number of cellular phosphatases on the order of several hundred, only three phosphatases have thus far been implicated in normal growth control and tumorigenesis: protein phosphatase 2A, Cdc25, and PTEN (Phosphatase and TENsin homolog deleted on chromosome 10).

Protein phosphatase 2A (PP2A) is a serine/threonine-specific trimeric protein complex that has been implicated in growth control based upon the effects of okadaic acid, a tumor promoter and potent inhibitor of PP2A (Schönthal 2001). In particular, targeting of PP2A with okadaic acid allows constitutive signaling through the MAPK pathway and activation of AP-1, suggesting that this phosphatase functions as a tumor suppressor. However, the loss of PP2A also is an embryonic lethal lesion (Götz et al. 1998), so it provides an essential function in growth and development. Studies of lung and colorectal tumors have identified mutations in small percentages of tumors (Wang et al. 1998), but the significance of these alterations remains equivocal. Therefore, it is currently unclear if PP2A functions in human tumorigenesis.

Cdc25 is a dual-specificity protein phosphatase capable of dephosphorylating CDK proteins containing inhibitory phosphothreonine and phosphotyrosine residues. In so doing, it stimulates CDK activity and cell cycle progression with the following target specificity (Gautier et al. 1991): Cdc25A on CDK2 and 4 and Cdc25B and C on CDK1. Because these molecules participate in cell cycle activation and cooperate with other oncogenes for transformation (Galaktionov et al. 1995), they behave as proto-oncogenes. Consistent with this idea, they are overexpressed in a variety of sporadic human neoplasms (Pestell et al. 2000) and are associated with a poor prognosis (Hernandez et al. 2000). Thus, phosphatases may behave as either oncogenes or tumor suppressors, depending upon whether phosphorylation of a target protein results in catalytic activation or suppression.

While PTEN has tyrosine phosphatase activity, it appears to act principally as an antagonist of PI3K activity through dephosphorylation of the PI3K product, the phosphoinositide lipid PtdIns(3,4,5)P3, which activates Akt (Yamada and Araki 2001) (see Figure 5–4). By metabolizing this substrate, cell proliferation and motility can be targeted, so PTEN functions as a tumor suppressor. In addition to metabolism of phospholipids, it also may behave as a tyrosine phosphatase to dephosphorylate and inactivate the adapter protein Shc and Focal Adhesion Kinase (FAK), a protein that mediates the extracellular signaling of integrins and activates

PI3K (Tamura et al. 1998). PTEN expression is deficient in a variety of tumors (Simpson and Parsons 2001), including brain, breast, endometrial, lung, and prostatic cancers, and overexpression of PTEN in tumor cell lines can induce apoptosis (Li et al. 1998), especially a form called anoikis, which is induced by loss of cell contact with the extracellular matrix (Davies et al. 1999). It can inhibit cell migration, and therefore its absence may play a significant role in tumor cell invasion (Tamura et al. 1999).

In addition to the demonstrated roles of dephosphorylating agents in growth control, regulation occurs at many other levels as well. The list of signal transduction pathways and the ability of these pathways to 'cross-talk' or interact with each other is continuously expanding. Thus, the PI3K/PDK1/Akt pathway talks to the Raf/MEK/ERK pathway through PDK1, which phosphorylates MEK1 and 2 on the same serine residues as the Raf kinases (Sato, Fujita, and Tsuruo 2004) to activate MAPK signaling. In addition, the MAPK and Wnt pathways both regulate the expression of the oncogene cyclin D1 independently of one another, so there is redundancy in pathways as well. There is also redundancy in pathway components. For example, there are at least twelve related enzymes or isoforms of protein kinase C (Tortora and Ciardiello 2003), adding another level of complexity to an already complex picture of positive/negative regulatory elements. However, as specific tissues are evaluated for these transducing pathways, it is possible that the number of active signaling pathways is quite limited and may, in fact, define the differentiation and maintenance of a given tissue.

5.8 Mechanisms of oncogene activation

As stated earlier, more than a hundred genes have been identified, which, under certain circumstances, can transform cells when a single gene copy is altered (a dominant transforming gene). The mechanisms responsible for oncogenic activation of these genes are quite varied but the principles are limited to three basic processes: misexpression of a transduced or translocated sequence, overexpression of a normal gene, or gain-of-function mutations. The reality is that a combination of these processes may apply in any given circumstance. In human cancers, misexpression of a normal gene can result from hypomethylation of a proto-oncogene's promoter/enhancer sequence (Esteller and Herman 2002). Such epigenetic mechanisms have been implicated in the dysregulated expression of the growth factor Igf-2 in Wilms' tumor or Ras in gastric cancers. Gene expression is partially regulated through methylation of nucleotides in CpG islands, which inhibits their expression. If this methylation is disrupted, the expression of a proto-oncogene may be inappropriately upregulated, initiating tumorigenesis. Overexpression in human tumors can also result from gene amplification, such as occurs for *erb*B-2/HER-2 in some breast tumors. For retroviral-mediated tumors, viral insertion promoter sequences near a proto-oncogene or transduction of cellular sequences by a retrovirus can lead to tumor development. In each circumstance, the cellular proto-oncogene would be placed in tandem with a constitutively active promoter/enhancer, resulting in its misexpression. Under experimental conditions, a proto-oncogene may be placed under the regulation of virtually any cellular or retroviral promoter/enhancer sequence, permitting the unnatural constitutive expression of that gene. Thus, misexpression of proto-oncogenic forms of *tgf-α* or the RTK c-*erbB-2* under the regulation of the mouse mammary tumor virus (MMTV) promoter in a transgenic mouse yields preneoplastic and neoplastic mammary lesions (Davies et al. 1999), while liver and breast lesions arise when *tgf-α* is regulated by a metallothionein promoter (Jhappan et al. 1990).

Studies of chemically induced and naturally occurring tumors demonstrate that oncogene activation can occur through gain-of-function mutations in the encoded sequences of the proto-oncogenes. The most frequently observed mechanism involves the incorporation of point mutations causing activation of the oncogene proteins. In both naturally occurring human tumors (Minamoto, Mai, and Ronai 2000) and chemically induced rodent tumors (Sills et al. 1999), a *ras* oncogene family member is frequently activated by point mutation in codons 12, 13, or 59 to 61 in a variety of tumors. In humans, nearly one-third of all tumors carry a *ras* mutation of which K-*ras* mutations in codon 12 occur at high frequency in adenomas and adenocarcinomas of the colon (Forrester et al. 1987) and in pancreatic adenocarcinomas (Almoguera et al. 1988).

It is also possible to induce a high frequency of mutations in *ras* genes of certain rodent tumors using specific carcinogens, suggesting that *ras* activation plays a significant role in carcinogenesis (explained in more detail later). Eighty-three percent of mammary tumors, initiated in rats with the alkylating agent N-methyl-N-nitrosourea (MeNU), contained activated H-*ras* sequences, and activation resulted predominantly from GGA to GAA transition mutations in codon 12 (Zarbl et al. 1985). Such a mutation causes a glycine residue to be replaced by glutamic acid in p21*ras*, which alters the conformation of the protein and permits continuous signaling through p21*ras*. 7,12-Dimethylbenz[a]anthracene (DMBA) also initiates mammary tumors in rats but does not induce mutations in codon 12 (Zarbl et al. 1985). Instead, codon 61 is affected. This is also the case for DMBA induction of mouse skin tumors in which 90 percent contained mutations (CAA to CTA) in codon 61 (Quintanilla et al. 1986). X-ray crystallographic analysis and mutagenesis studies of p21*ras* show that mutations in these hot spot regions prevent an interaction between mutated p21*ras* (codons 12, 13, 59, and 61) and the negative regulator GAP, which converts p21*ras*-bound GTP to GDP (Der, Finkel, and Cooper 1986; Milburn et al. 1990). These alterations result in the accumulation of GTP-bound protein, which is the activated and signaling form of p21*ras* (Cox and Der 2002).

The *neu/erb*B-2/HER-2 oncogene is also frequently activated in rodents by point mutation (Bargmann, Hung, and Weinberg 1986). The proto-oncogene sequence encodes a receptor tyrosine kinase with considerable homology to the EGFR (*erb*B-1). GTG to GAG (valine to glutamic acid) mutations in the putative transmembrane region of the encoded protein were detected in more than 90 percent of N-ethyl-N-nitrosourea (EtNU)–induced schwannomas, tumors of the peripheral nervous system, in rats but in none of the tumors of the central nervous system elicited in the same protocol (Perantoni et al. 1987). Such a mutation yields dominant transforming *neu* sequences. These mutations are specifically associated with the carcinogen EtNU because they occur infrequently in spontaneous schwannomas (Perantoni et al. 1994). They also appear very early in the neoplastic process, suggesting they could represent initiating mutations (Nikitin, Ballering, and Rajewsky 1991).

Point mutations are not the only mechanism whereby oncogenes acquire their oncogenic potential. In human tumors, *erb*B-2/HER-2, the human equivalent of *neu,* does not incur a similar mutation; however, as already mentioned, gene amplification, which causes an elevated expression of the *erb*B-2 gene product p185, is frequently associated with specific human tumors (Menard et al. 2001). Notably, a poor prognosis for women with mammary (Slamon et al. 1987) and ovarian tumors (Berchuk et al. 1990) correlates directly with the extent of overexpression of *erb*B-2 in tumor tissues. Because overexpression of the normal p185*erb*B-2 can transform cells in culture (Di Fiore et al. 1987) and provide cells with a selective growth advantage (Nikitin et al. 1991), overexpression of this receptor appears to be a viable mechanism for oncogene activation as well. Since this tyrosine kinase receptor is localized to the cell membrane, it is a readily available target for immunotherapy, and, as mentioned earlier, one based upon neutralizing antibody fragments is currently in clinical trials.

Gain-of-function point mutations have been observed in RTKs from human tumors as well (reviewed in Blume-Jensen and Hunter 2001; for *ret* specifically, Alberti et al. 2003). As mentioned earlier, the receptor Ret is critical to development of the kidney, small intestine, and neuroectoderm, and mutations in *ret* are characteristic of the familial tumor syndromes, Multiple Endocrine Neoplasia 2A (MEN2A) and Familial Medullary Thyroid Carcinoma (FMTC). The *ret* mutations occur in one of the extracellular cysteine residues and promote constitutive dimerization and activation of the receptor in the absence of its ligand glial-derived neurotrophic factor (Santoro et al. 1995). Alternatively, in MEN2B, a mutation in the kinase domain of Ret (Met918Thr) enhances constitutive kinase activity, allows novel substrates access to phosphorylation, and activates PI3K. That mutated *ret* functions as an oncogene in tumor formation has been well documented in a transgenic mouse model for this disease (Smith-Hicks et al. 2000).

Gain-of-function chromosomal rearrangements provide another mechanism for oncogene activation and may be a common pathway for induction of leukemias in humans. Chronic myelogenous leukemia (CML) is often associated with the appearance of the Philadelphia chromosome (Nowell and Hungerford 1960), which results from a reciprocal translocation between chromosomes 9 and 22 [t(9;22)(q34;q11)](Rowley 1973). This rearrangement juxtaposes the N-terminal exons of *bcr,* a gene that encodes a coiled-coil domain for protein oligomerization, a Grb2-binding site (SH2 domain – Tyr177), and a serine/threonine kinase from

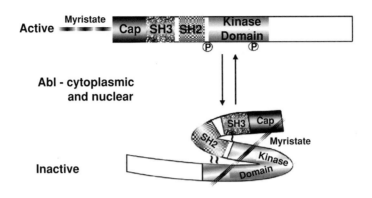

Figure 5–7. Allosteric regulation of Abl kinase activity. In the absence of phosphorylation, the Abl protein folds upon itself through interactions involving at least three different domains: the SH3 domain with the SH2-kinase linker region and the SH2 domain and the amino terminal myristate chain with the carboxy-terminal aspect of the kinase domain. Phosphorylation alters the configuration of the protein, activating the kinase domain. Abl is homologous to other nonreceptor kinases such as the Src family members. In CML, a chromosomal translocation often results in a fusion protein involving Bcr and Abl. The resulting protein contains a constitutively active Abl kinase, which localizes to the cytoplasm. The coiled-coil domain (CCD) of Bcr facilitates oligomerization and autophosphorylation of the protein for Abl kinase activation.

chromosome 22 with an exon 1-deficient c-*abl* proto-oncogene from chromosome 9 (Kurzrock, Gutterman, and Talpaz 1988; Ren 2002). Normal nuclear-localized c-Abl protein has tyrosine kinase activity that induces apoptosis following DNA damage, and this function is abrogated by localization of the Bcr-Abl fusion protein in the cytoplasm (Vigneri and Wang 2001). The coiled-coil domain of Bcr is critical for transformation, bringing the truncated Abl proteins together to 'autophosphorylate' for constitutive activation of the tyrosine kinase domains (Figure 5–7). The SH2 domain in Bcr also participates in transformation (Pendergast et al. 1993), probably through regulation of Ras signaling by providing a docking site for the Grb2 adapter protein as described above. The Abl portion of the fusion protein contributes principally through its tyrosine kinase activity. Normally, the Abl protein is negatively regulated by its amino-terminal sequences, which include a 'Cap' (Pluk et al. 2002), an SH3 domain, and an SH2 domain in that order. The SH3 domain interacts with amino acids that link the SH2 domain with the kinase, and the SH2 domain binds the C-terminal aspect of the kinase, creating an SH3-SH2 clamp that blocks the active site of the kinase (Courtneidge 2003). This mechanism is not unlike that described previously for c-Src. In any event, removal of the clamp sequences either through genetic engineering (Jackson and Baltimore 1989) or viral transduction yields a tranforming protein. The tyrosine kinase activity of the Bcr-Abl protein causes phosphorylation of STAT-5, which in turn upregulates the expression of the antiapoptotic factor Bcl-x$_L$ (Horita et al. 2000). This mechanism may in part define its role in Bcr-Abl-mediated tumorigenesis.

The *bcr-abl* fusion sequence can cause a CML-like syndrome when introduced in bone marrow cells of mice using a retroviral construct (Zhang and Ren 1998), and such models have helped identify functional determinants of tumorigenicity and cell signaling and have provided effective models for developing target-specific chemotherapeutic interventions. In particular, the cancer drug imatinib mesylate/Gleevec/STI-571 targets the tyrosine kinase domain of the Abl protein. Since kinase activity is at least partially responsible for cytoplasmic retention of Bcr-Abl, inhibition allows the fusion protein to translocate to the nucleus, stimulating apoptosis (Druker et al. 1996; Hantschel and Superti-Furga 2004).

Although the exact downstream signaling mechanism(s) responsible for Bcr-Abl-induced transformation has not been determined, several pathways have been implicated that enhance cell survival by inhibiting apoptosis and stimulating cell proliferation. These include activation of MAPK through Ras/Raf, Jnk, PI3K, as well as transcription factors c-Jun, c-Myc, and STAT-5 and antiapoptotic factors Bcl-2 and Bcl-x$_L$ (Scheijen and Griffin 2002). Furthermore, apoptotic and antiproliferative factors such as Bad, caspase 3, and p27/Kip1 are also inhibited. It is unclear whether the loss of nuclear localization and autoinhibitory sequences are both important for tumorigenesis. It is intriguing, though, that mislocalization may offer new protein targets for phosphorylation and thus provide other possible mechanisms for escape from growth control.

Studies of B cell tumors such as Burkitt's lymphoma have also implicated proto-oncogene rearrangements in tumorigenesis (Taub et al. 1982; Cory et al. 1999). Burkitt's lymphomas are characterized by reciprocal translocations of chromosomes 8 and 2 (5%), 8 and 14 (80%), or 8 and 22 (15%). The result of t(8;14) is the deregulation of transcription from the c-*myc* locus (Dalla Favera et al. 1982). This reciprocal recombination places the c-*myc* gene under the regulation of an immunoglobulin heavy chain enhancer (*Igh* locus), which is constitutively active in differentiating B cell populations. This causes the continuous expression of the c-*myc* gene, which encodes a transcription factor important in cell proliferation. The transforming potential of such a construct has been verified in transgenic mice, which develop B cell tumors (Adams et al. 1985). However, the fact that all B cells are not tumorigenic in transgenic animals, although all B cells carry the rearrangement, indicates that this rearrangement predisposes to tumor formation but by itself may not be sufficient for tumorigenesis. Similarly, as many as 90 percent of follicular lymphomas manifest a 14;18 translocation, again involving the *Igh* locus but this time bringing the *bcl-2* (B-Cell Lymphoma gene 2) locus under its positive regulation (Tsujimoto et al. 1984; Cleary, Smith, and Sklar 1986). In addition, transgenics expressing a similar construct show an expansion of lymphoid populations and lymphoid resistance to cell death (McDonnell et al. 1989). Bcl-2 has since been characterized as a member of a complex family of cell survivor proteins that regulate apoptotic pathways. By itself, Bcl-2 is not strongly oncogenic, and tumors arising in *bcl-2* transgenic mice frequently exhibit rearrangements involving the *myc* gene as well (Strasser, Harris, and Cory 1993). Furthermore, *bcl-2* can cooperate with *myc* under the control of immunoglobulin enhancer elements to rapidly transform B cells in transgenic animals (Strasser et al. 1990).

5.9 Carcinogens and oncogene activation

Although certain proto-oncogenes, notably *ras* gene family members, are readily mutable with carcinogen exposure, do the activating mutations necessarily function in multistage carcinogenesis? Are oncogenes involved in the various stages of tumorigenesis (i.e., initiation, promotion, and progression)? In the rat, transforming mutations found in the *neu/erbB-2/HER2* oncogene from chemically induced schwannomas were detectable in rat Schwann cells within one week of exposure to the carcinogen EtNU (Nikitin et al. 1991). DMBA-induced skin tumors in mice have a high incidence of H-*ras* mutations in papillomas as well as in carcinomas, and these mutations are detectable early in tumorigenesis as well (Nelson et al. 1992). MeNU, which induces codon 12 mutations of H-*ras* in rat mammary tumors, may act as an initiator since mutations were also present prior to tumor formation (Kumar, Sukumar, and Barbacid 1990). However, in nitrosamine-induced rat kidney tumors, different regions of the same neoplasm were found to contain different K-*ras* mutations (Higinbotham et al. 1996), suggesting that activation occurs after neoplastic conversion in these tumors. In human studies, *ras* mutations arise with increasing frequency as colon tumors progress, but they are observed in adenomas at a significant frequency once the tumors exceed 1 cm, suggesting they represent an event in tumor progression and not initiation (Vogelstein et al. 1988). Closer examination of various stages in colorectal tumorigenesis by microdissection has revealed that the number of tumors bearing a K-*ras* mutation increases with progression, the number with multiple K-*ras* mutations (heterogeneity) also increases with stage, and a poor prognosis is associated with these mutations (Span et al. 1996; Al-Mulla et al. 1998). These observations are consistent with ras involvement, not in the initiation stage of colorectal tumorigenesis in humans, but rather subsequent to that. They also suggest that oncogenic K-*ras* may be a poor choice as a chemotherapeutic target. For human cutaneous malignant melanoma, only N-*ras* mutations occurred, but these could be detected in preneoplastic nevi as well and were maintained with neoplastic progression, suggesting a role in initiation (Demunter et al. 2001). However, the frequency of mutations did increase with progression, and codon 61 mutations, which are associated with uV-induced cyclobutane dimers, predominated in the later stages. In human lung, K-*ras* activation was not detected in preneoplastic lesions but occurred in a large proportion (32 percent) of adenocarcinomas (Sugio et al. 1994; Keohavong et al. 1996). Additionally, for human prostate tumors, intratumor heterogeneity has been reported for *ras* alterations (Konishi et al. 1995). Although oncogene activation is clearly associated with tumorigenesis in these studies, the exact stage at which it functions seems to be variable.

5.10 Oncogene cooperation

The ability of oncogenes to transform cells has been tested directly in cultured mammalian cell lines using transfectable precipitated preparations of genomic DNA or cloned genetic sequences in plasmids. Single activated oncogenes are generally nontransforming for most cells, with the notable exception of mouse NIH 3T3 cells, which are particularly sensitive to transfection with certain oncogenes such as activated *ras* or *neu* sequences. However, this immortalized cell line is insensitive to transformation by most oncogenes unless they are introduced in special DNA constructs that provide high levels of expression. In cell preparations from freshly dissociated tissues (primary cultures), individual oncogenes are invariably nontransforming. Instead, combinations of oncogenes are required for inducing transformation. For example, when *ras* and *myc* are simultaneously transfected into primary cultures of rat embryo fibroblasts, transformation occurs (Land, Parada, and Weinberg 1983). Individually, mutated *ras* is only weakly transforming, but *myc* complements *ras* either by immortalizing cells or markedly reducing cell senescence. Similarly, in reconstituted prostate cultures, cells containing activated *ras* oncogenes were dysplastic, whereas those containing *myc* oncogenes were hyperplastic; however, only cells containing both *ras* and *myc* were neoplastic (Thompson et al. 1990). Similarly, pre-B lymphoid cells expressing an *Igh-Myc* fusion gene were nontumorigenic unless simultaneously expressing v-*ras*, which facilitated malignant transformation (Alexander et al. 1989). This collaboration is believed to be due at least in part to the ability of Ras to stabilize and stimulate the accumulation of c-Myc (Sears et al. 1999). This is not sufficient, though, since constitutive overexpression of c-Myc will induce apoptosis, not neoplasia. Thus, Ras probably also provides a survival signal through the PI3K/Akt pathway described above (Kauffmann-Zeh et al. 1997). One might envision then an oncogenic mechanism whereby Myc promotes cell transit through the cell cycle, while Ras redirects a cell from its inexorable death march (Sears and Nevins 2002). The studies of tumorigenesis in transgenic mice bearing individual oncogenes are consistent with results of cell culture studies. It is the exception that an individual oncogene can cause rapid tumor formation when introduced globally in the mouse germline (Hunter 1991).

These molecular observations support the view that carcinogenesis requires multiple genetic alterations. Through studies of complementation groups of oncogenes, it has been found that growth control remains operative when members of the same grouping are simultaneously expressed, and members of different signaling pathways or at different levels of signaling are more likely to be transforming in combination. For example, an activated cytoplasmic tyrosine kinase may combine with a nuclear oncogene to transform a normal cultured cell (Ruley 1990). Certainly the combinations, however, must cooperate at some level, such as described for *ras* and *myc* for transformation to occur. Studies of signal transduction pathways are beginning to unravel the mysteries of these interactions through an understanding of how these complementary pathways interconnect to regulate or dysregulate proliferation and cell death and promote autonomous growth (Sears and Nevins 2002).

5.11   Normal cells suppress tumor growth

As to why individual oncogenes are incapable of causing transformation, may it be due in part to the fact that normal cells, including stromal populations that surround initiated cells or the extracellular matrix components that they elaborate, may suppress the outgrowth of the oncogene-transformed cell. This was first demonstrated using v-*src*–infected avian embryonic cells, which were nontumorigenic in embryos despite active expression of the oncogene but were morphologically transformed when dissociated and placed in culture (Stoker, Hatier, and Bissell 1990). It appears that the controls placed on the aberrant cells by adjacent normal cells can be abrogated by wounding or treatment with tumor promoters (Kenny and Bissell 2003). For example, TPA induces the selective outgrowth of cultured *ras*-transformed rat embryo cells, resulting in their clonal expansion and transformed focus formation in culture (Dotto, Parada, and Weinberg 1985). In v-*src*-infected chicks, simple treatment with TGF-β, a factor involved in wound healing, was sufficient to induce tumor formation (Sieweke et al. 1990). Again, these findings are consistent with the models of multistage carcinogenesis described in Chapter 3. One might speculate that cells containing an initiating mutation in an oncogene such as *ras* remain quiescent and regulated by adjacent cells until released by immortalization, clonal selection, or death of the surrounding cells. In any of these scenarios, more than one event would be required for tumorigenesis.

5.12   Angiogenesis and tumor development

One restriction placed upon tumor growth by the surrounding microenvironment involves availability of nutrients and oxygen. Normal cells reside in close proximity to a continuous bath of nutrients provided by the microvasculature; however, tumor growth eventually precludes cell access to this network unless the tumor cells become able to induce neovascularization. Without this ability, tumor growth is limited, generally yielding microscopic lesions, since tumor cells rapidly achieve a balance between proliferation and apoptosis. To expand beyond these limits, tumor cells shift the balance by recruiting endothelial cells from the vasculature and inducing them to proliferate through the elaboration of angiogenic factors. In addition, tumor disruption of the inhibitory matrix surrounding a tumor mass by secretion of proteolytic enzymes permits endothelial cell invasion of the tumor and causes the release of matrix-bound angiogenic factors. Several reported events can mediate the activation of the angiogenic switch in tumor cells. A low oxygen potential created by tumor growth is itself a signal due to transcriptional activation of the transcription factor Hypoxia-Inducible Factor-1α (HIF1α). The HIF1α protein is frequently overexpressed in a variety of tumors and has been shown to regulate the expression of more than sixty genes, including several angiogenic molecules such as vascular endothelial growth factor (VEGF), its receptor VEGFR2, TGF-β3, and leptin (Semenza 2003). Constitutively active oncogenes, such as the *ras* and *raf* family members, *src*, *bcl-2*, *HER2/neu*, and *fos*, also induce the expression of VEGF along with other cytokines (e.g., Fgf-2, Igf-2, Tgf-α, and Pdgf) and matrix-degrading proteases (Folkman 2003a). On the other hand, the loss of p53 suppressor activity in tumors can also contribute to angiogenesis by permitting HIF1-α and VEGF expression, which p53 normally suppresses, and eliminating expression of thrombospondin-1 (TSP-1), a secreted cell adhesion molecule and tumor suppressor. Thus, the environment as well as tumor cell alterations contribute to angiogenic conversion.

A series of antiangiogenic factors, such as endostatin and angiostatin, are produced by dormant tumor cells and modulate neovascularization and tumor growth (Tonini, Rossi, and Claudio 2003). It is the balance then between proangiogenic and antiangiogenic factors that regulates vascular development and apparently also tumor growth. For endostatin, the protein interacts with VEGF receptors to block signaling, while angiostatin interferes with matrix degradation to inhibit tumor cell invasion. Because angiogenic behavior is clearly a prerequisite for tumor growth and invasion, considerable effort has been focused on the development of antiangiogenic therapies and several are currently in clinical trials (Folkman 2003b). In vivo studies of tumor xenografts have demonstrated that systemic application of agents such as endostatin can significantly impact tumor growth (Blezinger et al. 1999; Kisker et al. 2001). Such studies lend credence to the future possibility that these tumor static agents may be useful either in preventing tumor formation in predisposed populations or in blocking their spread after they are detected.

5.13   Tumor Suppressor genes

Some of the earliest experiments using in vitro cell membrane fusion techniques involved formation of cell hybrids from tumor and normal cells in culture to establish

Table 5–2. *Cloned tumor suppressor genes with known functions*

Gene	Consequence of gene loss in humans	Function of encoded protein
APC	familial adenomatous polyposis colorectal tumors	interacts with β-catenin to target for degradation
BRCA1	breast & ovarian cancers	? DNA repair, transcription, cell cycle regulation?
INK4a/ARF	melanoma & lung cancer	inhibits CDK4 and Mdm2
MSH2	hereditary nonpolyposis colorectal cancer	mismatch repair processes
NF1	von Recklinghausen's disease neurofibromatosis type 1 schwannoma and glioma	GTPase-activating protein (GAP), which regulates signal transduction through $p21^{ras}$
NF2	neurofibromatosis type 2 acoustic nerve tumors & meningiomas	locus encodes merlin (Moesin/Ezrin/Radixin-Like protein), involved with cytoskeleton and Mdm2 inhibition
p53	Li-Fraumeni syndrome multiple tumor types	transcription factor pivotal in progression of cell cycle/surveillance factor for DNA damage
PTEN	glioblastoma/prostate & endometrial cancers	dual-specificity phosphatase that inhibits PI3K signaling
Rb	retinoblastoma & osteosarcoma	binds and sequesters transcription factor E2F to maintain cells in G_0 of cell cycle
VHL	von Hippel-Lindau syndrome & renal cell carcinoma	inhibits expression of HIF-1α transcriptional targets
Wt1	Wilms' tumor	transcription factor required for renal development

genetic dominance of the neoplastic phenotype. Invariably, the hybrids were less tumorigenic (Harris 1988), and reversion to malignancy occurred with the loss of specific chromosomes in the unstable hybrids (Jonasson, Povey, and Harris 1977). For some tumors, malignancy could be suppressed with the introduction of a normal chromosome to replace lost sequences (Tanaka et al. 1991). These studies suggested that transformation results from a loss of function required for normal cell regulation rather than the gain of function of a dominant transforming sequence, thus implicating tumor suppressor genes in transformation.

More than twenty suppressor genes have now been cloned and characterized (Table 5–2), and several more have been localized in the genome. These genes encode proteins that negatively regulate the growth of cells and function in a small number of signaling pathways that modulate apoptosis, the cell cycle, or cell differentiation. As described above, some tumor suppressors, such as PTEN or the INK4 family members, inhibit the same pathways that are activated by proto-oncogenes. Although studies of experimental models of chemical and viral carcinogenesis have delineated the majority of dominant transforming genes or oncogenes, genetic analysis of specific types of human tumors has led to the discovery of most recessive suppressor genes. Families predisposed to certain types of cancers often exhibit the loss or deletion of one copy of a specific chromosomal region that contains a suppressor gene. This loss occurs in the germline, allowing it to be passed to subsequent generations. Although the absence of one copy (heterozygosity) is often silent, tumorigenesis may occur with the loss of the second copy (homozygosity or loss of heterozygosity). Because the probability of knocking out a single copy of a suppressor sequence is considerably greater than knocking out two copies of the gene, the likelihood of tumor development in families with a germline mutation already present is likewise considerably greater than in the general population.

5.13a The *Rb* locus

Retinoblastoma is a highly malignant tumor of the eye that as a genetic disorder is characterized by a loss of heterozygosity in chromosome 13q14. The gene implicated, the *Rb* locus, is either missing or altered (often by point mutations) in nearly every retinoblastoma analyzed, whether familial (germline) or sporadic (somatic) (Friend et al. 1987; Yandell et al. 1989). In addition, osteosarcomas, (Friend et al. 1986) and small cell lung (Harbour et al. 1988), bladder (Horowitz et al. 1989), breast (Lee et al. 1988), and pancreatic (Ruggeri et al. 1992) carcinomas all have altered Rb proteins to varying extents. Because reintroduction of normal *Rb* sequences into retinoblastoma cells reduces tumor cell malignancy (Huang et al. 1988), alterations in *Rb* appear to be essential for tumorigenesis.

The Rb protein belongs to a family of three members and includes p107 and p130. All three function in negatively regulating cell entry into the cell cycle through interaction with members of the E2F family of transcription factors. The Rb protein can interact with more than 100 other proteins, including c-Abl, but its principal role appears to involve E2F. It has been estimated that Rb signaling is targeted in more than 80 percent of all sporadic human tumors (Sherr 1996), although this generally occurs not through mutations in Rb, as described above, but rather through alterations of other suppressors such as INK4 family members which regulate Rb activity (Sherr and McCormick 2002).

The *Rb* gene encodes a nuclear protein, 110 kD, which inhibits cell proliferation. When functioning normally, cells express Rb in a hypophosphorylated or active form at the G_0/G_1 phase (see Figure 5–6). It becomes inactive through the phosphorylation of multiple serine/threonine residues (Grafstrom et al. 1999) by an active CDK/cyclin complex, for example, CDK4/cyclin D1 and CDK2/cyclin E, as G_1 progresses and is reactivated by a phosphatase, possibly protein phosphatase 2A (Feschenko et al. 2002), at the conclusion of mitosis (Buchkovich, Duffy, and Harlow 1989). In the hypophosphorylated state, Rb can interact with several proteins. As mentioned, it binds to the transcription factor E2F and thereby represses expression from promoter/enhancers that contain an E2F-binding motif (Harbour and Dean 2000). In addition, several oncoproteins from viruses bind to Rb. The protein EBNA-5 from the Epstein-Barr virus (Szekely et al. 1993), E7 protein from human papilloma virus (Munger et al. 1989), adenovirus E1A protein (Whyte, Williamson, and Harlow 1989), and SV40 large T antigen (DeCaprio et al. 1988) are all capable of sequestering Rb and thus releasing E2F for transcriptional activation. Mutations in viral proteins that eliminate transforming activity are associated with a loss of ability to interact with Rb. Thus, viruses that can repress *Rb* function may be transforming as a result of this activity, because cells would be released from the G_0/G_1 block in the cell cycle. Conceivably, certain nuclear proto-oncogenes, when activated, might similarly overwhelm the ability of Rb to regulate the cell cycle due to excesses in cell cycle–promoting molecules or altered affinities of effectors for Rb interaction and suppression. This may explain the transforming ability of cyclin D overexpression in cells or activation of the nuclear tyrosine kinase c-Abl, which is normally inhibited when Rb binds to its catalytic domain (Weinberg 1995).

In addition to its role in regulation of the cell cycle, Rb also has been identified as an antiapoptotic factor, a seemingly odd function for a tumor suppressor but one which places it perhaps in cellular decisions of life or death (Chau and Wang 2003). Such a function was first described in genetic studies in which Rb null homozygotes were shown to experience excessive levels of apoptosis in the lens of the eye, nervous system, and skeletal musculature, albeit with an increase in pituitary tumors (Jacks et al. 1992). Normally, it is targeted for degradation by caspase 3 (Fattman et al. 2001), an activator of apoptosis which is also upregulated when E2F is released by Rb. Furthermore, in mice transgenic for an undegradable form of Rb, apoptosis-inducing challenges are significantly inhibited (Chau et al. 2002). Although the selective basis for this dual function is not known, one might speculate that Rb may provide a critical block in cell cycle progression and if lost, without a fail-safe function, would readily lead to unregulated growth. However, if its loss also directs the cell to apoptosis, this would provide another level of regulation counter to tumor development.

5.13b   *p53* suppressor gene

The most frequently mutated gene associated thus far with human tumors (>50 percent) is the *p53* suppressor gene (Olivier et al. 2002). The loss of function of this nuclear protein is associated with colorectal carcinoma and the Li-Fraumeni syndrome (LFS), a familial condition characterized by a high frequency of a diverse group of neoplasms. In families of children with rhabdomyosarcomas, family members also presented with a high incidence of breast and brain cancers, osteosarcomas, and leukemias. These individuals often lose one *p53* allele from chromosome 17 in the germline, and tumors sustain a point mutation in the remaining allele (Malkin et al. 1990; Malkin 1993). Similarly, colorectal tumors present with genetic alterations involving both copies of the *p53* gene, and the return of a wild-type copy to tumor cells represses its neoplastic behavior (Baker et al. 1990). Mice harboring a knockout deletion for p53 are grossly normal at birth but eventually develop a variety of tumors with a shortened latency (Donehower et al. 1992). Aggressive choroid plexus tumors with inactive p53 increase rapidly in size in mice, unlike tumors with an active normal form of p53 (Symonds et al. 1994). When these tumors were

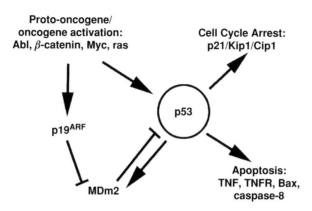

Figure 5–8. The p53 suppressor protein plays a pivotal role in the cell's decision to arrest growth or become apoptotic. Cell cycle arrest is mediated through the Cip/Kip family members and apoptosis through the tumor necrosis factors (TNF), receptors, or their downstream signaling molecules. Mdm2 negatively regulates p53 through ubiquitination, while p19ARF inhibits Mdm2/Hdm2 expression. These factors, whether oncogenes or tumor suppressors, have been implicated in a variety of human neoplasms.

evaluated for growth, they exhibited comparable rates of proliferation; however, the number of apoptotic cells in the slow-growing tumors was markedly higher; so p53-mediated apoptosis can suppress tumor growth. In addition, germline *p53* mutations can cooperate with *Rb* mutations to accelerate tumorigenesis in general and to induce tumor formation in tissues for which altered *p53* or *Rb* genes are incapable of inducing individually, for example, pinealoblastomas and pancreatic islet cell tumors (Williams et al. 1994). This may explain why mutations in both genes often occur in human tumors.

The p53 protein is a homotetrameric transcription factor which becomes active when a cell is stressed by hypoxia, irradiation, DNA damage, or oncogene activation (Vousden 2002). Once activated, p53 serves as a "checkpoint control" to direct cells either to cell cycle arrest in G1 phase for repair or to apoptosis if the damage is extensive (Figure 5–8). As a cell cycle inhibitor, its effects are associated principally with transcriptional activation of the tumor suppressor p21/Waf1/Cip1, a potent inhibitor of cell cycle progression. When expressed, p21/Waf1/Cip1 inhibits virtually all of the CDK/cyclin complexes. It also interferes with Proliferating Cell Nuclear Antigen (PCNA)–induced DNA replication, directly binds and inhibits E2F, and blocks Myc and STAT-3 activation (Roninson 2002). Thus, p53 activation has multiple indirect effects on cell replication and senescence.

The role of p53 seems pivotal in that it can be regulated at so many levels. In the proliferating cell, p53 levels are controlled primarily by protein degradation through a process called ubiquitination. This involves the covalent addition of a ubiquitin molecule, a 76–amino acid protein that facilitates protein target interaction with a proteasome for degradation. This process is common to the regulation of several oncogenic proteins, including β-catenin, c-fos, c-jun, and the cyclins. The ubiquitin ligase responsible for p53 degradation is the oncoprotein Murine Double Minute 2 (Mdm2 or Hdm2 in humans), which interestingly is also a target of p53 transcriptional regulation and, therefore, provides a feedback loop for the negative regulation of p53 activity. Mdm2 binding to p53 also inhibits p53-induced transcription, so it regulates p53 at more than one level. The importance of this relationship has been demonstrated in mice with a targeted deletion of Mdm2, which is embryonically lethal unless p53 activity is also lost, because unregulated p53 suppresses growth and induces apoptosis (Jones et al. 1995). The serine/threonine kinase Akt apparently works in concert with Mdm2 to stimulate p53 degradation. It does this by inhibiting p53 transcription (Yamaguchi et al. 2001) as well as by phosphorylating Mdm2 to promote its nuclear localization for p53 interaction (Mayo and Donner 2001).

Multiple signaling pathways converge to positively regulate p53 through Mdm2. The Alternative Reading Frame (ARF or p14^{ARF}/p19^{Arf}) protein interacts with Mdm2 to antagonize its inhibition of p53 (Sherr 2001). The ARF protein is generated from a sequence that overlaps the INK4a protein but in an alternate reading frame, as indicated by its name. As shown for p53, ARF-null mice are predisposed to neoplasia (Kamijo et al. 1999), suggesting it functions in a critical manner as a tumor suppressor in sustaining cellular p53 levels. Hypophosphorylated and active RB can also interact with Mdm2, preventing its binding to p53 and providing feedback for growth suppression from the cell cycle (Hsieh et al. 1999). Furthermore, unbound active E2F induces expression of the ARF protein to generate a growth inhibitory signal from an active cell cycle (DeGregori et al. 1997).

In addition to regulation by ubiquitination, p53 is also regulated by other posttranslational modifications, including phosphorylation in various domains within the protein. For example, N-terminal serine phosphorylation of p53 blocks the Mdm2-binding site and a nuclear export signal, causing nuclear retention and stabilization of p53 (Zhang and Xiong 2001). p53 activity is also regulated by transcriptional coactivator or corepressor proteins. These are proteins that bind with p53 to stimulate or repress gene expression from specific target binding motifs. The Apoptosis-Stimulating Protein for p53 (ASPP) family members are coactivators that interact with p53 to induce the transcription of apoptotic targets and cause cell death (Samuels-Lev et al. 2001).

As a key regulatory component of the cell cycle and apoptosis in response to cellular stress, it is not surprising then that p53 is frequently mutated in tumors. Even in human tumors with a normal *p53* gene, other factors that regulate its activity are invariably involved. For example, Hdm2 or ARF may be mutated (Sherr and McCormick 2002). If normal p53 activity can somehow be targeted and restored in tumor cells either through genetic manipulation or chemotherapy, it may be possible to redirect these cells to an apoptotic fate and eliminate a wide variety of tumors.

It has long been known that tumor cells are genetically unstable and this instability precedes tumorigenesis (Hartwell 1992). Accordingly (as discussed in Chapter 3), there has been considerable speculation regarding the existence of a mutator gene, which, once altered itself, could accelerate the rate of endogenous mutations in the genome and thus eventually mutate critical targets in growth control. DNA polymerases are likely candidates and have been observed to be mutated in some human colorectal tumors, which may account for the high mutation rate found in these tumors (Wang et al. 1992). The *p53* gene might also provide such a function. When altered, the cell loses the ability to growth-arrest damaged cells and thus repair DNA mutations prior to their fixation in the genome. The resulting accelerated accumulation of mutations in a variety of loci could readily expedite tumorigenesis in several tissues, as appears to be the case for Li-Fraumeni syndrome.

5.13c   Other tumor suppressors

In Table 5–2, several other suppressors are listed. While it is beyond the scope of this text to describe all of the tumor suppressors identified and their putative mechanisms of action, it is useful to at least briefly describe two more that have a demonstrated impact upon human tumorigenesis. Germline mutations in the Adenomatous Polyposis Coli (APC) suppressor, an inhibitor of the Wnt/β-catenin signaling pathway, were first implicated in a disease syndrome called familial adenomatous polyposis (FAP), which is characterized by multiple adenomatous polyps and colorectal neoplasms (Kinzler et al. 1991). Since then, somatic APC mutations have been detected in most sporadic colorectal cancers (Aoki et al. 1994). The APC protein forms a complex with axin and glycogen synthase kinase-3β (GSK3β) to phosphorylate β-catenin, marking it for ubiquitination and proteasomal degradation. Truncation mutations of APC prevent targeting of β-catenin, which then accumulates and functions constitutively as an oncogene to stimulate transcription of survival- and growth-promoting genes (Korinek et al. 1997). Furthermore, mouse genetic models clearly demonstrate the involvement of truncated forms of APC in the development of spontaneous and chemically induced colorectal tumors (reviewed in Perantoni and Rice 1999).

Another suppressor found to play a critical role in human tumorigenesis is the membrane receptor Patched (Ptch), which interacts with members of the hedgehog (Hh) family of secreted signaling ligands. Hedgehog signaling, especially from sonic hedgehog (Shh), regulates patterning, proliferation, and survival in many tissues during embryogenesis. In the absence of Hh, Ptch represses activation of the membrane coreceptor Smoothened (Smo). When an Hh ligand binds to the Ptch–Smo membrane complex, members of the Gli family of transcription factors are stabilized sufficiently for nuclear translocation and transcriptional activation (Ruiz i Altaba, Sanchez, and Dahmane 2002). In the familial autosomal dominantly inherited disorder called basal cell nevus syndrome, inactivating germline mutations in Ptch result in constitutive activation of the Smo–Gli pathway (Johnson et al. 1996). Furthermore, somatic mutations in Ptch have been detected in sporadic human basal cell carcinomas (BCC) and medulloblastomas (Goodrich et al.

1997). A genetic model to study this process has been devised using mice that are heterozygotic for Ptch. When exposed to ultraviolet (Aszterbaum et al. 1999) or ionizing radiation (Mancuso et al. 2004), animals deficient in one copy of Ptch are hypersensitive to the induction of BCC. Similarly constitutive overexpression of other components of the signaling pathway, for example, Smo (Xie et al. 1998) or Shh (Oro et al. 1997), causes BCC or BCC-like lesions, consistent with Ptch functioning as a tumor suppressor.

5.13d Apoptosis and its role in growth regulation

Apoptosis, or programmed cell death, is the negative regulatory mechanism whereby superfluous tissues are eliminated during differentiation, proper cell numbers are maintained in differentiated tissues, or irreparably damaged cells are eliminated. Simply stated, it is genetically determined cell suicide. Stem cell proliferation ensures continuous production of healthy functional cells, and apoptosis regulates the rate at which the cells die, so a balance between cell replacement and cell loss can be maintained. In the developing embryo, this process is crucial to tissue restructuring (e.g., digit formation in the developing limbs). In adults, it accompanies the dynamic process of normal tissue renewal as well as tissue degeneration, as in the postlactating breast. The process even functions in tumor cells, albeit at a reduced rate. As described earlier, it can provide a mechanism for removal of cells under severe stress, such as follows extensive radiation-induced DNA damage or growth factor or extracellular matrix deprivation (Frisch and Francis 1994). In tumor development, however, resistance to or inhibition of apoptotic signaling is an apparent prerequisite for the initial stages of the neoplastic process. After all, enhanced cell survival, which allows clonal selection in preneoplastic lesions, would have to precede malignant conversion.

There are two principal pathways that mediate apoptotic signaling (Figure 5–9): one involving external signals and the cell membrane (also called Activation-Induced Cell Death or AICD) and the other initiated by internal signals and changes in mitochondrial permeability (Hersey and Zhang 2003). Regardless of the preliminary signal, both are initiated and driven by cascades of the Cysteine Aspartyl Proteases (Caspases), which cleave protein targets after specific aspartic acid residues. For external signals, the cascade is initiated by caspases 8 and 10, while internal signals activate caspases 2 and 9.

External apoptotic signals are propagated through a superfamily of membrane-bound proteins called Tumor Necrosis Factor (TNF) receptors, for example, lymphocyte-expressed CD95/Fibroblast Associated (Fas) which is activated by a soluble or membrane-bound form of the Fas Ligand (FasL). The cytoplasmic sequence of the TNF receptors includes a Death Domain (DD), a 65–amino acid tail that recruits Fas Associated Death Domain (FADD) protein to the membrane. FADD is an adapter protein with a caspase-interactive Death Effector Domain (DED). FADD

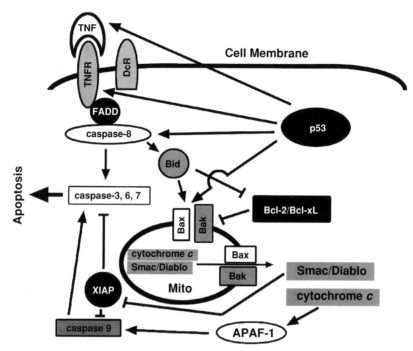

Figure 5–9. Two principal pathways mediate apoptosis. The first involves membrane signaling through the tumor necrosis families of ligands (TNF) and their receptors (TNFR) and is initiated by activation of caspase 8. The second pathway is mediated through the mitochondrion (Mito), which, with p53-induced heterodimerization of Bax and Bak, releases cytochrome c and Smac/Diablo to activate caspase 9. This pathway is negatively regulated by Bcl-2 and Inhibitor of Apoptosis protein (IAP). DcR – decoy receptor, FADD – Fas Associated Death Domain, APAF-1 – Apoptosis Activating Factor-1, XIAP – X-linked IAP.

therefore brings an initiator procaspase, such as procaspase 8, to the activated TNF receptor to form a Death-Inducing Signaling Complex (DISC). DISC formation facilitates the proteolytic activation of the initiator caspase, which in turn activates the effector caspases 3, 6, and 7 (Ashkenazi 2002).

Internal signals in the form of proapoptotic molecules emanate from the mito-chondria under the regulation of B cell lymphoma gene 2 family (Bcl-2) members, which were mentioned previously. The proto-oncogene *bcl-2* was discovered as part of a translocation event in B cell lymphomas (Cleary, Smith, and Sklar 1986), and it induces similar tumors in mice when overexpressed (McDonnell et al. 1989). Inter-estingly, it functions not by enhancing cell proliferation, but instead by prolonging stem cell survival, that is, inhibiting apoptosis and "immortalizing" the altered cells. Several additional homologous sequences have been identified, including Bcl-x_L and Mcl-1, which are antiapoptotic factors like Bcl-2, but others, such as Bad, Bid, and Bcl-x_S are proapoptotic and act as sensors of DNA damage (Zörnig et al. 2001).

These proapoptotic factors interact directly with Bcl-2 and Bcl-x_L to inhibit their activities. Two other proapoptotic family members Bax and Bak function through their recruitment to the outer mitochondrial membrane to alter the membrane potential and induce the release of other proapoptotic mediators. Perhaps the best understood pathway thus far involves TNF-mediated signaling. Caspase 8 activates Bid through proteolytic cleavage. The truncated Bid protein causes oligomerization of both Bax and Bak, which then incorporate into the mitochondrial membrane, allowing the release of apoptotic triggering factors, cytochrome c (Eskes et al. 2000) and Smac/Diablo (Du et al. 2000). Current speculation suggests that the membrane incorporations of Bax and Bak create pores in the mitochondrial membrane, facilitating the release of the apoptotic triggers (Hersey and Zhang 2003). Furthermore, Bcl-2 and Bcl-x_L may inhibit this by binding Bax or Bak directly to prevent their oligomerization (Mikhailov et al. 2001). The importance of Bax and Bak in the apoptotic mechanism has been demonstrated using Bax and Bak double-mutant mice, which are deficient in apoptotic responses to a number of stress stimuli (Zong et al. 2001).

Once cyctochrome *c* is released into the cytoplasm, it interacts with Apoptosis Activating Factor-1 (APAF-1). This complex triggers caspase 9 activation through a conformational change, and caspase 9 in turn activates caspases 3, 6, and 7 (Slee et al. 1999). Alternatively, apoptotic signaling can be triggered through Smac/DIABLO protein. This protein associates with one of the Inhibitor of Apoptosis Protein (IAP) family members, which block apoptosis by directly binding to several of the caspase family members. X-linked IAP (XIAP), for example, inhibits caspases 3,6,7, and 9, and Smac/DIABLO directly competes with caspase 9 for interaction with XIAP. Thus, Smac/DIABLO prevents XIAP from inhibition of caspase 9-mediated apoptosis (Hersey and Zhang 2003).

In general, the type of stress imposed upon a cell determines the apoptotic pathway that will be invoked in response to that stress. Thus, if growth factors are no longer available to promote cell survival, antiapoptotic factors, such as bcl-2, are lost and apoptosis is mediated through a caspase 9–dependent mechanism. On the other hand, if the stress is due to DNA damage, p53 activates a different series of proapoptotic factors, which vary depending upon the type of cellular damage incurred or the tissue involved. However, most of the documented p53 proapoptotic activities involve the upregulation of specific proapoptotic pathway players: Bax (Miyashita and Reed 1995), TNF receptors DR5 (Takimoto and el-Deiry 2000) and CD95 (Aragane et al. 1998), mitochondrial-localized apoptotic factor p53–regulated Apoptosis-Inducing Protein [p53AIP1] (Oda et al. 2000), pro-apoptotic Bcl-2 family members Noxa and Puma (Villunger et al. 2003), and caspase-6 (MacLachlan and el-Deiry 2002) or – 8 (Liedtke et al. 2003). In addition, p53 can traffic to the mitochondria for induction of apoptosis by a nontranscriptional mechanism (Marchenko, Zaika, and Moll 2000). These observations indicate, as expected, that mechanisms of p53-dependent apoptosis are intertwined with the established apoptotic pathways.

Since acquired resistance to apoptotic signaling appears to be a prerequisite for neoplasia, it is not surprising that mutations in genes encoding apoptotic proteins have been reported in human tumors. These include both gain-of-function mutations in oncogenes and loss-of-function mutations in tumor suppressors. For example, Bcl-2 is upregulated due to a chromosomal translocation in follicular B-cell lymphomas, and the Bcr-Abl fusion protein induces expression of Bcl-x_L, as mentioned above. For proapoptotic factors, frameshift loss-of-function mutations in Bax are found in some colorectal cancers (Rampino et al. 1997), and Bax loss of function in mice accelerates mammary tumorigenesis and reduces apoptosis in these tissues (Shibata et al. 1999). Similarly, the TNF family of death receptors behave as tumor suppressors. A soluble form of CD95/Fas has been reported in some lymphoma patients, and this competes with the membrane-bound form for its ligand to block apoptotic signaling. In gastric tumors, missense mutations were detected in the DD of CD95/Fas, causing its inactivation, and in lung and colorectal tumors and EB virus-associated lymphoma, a decoy receptor DcR3 is overexpressed (Zörnig et al. 2001). In this circumstance, the tumor produces a cell membrane–bound receptor form of CD95/Fas which retains the ligand-binding portion of the protein but lacks the cytoplasmic aspect. This effectively competes with the intact receptor, blocking apoptotic signaling. Mice with loss of function for CD95/Fas are not predisposed to tumor development unless linked with an oncogene such as *bcl-2* (Traver et al. 1998); however, the loss of function does induce a severe accumulation of lymphocytes. Other mutation sites in apoptotic signaling include the oncoprotein IAP family members (particularly survivin) and the suppressor Apaf proteins. The presence of mutations in various factors in apoptotic signaling suggests that these pathways may provide useful target sites for chemotherapeutic intervention, and indeed substantial efforts are currently directed to that possibility.

5.13e   Senescence

Among the challenges of early cell culture studies was the inability to propagate primary cells beyond a limited predictable replicative potential. Termed the 'Hayflick' phenomenon for its discoverer, normal nontransformed cells were observed to eventually become senescent, that is, quiescent nonreplicating populations, that would progress to an apoptotic crisis, the survivors of which were immortalized (Hayflick and Moorhead 1961). Since then, the phenomenon has been linked to the degradation of telomeres, lengthy regions of highly repetitive DNA at the ends of chromosomes that stabilize and protect the chromosome from genetic loss. With each cycle of DNA replication, the terminal 50–200 base pairs of a chromosome are lost due to the inefficiency of the DNA polymerases (Harley, Futcher, and Greider 1990). As a result, there is a limited number of replicative cycles that can occur before the 'molecular clock' runs out and cells are forced into senescence when important gene functions are lost. As the damage accumulates, chromosomes eventually develop end-to-end fusions, which directly precede cellular apoptosis (Hahn et al. 1999), thus eliminating 'aged' cells from the proliferative pool.

The cell is not without a repair mechanism to help manage accumulating damage to the telomere. In fact, an enzyme called telomerase was detected in tumor cell lines and found to help maintain constant telomere length despite repeated cell replication of immortalized cell lines (Kim et al. 1994). The gene that encodes the catalytic subunit of the telomerase holoenzyme – a reverse transcriptase - was subsequently cloned (Lingner et al. 1997) and used in ectopic expression studies to determine the effect of its activity on senescence, immortalization, and tumorigenesis. In many cases, cultured cells expressing telomerase showed a normal karyotype, elongated telomeres, and vigorous growth in excess of their normal culture lifespan (Bodnar et al. 1998). In other cases, expression was not sufficient for immortalization but required the loss of Rb/INK4 activities as well (Dickson et al. 2000). A link with tumorigenesis is not quite as clear since overexpression of telomerase is not required for cell transformation. In mice deficient for telomerase activity, transformation efficiencies did not decrease (Blasco et al. 1997), although evidence of diminished cellular lifespan was demonstrable. Furthermore, levels of *myc/ ras*-induced transformation of cells from wild-type or telomerase-negative animals were the same, showing that it may not be a prerequisite for cell transformation.

5.14 Where pathology meets molecular biology

As described in an earlier chapter, pathologists have long known that cells pass through a series of preneoplastic stages on the way to becoming tumors, and cytogeneticists have shown that specific tumor types contain consistent patterns of chromosomal abnormalities. With the identification and characterization of the dominant (oncogenes) and recessive (tumor suppressors) elements in tumors, we have reached the point where we can now link certain pathologic changes or chromosomal aberrations with specific molecular alterations. Through exhaustive evaluations of human colon tumors, Vogelstein's laboratory described a series of events associated with colon tumorigenesis (Fearon and Vogelstein 1990; Vogelstein and Kinzler 1993; Kinzler and Vogelstein 1996). Their now classical studies have shown that most colorectal tumors, whether familial or sporadic, suffer a loss of the suppressor sequence *apc* (adenomatous polyposis coli). As mentioned, this gene encodes a cytosolic protein with β-catenin-binding domains and inhibits signals from proliferation-inducing Wnt proteins. Furthermore, when wild-type Apc replaces a mutated form in colonic epithelial cells, the cells undergo apoptosis, suggesting that Apc controls a decision to initiate cell death. Mutations in Apc occur early in tumorigenesis and may be essential initiating events in tumor formation due to the nature of colonic differentiation. The stem cells are maintained in crypts of the colonic villi, and as they differentiate they move away from the crypt, eventually die, and are sloughed in the gut. For transformation to occur, an altered cell would have to be retained as a viable entity. Inhibition of apoptosis would provide an efficient mechanism for achieving this. Once "immortalized," time and probability would then determine neoplastic development in these initiated cells. A second common lesion in colorectal cancer involves a group of genes responsible for a type of DNA repair, namely mismatch repair (*mmr*). Mutations in these genes predispose cells to a high frequency of mutation in the genome in general and can result in the accelerated accumulation of mutations in a variety of genes, including those involved in signal transduction or cell cycle regulation. Thus, the frequently observed alterations in these *mmr* genes may increase the likelihood that the other commonly found mutations in colorectal cancer occur (i.e., K-*ras* activation in the larger adenomas and *p53* inactivation late in the neoplastic process). These studies demonstrate that no single lesion is generally transforming and that neoplasms probably result from an initiating event involving a block in apoptosis followed by a progression event that causes genomic instability such as occurs with mutations in *mmr* and generates the multitude of alterations attributed to malignant cells.

5.15 Summary

Since the first writing of this chapter in 1995, our understanding of the molecular events that regulate cell proliferation, differentiation, and apoptosis and how they are subverted in tumorigenesis has leaped forward. It is now clear that cells exist in homeostasis due to a critical balance between activation and inhibition of each of these cellular processes. Thus, when PI3K signals for cell proliferation, PTEN keeps it in check by removing the PI3K messenger. When proto-oncogenic proteins Myc or Bcl-2 signal for proliferation, they also induce so-called 'failsafe' pathways involving *p53* or *Ink4a* expression, which counter this signal (Schmitt 2003). For transformation to occur, homeostasis must be disrupted, and there are many documented examples of this presented in the chapter. With this understanding though, it is now possible to target pathways that play a critical role in this regulation, and current efforts are directed toward the development of inhibitors of proliferation-inducing proteins such as Akt or stimulants of apoptotic activity such as p53. These efforts are all the more important since it is known that tumor relapse following treatment with the standard arsenal of cytotoxic agents is often associated with cellular loss of apoptotic activities mediated by factors such as p53. We have now entered a phase of cancer research where the basic mechanisms of tumorigenesis are understood and successful targeted therapeutic interventions exemplified by agents such as Gleevec or Herceptin are now a matter of time. The mere identification of targets specific to tumors provides potential markers for detection in cancer diagnosis, where the mantra is "the earlier we can detect tumors, the more likely that intervention will be successful." Furthermore, armed with knowledge of the specific somatic sites, it may soon be possible to assess an individual's risk for a particular type of cancer through measurements of relevant mutations such as has been reported for the incidence of *p53* mutations in ultraviolet-damaged skin (Nakazawa et al. 1994). Through studies of individuals who develop specific types of cancers, it may be possible to delineate more germline lesions that predispose to cancer. In fact, it is conceivable that as we understand the interactions among the various signaling pathways and their relationship to tumorigenesis, we may find that every individual who develops cancer actually belongs to a predisposing group based on some germline mutation. Such a circumstance could lead to early identification of these lesions in predisposed individuals and even to the development of therapies for tumor prevention once we learn to manipulate cellular signaling pathways effectively. It is indeed remarkable that we can now envision such possibilities.

6

Cancer in nonhuman organisms

Robert G. McKinnell

The archived specimens (in the Registry of Tumors in Lower Animals) provide historical data to establish epidemiological associations, geographic ranges and prevalence trends. It provides reference material for study by students and research scholars and provides residual material for retroactive genetics studies as new tools in molecular genetics are developed. Further, whole tumor displays in the Registry impart a convincing reality of the disabling effect of this disease on endangered species to visiting reporters, government representatives, scientists and lay public. If one picture is equal to a thousand words, an actual specimen is worth at least a hundred pictures.

J. C. Harshbarger 2002

Workers who deal with cancer problems only in man and his closest relatives among the homeothermic vertebrates may look disdainfully at those who invest some of their energies in studies of neoplasms in creatures such as fish, frogs, snakes, and the numerous species of spineless, often slimy animals that were, in the days of the ancients, lumped together as "vermin." After all, Alexander Pope, a poet of some wisdom, admonished: "The proper study of man is man."

So it may be. But Aristotle, long before modern ecologists, recognized that man is an organism within a greater organism, the earth ecosystem. The ecosystem is not dependent on man, though man is dependent on the ecosystem. Ought not the study of man be extended to include the other animals that coexist with and support him on this motherly planet? They share with man many diseases, including cancers, and it is pertinent to ask: In what ways do neoplasms of animals at various phyletic levels differ from or resemble those of man in relation to etiology, natural history, immune factors, biochemistry, morphology, molecular biology, and the rest? This is the basic question to which comparative oncology addresses itself.... [A]s if in reply to Pope, Kipling asked: "What do they know of England who only England know?"

C. J. Dawe 1969

6.1 Introduction

Phyletic aspects of cancer (Dawe et al. 1981) are discussed in this chapter. This means that descriptions of the distribution, prevalence and etiology of malignant tumors in organisms other than humans are considered. Why? Are some biologists spending scarce funds in an endeavor to cure cancer in amphibians and reptiles (and other "lower" beasts)? Of course not! Research for the prevention and cure of human cancer has a higher priority than comparable studies of lower organisms and this is entirely proper. However, there is profound need for the study of cancer in creatures other than humans and the eloquent prose of John C. Harshbarger and the late Clyde Dawe (see epigraphs) relates to that need.

A rationale that may not appeal to practical-minded students is that there is a certain intellectual fulfillment in understanding natural phenomena. Which plants have tumors and why? What animals are commonly afflicted with cancer? Learning about phyletic aspects of cancer is an end in itself for some scholars.

There are, however, several very practical reasons for seeking knowledge concerning cancer of plants and animals. For instance, if fish tumors are historically scarce in a particular area but those tumors are now being found in increasing abundance, one should seek the cause of the increased rate of neoplasia. Fish live in the water that drains our agricultural and industrial lands, and afflicted fish, as sentinel organisms, may serve as an early warning signal that the level of carcinogens is increasing in that environment. Humans share that environment. The reduction in cancer prevalence in natural populations, when it occurs, comes as a welcome signal that cleanup of a particular site or region has been successful.

A significant motive for the scrutiny of lower organisms is that their cancers provide information directly useful in understanding human biology. For example, the renal adenocarcinoma of the Northern Leopard Frog, *Rana pipiens*, is caused by a herpesvirus. This frog cancer was the first malignancy associated with a herpesvirus (see discussion in this chapter and Viral Carcinogenesis in Chapter 3). More recently, other animal cancers and human malignancies have been associated with herpesviruses. Whether or not the herpesvirus association in the human cancers proves to be etiologic, certainly the existence of a frog cancer caused by a herpesvirus (and the many other animal cancers caused by viruses) adds credence to that possibility. Moreover, experiments that would be inappropriate in humans relating to viral etiology and cancer treatment can be carried out on nonhuman creatures. Certainly tumors of lower animals have added immensely to the understanding of cancer in humans.

Another basis for the phyletic approach to cancer relates to the finding of organisms that are relatively resistant to cancer. For example, crayfish were thought *not* to get tumors (Dawe 1964). Similarly, sharks were once thought to be cancer free. That notion is now untenable. Sharks and related cartilagenous fish (chondrichthyes, elasmobranchs) are indeed afflicted with neoplasms (Harshbarger and Ostrander 2000) much as all other organisms are thought to be. However, even if crayfish and sharks do have cancer, it may be that their tumors occur at a low rate. Why? Do these animals live in an environment with little or no pollution? Or, do they have a peculiar biology that protects many of them from cancer-causing agents? If the latter is the case, what is that biology and how does it differ from the biology of organisms with greater susceptibility to cancer? It is possible that we could learn much from animals such as crayfish and sharks concerning protection of humans from cancer.

Yet another rationale for phyletic studies relates to cancer prevention and treatment. Experimental organisms permit novel and unusual approaches to this aspect of cancer research. Consider that an economically important plant cancer can be prevented with an antibiotic substance already in commercial use. Cancer vaccines have kept the price of chicken well within affordable limits for many years. With existing technology, a cat can be protected against a particularly virulent and deadly form of leukemia. These are not pipe dreams and the plant cancer, the chicken neoplasia, and the cat leukemia are discussed later in this chapter. Perhaps success stories such as these encouraged workers seeking to prevent cervical cancer with a vaccine against human papillomavirus 16 (HPV-16) (Koutsky et al. 2002, Brown et al. 2004), an achievement that may be the "beginning of the end" for cervical cancer (Crum 2002).

Finally, plants and lower animals have been the source of substances that may be useful in the treatment of human cancer. It would be naïve to believe that more naturally occurring drugs cannot be found. Pharmaceuticals derived from nature range from drugs under study (examples: bryostatin 1 derived from the marine bryozoan *Bugula neritina*, didemnin B obtained from the marine tunicate *Trididenmun solidum*, Neovastat AE 941 derived from shark cartilage, and discodermolide from the Caribbean sponge *Discodermia dissoluta*) to well-established derived-from-nature chemotherapeutic agents (examples: the vinca alkaloids derived from *Catharanthus roseus* the plant with pretty pink flowers and paclitaxel derived from the bark of the Pacific yew, *Taxus brevifolia*) (Schwartsmann et al. 2001; Dupont et al. 2002; Honore et al. 2004). The toxic side effects of human-made chemotherapeutic agents are well known (Chapter 8). Some readers may hope that pharmaceuticals derived from nature will be less toxic than chemicals designed by humans. The author of this chapter wishes to disabuse anyone who believes there is less toxicity in drugs from nature – for such is not the case.

It is hoped that this somewhat inordinately lengthy introduction will prepare the student for a perusal of plants and animals of interest to cancer researchers. Let no one minimize the usefulness of nonhuman organisms in the study of cancer in humans.

6.2 Plant growths

Insect galls, which occur in plants in a diversity of forms, are caused by fungi, bacteria, and round worms (nematodes) but the most common cause of galls are insects and most galls occur in oak trees. While insect galls are abnormal growths resulting from cell proliferation with a tumor-like appearance, they are not true tumors. Malignancy in humans and other animals is characterized by inappropriate mitotic activity (growth), plus anaplasia, invasion, and metastasis (Chapters 1 and 2). Insect galls form characteristic lumps that permit identification and classification and have a well-circumscribed growth potential. Their abnormal growth follows specific patterns for specific galls. Unlike the cells of animal tumors however, normal growth and differentiation returns to the cells of insect galls upon removal of the causative insect or its secretions. True tumors in animals, with very few exceptions, do not revert to normal with the removal of the etiological agent. As ancient tumors have been studied in fossil animals and humans (Chapter 7), so too paleobotanists studied galls of fossil plants. Recently, evidence has been presented that insects were galling tree-fern fronds during late Pennsylvanian times (320 million years ago) (Labandeira and Phillips 1996; see also Labandeira, Johnson, and Wilf 2002) indicating the immense antiquity of these plant growths.

In contrast to insect galls, which impress some people as tumors but which are not, some plant growths are true tumors. "True" tumors here refer to plant growth with a persistently altered condition known as autonomy. The late Armin Braun of The Rockefeller University, New York City, described autonomy in plant tumors as resulting in unrestrained growth. Further, autonomy permits the transplantability of plant tumors. Autonomy ensues from a "transition from a normal cell to a tumor cell (with) a profound and heritable change... which allows a tumor cell to determine its own activities largely irrespective of the laws that govern so precisely the growth of all normal cells of an organism. This newly acquired property is the most important single characteristic of tumor cells since without it there would be no tumors (Braun 1974)." The Rockefeller professor went on to state that true tumors in plants are transplantable into appropriate hosts of the same species and continue with their growth identical to the tumor from which they were derived. Transplantability itself is derived from the autonomy of the plant cells. Hence, if one uses autonomy as a characteristic of the tumorous state, then clearly crown galls are "true" tumors. This statement is necessary because in most animal and human cancer, metastasis occurs. Plants, being deprived of a vasculature, have no blood-borne metastasis. Lacking this fundamental aspect of cancer, some may say that true tumors cannot occur in plants. However, the author of this chapter sides with Armin Braun with the notion that plants do indeed have true tumors and that a consideration of them is useful for an understanding of tumor biology.[1]

---

[1] Although neoplasms are discussed in this section, the literature of plant cancer is brief. Plants are remarkably resistant to neoplastic change despite their exposure to massive doses of ultraviolet radiation, which is known to cause DNA damage and cancer in animals and humans (see discussion of skin cancer in Chapter 7; also Doonan and Hunt 1996).

Perhaps the most studied of plant tumors is the crown gall, which afflicts a diversity of dicotyledonous plants with global distribution. Some of the plants vulnerable to crown gall include apple, almond, walnut and pecan trees, various stone fruits, roses, grapevines, raspberry and blackberry bushes, tobacco, chrysanthemums, and willows. Crown galls presents as large cauliflower-like swellings, usually where lateral roots branch from the underground stem but also on stems and leaves. The crown gall was originally thought to be caused by a gram-negative soil bacterium, *Agrobacterium tumefaciens* (Smith and Townsend 1907), but later it was shown that it was not the bacterium *per se* that caused the tumor; the disease was found to be mediated by the T_i (tumor-inducing) plasmid, a closed circle of nonchromosomal DNA. A small portion of the plasmid DNA, known as T-DNA (transferred DNA) becomes stably integrated (inserted) into the plant cell genome (Binns 2002). The interkingdom (so-called because of the insertion of bacterial DNA into plant DNA) inserted T-DNA contains genes that activate the host plant's synthetic activities, resulting in the formation of the crown gall. Oncogenomics seeks to understand cancer through genome research – hence, it is no surprise to learn of the sequencing of the tumor-inducing plasmid (Suzuki et al. 2000) and the *A. tumefaciens* genome (Wood et al. 2001).

Crown galls of tobacco grow well in vitro as chaotic arrangements of neoplastic tissues and organs. Abnormal shoots and leaves may grow from the neoplastic mass. In an elegant series of experiments, the late Armin Braun showed that this abnormal tissue could give rise to normally differentiated plants that flowered and set seed. His procedure involved serial grafting of neoplastic tissue to normal hosts (Braun 1981). The grafted tumor-derived cells ultimately gave rise to normal plant tissue providing unequivocal evidence that neoplastically transformed tobacco cells undergo no loss or irreversible change of their genome that would preclude the formation of normal plant tissues.

At the time of Braun's experiments, it was generally believed that cancer cells could give rise only to more cancer cells. Thus, when Braun reported to a scientific audience that his plant cancer cells gave rise to normal mitotic progeny, a Nobel laureate in the audience remarked that although the experiments on plant cell differentiation were interesting, they could have no relevance to cancer. The Nobel laureate stated a longheld dictum: cancer cells can give rise only to more cancer cells. Inasmuch as Braun's cancer cells produced normal cell progeny, it was thought that in no way could the plant cells be malignant. But because of Braun and others, the dictum "cancer cells can give rise only to more cancer cells" is no longer tenable.

Malignant transformation persists after removal of the carcinogenic agent (Chapters 1 and 3). This principle distinguishes temporary perturbations of cell structure and activity from true malignancy. However, the crown gall studies of Braun provided indisputable evidence that the phenotypic expression of a malignant cell may be altered. In this instance, there is a complete remission of the malignant state resulting in normal mitotic progeny. Thus, the principle enunciated earlier should be modified by insertion of the word "ordinarily" in the phrase to read that "malignancy *ordinarily* persists even after the removal of the inducing agent." Other examples of induced differentiation of cancer cells, including the Lucké renal adenocarcinoma of frogs, are discussed below.

Botanists who study the crown gall have contributed to another important area of cancer research. Knowledge of etiology frequently helps in devising rational treatment. Such is true for the crown gall. As indicated earlier, crown galls ensue after the integration of T-DNA into the recipient cell. Thus far, chemical control of the disease seems ineffective. However, it was reported that *Agrobacterium radiobacter* strain K84 carries a plasmid designated pAgK84 (Kerr 1980), which codes for the production of an antibiotic designated Agrosin 84. That antibiotic, a substituted adenine nucleotide, breaks down the opine nopaline necessary for growth of the oncogenic *A. tumefaciens* bacterium. Agrocin 84 then appears to interfere with DNA synthesis of the *A. tumefaciens,* resulting in lethality to the pathogen. Agrocin 84 is used commercially for protection of plants vulnerable to crown gall (Pesenti-Barili et al. 1991). Although Agrocin 84 protects plants from infection and subsequent malignant transformation, it does not cure already existing crown gall.

At the risk of seeming overly enthusiastic about developments in plant pathology, we should stand in awe considering the results reported here. In the case of crown gall, a disease with similarities to cancer in animals, the cause of the neoplasm, an infection with a bacterium, is known. A portion of the bacterial plasmid integrates with the host cell genome. The DNA of the noxious bacterium and its tumor-inducing plasmid has been sequenced, malignant cells can be induced to give rise to normal mitotic progeny, and a means of preventing the tumor is already in hand. Biomedical scientists who work with animals and humans should be appropriately impressed.

6.3   Invertebrate animals

As late as 1930 it was asserted by some workers that invertebrates did not, or even for alleged reasons could not, develop tumours; and even today it is sometimes stated that invertebrate tumours are not "really" neoplasms, or are irrelevant to the study of vertebrate cancer. Such statements appear quite ill-founded.

J. Huxley 1958

Are the tumors of invertebrates "true" cancers? That same question was asked of the plant neoplasms already discussed. While differences doubtlessly exist, it is likely that genetic and molecular mechanisms that control normal and neoplastic growth are highly conserved in all eukaryotic organisms. It follows from this that tumors of invertebrates, properly diagnosed and characterized, may well be a lode awaiting mining by comparative oncologists.

Neoplasms, or what appear to be neoplasms, have been reported to occur in mollusks, curious marine worms known as sipunculids, arthropods, ascidians (which are primitive sessile chordates with a free-living tadpole stage), and annelids (Dawe 1969; Dawe et al. 1981) and probably afflict all metazoans.

Drosophila melanogaster. The fruit fly, *Drosophila melanogaster,* as is universally recognized, has been the "most important model organism for genetic studies since early in the twentieth century (Snustad and Simmons 2003)." Because of the genetic component in most, if not all cancer (Chapter 4), studies of tumors in the fruit fly are of obvious importance. Melanotic "tumors" in fruit flies, known almost since the beginning of *Drosophila* genetics (see pages 150 and 151 of Bridges 1916; Stark 1918), should not be considered to be true malignancies because of their paucity (or absence) of mitosis, their failure to manifest invasion, and their encapsulation by blood cells, coupled with their destruction at metamorphosis. Melanotic tumors are probably analogous to granulomas, or pseudotumors, or simply nonmalignant aggregations of cells (Gateff 1981); they are cited here to indicate a long history of fly-workers' interest in neoplasms.

However, cancers do exist in *Drosophila.* Because of the "high level of gene and pathway conservation, the similarity of cellular processes and the emerging evidence of functional conservation of tumor suppressors between *Drosophila* and mammals," it is clear that studies of tumorigenesis in flies can directly contribute to the understanding of human cancer (Potter, Turenchalk, and Xu 2000).

Malignant neuroblastomas can be recognized in sixteen-hour fly embryos. In these cases, definitive differentiation of adult optic neuroblasts fails to occur, and the dividing neuroblasts invade and destroy normal portions of the brain. Six independent mutations are known that arrest differentiation of the neuroblasts and give rise to these malignant neuroblastomas. Further, imaginal disc (larval cells that will participate in adult development) tumors of *Drosophila* have been described which have lost the capacity to differentiate and which grow in an autonomous and lethal manner. Mutations to specific tumor suppressor genes result in the imaginal disc tumors (Kurzik-Dumke et al. 1992). Transplanted tumors of *Drosophila* reveal their similarities to malignancies of humans and other mammals by their growth potential, invasiveness, and metastastic competence (Woodhouse, Hersperger, and Shearn 1998). Not unlike tumors of higher organisms (Chapter 2), the *Drosophila* tumors elaborate a proteolytic enzyme that is thought to be a Type IV collagenase (Woodhouse et al. 1994). These studies provide hope that the "correlation between the primary genetic change and the malignant neoplastic transformation of cells" may be identified and characterized in that organism which is best known genetically (Gateff 1981). That hope is furthered by the sequencing of most of the euchromatic portion of the fruit fly genome (Adams et al. 2000).

Blood vessels are found in all vertebrates. Malignant cells are moved from a primary tumor to distant sites via vascular dissemination. Intravasation, movement in the vasculature, and extravasation by cancer cells is a major component of metastasis (Chapter 2). Flies do *not* have arteries, veins, or capillaries. Thus, this aspect of metastasis cannot be studied in *Drosophila*.

Mya arenaria. Gonadal tumors (Van Beneden, Rhodes, and Gardner 1998) and leukemia (Kelly et al. 2001) have been studied in Maine soft-shell clams, *Mya arenaria.* Leukemia-like disseminated neoplasia is known to occur in certain populations of the soft-shelled clams at an astonishing 76 percent frequency (Dungan et al. 2002).

Masui's Study of Oocyte Maturation in the Northern Leopard Frog, *Rana pipiens*, Demonstrated Cell Division Controls that are now Recognized as Universal Among Eukaryotes.

Cell division is an essential aspect of life and therefore one would predict that its molecular pathways would be evolutionarily conserved. Indeed, the pioneering work of an extraordinarily gifted scientist, Yoshio Masui, permitted molecular biologists to decipher the molecular pathways of cell division in eukaryotes. Professor Masui identified the cytoplasmic components controlling oocyte maturation in the Northern Leopard Frog, *Rana pipiens*. His studies, though performed on *normal* oocytes in a frog, provided significant insight into the controls of cell division in all eukaryote creatures – in both normal *and* malignant cells.

Frog oocytes have a long growth period. When the primary oocytes are fully grown, they arrest at a stage containing germinal vesicles (egg nuclei) that are meiotically immature. At ovulation, meiotic arrest is broken and maturation occurs with the breakdown of germinal vesicles and the release of first polar bodies. Polar body release is, of course, a form of cell division.

Dr. Masui wanted to know what controls these events in oocyte maturation. First he treated oocytes from mature *R. pipiens* with progesterone which induced maturation. Then he injected cytoplasm from the progesterone-treated oocytes into fully grown but immature (meiotic-blocked) oocytes. Those immature oocytes underwent meiotic *maturation*, the release of the first polar body, and development to metaphase of the second meiotic division. The cytoplasmic substance that induced this complex response was termed "maturation promoting factor" (MPF) (Masui and Markert 1971).

What is truly impressive in the context of evolutionary conserved mechanisms is that MPF *controls not only meiotic maturation but cell division in all eukaryotic organisms* from yeast through frogs up to and including humans. It was later shown by others that Masui's MPF is a molecular complex of two proteins; cyclin-dependent kinase (CDK) and cyclins. CDKs regulate other proteins by phosphorylation and are inactive without the presence of cyclins. The mechanisms that drive cell division are complex indeed and when they were deciphered in yeast and in frogs, it soon became apparent that cell division controls were essentially identical in all eukaryotic organisms. Cell division in neoplastic cells is inappropriately regulated and many workers are seeking insight into cell division controls as a means of developing new generations of cancer drugs. A study in an ectothermic vertebrate, the frog, elucidated cell division controls, an understanding of critical importance to cancer.

Yoshio Masui was born in Kyoto, Japan and he received his Ph.D. degree from Kyoto University in 1961. After an academic appointment at Yale University, Dr. Masui moved to the University of Toronto where he is now a Professor Emeritus. In 1998, Dr. Masui received the Lasker Basic Medical Research Award for pioneering genetic and molecular studies that revealed the universal machinery for regulating cell division in all eukaryotic organisms from yeast to frogs to humans. Dr. L. H. Hartwell and Dr. P. Nurse were co-winners of the award. In 2003, the Governor General of Canada appointed Professor Masui "Officer of the Order of Canada," Canada's highest honor for lifetime achievement.

6.4 Cancer in selected ectothermic (cold-blooded) vertebrates

Examples of abnormal growth have been encountered in many taxonomic groups of animals and at every level of biological complexity. However, except for the mammals, the study of neoplasia, especially in the lower vertebrates, is to a large extent descriptive, haphazard, and incomplete.

D. G. Scarpelli 1975

Although significant progress in the study of neoplasms in poikilotherms has been made since the comment of Scarpelli of many years ago, it is still true that the oncology of lower vertebrates is, to some extent, "descriptive, haphazard, and incomplete." The deficiency in knowledge is changing rapidly as can be seen in the following section.

6.4a   Fish

There are about 20,000 species of fish that vary enormously in size, habitat, and life expectancy. It should come as no surprise, therefore, that much is known about tumors of these creatures (Masahito, Ishikawa, and Sugano 1988). Fish are of greater economic importance than any other group of ectothermic (poikilothermic) vertebrates. Fishes harvested for commercial purposes have provided neoplasms which include all of the major histologic types recorded for mammals, viz., epithelial, mesenchymal, pigment, and nerve cell cancers. The Registry of Tumors of Lower Animals, formerly located at George Washington University in Washington, D.C., reported that almost three-fourths of the tumors of chordates in its collections were neoplasms of fish (Harshbarger, Charles, and Spero 1981). Most fish neoplasms are found in teleosts (bony fishes).

Do tumors occur in fossil fish?

"The study of fossil fish for neoplastic development is a field lying completely fallow"
                                                                                          Masahito et al. 1988

Fossil galls (plant tumor-like growths) were discussed earlier. What about fossil fish? Fossil fish are more abundant and better known than fossils of any other vertebrate class. Yet no scientific record of tumors appears in these fossils. Fortunately, contemporary fish are found that are similar to the fossils. These contemporary animals have changed little during their immense evolutionary history and are thus sometimes referred to as "living fossils" "Living fossil" examples from the Dipnoi (Dipnoi are primitive fish with a persistent notocord, a largely cartilagenous skeleton, and lungs as well as gills) include the tropical freshwater African lungfish, *Protopterus amphibius* and the South American lungfish *Lepidosiran paradoxa*. Dipnoi appeared during the "age of fishes," also known as the Lower Devonian, about 400 million years ago. In contrast to their fossil ancestors, which appear to be tumor-free, contemporary "fossil" fish are afflicted with tumors and they may occur at a relatively high prevalence (Masahito et al. 1988). An example: An autopsy series of fourteen lungfish revealed three cases of hepatocellular carcinoma, one of which had multiple metastatic colonies (Masahito et al. 1986). A spontaneous spermatocytic seminoma (Masahito, Ishikawa, and Takayama 1984a) and a huge spontaneous neurinoma (Ishikawa et al. 1986) have also been described in lungfish. Gar are ganoid fishes of North and Central America of the genus *Lepisosteus* and they are also considered to be representative of an ancient group of fish. Epitheliomas have been described in two species of gar (Takayama et al. 1981). One may ask whether the environment of contemporary fish has deteriorated with resultant induction of neoplasms by noxious substances, or is it that we simply have not studied (or cannot study) fossil fish adequately? While that question remains unanswered, it is clear that contemporary "fossil fish" may become afflicted with tumors. Clearly, this intriguing field of investigation is *not* completely fallow due to the enlightened efforts of Masahito and his colleagues.

Environmental studies. Many studies associate selected malignancies with noxious chemicals in the environment. While an association cannot establish a cause and effect relationship, the putative cancer-causing chemical becomes more plausible as an etiological agent if changes in both the chemical and the malignancy are correlated. Such is the case in a study of hepatic cancer in wild-caught brown bullhead catfish, *Ameiurus nebulosus*. A coking facility located in Lorain County, Ohio was closed and subsequently, polycyclic aromatic hydrocarbons (PAHs) in the environment plummeted. So, too, did liver cancer in the catfish plummet, suggesting a causal relationship (Baumann and Harshbarger 1995).

Another mode of seeking insight into the effects of noxious chemicals on cancer is to examine one vulnerable species at several locations with diverse levels of contamination. Liver and skin tumor prevalence in *A. nebulosus* collected at several sites in the tidal Potomac River watershed, near Washington, D.C., was examined. The sites were contaminated with polychlorinated biphenyls (PCBs), PAHs, and organochlorine pesticides. Significant differences were observed in liver tumor prevalence (a selected example: fish from the Anacostia river site had a liver tumor frequency in the autumn of 60 percent versus the frequency of liver tumors in fish from the Tuckahoe River of 10 percent). The Tuckahoe River site, not a part of the Potomac River watershed, was chosen because it drains a forested and agricultural area on the east shore of Chesapeake Bay and thus would serve as a control for the contaminated sites on the Potomac River (Pinkney et al. 2001).

Sometimes the environment that leads to new insights is a man-made environment. Such is the case of cancer in the rainbow trout (*Salmo gairdnerii*). It has been known for many years that rainbow trout may be found in nature with liver cancer (Haddow and Blake 1933). Epizootic outbreaks of liver cancer in rainbow trout occurred in 1937 (Wales and Sinnhuber 1966) and a number of subsequent years. The abundant neoplasms were traced to fish food rations. Eventually it was shown that the noxious ingredients in the rations were aflatoxins derived from the mold *Aspergillus flavus* (see discussion of these potent carcinogens in Chapter 3 as well as in the Appendix under "liver cell carcinoma"). We should be appreciative of rainbow trout, which with commercially grown turkeys (Chapter 3), made it evident that aflatoxins were powerful liver carcinogens.

Germ cell tumors in the brain. Malignant cells have the competence to translocate, that is, metastasize, to sites in the host that are anatomically distant from the primary tumor. In fact, that competence has been considered to be absolutely decisive in categorizing a tumor as malignant. The multitude of steps involved in the translocation of cancer cells has been considered previously (Chapter 2). Biologists in general, and embryologists in particular, are well aware that translocation is not a property limited to malignant cells. The migration of normal primordial germ cells (PGCs) is perhaps the classic example among the many kinds of migrating embryonic cells. PGCs have a nongonadal origin and achieve their definitive differentiation after arrival at the anlage (primordium) of the gonad. What happens if migrating PGCs get "off track" and become lodged in a nongonadal site? At least a partial answer to this question is suggested by a study of three germinomas that were found in wild-caught lake whitefish (*Coregonus clupeaformis*).

Germ cell tumors in fish are relatively rare to begin with (Masahito et al. 1984b). Add to their being uncommon the peculiar location of a germ cell tumor in the brain and indeed, an interesting story of fish neoplasia emerges. Germinomas were found at the ventral surface of the brain near the pituitary in the lake whitefish. The pituitary is partially encased in the sella turcica and the germinomas were just above the sella; hence, the designation of these neoplasms as "suprasellar germinomas." It was thought unlikely that the germ cell tumors were metastatic from the host gonads because the gonads were entirely free of malignancy. Yet the histology of the germinomas was identical with germinomas of the male and female gonads; that is, testicular seminomas and ovarian dysgerminomas. The authors concluded that *normal* migrating cells became misplaced and ultimately expressed the malignant phenotype of germ cell tumors (Mikaelian et al. 2000).

Cancer in Sharks and Other Elasmobranchs. Shark cancer was mentioned briefly in the introduction to this chapter. While it was formerly thought that elasmobranchs in general, and sharks in particular, are not afflicted with cancer, it is now known that sharks do indeed become afflicted with cancer. Over forty cases of tumors in elasmobranchs (sharks, skates, rays, and ratfish) were reported, about half of which involved sharks. These tumors included squamous cell carcinoma, fibrosarcoma, melanoma, renal adenocarcinoma and chondroma. All organ systems were represented except for respiratory. This report should set to rest the notion that sharks and other elasmobranch fish are tumor-free (Harshbarger and Ostrander 2000; see also Borucinska et al. 2004).

Pigment cell tumors. Pigment cell tumors occur more commonly in fish than in mammals, including humans. These include tumors of erythrophores (red or yellow pigment cells producing erythrophoromas), iridophores (iridescence-producing cells resulting in iridophoromas) and melanocytes (melanin-producing cells that result in melanomas). Only melanomas occur in humans.

Next to melanomas, erythrophoromas are the most common pigment cell tumor of goldfish, *Carassius auratus*. An extensive collection of fifty-two erythrophoromas from thirty-eight goldfish provided material for characterization of these fish neoplasms (Color Plate 14). The tumors occur in the dermis of fish as well as amphibians and reptiles. The erythrophoromas are found on the eyes, head, operculum, flank, fins, and other parts of the fish body. They are deep red to orange in color and can be grown in cell culture for prolonged periods (Masahito, Ishikawa, and Sugano 1989). These beautiful tumors have invasive patterns of growth following nerves, vasculature, and mesothelial coverings of body cavities. So, too, do normal pigment cells, which makes it difficult to distinguish normal invasion from malignant invasion. It is not uncommon for cancer cells to exhibit characteristics that are identical or similar to those of normal cells – this issue is discussed in Chapter 2.

Normal goldfish erythrophores were transfected with *ras*, *src*, and *myc* oncogenes. The transfected cells replicated with a tumor phenotype as evidenced by a piling up of cells in a monolayer culture and growth of colonies in soft agar. In time, the transfected cells ceased to divide and they spontaneously differentiated as normal-appearing erythrophores. The normal-appearing, formerly transformed, cells produced sepiapterin, 7-hydroxybiopterin and isoxanthopterin, all of which are found in normal erythrophores but *not* found in the oncogene-transfected mitotically active cells (Masahito et al. 1989; Matsumoto et al. 1993). The controlled differentiation of tumor cells (as discussed elsewhere in this chapter and in Chapter 1) is a potential therapeutic procedure that has the promise of yielding cancer treatment with little or no cytotoxic side effects.

Oncogenes. Of course, oncogenes (Chapter 5) have been isolated and sequenced from fish (Nemoto et al. 1986; Van Beneden et al. 1990). These and other molecular studies permit a characterization of the evolutionary significance of DNA sequences thought to be important in oncogenesis.

6.4b Amphibia

Pancreatic Cancer in Frog Hybrids. In the United States, cancer of the pancreas is the fourth leading cause of cancer death. Humans with pancreatic cancer have an extremely short survival time after diagnosis. Pancreatic carcinomas of Japanese, Chinese, and Korean pond frog (*Rana nigromaculata* group) hybrids (Masahito et al. 1995) were reported (Figure 6–1). The study of spontaneous pancreatic cancers in frogs offers the possibility of enhanced understanding of the very malignant human equivalent.

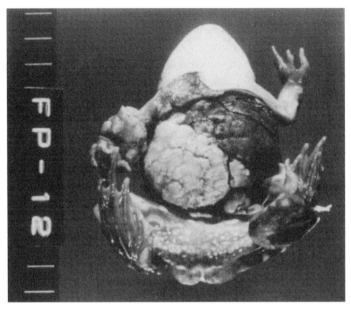

Figure 6–1. Pancreatic carcinoma in a hybrid frog. (From Masahito et al. 1994. Frequent development of pancreatic carcinomas in the Rana nigromaculata group. *Proc 8th Int Conf Int Soc Diff*, pp. 183–6, Hiroshima, Japan: International Society of Differentiation, Inc.); see also Figure 1 of Masahito et al. 1995.

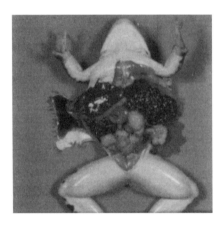

Figure 6–2. Sexually mature female northern leopard frog, *Rana pipiens*, with Lucké renal adenocarcinoma. The renal adenocarcinoma appears as a light colored nearly spherical mass just to the right of the midline posterior to the ovaries. Just posterior to the spherical mass are several other lobes of the tumor. (Photograph from the author's files).

Kidney Cancer in Toad Hybrids. Renal cell carcinomas develop in hybrids of female Japanese toads, *Bufo japonicus*, and male Chinese toads, *Bufo raddei*. Pathogenesis of these spontaneous tumors in toad hybrids occurs with the formation of polycystic kidneys, then papillary lesions followed by renal cell carcinoma. Kidney cancer has never been reported in nonhybrid toads of these species (Masahito et al. 2003). If the toad-hybrid tumors are virus-free, then there is a possibility that they may shed insight on genetic induction and development of renal cancer.

The kidney cancer of the Northern Leopard Frog, Rana pipiens, and RaHV-1. The renal adenocarcinoma of the Northern Leopard Frog, *Rana pipiens* (Figure 6–2), was originally described by Lucké (1934). Scarpelli (1975), when writing about tumors of lower animals, said the Lucké tumor deserves "special mention" because of its major contributions to knowledge concerning the biology of cancer. The frog tumor was the first cancer associated with a herpesvirus.

Herpesvirus species are found in many vertebrates and each virus is associated with a single host species. Because of this, Ranid Herpesvirus 1 (RaHV-1) is thought to occur only in *R. pipiens.* The herpesviruses are among the largest and most complex of viruses. Genomic sequences are known in at least twenty-six herpesviruses, which permits the construction of an evolutionary tree. The tree shows, among other things, the relationship of RaHV-1 with other herpesviruses (Davison 2002).

A virus was thought to be associated with the frog tumor in the earliest studies (Lucké 1938). Of course that virus, RaHV-1, is now recognized as the etiologic agent of renal carcinoma. Electron micrographs of tumor reveal RaHV-1 (Figure 6–3). Field inversion gel electrophoresis of purified viral DNA revealed the viral genomic size to be approximately 220 kbp (Figure 6–4) (Sauerbier et al. 1995). A subsequent estimate of genomic size, made by summation of restriction fragments, was 217.14 kbp (Davison et al. 1999). The genomic size calculations, produced by different procedures, are remarkably similar and provide confidence that the size estimates are reliable and essentially correct (McKinnell and Carlson 1997). Knowledge concerning the oncogenicity of herpesviruses began with frogs. Perhaps knowledge of virus/cancer cell interaction in frogs may provide insight into herpesviruses and their causative relationship to other cancers.

Nuclear Transplantation (Cloning) of a frog cancer. Cloning of metazoans by nuclear transplantation was developed in the Northern Leopard Frog, *R. pipiens,* as a means for characterizing the developmental potential of somatic cell genomes (Briggs and King 1952; McKinnell and DiBerardino 1999; DiBerardino and McKinnell 2004). Successful nuclear transplantation experiments mandate a normal or near-normal complement of chromosomes. For that reason, chromosomes of the Lucké tumor were studied, and most cells were found to be euploid or nearly euploid (DiBerardino et al. 1963; see Figures 4–1 through 4–3, Chapter 4). A study of the capacity of the genome of the Lucké renal adenocarcinoma to promote development and give rise to different cell types was undertaken (King and McKinnell 1960). Instead of forming a small population of cancer cells, tumor nuclei transplanted into enucleated ova (ova or eggs from which the maternal genes have been removed) formed young embryos (Figure 6–5) that were abnormal primarily because they failed to reach maturity (McKinnell, Deggins, and Labat 1969).

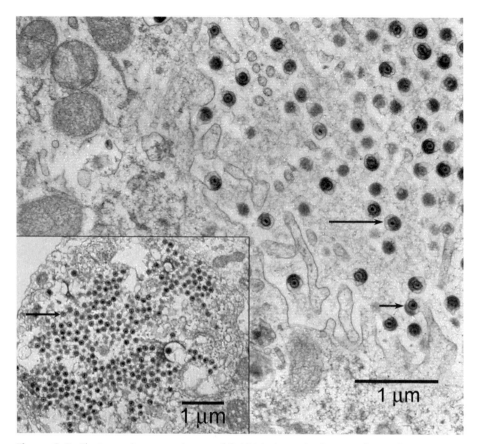

Figure 6–3. Electron microscope image of RaHV-1 viruses in the cytoplasm of an induced Lucké renal adenocarcinoma incubated at 4 degrees C. Long arrow indicates mature virus with a cell membrane surrounding the viral particle. Short arrow indicates a mature virus on a stalk presumably budding to the exterior. Inset: Micrograph of a spontaneous primary tumor incubated at 4 degrees C. Arrow indicates one of many mature viruses inside the cell. Images courtesy of Carol S. Williams and John W. Williams III, Center for Biomedical Research/RCMI (5 G12RR03059), Tuskegee University, Tuskegee, Alabama.

A significant enhancement of the differentiation and growth of a number of tissue types was reported in grafts of tumor nuclear transplant tissue (Lust et al. 1991; McKinnell et al. 1993; McKinnell 1994) (Figure 6–6). The investigators believed that failure to mature with resulting death of the nuclear transplant larva might be due to improper activation of one or a few genes of the embryo but that other genes, necessary for normal development, were present and could be expressed if the tissue were placed in an optimal environment (see discussion of gene activation and reprogramming in cloned animals in Hochedlinger and Jaenisch 2002). The "optimal environment" chosen was the tail of a normal embryo, where

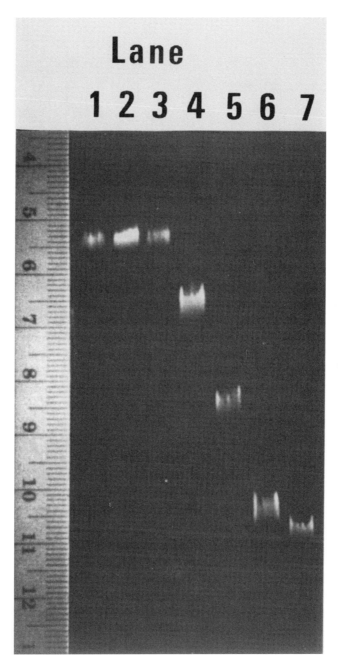

Figure 6–4. Field inversion gel electrophoresis of RaHV-1 DNA on agarose (lane 2). Lanes 1 and 3: DNA from chromosome 1 of *Saccharomyces cerevisiae*, with a molecular size of 220,000 bp. The *S. cerevisiae* DNA flank and comigrate with the RaHV-1 DNA, providing a size estimate for RaHV-1 DNA of 220,000 bp. Lanes 4, 5, 6, and 7 are other size markers. (From Sauerbier, W., Rollins-Smith, L. A., Carlson, D. L., Williams, C. S., Williams III, J. W., and McKinnell, R. G. 1995. Sizing of the Lucké tumor herpesvirus genome by field inversion gel electrophoresis and restriction analysis. *Herpetopathologia* 2:137–43.)

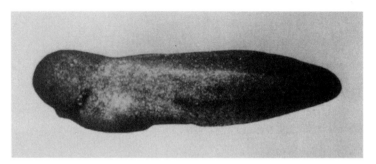

Figure 6–5. Swimming tadpole produced from the insertion of a Lucké renal adenocarcinoma nucleus into an activated and enucleated ovum of the northern leopard frog, *Rana pipiens*. The tadpole swims, has a beating heart, with well-formed body and tail. Abnormal primarily because it fails to feed, the tadpole dies shortly after reaching the swimming stage. The specialized cells of the tadpole all descended from a cancer cell genome (From McKinnell, R. G. 1973. Nuclear transplantation. In Seventh National Cancer Conference Proceedings, pp. 65–72. Philadelphia: Lippincott-Raven Publishers.)

a fragment of the tumor nuclear transplant embryo was grafted. Exquisite differentiation occurred in many of the grafts (Figure 6–7). The nuclear transplant studies reveal clearly that the genome of the Lucké renal adenocarcinoma has the competence to direct the differentiation of a number of cell types in addition to neoplastic cells.

Normal cell progeny from a neoplastic genome is an inadequately exploited phenomenon in cancer research. If the phenomenon is of general biological significance, it would be expected (and hoped) that other tumor types could be manipulated to give rise to normally differentiated cells and tissue: they do (McKinnell 1989a, 1989b; see differentiation of malignant stem cells, Chapter 1, and Hochedlinger et al. 2004).

6.4c   Reptiles

Laboratory animals tend to be small, inexpensive, and easy to manage. Perhaps because reptiles are *not* commonly used as laboratory animals and therefore may be scrutinized for cancer less often than other animals, relatively few cancers have been reported. Nevertheless, neoplasms have been reported in the Boa constrictor (*Boa constrictor*), the Komodo dragon (*Varanus komodoensis*), the California king snake (*Lampropeltis getulus*), the marsh terrapin (*Pelomedusa subrufa*), the rattlesnake (*Crotalus horridus*), the rock python (*Python molurus*), and the box turtle (*Terrapene carolina*) (Harshbarger et al. 1981; Zwart and Harshbarger (1991). Tumors of five marine turtle species, obtained from North and South America, Europe, Asia, and Oceania, are described by Harshbarger (2002).

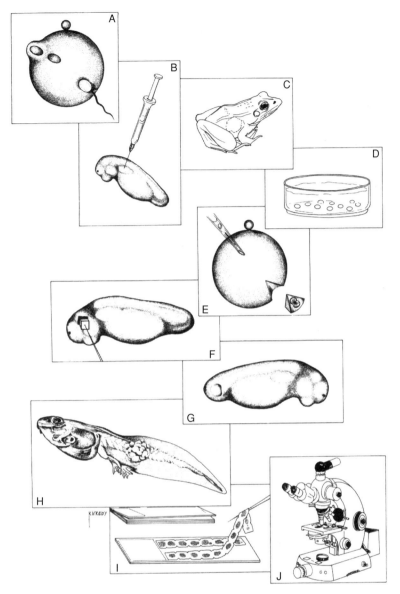

Figure 6–6. Schematic protocol for allografting experiments designed to reveal differentiation potential of tumor nuclear transplant embryos. **(A)** Hydrostatic pressure applied to eggs shortly after insemination results in triploidy (two maternal haploid nuclei plus the haploid sperm nucleus fuse to form a triploid zygote and subsequent triploid embryo). Triploidy will subsequently serve as a nuclear marker in the grafting experiments. **(B)** Triploid embryo injected with Lucké tumor herpesvirus-containing preparation, which induces a renal adenocarcinoma (dotted line) in a juvenile frog **(C)** A small fragment of the tumor is dissociated **(D)** and cells are transplanted into an enucleated oocyte **(E)** A triploid embryo ensues from the nuclear transplant procedure **(F)** and a fragment, in this case an eye primordium, is allografted to the tail of a diploid host **(G)** Extensive growth of the transplant occurs in the next 40 days **(H)** after which the graft is harvested, examined histologically **(I,J)** Triploidy, as ascertained by quantitative photometric evaluation of Feulgen stained nuclei of the allograft, excludes the unlikely possibility of development induced by nuclei other than the transplanted tumor nucleus (From McKinnell et al. 1991).

213

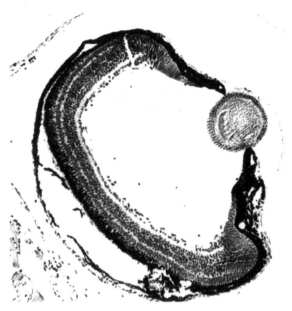

Figure 6–7. Histology of an eye primordium, obtained from a tumor nuclear transplant tadpole similar to that shown in Figure 6–5, allografted to the tail of a normal tadpole. The grafted eye rudiment differentiates a normal-appearing eye with lens and retina. (Courtesy of Dr. J. M. Lust.)

Harshbarger and a Lifetime Study of Tumors of Lower Animals

"If one picture is worth a thousand words, an actual specimen is worth at least a hundred pictures." This is the view of Dr. John C. Harshbarger (see the first epigraph of this chapter) who has devoted his scientific life to the study of cancer in lower animals. His studies encompass neoplasms of both invertebrates and ectothermic vertebrates. Lumps and lesions of a bewildering diversity of creatures with neoplasms have come under his expert eye as a Professor of Pathology, shown here at his microscope, at George Washington University Medical Center in Washington, D.C. The studies would be useful simply as a catalog of neoplasia in lower forms – he has indeed diagnosed and described tumors in marine turtles, an Ozark Hellbender, platyfish, the giant gourami (freshwater food fish from SE Asia), brown bullhead fish, channel catfish, the zebrafish, a day gecko, sharks and rays, axolotls, snakes, lake whitefish, marine bivalve molluscs and, of course, a host of other kinds of animals. However, as is evident to the readers of this chapter (see Section 6.4a), Dr. Harshbarger is not content to only catalog; he seeks understanding of the etiology, geographic distribution and epidemiology of these tumors of lower animals. As was pointed out, fundamental studies of this nature have the potential of protecting human populations from toxic substances introduced into the environment. Many people who have never heard of John C. Harshbarger are beneficiaries of his efforts.

Dr. Harshbarger received his Ph.D. from Rutgers University. From 1967 through 1995, he served as Director, Registry of Tumors of Lower Animals, Smithsonian Institution, Washington, D.C. He became Professor of Pathology at George Washington University Medical Center (1995) and continued as Director of the Registry until his retirement. It is appropriate that he and his work have been recognized by a cover photograph on the journal *Cancer Research* (Volume 40, Number 9, 1980). His long efforts in the study of tumors in lower organisms was recognized by his selection as the first recipient of the Prince Hitachi Prize for Comparative Oncology of the Japanese Foundation for Cancer Research (the Cancer Institute) in Tokyo, Japan. The award was personally presented to Dr. Harshbarger by His Imperial Highness Prince Hitachi on 27 May 1996.

6.5 Cancer in selected warm-blooded vertebrates

6.5a Birds

Rous Sarcoma of Chickens. The Rous Sarcoma occurs in Barred Plymouth Rock chickens (Rous 1911). Peyton Rous of the Rockefeller Institute showed that the sarcoma of chickens could be transmitted by means of a cell-free filtrate. The studies of Rous were difficult to ignore because they were carefully done, but they were not in harmony with the then current beliefs concerning the origin of "spontaneous" tumors. Rous showed that something (a virus) was in his cell-free preparations, and it was this "something" that caused the sarcoma of chickens. This was the first demonstration that a virus, the Rous Sarcoma Virus (RSV) could cause a cancer (see Chapter 5 for a discussion of oncogenic retroviruses and the RSV). Even after a half century, Rous was thought to have misled cancer researchers by insisting that cancer could be caused by a virus. However, at age 87, the views of Rous were vindicated by the award of the Nobel Prize.

Marek Disease (MD) of Chickens. MD is an acute transmissible lymphoproliferative disorder of domestic chickens, *Gallus domesticus*, that is worldwide in distribution. A significant portion of the infected chickens manifest lymphoid tumors, which is why MD is regarded as a neoplastic disease. It is caused by a herpesvirus that is spread to susceptible chickens from infected feather follicles via airborne dander, dust, and animal house rubbish. It is of considerable importance to farmers because of the potential for economic loss. Marek kills more chickens than any other disease. Chickens can be protected with a vaccine produced from a herpesvirus of turkeys or from attenuated strains of the Marek herpesvirus, with a resultant economic gain (see Biggs 2001 and other papers in the volume devoted entirely to MD). The efficacy of the vaccines led a biologist to exclaim with gusto that the protection to chicken flocks was "worth crowing about!" It could be added that development of the first effective vaccine that protects animals against cancer is a milestone of incredible importance!

6.5b   Mammals

It has been stated that the development of a multicellular body results in vulnerability to cancer. There probably is much truth to that statement, which applies to mammalian species as it does to all other multicellular creatures. Thus, a complete review of malignancy in mammals would require a review of all mammals. Because that is not possible here, brief mention is made of neoplasms of special interest.

Marsupials. Cancer of the Tasmanian Devil. Marsupials are primitive mammals with pouches in which the young develop after what would be considered for other mammals a premature birth. A dog-sized carnivorous marsupial, *Sarcophilus harrissii*, is found only in Tasmania and is popularly known as the "Tasmanian Devil." A previously unknown neoplasm is believed to have first been noted around 1995. The cancer of the Tasmanian Devils occurs only in adults and usually starts around the mouth or face. The mouth tumors grow into masses that occlude the oral cavity, resulting in starvation or impaired vision. Of course, cancer occurs in many if not all mammals. What is of special note about this facial affliction of Tasmanian Devils is that the cancer seems to be contagious, with an epidemic that has in less than a decade destroyed a third to half of the population. The tumor resembles a malignant lymphoma (Dr. John C. Harshbarger, personal communication) but histopathological studies are continuing (Dr. Richmond Loh, personal communication). The neoplasm

Figure 6–8. Mouse with a spontaneous (i.e., not induced by humans) mammary carcinoma. (From Gross, L. 1970. *Oncogenic Viruses*. Oxford, UK: Pergamon Press.)

may be spread by an infection of an intact cell from the bite of another Tasmanian Devil. One would ordinarily expect rapid immunological rejection of a foreign cell. However, because of inbreeding of an initially small population, contemporary Tasmanian Devils are like clones and are immunologically tolerant of cancer cells from others of the same species. The cancer epidemic continues (Sandra Blakeslee, *New York Times*, 31 May 2005).

Marsupials. Melanoma. The South American opossum *Monodelphis domestica*, develops melanoma on exposure to ultraviolet (UVB) radiation. This marsupial is the only species of mammals, other than humans, known to develop melanoma (Wang et al. 2001). Heightened interest in this animal melanoma relates to the burgeoning prevalence of this lethal form of human skin cancer that is related to increased UVB and ozone depletion (see human melanoma in Chapters 7 and 8).

Mouse Studies. Mice are better known with respect to their genetics than any other mammal (e.g., see Rader 2002). The creatures are small, well adapted to laboratory husbandry, and are studied extensively. For example, the mouse mammary carcinoma (Figure 6–8) was shown to be transmitted via an agent (a virus) in milk (Bittner 1937), marking a historical landmark in mammalian viral oncology. The mouse mammary carcinoma has been used in experimental studies of metastasis in which the production of the metalloproteinase collagenase was shown to be an important factor (Chapter 2 and Tarin 1992).

Teratocarcinomas. This "germ cell tumor" (see Appendix), occurs spontaneously or may result from grafting embryos to adult structures (Stevens 1984). Mouse teratocarcinomas arise from germ cells that first develop parthenogenetically as

Figure 6–9. Albino mouse with pigmented hair at anterior base of left ear and several locations on the right flank and thorax. The normal pigmented hair developed from embryonal carcinoma cells of agouti (pigmented) genotype. This is a now-classic example of normal cell differentiation ensuing from malignant cell mitotic progeny. (From Brinster, R. 1993. Stem cells and transgenic mice in the study of development. *Int J Dev Biol* 37:89–99.)

teratomas. They are comprised of many kinds of cells and tissues, including nervous system, cartilage, bone, skin, striated muscle, and respiratory and gut mucosa. The arrangement of these tissues, seemingly without regard to each other, was described as an "approximation to chaos" (Needham 1942). Mouse teratocarcinoma cells have been used in a great diversity of experiments demonstrating differentiation competency (Figure 6–9) (see discussion of differentiation in mitotic progeny of plant, fish, and frog tumors presented earlier in this chapter; Chapter 1; Sell and Pierce 1994; Andrews 2002).

Canine Cancer. Tumors occur in dogs at about twice the rate of cancer in humans. Tumors of dogs include skin and soft tissue tumors (most common) followed by gut, mammary, urogenital, lymphoid, and other malignancies (Dobson et al. 2002). Although rodents are small and relatively easy to rear in the laboratory, dogs have a longer life expectancy, share food and lodging with their owners, and when they become afflicted with neoplastic disease, they often are treated. The response of a dog to chemotherapy, which may lead to better treatment options for other dogs and humans, could be interpreted as a last friendly act from a devoted companion.

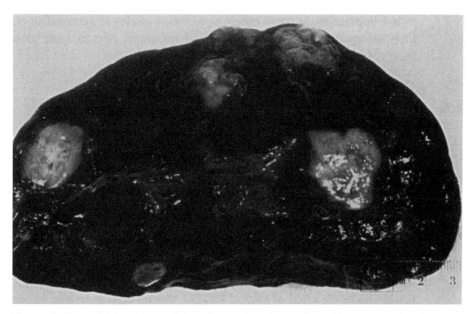

Figure 6–10. Multiple white nodules of a renal carcinoma found in a mature female California sea lion, *Zalophus californianus*. The malignancy was anaplastic and widely metastatic. (Reprinted with permission from Howard, E. B., Britt, Jr., J. O., and Simpson, J. G. 1983. Neoplasms in marine mammals. In E. B. Howard, ed., *Pathobiology of Marine Mammal Diseases*, Vol. II, p. 146. Copyright © 1983 CRC Press, Boca Raton, FL.

Feline Leukemia. Cats are vulnerable to a lethal disease caused by an infection of a retrovirus, feline leukemia virus (FeLV). The virus is found in saliva and transmission occurs by transfer of infected saliva associated with long term close contact of non-infected cats with infected cats. Mutual grooming, cat fights, and sharing food dishes are thought to be involved in contagion. The FeLV infection causes many clinical conditions in cats in addition to leukemia. The non-neoplastic illnesses include anemia, gut problems of diarrhea or constipation, respiratory distress, excessive thirst, and a loss of immunity to infectious diseases. There is no cure for a cat with FeLV disease. However, several vaccines provide significant protection for cats against the feline disease (e.g., see Harbour et al. 2002). If the development of a vaccine against Marek disease, a cancer of chickens, "is worth crowing about," then one might observe that protective vaccines against cancer in felines should be referred to collectively as "the cat's meow."

Tumors of other mammals. The autopsy of stranded dead beluga whales (*Delphinapterus leucas*) revealed a cancer prevalence (27 percent) much "higher than that reported for any other population of cetacean." (Martineau et al. 2002). A kidney carcinoma of a California sea lion, *Zalophus californianus* is illustrated in Figure 6–10. Metastatic mammary carcinoma was reported in a horse (Munson 1987), and a diversity of neoplasms are known from cattle and other domestic animals (Naghshineh et al. 1991; Rostami et al. 1994).

It is obviously impractical, as stated earlier, to provide a catalog of mammalian cancer. For more information on the malignancies of mammals, refer to other sources (e.g., Moulton 1978; Turusov and Mohr 1994).

6.6 Summary – But try anyway!

The rationale for a study of cancers in organisms other than humans is explained nicely in a paragraph taken from a symposium preface by Clyde Dawe et al. (1981). This chapter began with Dawe and it ends with Dawe – only because he phrased it so well.

"By retracing the steps followed in the phylogenetic development of regulatory mechanisms, are we not likely to find understanding of these mechanisms, just as developmental biologists expect to reach such understanding by retracing ontogenetic development? Regrettably, it does not necessarily follow that with understanding will come control of neoplasia. An escape from this uncomfortable thought lies in the advice of Michael Faraday: "But try anyway; no man knows what is possible."

7

Epidemiology

Robert G. McKinnell

The study of occupational, dietary, habitudinal, and other environmental cancers can point the way to measures of cancer prevention.

Julian Huxley 1958

7.1 Introduction

Epidemiology is the study of the distribution, prevalence, and etiology of disease and for this book the disease is, of course, cancer. Epidemiology is studied both by observation and with experiments. This chapter is confined to observational investigations. Risk factors identified in epidemiological studies have the potential, if put to use, to prevent or lower the incidence of certain cancers. These studies include an examination of the effects of occupational exposure to carcinogens as well as lifestyle activities that tend to make certain individuals prone to cancer (lifestyle is examined in greater detail in the following chapter). A practical example, already widely recognized, relates to lung cancer and smoking. Lung cancer does not randomly afflict individuals within a population. Rather, it appears disproportionately in people who smoke cigarettes. Lung cancer differences in different groups can be, and is, correlated with different exposure to cigarette smoking. When smoking increased, so too did the incidence and deaths due to lung cancer. In the United States, smoking-induced lung cancer remains the greatest cause of death by malignancy for men and women (Figures 7–1 to 7–3; Color Plates 5 through 10). Identifying tobacco as a risk factor for lung cancer has, inherent with that knowledge, the potential for prevention of much of the disease. Doll has referred to this as the "acid test of practice," that is, ascertaining if cancer can be prevented by changing the exposure to the presumed etiological agent – in this case, tobacco smoke (see "box" on Doll). Today, there is every reason to believe that lung cancer prevalence and mortality will

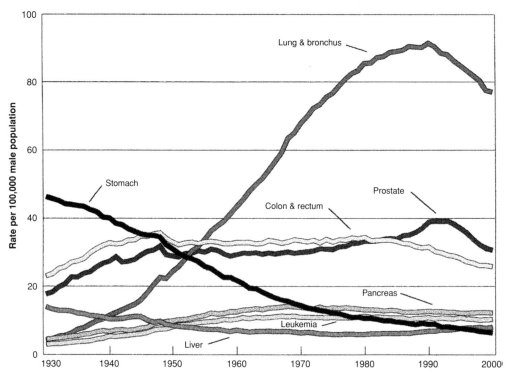

Figure 7–1. Age-adjusted cancer death rates, males by site, United States, 1930–2000. Rates per 100,000 are indicated on the ordinate, and years are indicated on the abscissa. (Source: American Cancer Society's *Cancer Facts and Figures 2004*. Copyright 2004 American Cancer Society, Inc. All rights reserved).

be reduced if tobacco consumption is decreased. Tobacco and lung cancer illustrate well the potential value of epidemiological studies.

A note of caution is necessary with descriptive or observational epidemiology. Strictly speaking, retrospective observations on a population correlating suspected cause with a particular effect do not merit a conclusion of etiology; the studies can only suggest a presumptive cause. For practical purposes, one must use the rule of reasonable explanations; tobacco smoke is more reasonably related to lung cancer than the sale of silk stockings – despite the fact that both tobacco and silk stockings were at one time correlated with increased prevalence of lung cancer (see "box" on Ochsner). Prospective epidemiological studies and experimental investigations enhance the understanding of the causes of cancer.

The nonrandomness of cancer. Epidemiologists are aware that much cancer strikes neither capriciously nor randomly. The risk for specific cancers varies with geography, and maps have been constructed that identify populations at differing risk for specific malignancies (Color Plates 15 and 16). The differences in cancer prevalence suggest that perhaps 75 percent to 80 percent of cancers in the United States

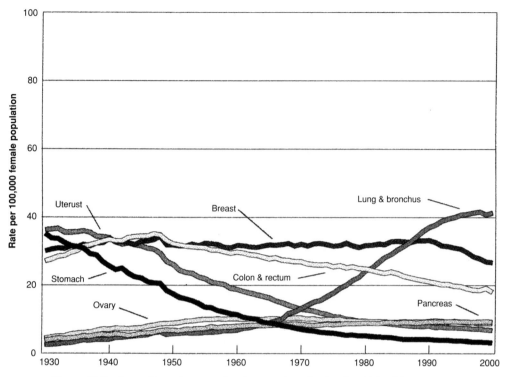

Figure 7–2. Age-adjusted cancer death rates, females by site, United States, 1930–2000. Rates per 100,000 are indicated on the ordinate, and years are indicated on the abscissa. (Source: American Cancer Society's *Cancer Facts and Figures 2004*. Copyright 2004 American Cancer Society, Inc. All rights reserved).

are due to factors in the environment (Doll and Peto 1981). "Environment" for the purposes of this chapter includes the personal environment of diet, exercise, occupation, and other factors such as exposure to the sun and tobacco use.

Geography is important. Breast cancer is seven times more frequent among Hawaiian women than among non-Jewish women of Israel. Prostate cancer in African American males of Atlanta is seventy times more common than in males of Tianjin, China (Fraumeni et al. 1993). Perhaps of greater impact are the studies of immigrants who rapidly assume the cancer rates of their adopted country. For example, gastric cancer is common in Japan. Japanese who immigrate to the United States undergo a marked dietary change and experience a decrease in stomach cancer rate. The offspring of the immigrants have an even greater decrease in stomach cancer mortality. In contrast, Japanese who move to California from Japan experience an increased colon cancer incidence presumably associated with dietary change (Dunn 1975; Wogan 1986). For these reasons, the geographic "environment" is considered to be a major factor in assessing risk for cancer.

Estimated New Cases*		Estimated Deaths	
Male	**Female**	**Male**	**Female**
Prostate 230,110 (33%)	Breast 215,990 (32%)	Lung & bronchus 91,930 (32%)	Lung & bronchus 68,510 (25%)
Lung & bronchus 93,110 (13%)	Lung & bronchus 80,660 (12%)	Prostate 29,500 (10%)	Breast 40,110 (15%)
Colon & rectum 73,620 (11%)	Colon & rectum 73,320 (11%)	Colon & rectum 28,320 (10%)	Colon & rectum 28,410 (10%)
Urinary bladder 44,640 (6%)	Uterine corpus 40,320 (6%)	Pancreas 15,440 (5%)	Ovary 16,090 (6%)
Melanoma of the skin 29,900 (4%)	Ovary 25,580 (4%)	Leukemia 12,990 (5%)	Pancreas 15,830 (6%)
Non-Hodgkin lymphoma 28,850 (4%)	Non-Hodgkin lymphoma 25,520 (4%)	Non-Hodgkin lymphoma 10,390 (4%)	Leukemia 10,310 (4%)
Kidney 22,080 (3%)	Melanoma of the skin 25,200 (4%)	Esophagus 10,250 (4%)	Non-Hodgkin lymphoma 9,020 (3%)
Leukemia 19,020 (3%)	Thyroid 17,640 (3%)	Liver 9,450 (3%)	Uterine corpus 7,090 (3%)
Oral cavity 18,550 (3%)	Pancreas 16,120 (2%)	Urinary bladder 8,780 (3%)	Multiple myeloma 5,640 (2%)
Pancreas 15,740 (2%)	Urinary bladder 15,600 (2%)	Kidney 7,870 (3%)	Brain 5,490 (2%)
All sites 699,560 (100%)	All sites 668,470 (100%)	All sites 290,890 (100%)	All sites 272,810 (100%)

Figure 7–3. Estimated new cancer cases and estimated cancer deaths in 10 leading sites for male and females of the United States, 2005. (Source: American Cancer Society's *Cancer Facts and Figures 2005.* Copyright 2005 American Cancer Society, Inc. All rights reserved).

Mortality of heart disease and cancer compared. It is estimated that 1,372,910 new cases (710,040 male and 662,870 female, Figure 7–3) and 570,280 deaths (295,280 male and 275,000 female, Figure 7–3) from cancer will occur in the United States in the year 2005 (Jemal et al. 2005). More people will die of heart disease. Thus, to the uninitiated, heart disease is more important than cancer as a cause of death. However, a disproportionate number of older people die of heart disease. Cancer, in contrast, may afflict the very young (as well as older people, of course). Cancer causes more deaths than heart disease in females in the United States at all ages until eighty years and above. Over twice as many males of the United States ages one to nineteen die of cancer than die of heart disease. Compare years of life lost by a young child with leukemia with years of life lost by an elderly person who dies of a heart attack. This is not to trivialize heart disease as a major medical problem but the comparison with cancer is offered to provide perspective to two very important medical scourges.

Changing prevalences and mortality. Cancer rates may change for reasons unrelated to occupation or environment or the other factors discussed here. One reason for change is the increased life expectancy experienced in most people of western Europe, parts of Asia, and North America. Many but not all cancers become more common in older individuals. While lifespan (years of life under optimal conditions) has remained unchanged, life expectancy (statistical estimate of years of life a person or populations is expected to live) increased greatly during the twentieth century and will probably continue to do so for some time. That means that instead of dying of contagious disease during childhood or early adulthood, there is a much greater likelihood of reaching old age. With more elderly individuals, it follows that an increased incidence will occur with those cancers associated with old age. Hence, the recent increased prevalence of some malignancies (and heart diseases) may not reflect a deterioration of the environment or improper attention to a healthy lifestyle but instead, the increased prevalence may reflect success of public health measures designed to eliminate contagious diseases which, in turn, permits a greater proportion of a population to reach old age.

The reduced prevalence, or absence, of a number of diseases in western countries, other than cancer and heart disease, speaks eloquently of the efficacy of control by prevention. Examples of diseases particularly appropriate in this context include pellagra, smallpox, scurvy, cholera, typhoid fever, and typhus. Many other examples could be cited, of course, but these are striking both because of their lethality and their present rarity. These diseases are currently so infrequent that few Americans of college age have had any experience with them. The absence or very low prevalence of these and a number of other serious illnesses results from their prevention. Public health measures such as vaccinations and attention to the purity of the food and water supply have been extraordinarily effective in disease control. Probably because of their complex and poorly understood etiologies, cancers are almost alone as a group of diseases for which preventive measures have received scant attention. This neglect of the area of prevention is rapidly being remedied, and there is now much research into cancer prevention.

Prognostication of cancer rates: A difficult and imprecise art. It is unlikely that cancer cures by surgery, radiation, or chemotherapy (Chapter 9) will improve sufficiently in the near future to greatly reduce the cancer death rate. Related to the unlikely event of a major change in cancer cure is the failure to meet the stated goal of the National Cancer Institute of a 50 percent reduction in cancer deaths in the United States by the year 2000 (Butrum et al. 1988). Indeed, there is a dismal prognostication in the World Health Organization's (WHO) *World Cancer Report* that new cancer cases will increase 50 percent by the year 2020. About 10 million people were diagnosed with cancer worldwide in 2000. That figure is estimated to grow globally to 15 million new cases two decades later (Stewart and Kleihues 2003). If not through cure, then by what means will an impressive decrease in cancer mortality occur? Almost alone, prevention has the possibility and indeed the likelihood of leading to much of the sought-for reduction in the cancer death rate. Epidemiology provides insight into which cancers are most likely to be prevented. With information obtained from epidemiology, modification of lifestyle (Chapter 8) may indeed bring about the hoped-for decrease in cancer deaths.

7.2 Cancer in fossil humans: A brief digression concerning paleopathology

It is sometimes suggested that cancer is caused by pollution of food, air, and water due to *modern civilization*. Some argue that we could avoid cancer altogether if we would only go back to a more "simple time." Cancer detected in fossil humans informs us, however, that not all cancer is attributable to carcinogens produced by chemical industries, automobile exhaust, and power plants of the late twentieth and early twenty-first century. Such a discussion belongs in a chapter on epidemiology because it brings a perspective to the considerations of what causes cancer and who is vulnerable.

Multiple myeloma is a malignant disease with an ancient past. It is a disease of bone marrow cells that has the competence to invade and destroy calcified bone. When it does so, the osteolytic lesions leave a lasting record in the bones of those unfortunate individuals who manifest the disease. Holes in the skull and other bones are formed with sharply defined borders that have been described as appearing "punched out." Of course, after several thousands of years most people leave nothing of their mortal remains. However, in exceptional cases, fossil bones remain and are discovered. Because multiple myeloma may leave its mark in bones, researchers seek evidence of this neoplasm in fossil human bones. Evidence of multiple myeloma has been detected in bones from such diverse locations as Kérpustza, Hungary; Ipswich, England; Indian Knoll, Kentucky; and Kane Mound, Missouri. These paleopathologic specimens vary in age from 3,000 to 5,000 years (Morse, Dailey, and Bunn 1976). They witness to the antiquity of human cancer in general and to multiple myeloma in particular.

Other malignancies have been reported in fossils, including nasopharyngeal carcinoma in a mummy, osteosarcoma in Celtic warriors (Ortner and Putschar 1981), and metastatic melanoma. Paleopathology is more than just the examination of "dry bones" (Anderson 1994) and provides insight into diet, parasitism, and wounds from combat, as well as the potential for the diagnosis of cancers. However, great care must be taken in that diagnosis. It is often made from fragments of skeletal material thousands of years old and as such, pathological diagnosis for the most part is conjectural and/or tentative. Be that as it may, let the skeletal remains and evidence of cancer described in these few paragraphs remove the naïve notion that cancer is only a disease of contemporary society.

7.3 Epidemiology of selected human cancers

Clearly, only a few cancers can be considered here. Data on many more cancers can be accessed by reference to Jemal et al. 2005 or checking an American Cancer Society web site (http://www.cancer.org/statistics). The selected cancers were judged to merit consideration for differing reasons. Lung (and bronchus cancer) in the

United States is the number one killer by cancer among women and exceeds breast cancer as a cause of death. Skin cancer is the most common malignancy in the United States and deaths from one form, malignant melanoma, are increasing faster than any other lethal cancer. The reason for inclusion of other cancers in this section will become obvious upon reading.

7.3a   Lung cancer

> A cuſtome lothſome to the eye, hatefull to
> the Noſe, harmefull to the braine, daungerous to
> the Lungs, and in the blacke ſtinking fume there-
> of, neereſt reſembling the horrible Sti-
> gian ſmoke of the pit that is
> bottomeleſſe.

Source: Counterblaste to Tobacco, first published anonymously in 1604 by James VI of Scotland who was also known as James I, King of Great Britain and Ireland. A "counterblaste" is an energetic argument or polemic against something, in this case, tobacco[1].

The magnitude of the drop in cancer mortality that would occur if Americans (and all other citizens of the world) ceased smoking could not be matched by any treatment of surgery, radiation, or chemotherapy currently available, in development, or even imagined by cancer research workers.

Pandemic. Lung cancer is the leading cause of cancer death in both males and females in the United States (Figures 7–1, 7–2, and 7–3; Color Plates 5 through 10). "Bronchial carcinoma became the most common cause of death from cancer in the *world* (italics added) in the early 1980s" (Doll 1992) and has been described as being "pandemic." More American women *die* of lung cancer than the combined total deaths due to breast cancer and all gynecological cancers – the epidemic of lung cancer deaths in women increased by an incredible 600 percent between 1930 and 1997 (Patel, Bach, and Kris 2004). Almost all lung cancer deaths are preventable if people do not smoke cigarettes. Prior to World War I, lung cancer was so uncommon that Adler (1912) wrote "On one point, however, there is nearly complete consensus of opinion, and that is that *primary neoplasms of the lungs are among the rarest forms of disease* (italics added)." Deaths due to lung cancer soared among males in the United States after World War I reaching a peak in death rate about 1990 (Figure 7–1). More recently, death rates increased in American women where a peak was reached about a decade later (Figure 7–2).

---

[1] The font and spelling of 400 years ago may be difficult for some students to read. For that reason, the quotation is herewith rendered in a contemporary font with some added explanation within parentheses: "A custome lothsome to the eye, hatefull to the Nose, harmefull to the braine, dangerous to the Lungs, and in the blacke stinking fume thereof, neerest resembling the horrible Stigian smoke of the pit that is bottomelesse." (stigian is an obsolete spelling of stygian which refers to the River Styx of the lower world, Hades, of Greek Mythology; the adjective pertains to something that is dark or gloomy or *hellish*). In brief, one might surmise that James VI of Scotland did not find the consumption of tobacco either pleasing or healthful.

Consumption of tobacco in the form of cigarettes surged during and after World War II. Probably 95 percent of lung cancer in men is attributable to smoking, and the 90,490 American men who will die of the disease in 2005 comprise about 31 percent of all male cancer deaths (Jemal et al. 2005). An additional 73,020 American women will die of lung cancer in 2005, making a total of 163,510 largely unnecessary American deaths. About two-thirds of Chinese males smoke; they also smoke at an increasingly younger age, and despite their knowledge of the hazards of tobacco consumption, they evidence little interest or desire to quit (Gong et al. 1995). China, the world's leading tobacco grower and most populated country, clearly has an incipient health problem of colossal proportions.

In the United States, lung cancer causes the loss of about 1 million years of life each year (i.e., if the loss of life, measured in years, of each smoker who dies of lung cancer is added together, the toll would be about 1 million years).

A long history. The concern for tobacco as it affects health is not new (note James IV's *Counterblaste to Tobacco* at the beginning of this section). In 1761, John Hill of London admonished against the use of snuff, reporting that it caused cancers of the nose. This warning was followed by the report of carcinoma of the lips "where men indulge in pipe smoking" by Sammuel Thomas von Soemmering of Mainz in 1795 (Shimkin 1977).

Lung cancer became increasingly common after World War I, and observers correctly noted a probable relationship between cigarette smoking and the increased rate of lung cancer (for example, see Tylecote 1927). Worthy of special note were Ochsner and DeBakey (1939, 1940, 1941) (See Box). Large epidemiological studies followed (Wynder and Graham 1950; Doll and Bradford-Hill 1950) that showed lung cancer death rate to be dose dependent; that is, it is directly related to the number of cigarettes smoked, and lung cancer incidence would be reduced by 80 percent to 90 percent if smoking could be eliminated. Most epidemiological studies focused on those who already manifested lung cancer and sought etiology by comparing them with disease-free individuals. These were "retrospective" studies. Another study approach, known as "prospective" was used by Doll and his associates. This involved a "forward looking" investigation of a normal population. Prospective studies minimize the difficulty of obtaining appropriate control groups and eliminate, to some extent, bias due to recall of past exposure. A group to be studied is selected in the present time and these individuals are followed prospectively into the future. In time, some of these study individuals will develop the disease, in this case lung cancer. Of course, the epidemiological study cannot identify molecular or cellular mechanisms but it can produce convincing evidence concerning causality. Doll and his colleagues started a prospective study of 34,439 British male doctors in 1951. Forty years later, 20,000 of the doctors had died. The absolutely astonishing study showed that not only did smoking cause lung cancer, it also was associated with an excess mortality for cancers of the mouth, esophagus, pharynx, lung, pancreas, and bladder, as well as to deaths due to respiratory disease (other than cancer), vascular disease, peptic ulcer and

Nylon stockings sales are increasing at a rate parallel to that of lung cancer

Dr. Edward William Alton Ochsner (1896–1981) did not refute the statement concerning nylon stockings nor did he agree with it. The statement came from a fellow physician who was a heavy smoker while Ochsner was making one of his regular polemical tirades concerning the hazards of smoking. Subsequently, the doctor who smoked died of lung cancer. So much for nylon stockings and lung cancer. It was not the end of ridicule however. When Dr. Ochsner entered a classroom of medical students once, the entire class lit up cigarettes simultaneously in contempt for the crusading surgeon. Life is not necessarily easy for one in the early stages of a struggle.

Alton Ochsner was better known by his last two names. He was extraordinarily well trained in thoracic surgery and because of his great skill in that area, he saw many lung cancer patients during his long career. Seeing many cases was not the way it should have been. While he was yet a junior in medical school, he was required to attend the autopsy of a man who had died of lung cancer. The year was 1919. Young Ochsner's attendance at the autopsy, and the rest of his class, related to the extreme rarity of the malignancy. The professor feared that this would be the only opportunity for Ochsner and his fellow students to see lung cancer. The professor was almost correct. Alton Ochsner did not see another case of lung cancer for seventeen years. But when he did, he saw not only one case but eight cases during a period of several months! He thought it was an epidemic.

Good doctors study their patients. To this end, Dr. Ochsner asked questions and listened to his patients. He found that many of his lung cancer patients were male veterans of World War I. They had taken up cigarette smoking while serving their country. They did not just puff, they inhaled deeply. Dr. Ochsner became convinced that cigarettes were causing the epidemic of lung cancer. He deplored the fact that this indulgence resulted in a self-induced lethal cancer. He referred to lung cancer as a "preventable disease." He deplored it all the more because treatment was "dismally unsuccessful." In a paper in 1941, he and his coauthor Michael DeBakey stated "It is our definite conviction that the increase in the incidence of pulmonary carcinoma is due largely to the increase in smoking, particularly cigaret smoking, which is universally associated with inhalation." Ochsner also wrote "It is the one cancer which is almost entirely preventable, and if there could be a complete cessation of tobacco use, the incidence of cancer of the lung would be as rare as when I was a student, and would become of no clinical significance." Dr. Ochsner gave many talks at medical meetings and wrote extensively on the subject. Because of his immense stature in surgery, he became well known in the fight against lung cancer. In this case, a surgeon made medical history in the field of epidemiology – his contribution to this field is much admired and will be long appreciated.

Alton Ochsner was born in a sod house near Kimball, South Dakota. He graduated from the University of South Dakota and earned his medical degree from Washington University in Saint Louis. He was a resident in surgery at four hospitals when none was customary at that time. At the age of 31, he became Professor and Chair of Surgery at Tulane University, New Orleans, Louisiana. Dr. Ochsner served as President of the American College of Surgeons, the American Association for Thoracic Surgery and the American Cancer Society. In 1937, he co-founded with Dr. Owen Wangensteen of the University of Minnesota the journal *Surgery* and was co-editor with Dr. Wangensteen for many years.

other medical conditions. The authors stated "It now seems that about half of all regular cigarette smokers will eventually be killed by their habit (Doll et al. 1994)." These and other major studies were reviewed by the Advisory Committee to the Surgeon General (1964), which reported that "the risk of developing lung cancer increases with duration of smoking and the number of cigarettes smoked per day, and is diminished by discontinuing smoking. In comparison with nonsmokers, average male smokers have approximately a nine- to ten-fold increased risk of developing lung cancer and heavy smokers at least a twenty-fold increased risk."

The original 1954 study of Doll and Hill was updated in 2004, precisely fifty years after its onset. There is no reason to doubt any of the earlier conclusions, but more recent data provide even more dreadful information. The authors appallingly predict that about one billion tobacco deaths will occur in the present century if current worldwide smoking patterns persist (i.e., about 30 percent of young people smoke). Further, all medical progress made to extend life expectancy during the twentieth century is wiped out by the smoking habit. For example, the chances of reaching ninety of a 70- year-old male smoker was a meager 7 percent; the non-smoker of the same age group had a 33 percent chance of reaching that age. A fragment of good news in the epidemiological study of Doll and his colleagues is the information that smokers who give up the tobacco habit between the ages of twenty-five and forty-four have almost the same life expectancy as nonsmokers (Doll et al. 2004).

Cigarette smoking greatly increases risk when combined with other hazards. For example, asbestos workers who smoke are far more vulnerable to lung cancer than their nonsmoking coworkers or smokers who do not work with asbestos (Selikoff, Seidman, and Hammond 1980).

Statistical studies can contribute to the prevention of cancer – prevention is no longer considered to be "faintly immoral."

A bit after the middle of the twentieth century, William Richard Shaboe Doll (far better known as "Richard Doll" or "Sir Richard") was advised by a professor of surgery that research on the prevention of cancer was not only a waste of time but it was "faintly immoral"; immoral because cancer was regarded as part of the normal aging process. Times have changed and a major motive force in that change is Richard Doll.

The increasing prevalence of lung cancer that had its origin shortly after World War I was thought to be caused by, among other things, diesel exhaust fumes, motor cars, atmospheric coal smoke, etc. However at midcentury, it was shown it was not the exhaust of diesel lorries or other suspected noxious fume producers, but cigarette smoke. This relationship of tobacco smoke and lung cancer, now firmly an irrefutable link, was revealed by Sir Richard Doll (see photograph) and his statistical colleague Sir Austin Bradford Hill. Doll and Hill were not the first. Other investigators both before and contemporarily also suspected that tobacco was the major culprit of the lung cancer epidemic. In contrast to several retrospective studies, Doll and Hill pioneered in prospective studies, which resulted in a clear link between cigarette smoking and lung cancer. Doll will always be remembered and honored for his epidemiological studies, which showed unequivocally that about half of all regular cigarette smokers will be killed by their habit.

Richard Doll was born 28 October 1912 in Hampton, located in the Greater London outer borough of Richmond upon Thames. He received his medical training at Saint Thomas's Hospital and Medical School at the University of London with graduation in 1937. He served in the Royal Army as a medical officer during World War II. Many years after the end of the war, Sir Richard wrote an utterly fascinating day by day account of the retreat to and evacuation of Dunkirk (*BMJ* 300:1183,1990 with subsequent installments in later issues).

After the war, he held an appointment in the Medical Research Council's Statistical Research Unit where he began his studies of tobacco and where he rose to the position of Director. He was appointed by the Prime Minister to the prestigious rank of Regius Professor of Medicine at Oxford in 1969. He was instrumental in the founding of Green College, Oxford, where he served as the first Warden (for American university students: the title of "Warden" is more or less equivalent to the American title of "Dean"). He received the Order of the British Empire in 1956, he was elected to the Royal Society of London in 1966, and he was knighted in 1971. Sir Richard worked until he died at age 92 in 2005.

Sir Richard was the author of a brief and very informative book with the title *Prevention of Cancer: Pointers from Epidemiology*, The Nuffield Provincial Hospitals Trust, 1967.

Involuntary Smoking. The association between passive breathing of other people's smoke and lung cancer has been shown in most studies to be positive and dose dependent (Boffetta 2002). Nonsmokers exposed to heavy smoke (e.g., bartenders) are reported to have elevated levels of carcinogens in their blood (Maclure et al. 1989) and the urine of such individuals is mutagenic (Sora et al. 1985). An editor, when contemplating secondhand smoke, wrote "Given the strangeness of the British, the final push (to eliminate smoking) may come because passive smoking kills pets as well as people (Anonymous 1994)."

All organs are composed of a diversity of kinds of cells. This generalization obviously includes lungs. Thus, the term "lung cancer" includes various types of neoplasms such as squamous cell carcinoma, small cell lung carcinoma ("oat cell cancer"), and adenocarcinoma. These types of lung cancer are discussed in the Appendix.

7.3b Breast cancer

Breast cancer in women is diagnosed more frequently than any other non–skin cancer in women and is second only to lung cancer as a cause of death by malignancy in the United States (see Figures 7–2 and 7–3). Breast cancer is the leading cause of cancer death in women aged twenty to fifty-nine after which lung cancer becomes the primary cancer killer from age sixty through eighty and above. It is estimated that 40,410 women will die of breast cancer (see Figure 7–3) in the United States in 2005. Men will suffer from about 1,690 of the 212,930 new breast cancer cases expected in 2005 (Jemal et al. 2005).

Breast cancer is uncommon before age thirty but rises thereafter. In the United States, the likelihood of developing breast cancer is 1 in 207 prior to forty years of age. The likelihood of developing breast cancer increases to 1 in 24 between the ages of forty and fifty-nine and climbs to 1 in 13 during the years sixty to seventy-nine. Hence, increased age is associated with increased risk for breast cancer. The lifetime (birth to death) risk for breast cancer is 1 in 7 (Jemal et al. 2005).

Contrary to a widely held notion, the United States does not lead in breast cancer mortality. Women of Denmark, the Netherlands, Ireland, Hungary, and New Zealand (in descending order; 2002 data) all suffer from a higher breast cancer mortality than U.S. women. In contrast, China has significantly less breast cancer. The breast cancer mortality rate of China is about one-fifth that of Denmark and one-third that of the United States (American Cancer Society 2005).

Breast cancer and history. Bernardino Ramazzini noted almost three centuries ago that breast cancer was more common among nuns than among other women (Shimkin 1977). His observations were correct, but not because the religious life renders nuns vulnerable to breast cancer. A "protective" role of marriage was suggested. However, it is not marriage per se that is protective. Rather, not having children (nulliparity) is associated with increased prevalence of breast cancer among women. The risk factor of nulliparity, or of having children after thirty-five years of age, is as real now as it was when Ramazzini first wrote about the peculiar vulnerability of nuns to breast cancer. Most of the increased risk associated with nulliparity is for postmenopausal breast cancer.

Risk factors. As stated earlier, epidemiologists search for risk factors in the hope that they may give insight into the causes of cancer and this, in turn, may suggest means of reducing cancer prevalence. In this discussion, cause is used in the sense of that which produces a consequence. A risk factor has the potential of inducing a consequence (in this instance, cancer). Cause is not used here in the sense of a change in a nucleotide sequence or altered transcription. Risk factors are unlikely to reveal mechanisms of fundamental changes in cell activity. An individual at increased risk will not necessarily succumb to cancer. Rather, the risk factor simply identifies individuals who are at higher risk than others who are not in that category. Consider breast cancer and the nuns studied by Ramazzini. Although nuns had (and still have) a higher prevalence of breast cancer than women who are not nuns, it is well to remember that the great majority of nuns die of conditions other than breast cancer. And so it is with other risk factors.

Reproductive History and Breast Feeding (Nursing). Reproduction and hormonal factors are important in the genesis of breast cancer. Age at the birth of the first child is a risk factor. Women who have children prior to age twenty have a reduced risk of breast cancer when compared to women who have their first child after age thirty. Associated with reproduction is lactation and breast feeding. Breast feeding is protective. There would be a worldwide significant reduction in breast cancer in developed nations if women breast feed their babies as their compatriots do in developing countries (Collaborative Group on Hormonal Factors in Breast Cancer 2002). Early menarche (the first menstrual period) and late menopause result in a long menstrual history with an increased risk for breast cancer. There may be an increased risk for premenopausal breast cancer with long-term use of oral contraceptives (Henderson et al. 1997). Postmenopausal hormone replacement therapy with estrogen and progestin are also associated with increased risk.

Dietary fat. On a worldwide basis, fat in the diet correlated to breast cancer (Figure 7–4). The availability of fats, and other nutrients, leads to increased height during the growing years, and the increased height (and weight) is correlated positively to breast cancer risk. Further, certain diets may lead to obesity – overweight women are more likely to die of breast cancer. Dietary fat as it relates to breast and other cancers is discussed at greater length in Chapter 8.

Cigarettes and alcohol. Tobacco use, although extraordinarily important in the genesis of lung cancer (Section 7.3a), probably has little effect on breast cancer prevalence. Alcohol consumed at the rate of one or more drinks per day is related to breast cancer. There is speculation that individuals who frequently drink also are likely to smoke, and there may be a synergistic effect of the two.

Maps and breast cancer. Geographic pathology is concerned with mapping distributions of cancer prevalence. Breast cancer is far from random in its geographic

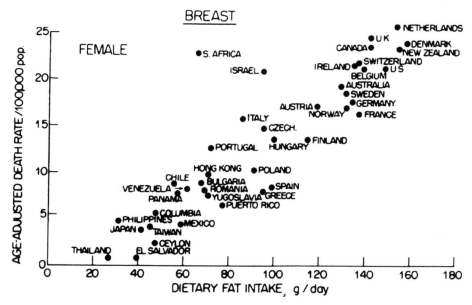

Figure 7–4. The age-adjusted death rate for breast cancer (ordinate) shows a positive correlation with the average amount of fat consumed per day per person (abscissa). (From Carroll, K. K., and Khor, H. T. 1975. Dietary fat in relation to tumorigenesis. *Prog Biochem Pharmacol* 10:331. Reprinted by permission of Karger, Basel.)

distribution. Women of the northeast United States, Chicago and vicinity, and the metropolitan areas of California are more vulnerable to breast cancer than women who live in the South, Midwest, and West (except for parts of California previously mentioned) (Color Plate 15) (Devesa et al. 1999). Affluent white women in the United States have a higher prevalence of breast cancer than women of lower socioeconomic status. The higher prevalence of breast cancer in the North may well be related to the historically greater affluence of the North compared with other parts of the United States.

Family history and breast cancer. A woman with a mother or sister with breast cancer is at an increased risk for this malignancy. For example, a woman whose mother or sister experienced breast cancer before age forty-five (premenopause) is at an increased risk compared to women of the general population. While only about one in twenty breast cancers ensue from the inherited susceptibility genes, BRCA1 and BRCA2, women who have these genes are at a greatly increased lifetime risk (see discussion in Chapter 4).

Concerning the fear of breast cancer and hope for the future. The majority of women in the highest risk categories do not develop breast cancer (Madigan et al. 1995), and the risk of dying of breast cancer is one-third the risk of developing that awesome malignancy. Surely students of epidemiology will, in time, seize on the nonrandomness of one or more risk factors (discussed above) and develop with clinicians a method of behavior modification or treatment that will lead to the ultimate control of this feared malignancy. "In the meantime, perhaps the best advice to reduce the risk of breast cancer is that of the American Cancer Society: avoid or limit intake of alcohol, partake in vigorous exercise, and minimize weight gain. While the evidence may be weak, it is not without merit to eat a diet rich in fruit and vegetables (Byers et al. 2002)."

7.3c   Skin cancer

Skin cancer is increasing in all populations of white people throughout the world. White people in this context are individuals of fair complexion, like those of Celtic or Nordic origin (Bevona and Sober 2002). About 1.3 million people in the United States will develop skin cancer each year. Almost all of those skin cancers will be nonmelanoma skin cancer. While the incidence among white people in the United States is high, it is by no means the greatest. The highest incidence of skin cancer in the world is found in Australia and New Zealand. Male citizens of Bombay, India, experience less than 1 percent of the skin cancer that afflicts the white males of Australia (compare 1.5 per 100,000 in the former country to 167 per 100,000 in the latter). Obviously, the indigenous people of Bombay are dark skinned, and the whites of Australia (and Tasmania and New Zealand) are descendants of British and other northern European migrants. As expected, nonwhite Americans have less basal cell and squamous cell skin cancer than their fellow white Americans. Of course, being white does not cause skin cancer. Excessive exposure to ultraviolet (UV) radiation causes this neoplasm (see Chapter 3 for UV as a cause of cancer; prevention of skin cancer is discussed in Chapter 8; see discussion of prostate cancer later in this chapter for an incongruous account of a postulated protective effect of ultraviolet radiation).

Of the three skin cancers, the prevalence of basal cell carcinoma greatly exceeds that of the less common squamous cell carcinoma. Malignant melanoma is the least common of the three. In the United States white population, about 75 percent to 80 percent of skin cancer is basal cell carcinoma; squamous cell carcinoma accounts for 20 percent to 25 percent. The nonmelanoma skin cancers are the most treatable forms of skin malignancy, with a high cure rate of 95 percent (probably due to the fact that they are easily noticed and early treatment is generally sought).

Basal cell carcinoma (Figure 7–5) is a skin cancer that has its origin in the basal layer of the skin epidermis. It is somewhat more common in men than in women. Occurring usually on the upper part of the face, it was formerly associated with aging. With increased skin exposure to ultraviolet radiation, it now occurs in younger individuals. It is slow growing and extraordinarily late in its metastatic behavior (the metastatic potential has been described as "practically nil"). Nevertheless, the malignancy has the competence to invade and destroy normal tissues. Tumors that persist for a long time may form lesions known as "rodent ulcer." Basal cell carcinoma has the distinction of being the most common cancer in the United States.

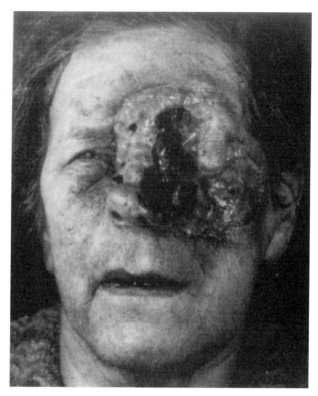

Figure 7–5. Basal cell carcinoma in a female patient. (From Walter, J. B. 1982. *An Introduction to the Principles of Disease*, p. 245. Philadelphia, PA: Saunders.)

Squamous cell carcinoma arises in the epidermis from cells that form keratin. As with basal cell carcinoma, it occurs in men more often than in women and it ranks as the second most common cancer in the United States. Squamous cell carcinoma arises primarily in skin exposed to the sun in a dose-dependent manner in individuals vulnerable to sunburn, those who tan poorly and are of light complexion. Individuals who have had organ transplantation and who are on immunosuppressive medication are at increased risk for squamous cell carcinoma. Metastasis is relatively uncommon with this skin tumor but the risk increases with the diameter and depth of the squamous cell carcinoma. Metastasis is also more common in immunosuppressed individuals.

Malignant melanoma is expected to account for 7,770 skin cancer deaths in the United States in 2005. More than half of these deaths (4,910) will be of males (Jemal et al. 2005). Malignant melanoma is unquestionably the most lethal of the skin cancers.

Malignant melanoma arises from pigment cells in the skin, and the prognosis is based, in part, on the thickness of the tumor and whether it has transgressed the epithelial basement membrane at biopsy. It has been reported that almost all patients

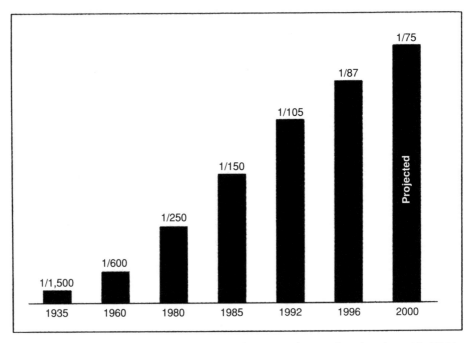

Figure 7–6. The lifetime risk of developing malignant melanoma has risen from 1 in 500 in 1935 to 1 in 87 by 1996. It was projected to increase to 1 in 75 by the year 2000, an increase of over 600% in two-thirds of a century. (From Rigel, D. S. 1996. Malignant melanoma: Perspectives on incidence and its effects on awareness, diagnosis, and treatment. *CA Cancer J Clin* 46:196. © American Cancer Society.)

with melanoma less than 1 mm thick survive compared with less than 50 percent survival among those with tumors over 4.6 mm thick (see the letter in Introduction). Alternative expressions of survival with lesions of varying thickness still convey the same message. For example, another author states that patients with melanomas less than 0.75 mm thick have a 97 percent to 99 percent survival rate compared with less than 50 percent survival rate if the lesion is more than 3 mm. Regardless of which set of figures are used, the message is the same: survival is inversely related to lesion thickness (Rigel 1996).

Malignant melanoma was once uncommon, but it is now increasing rapidly (Figure 7–6) with new cases estimated in the United States to be 59,580 in 2005 (33,580 male and 26,000 female, see Figure 7–3) (Jemal et al. 2005). Although susceptibility increases with age, a significant proportion of relatively young individuals die of malignant melanoma. As in basal cell and squamous cell skin cancers, solar exposure is believed to be of great importance in the induction of melanoma. Because solar radiation is more intense in southern latitudes than in the north, melanoma mortality is found predominantly in the south-central and southeastern states as well as parts of sunny California. There is no measurable ultraviolet radiation (due to overcast skies and the low angle of the sun with the horizon) during much of Minnesota's winter. The salutary effects of the sky and sun angle is apparent by scrutiny of the map of melanoma rates (Color Plate 16) (Devesa et al. 1999). The take-home message is that lethality (especially of melanoma) can be largely eliminated by protection from sun exposure (Chapter 8).

Ozone layer depletion. An ominous fear further adds to the already appropriate concern for ultraviolet exposure: the possibility of increased ultraviolet radiation associated with a diminished ozone layer (de Gruijl et al. 2003). The ozone layer of the atmosphere, six to thirty miles above the earth's surface, provides limited protection from skin-damaging ultraviolet radiation. This limited protection was diminished by a hole in the ozone layer which affects the extreme tip of South America. Here citizens of Tierra del Fuego have been exposed to extraordinarily high levels of ultraviolet radiation. The damage to the ozone layer has been attributed to release of chlorofluorocarbons, formerly used as refrigerants. Damage is also attributed to the pesticide methyl bromide and halons used in fire extinguishers.

Mutagenesis and UV Radiation. Studies of animals and of the epidemiology of skin cancer have revealed it is the ultraviolet portion of solar radiation that causes skin cancer (Wang et al. 2001; de Gruijl et al. 2003). Ultraviolet light consists of longer wavelengths of 320 to 400 nm, known as ultraviolet A (UV-A) and shorter wave lengths of 280 to 320 nm, known as ultraviolet B (UV-B). DNA absorbs light in the UV-B wavelengths, and it is these wavelengths that have been associated with skin cancer (see discussion of ultraviolet carcinogenesis in Chapter 3).

Ultraviolet radiation can break DNA strands and induce the formation of cyclobutane pyrimidine dimers between adjacent pyrimidines (Setlow 1974; Haseltine 1983). It is this DNA damage that presumably results in carcinogenesis. Fortunately, most of the damage is repaired by cellular enzymes, which probably accounts for a lower cancer prevalence and mortality than might otherwise be expected with UV-B exposure. Individuals who do not have the proper enzymes for DNA repair are particularly vulnerable to skin cancer (Cleaver and Kraemer 1989; Cleaver and Crowley 2002).

7.3d Prostate cancer

The prostate gland is important in reproduction because it provides most of the fluid that carries sperm to the exterior. The prostate is located at the neck of the bladder and surrounds the urethra, which is the passageway for urine and semen to exit the body. Cancer of the prostate is the most common deep-seated (internal) human malignancy in the United States, with about 232,090 new cases (33 percent of all new cancer cases in males) in 2005 with an estimated 30,350 deaths (10 percent of male cancer deaths) (Jemal et al. 2005) (see Figure 7–3). Actual new case rate may be even higher because many elderly men die of stroke, heart disease, lung, or colon cancer, or any of a number of other diseases, and with them dies a malignancy that had not yet manifested itself as a life-threatening disease. The small and undiagnosed prostate cancers are nevertheless real malignancies with the same lethal propensities of other cancers. Perhaps the new case rate would be even higher with increased use of biopsy (Doll 1992).

The author of this chapter, as many older American males probably are also, is pleased to note a declining rate in prostate cancer since about the mid-1990s (see Figure 7–1). Despite the overall decline of prostate cancer in the United States, huge differences in morbidity and mortality are found among various ethnic groups. African American males have the highest frequency of prostate cancer in the world, about half again higher than the frequency of white American males and five times that of American Indian and Alaska natives. African American males of Atlanta, Georgia, have an incidence of prostate cancer fourteen times greater than Japanese males. African American prostate mortality (deaths per unit time) is 500 percent higher than that in Asian Americans and Pacific Islanders living in the United States (American Cancer Society, 2005).

Just as ethnic differences in morbidity and mortality are found, so too are geographic differences. Uganda has over twice the rate of prostate cancer deaths compared with the United States. Other countries with prostate cancer death rates higher than the United States include Norway, Sweden, and Cuba. The death rate of prostate cancer in China is the lowest in the world (American Cancer Society, 2005).

Little information is available at the present time concerning the role of diet, occupation, or other identified factors in the genesis of this disease. Surely the epidemiologic differences in prostate cancer prevalence, especially the extraordinarily high rate of prostate cancer in African American men compared with the very low mortality of the disease in Japanese and Chinese men, will offer insight into the cause(s) of this very common malignancy and ultimately into its prevention.

An intriguing hypothesis concerns reduced ultraviolet radiation and prostate cancer. African American men in United States and western Europe have elevated rates of prostate cancer, as stated previously. Reduced ultraviolet radiation with northern latitudes is correlated to reduced conversion of 7-dehydrocholesterol to vitamin D in the skin, which in turn may cause the increased vulnerability of these darkly pigmented men (Studzinski and Moore 1995). Sunlight exposure is inversely proportional to prostate cancer mortality in other less pigmented populations as well (see review by Polek and Weigel 2002). The notion that ultraviolet radiation may protect against one form of cancer is paradoxical when one considers that ultraviolet radiation is associated with the etiology of another form of cancer (melanoma).

The antioxidant lycopene, found in tomatoes, has been advocated as protective against prostate cancer. Lycopene is discussed in Chapter 8.

Although the etiology of prostate cancer remains poorly known (a gene that increases risk for prostate cancer when mutated is described in Chapter 4), Charles Brenton Huggins received the Nobel Prize for his studies of this cancer. Huggins and Hodges (1941) showed that most prostate cancer cells are not autonomous and these androgen-dependent cells of the prostate undergo cell death (apoptosis) after androgen ablation. Castration (orchiectomy) of prostate cancer patients resulted in an immediate reduction in the elevated serum acid and alkaline phosphatases of patients and a regression of the tumor with a concomitant reduction of pain. Unfortunately, not all prostate cells respond to androgen ablation, and those cells that do not will eventually increase in number with an expanded tumor mass. The new cells are not androgen sensitive.

7.3e Colorectal cancer

About 104,950 new cases of cancers of the colon and 40,340 cases of rectal cancer are expected among men and women in the United States in 2005 (Jemal et al. 2005). Deaths in 2005, estimated to be 56,290, is third highest among cancers of both men and women in the United States (see Figure 7–3). Cancers of the small intestine have a much lower morbidity and mortality.

Why is the long small intestine (about twenty feet in length in an adult with 90 percent of the absorptive surface) almost invulnerable to cancer, whereas the short (five feet in length with only 10 percent of the absorptive surface) large intestine so susceptible to malignancy? Solving this enigma might point the way to preventing deaths from the third most lethal of human cancers. The nearly 57,000 American men and women who will die of cancer of the large intestine and rectum contrasts with the 490 women and 580 men who will die of cancer of the small intestine (Jemal et al. 2005). The answer to the query about differences in causation between small intestine and large intestine cancer is not known. The rhetorical question is posed because it provides insight into the "why" of epidemiology. If data were not collected concerning how many people die of what kind of cancer, it would be unlikely if anyone would ever know that cancer of the small gut has a distinctly different frequency than does cancer of the large gut.

Because both the small and large intestines are comprised of gut epithelium, smooth muscle, glands, and connective tissue, there is indeed a puzzle why the small intestine is "protected" and the large intestine seems not to be. What might influence this difference in cancer susceptibility? One may examine the bacterial flora as it changes along the length of the gut. Small intestine bacteria are found in lower numbers and are metabolically less effective in activating procarcinogens than the bacteria of the large intestine. Transit through the small intestine (the interval of time between material entering and leaving the small intestine) is less than transit through the large intestine. Transit time translates into how long epithelial cells in a particular portion of the gut are exposed to activated procarcinogens – and other noxious substances. The large intestine has a significantly longer exposure. Water is removed from the contents of the large intestine with resultant concentration of the activated procarcinogens. Combine a greater concentration and a longer exposure to carcinogens and an intellectual framework emerges for possible cancer causation. It seems likely that from this knowledge preventive measures (altered diets, for example) will be devised. If this occurs, we can

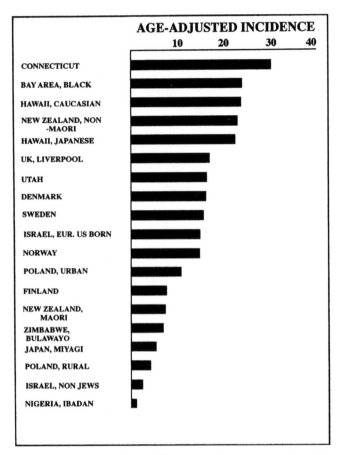

Figure 7–7. Cancer prevalence varies in different populations throughout the world. Illustrated here is age-adjusted colon cancer incidence by geographic area. (From Greenwald, P., Lanza, E., and Eddy, G. A. 1987. Dietary fiber in the reduction of colon cancer risk. *J Am Diet Assoc* 87:1179. Copyright The American Dietetic Association. Reprinted by permission of the Journal of the American Dietetic Association.)

thank the epidemiologists for their insight into differences in gut vulnerability to cancer.

As with most cancer, the rate of colorectal cancer differs in different parts of the world (Figure 7–7). The citizens of Connecticut[2] have ten times as much colorectal cancer as do the people of Bombay, India (Higginson et al. 1992). Men of the Czech republic are at least seven times more likely to die of cancer of the colon and rectum than men of Mexico (American Cancer Society 2004). People who live in the northeast part of the United States are more vulnerable to cancer of the large intestine than other Americans (Pickle et al. 1987). In the United States,

[2] The prevalence data cited here are old. The 1997 estimated colon and rectum cancer mortality rate in Connecticut is 690 per 100,000 (Parker et al. 1997), which indicates an increase in recent years.

more African Americans die of colorectal cancer (34.6 per 100,000) than do white Americans (25.3 per 100,000).

Age is a major risk factor for colorectal cancer. One cannot control becoming older so it is legitimate to ask if there is any way, short of seeking the fountain of youth, that a healthy person can reduce risk for colon cancer? Emerging evidence indicates that perhaps as much as 90 percent of all colorectal cancer cases and deaths could be prevented and these issues are discussed in Chapter 8.

7.3f Cervical cancer: "The Beginning of the End"

The uterus, a pear-shaped hollow organ about 3 inches long, 2 inches wide, and 1 inch thick, consists of an expanded upper portion known as the corpus and a more narrow lower portion, the cervix, that extends into the vagina. The cervix provides an opening to the interior of the uterus. The stratified squamous epithelium of the vagina changes abruptly to the columnar epithelium of the uterine interior at the cervix. Most cervical cancer is squamous cell in origin.

Although the death rate of uterine cancer (cervix and corpus combined) has declined markedly since the onset of World War II (see Figure 7–2), nevertheless about 10,370 new cases of cervical cancer with about 3,710 deaths (see Figure 7–3;) are expected from that malignancy in the United States in 2005 (Jemal et al. 2005). Cervical cancer is found infrequently in women younger than twenty, among celibate females (e.g., nuns), and among those who have had only one sexual partner who in turn has had only one sexual partner. It is uncommon among Jewish women. While uterine cancer has declined in the United States, it is the *leading cause of death* by cancer among women in developing countries, claiming an estimated 200,000 lives each year. While Mexico is low in overall cancer deaths, it is first in cervical cancer deaths (American Cancer Society 2004).

The changes in cell structure designated "cervical dysplasia," seen in pap smears and biopsies, are believed due to human papilloma virus (HPV) infection; HPV-infected individuals often show a significant increase in the number of dysplastic cells in a pap smear. The dysplastic cells contain HPV DNA and virus-specific antigens. Cervical dysplasia is in turn believed related to carcinoma in situ, which is ultimately linked with invasive squamous cell carcinoma of the cervix (Howley et al. 1997). The close association of the DNA-containing human papilloma viruses with cervical cancer, and the abundance of cervical cancer worldwide, led a German virologist to state at a scientific meeting that DNA viruses are now second only to tobacco smoke as a "cause" of human cancer.

A note on viruses as a "cause" of cancer is considered here. All cancer etiology is complex. Consider tobacco and lung cancer. Not all smokers, even heavy smokers, succumb to lung cancer. Obviously other factors affect causation. Individual vulnerability is affected by genetics as well as a diversity of environmental factors. However, tobacco smoke is so intimately linked with lung cancer that it is generally considered to be etiologic even while considering modulating factors. Viruses as a cause of cancer are perhaps similar. Because of this, it is not uncommon to refer to "virus-associated cancer." This designation of an "association" is more in harmony with the discussion of viral etiology in Chapter 3. It is wise to remember that cancer etiology is not simple.

The first edition of this book contained the statement: "If epidemiologic studies do indeed confirm one or more human papilloma viruses as the causal agent for cervical carcinoma, there will be abundant opportunity to devise ways to prevent the spread of the virus, which in turn should reduce the frequency of the cancer." To a very real extent, that prognostication has come true with the development of a vaccine against HPV-16 that holds promise in preventing cervical cancer (Koutsky et al. 2002). The vaccine was hailed as the "beginning of the end for *cervical* cancer" (Crum, 2002). If the end is in sight, it will be due to splendid epidemiological studies coupled with state of the art vaccine production.

7.3g   Hodgkin lymphoma

Hodgkin lymphoma (formerly known as Hodgkin's disease) is of particular interest because it is a potentially curable cancer. It was invariably fatal when first described in 1832 but can now be successfully treated in many cases. Mortality has been decreasing for the past third of a century with 1,410 expected deaths in the United States in 2005 (Jemal et al. 2005). The disease, a primary malignancy of lymph nodes, results in enlargement of the spleen and other lymphoid tissues. Multinucleated giant Reed-Sternberg cells are diagnostic of this neoplasm.

Hodgkin lymphoma increases in incidence until about twenty-five years of age, then decreases until age forty-five, after which it increases again (Doll 1992). Hodgkin lymphoma affects more males than females and it is more common among the educated than the uneducated, among the urban dweller than the rural person, and among siblings of a patient with the cancer. Americans and Italians have among the highest prevalence of Hodgkin lymphoma; Chinese and Japanese have the lowest (Higginson et al. 1992).

Infection with mononucleosis results in a small increase in vulnerability to Hodgkin lymphoma, which suggests a role for the Epstein-Barr herpesvirus in some cases of the disease (Thomas et al. 2002). Because the increased risk of Hodgkin lymphoma after infectious mononucleosis is small, it would seem that other important, and as yet unknown cofactors, are involved in etiology (Hjalgrim et al. 2003).

7.4   Occupational cancers

The 1775 description of cancer of the scrotum of chimney-sweep boys by Percivall Pott is the classic epidemiological study of occupational neoplasia. The plight of young boys, purposely kept small to permit them to fit into and clean hot and sooty chimneys is a compelling story. The story still elicits compassion in the hearts of readers (Pott 1775). The brutal treatment of these boys brings to mind the writings of Charles Dickens in the following century.

Pott was probably the first to associate a particular malignancy (the scrotal cancer) with a specific occupation (chimney-sweeping). He believed it was the coal soot caught in the skin folds of the scrotum that caused the cancer. Later, coal tar was indeed shown to cause experimental skin cancer in rabbits (Yamagiwa and Ichikawa 1918). Coal tar is a mixture of substances. Kennaway (1924) showed that coal tar contained polycyclic aromatic hydrocarbons, particularly benzo-(a)-pyrene, and it was these substances that conveyed a carcinogenic potential to soot. The chimney-sweeps were vulnerable to a particularly noxious and loathsome cancer because of their occupation during their youth.

Since those early days, there has been increased interest in occupational cancer. No one knows how much cancer is due to occupation, but Landrigan (1996) suggests that about 10 percent of cancer deaths are due to occupational exposure. More than 570,000 U.S. citizens are expected to die from cancer in 2005. Ten percent of that figure, 57,000, is the estimate therefore for deaths due to occupational exposure. That figure is much greater than the number of American military personnel, 36,940, who were killed during the entire Korean War (1950–3). Almost all occupational cancers can be prevented. Thus, it is a rich area to be explored in the quest for reduced cancer deaths.

Other substances and the cancers they cause include (but are not limited to) benzene with leukemia, asbestos with lung cancer and mesothelioma, and vinyl chloride with angiosarcoma of the liver (Landrigan 1996). Listings of human carcinogens, identified by occupational exposure and judged to cause cancer, are found in Stellman and Stellman (1996), Trichopoulos et al. (1997), and Chapter 3. It is gratifying to learn that "in the United States and other developed countries, exposure to established occupational carcinogens has been substantially reduced" (Trichopoulos et al. 1997). However, it is disquieting to note that fewer than 1,000 of the 50,000 chemicals regularly used in commerce have been examined for their cancer-causing potential (Stellman and Stellman 1996). The epidemiology of occupational cancer remains a potentially fruitful area for studies which seek to minimize unnecessary deaths.

7.5 AIDS-related Kaposi's sarcoma

An individual with the human immunodeficiency virus (HIV) infection has an increased risk for Kaposi's sarcoma (KS). KS is estimated to afflict acquired immunodeficiency syndrome (AIDS) victims at the incredible rate of 20,000 times more commonly than that of the general population (Beral et al. 1990). KS arises in vascular or lymphatic endothelial cells, causes pink, red, or blue spots on the skin of the trunk, arms, head, and neck, and in the mouth. The disease also spreads to the gut, lymph nodes, and other internal organs. KS increased in prevalence in the 1980s and is the most common neoplasm afflicting HIV/AIDS patients (Geraminejad et al. 2002; Martinel et al. 2002).

A herpesvirus known as Human Herpesvirus 8 (HHV-8), also known as Kaposi Sarcoma Herpesvirus, is associated with virtually all lesions of KS. Kaposi HHV-8 is closely related to the Epstein-Barr Virus. While HHV-8 is considered by many to be a human cancer virus, it is believed by others to be a necessary but not sufficient cause of KS.

Risk factors for KS, other than HIV infection, include male homosexual lifestyle, cytomegalovirus disease, infection with other sexually transmitted diseases, inhaled nitrates, and the number of sexual partners in HIV-endemic metropolitan areas. The neoplasm occurs less commonly among HIV-infected individuals who are drug users, hemophiliacs, and women and children. Kaposi sarcoma also arises in some transplant patients who are maintained on immunosuppressive drugs, as well as elderly men of Mediterranean or Eastern European origin.

Other AIDS-related malignancies. Non-Hodgkin lymphoma (NHL) is the second most common malignancy of AIDS patients with a rate about sixty times greater than in the general population. Persons with AIDS also have increased vulnerability of primary central nervous system lymphoma (PCL). The average age of non-HIV infected patients with PCL is about sixty; in contrast, the average age of AIDS patients with PCL is thirty-five (Safai et al. 1992). Of course, immunocompromised patients who are not AIDS patients also suffer from an increased risk for Kaposi sarcoma and non-Hodgkin lymphoma. Because of this, there is emerging interest in differences in the epidemiology of these tumors among AIDS patients and other immunocompromised patients as a means of studying pathogenesis of the neoplasms.

7.6   What is next?

Increasingly more complete and reliable data concerning activities during life, and cause of death, make for ever better descriptive epidemiological studies. This, however, is not an end in itself. The epidemiological data must be translated into information with the potential for reduction in cancer morbidity and mortality.

Biological factors may be factored into the puzzle in the continuing effort to understand cancer morbidity and mortality. Discussed above (section 7.3e) is the relatively lower frequency of cancer of the small intestine versus frequency of cancer of the large intestine. What can be learned from the epidemiological data? Another example that may be a rich lode suitable for mining is the report that Down syndrome individuals, while suffering from an excess of leukemia and testicular cancer, are afflicted with only one-tenth the abundance of solid tumors as other people (Yang, Rasmussen, and Friedman 2002). Down syndrome patients are trisomic for chromosome 21 (three of chromosome 21) and speculation has been made that the extra chromosome may result in enhanced expression of solid tumor suppressor genes located on chromosome 21. Whether it is this enhanced expression of specific genes or a combination of other factors, Down syndrome patients may yet provide the gift of greater understanding of cancer.

While some of the epidemiological studies make it clear what should be done to reduce risk factors for certain cancers, as discussed in the next chapter, human populations have been extraordinarily recalcitrant in their willingness to alter risky behavior. Perhaps studies in behavior modification should be the next big effort of those who seek to reduce the toll taken by cancer.

8

Lifestyle: Is there anything more important?

Robert G. McKinnell

Louis Pasteur (1822–1895), when considering a disease, was more interested in how can it be prevented, not, how can it be cured.

A. Ochsner 1954

The most desirable way of eliminating the impact of cancer in humans is by prevention.

L. Wattenberg 1985

The ultimate goal in the control of any disease is prevention, and so it is with cancer.
Y. Hayashi et al. 1986

Evidence suggests that one third of the more than 500,000 cancer deaths that occur in the United States each year can be attributed to diet and physical habits, with another third due to cigarette smoking.

T. Byers et al. 2002

8.1 Introduction

It has been estimated that ***two-thirds*** of all cancers can be prevented by lifestyle modification (see Byers epigraph above). In contrast to the abundance of cancer that is clearly preventable, it may come as a surprise to learn that there is an anticipated massive 50 percent increase worldwide in cancer incidence by 2020 (Stewart and Kleihues 2003). The title to this chapter is thus entirely appropriate and readers of this book, with the possible exception of those who are already afflicted with cancer, should take heed. Of course, those who already suffer with cancer have family and friends and hence, it is not inappropriate for them to read this for the benefit of those they love. Ochsner (1954) believed that education must play a role in risk reduction. It is hoped that readers who still smoke, are overweight, who do not exercise adequately, who bask in the summer sun, or do other things that put them at increased risk for cancer will take notice of the "educational" messages contained herein. As illustrated in the last chapter on epidemiology, cancer strikes neither capriciously nor randomly but at certain groups that are at risk. Modification of that risk is often possible. Hence, there is an urgency for all individuals who are not already doing so to make appropriate changes in their lifestyle in an endeavor to reduce their vulnerability for cancer.

An underappreciated aspect of cancer risk reduction is the simultaneously reduced risk for major medical problems such as heart disease, diabetes, and other chronic diseases (Curry, Byers, and Hewitt 2003). There is even further merit to cancer risk reduction by the increased feeling of well-being resulting after the escape from obesity and smoking with a healthy diet and with increased exercise. The recommendations of this chapter are consistent with those of the American Cancer Society, as well as the National Cancer Policy Board, Institute of Medicine of the National Academy of Sciences, the National Cancer Institute, the National Research Council, the U.S. Department of Agriculture/Department of Health and Human Services Guidelines, and the American Heart Association. They are similar to the recommendations made in Japan and also the European Organization for Cooperation in Cancer Prevention. The American Cancer Society, the American Diabetes Association, and the American Heart Association have adopted a common agenda that is designed to reduce the incredible toll in human suffering and financial loss due to disease – this chapter is in harmony with that agenda (Eyre et al. 2004). Although the uniformity of opinion of a multiplicity of agencies does not in itself establish truth, the fact that the agencies concur on their recommendations indicates that a number of thoughtful individuals have studied the evidence and have arrived at the same conclusions. Because of this uniformity of opinion, the author of this chapter believes that the recommendations of the organizations are reasonable and in all probability valid.

8.2   Lung cancer is a preventable disease

"Tobacco is the greatest contributor to deaths from cancer, and reduction in tobacco use offers the greatest opportunity to reduce the incidence, morbidity, and mortality of cancer (Curry et al. 2003)." "Smoking remains the most common preventable cause of death in the developed world, and is rapidly becoming an important cause of death in the developing world (Sargent and DiFranza 2003)." "It now seems that about half of all regular cigarette smokers will eventually be killed by their habit (Doll et al. 1994)." The three quotations that begin this paragraph should indicate to the reader why lung cancer leads in this chapter on cancer prevention. That lung cancer, caused by tobacco smoke, is largely a self-induced disease is discussed extensively in the preceeding chapter. If almost all lung cancer is due to tobacco, what measures are required to reduce its prevalence until it becomes *exceedingly rare* as it was during the early part of the twentieth century (Adler 1912). The answer to that question is both simple and extremely complex. The simple answer is if you do not smoke, do not start; and if you do smoke, quit now. The complex portion of the advice is that if a person already smokes, it is extraordinarily difficult to stop.

The habit of smoking cigarettes often begins during the early teenage years. Young people are influenced by their fellows at school and by their parents and, perhaps especially, by individuals whom they admire. Motion pictures play an important role because of the nearly ubiquitous portrayals of smoking as a sophisticated habit. One should note that the hazards of smoking are already well known to young people. Smoking is "cool" and begins despite well known dangers associated with the habit.

Not all people take up cigarette smoking and not all who do have the competence or willpower to quit when they wish to do so. There are factors other than nicotine addiction that make quitting difficult. Seeing smoke and smelling smoke are intensely pleasurable stimuli to the addicted. The pleasure is enhanced in social situations. It is difficult to avoid a smoking situation if one's co-workers are smokers – and this complicates smoking cessation.

Vigorous measures to inhibit the onset of smoking in young people include tax increases on cigarettes (young people respond to price more than adults and tax increases do indeed discourage smoking among the young), strictly enforced prohibition of tobacco sales to non-adults with tobacco-licensing for vendors of tobacco, making public buildings entirely smoke-free, and further restrictions on tobacco promotion and advertising (Sargent and DiFranza 2003; Curry et al. 2003; Eyre et al. 2004). It is obviously far more economical and humane to discourage young people from smoking than to endeavor to cure the diseases that ensue from that habit.

Much has been written about "second-hand smoke," also known as "involuntary smoking," that is, exposure of non-smoking people to the smoke of others. There is indeed increased risk for lung cancer from this type of exposure but the increase per individual is small. So, why the concern? The concern derives from second-hand smoke exposure to huge numbers of people. Cancer in a small fraction of the huge number who breathe other peoples' smoke results in much new cancer (Boffetta 2002; the entire issue of the journal containing the Boffetta paper is devoted to the health effects of environmental smoke, including increased risk of lung cancer, cardiovascular disease, and reproductive difficulties).

Tobacco use in other forms is hazardous. Children often think that "smokeless tobacco" is safe and thus begin routine use of snuff as early as seven years of age. Snuff use is strongly correlated with cancer of the cheek, gums, and pharynx. This is not new information, which suggests, among other things, that *Homo sapiens* is a slow learner. John Hill in 1761 wrote, after describing cancer of snuff users, "no man should venture upon Snuff, who is not sure that he is not so far liable to cancer: and no man can be sure of that" (Shimkin 1977).

The section on smoking in this chapter on lifestyle may seem strident to some readers – so be it. It is written as such because there is no other single lifestyle alteration known that would have the impact on cancer deaths as tobacco avoidance. Is anyone listening?

8.3   Ultraviolet radiation and that "healthy tan"

A long history relates exposure to the sun to skin damage and skin cancer (e.g., see Hyde 1906). Light-complexioned people who burn easily, tan poorly, and freckle, such as those of Irish, Scotch, English, and Scandinavian origins (these people were referred to as being of Celtic or Nordic origin in Chapter 7), and who are regularly exposed to the sun (note the letter to Jill from Amy in the Introduction) are much more vulnerable to skin cancer than those who have darkly pigmented skin or who, since childhood, eschew being in the open sunshine. Fishermen, farmers, ranchers, and other outdoor workers are frequently afflicted with skin cancers on the backs of their hands, necks, and faces which are, of course, anatomical sites with sun exposure.

Because the intensity of sunshine varies with latitude, southern people of the United States are more vulnerable to melanoma than people who live in the north (Color Plate 16). The depletion of protective ozone in the stratosphere has resulted in citizens of Tierra del Fuego having experienced perhaps the highest level of UV radiation in the entire world (Chapter 7).

The tanned body is deemed by many to be more attractive than a body of lighter hue. Cosmetic darkening of the skin by tanning lamps is used by more college women than college men and more than 90 percent of the users of the lamps are well aware of the hazards of premature skin aging and the risk of skin cancer. The users of the tanning lamps ignore significant risk for the pleasure of the moment. To change risky and unnecessary exposure to damaging UV will require a change in the harmful notion than tans are attractive and healthy (Knight et al. 2002).

It is a relatively recent view that tanning is beautiful. Perhaps this is because lifestyle during the Great Depression (ca. 1929 to 1939) precluded frequent sunbathing expeditions and prolonged exposure to the sun for most people. People were just too busy endeavoring to survive a difficult economic time to get out into the sun, thus white people were white. Actually, in the even more distant antebellum time (the decades prior to 1861) in the deep South of the United States, "refined" women carefully protected themselves from sun exposure. The southern belle when venturing out of doors was garbed in long sleeves, white gloves, and carried an elegant parasol. This attire was to maintain her concept of beauty with a fair complexion and to distinguish herself from weather-beaten workers with sun-damaged skin who labored in the fields. There were healthful benefits from this stylish (and probably snobbish) upper-class behavior. The Southern Belle story is notable because it suggests that the hue of skin for "white" people is strictly a matter of style and fortunately, styles can and hopefully will change.

Currently, ever more skin is exposed during the time spent in the outdoors, with entirely predictable results. Women who wear bikini bathing suits, or who swim nude, are almost 13 times more likely to develop malignant melanoma in the trunk region than women who wear one-piece bathing suits with high backs (Holman, Armstrong, and Heenan 1986). A combination of more time for recreation and clothing that exposes more skin, with a cultural value placed on the incorrect notion of a "healthy" tan in white people, and it is not surprising that the lifetime risk for malignant melanoma has increased about 20 times since 1935. (Author's note: does it convey more urgency to state that melanoma has increased 2000 percent since 1935?)

Inasmuch as UV-B (Chapter 7) is believed to be the culprit in skin damage, one might ask if exposure to UV-A would be safe and still result in the desired bronze color. For a variety of reasons that include damage to dermal elastic tissue, lens damage with possible cataract formation, and the failure of UV-A tanning to protect from UV-B damage (and even the production of UV-B by UV-A lamps), most health-care providers caution individuals against exposure to UV-A tanning lamps. If these considerations fail to convince the reader of potential hazard, it may be useful to note that UV-A causes tumors in hairless mice (Diffey 1987).

The fact that skin cancer among African Americans is relatively rare supports the view that melanin pigment protects against ultraviolet damage and subsequent skin cancer. This view is substantially strengthened by the observation that albino blacks, who lack skin pigmentation, are particularly vulnerable to skin cancer (Luande, Henschke, and Mohammed 1985).

8.3a How to minimize risk for skin cancer

Public health programs designed to prevent skin cancer focus almost entirely on protection from sunlight (IARC 2001). Thus, the best expedient to avoid skin cancer is to stay out of the sun, especially the sun at midday. Seek shade, not open sunshine. The next best expedient is proper choice of clothing if outdoor exposure is unavoidable. Hats with brims are protective, and the broader the brim, the more protection. The nearly ubiquitous ball cap of young men in the United States does *not* protect the ears, with predictably dangerous results. Ball caps worn backwards do not protect the eyes and forehead. Dark, closely woven cloth protects more than white loosely woven material. A simple test for protection is to see how dense a shadow the cloth produces in bright sun; the more dense the shadow, the more protection the cloth provides from ultraviolet radiation. Avoid or severely limit exposure to sunshine, especially at midday, and wear protective clothing – these are the golden rules of skin cancer protection.

8.3b The peculiar status of protection by sunscreens

Not mentioned in the discussion above is the use of protective sunscreens. As strange as it may seem, there is at the present time insufficient evidence to support the view that sunscreens protect from basal cell carcinoma or malignant melanoma. The reason for this is unknown. It may be that users are also people who sunbathe often. It may be that users are fairskinned individuals (Celtics and Nordics) and are already at increased risk. It may be that data available relate to the use of products with lower sun protection factors (SPF) than sunscreens currently available. However those facts may be, one should note that "as a pharmaceutical product marketed for melanoma protection (instead of the prevention of sunburns) sunscreens would have failed the tests of efficacy during the approval process (Gefeller and Pfahlberg 2002)." While the sunburn protective action of sunscreens can increase the amount of sunbathing without burning, the increased time in the sun may well increase the risk for melanoma (Vainio, Miller, and Bianchini 2000). Notwithstanding, it is reasonable and prudent (because of the lethality of melanoma) to seek sunscreen protection from UV when outdoor exposure is absolutely required. This is the recommendation of the American Academy of Dermatology and the American Cancer Society (Rigel 2002). And so, added to the two proven golden rules of skin cancer protection (stay out of the sun when possible and use proper protective clothing when exposed to the sun) may be added a third which is, when exposure to sunshine cannot be avoided, use a sunscreen. This advice is given with trepidation because of the uncertain status of cancer prevention by sunscreens and because the potential for abuse of sunscreens by increasing exposure to UV during the tanning process (IARC 2001).

8.4 Diet, nutrition, and cancer

For a number of years, there has been strong but indirect evidence that common neoplasms could be made less prevalent if suitable alterations were made to the diet (Doll and Peto 1981). It is commonly accepted that diet can affect cancer prevalence, either by the avoidance of food substances that increase risk or by selecting foods thought to protect. Unfortunately, because of many problems relating to the identification of just what is an optimal diet for cancer, little definitive information is available about which foods should be selected and which should be avoided. Even less is known about how some nutrients in these foods effect reduced risk of cancer. Nevertheless, major health organizations concur concerning dietary recommendations. The recommendations of the American Cancer Society (Byers et al. 2002) (Table 8–1) are given. Although information thought to be useful in reducing cancer risk is widely available, only a minority of Americans deign to consume healthful foods; thus, progress toward attaining compliance with dietary guidelines appears to be minimal. For example, of 11,658 adults studied, only 16 percent consumed high-fiber breads and cereals, and only 21 percent ate fruits and

Table 8–1. *American Cancer Society (2002) Guidelines on Nutrition and Physical Activity for Cancer Prevention (Byers et al. 2002).*

ACS Recommendations for Individual Choices

1. Eat a variety of healthful foods with an emphasis on plant sources.

   - Eat five or more servings of a variety of vegetables and fruits each day.
   - Choose whole grains in preference to processed (refined) grains and sugars.
   - Limit consumption of red meats, especially those high in fat and processed.
   - Choose foods that help maintain a healthful weight.

2. Adopt a physically active lifestyle.

   - Adults engage in at least moderate activity for 30 minutes or more on five or more days of the week; 45 minutes or more of moderate-to-vigorous activity on five or more days per week may further enhance reductions in the risk of breast and colon cancer.
   - Children and adolescents: engage in at least 60 minutes per day of moderate-to-vigorous physical activity at least five days per week.

3. Maintain a healthful weight throughout life.

   - Balance caloric intake with physical activity.
   - Lose weight if currently overweight or obese.
   - If you drink alcoholic beverages, limit consumption.

ACS Recommendation for Community Action

Public, private, and community organizations should work to create social and physical environments that support the adoption and maintenance of healthful nutrition and physical activity hehaviors.
   - Increase access to healthful foods in schools, worksites, and communities.
   - Provide safe, enjoyable, and accessible environments for physical activity in schools, and for transportation and recreation in communities.

vegetables high in vitamin A (Patterson and Block 1988). Little change for the better has been noted since the 1988 study.

Several major dietary components linked with cancer risk reduction as well as substances that occur in trace or limited quantities are discussed later. In addition to vitamins, a large number of nonvitamin, nonnutrient dietary components that occur in small quantities in certain foods may be important in protection against cancer. The existence of this array of protective substances adds credence to the admonition of dieticians to not rely on vitamin pills in lieu of a balanced diet. Several kinds of nonnutrient, nonvitamin, protective dietary substances not yet available in pills are discussed.

As stated, the information that follows is concerned primarily with cancer risk reduction by consumption of foods thought to protect. The other side of the coin is the avoidance of harmful substances. Alcohol is considered in this context. Other substances are thought by some to be potentially harmful, but many of these are of little day-to-day concern to most residents of the United States. Included in this category are aflatoxins produced by *Aspergillus* molds (Chapter 3). A direct relationship has been found between the level of food contamination by aflatoxins and human liver cancer in parts of Africa and East Asia. Esophageal cancer in several provinces of the People's Republic of China seems to be related to the consumption of pickled or moldy food.

8.4a   Dietary fiber and colorectal cancer

The historical importance of Burkitt. Denis Parsons Burkitt (see Box), a surgeon who spent considerable time in Africa, noted that individuals on a native diet did not suffer, or suffered less frequently, from a number of pathological conditions of the gut common in his home country. Most significantly, for the purposes of this chapter, was his observation that cancer of the colon and rectum occurred less frequently in Africa than in England or the United States (see Figure 7–7). The difference is striking when one compares colon cancer rate in Connecticut, 32.3 per 100,000, with Dakar, Senegal in West Africa, with an incidence of 0.6 per 100,000 (Weisburger and Wynder 1987). (Note: crude cancer rates may be misleading for comparison of countries where survival past fifty years of age is relatively rare compared to countries such as Sweden, Norway, Belgium, the United Kingdom, and the United States where survival past age seventy-five is common. Because of this, the age structure of a population is included in the computation of the age-adjusted rate for cancer prevalence or mortality that is used throughout this book.)

The cluster of digestive tract maladies correlated with bowel cancer suggested to Burkitt a common or related etiology. He noted that Africans consumed a diet with far more fiber than was found in contemporary English (and American) food and that there was an inverse relationship between total dietary fiber intake and colon cancer (Burkitt 1971).

However, while the observations of Burkitt show a correlation between fiber and colorectal cancer, etiology is not established by simple correlation (recall the nylon stocking and lung cancer correlation, see "Box on Ochsner," Chapter 7). Burkitt's views stimulated a spate of studies seeking understanding of the relationship between dietary fiber and colorectal cancer. A review of some of the literature suggests that a causal relationship between the two may not exist (Lawlor and Ness 2003, Giovannucci 2003; Curry et al. 2003). Nevertheless, an investigation of more than one-half million individuals, aged twenty-five to seventy, gave strong support to the protective role of fiber. The study concludes that in groups with low average intake of dietary fiber, a doubling of dietary fiber "could reduce the risk of colorectal cancer by 40 percent (Bingham et al. 2003)." Because prevalence and lethality of lower gut cancer is so high (Chapter 7), the possibility of a 40 percent reduction of that cancer in certain populations is indeed noteworthy.

Recommendations for reduced colorectal cancer risk. Even those who question the role of fiber in cancer prevention agree that with certain lifestyle alterations, including increased physical activity, avoidance of obesity, limiting alcohol consumption, limiting red meat and highly processed carbohydrate intake, the majority of colorectal cancers could be prevented (Giovannucci 2003; Curry et al. 2003). To these recommendations may be added that it is likely that folate (the anionic form of the B vitamin folic acid), may protect. People who smoke are urged to stop. There is discussion that foods with a high glycemic index (foods which are digested quickly with a sudden surge of glucose to the blood) may be related to an increased risk of colorectal cancer, but more studies are needed (Curry et al. 2003). The American Cancer Society recommends adding more vegetables and fruits to the diet and the avoidance of tobacco (Byers et al. 2002; Eyre et al. 2004). In summary and to repeat, "eating a diet rich in plant foods, in the form of fruit, vegetables, and whole-grain cereals, probably remains the best option for reducing the risk of colon cancer, and for more general health protection (Ferguson and Harris 2003)."

Regular colorectal screening does not prevent cancer, and thus is not discussed here. However, screening has the potential, when accompanied with proper medical treatment, of decreasing the lethality of the disease if and when it does occur.

The risk of colon cancer in the United States is significant. Hence, there is the possibility of an incredible savings of morbidity and mortality if recommendations for reduced risk are followed – we owe much to Mr. Denis Burkitt, who noted variant rates of colorectal cancer among the world's population and who was the prime motivator for many of the studies that followed.

Denis Parsons Burkitt (From: Jon A. Story and David Kritchevsky. 1994. Denis Parsons Burkitt (1911–1993). *J. Nutr.* 124:1551–4).

"Don't forget fibre in your diet"

The admonition that heads this essay is the title of a book written by the late Denis Parsons Burkitt. Charles Darwin was not the first to write on evolution and neither was Burkitt the first to consider the role of fiber in the diet. However, Darwin's name is inexorably associated with evolution as is Denis Burkitt's name with dietary fiber. The notion that fiber may have a role in cancer prevention is related to the medical career of this man from Northern Ireland. Denis Burkitt served in East Africa and Ceylon (now Sri Lanka) as a surgeon in the Royal Army Medical Corps during World War II. After the war, he sought for and obtained an appointment in Uganda with His Majesty's Colonial Service. He made two major contributions to medical science related to his experiences in Africa. The first was the description, distribution, and ultimately, the etiology of a pediatric cancer that bears his name, Burkitt lymphoma (Chapter 4). The second related to diet and diseases common in the western world.

Mr. Burkitt (British surgeons disdain the title "doctor") noted the rarity of diverticulosis, benign polyps, ulcerative colitis, appendicitis, hemorrhoids, other medical conditions, and most especially for this discussion, colon cancer, among the people he saw in his African clinic. These are of course common in developed countries such as England and the United States. He also noted that the native diet had far more fiber than that typical of the western world. Burkitt observed that with the high fiber diet was reduced "transit time", that is, the time it takes for feces to egress from the body after consumption of food. He suggested that if carcinogens were in the diet or produced by bacteria in the gut, colonic cells would be exposed to these substances for shorter times with the high fiber diet (because of reduced transit time). He studied fecal volume. Obviously, high levels of fiber result in high fecal volume which in turn dilutes noxious substances that may be in the gut. His thesis on fiber and its protective role was described in many publications, which gained world-wide attention and subsequent recognition for Burkitt.

While the jury is still out with respect to the role of fiber in prevention of colon cancer, the contribution of Burkitt to an awareness of the importance of diet to cancer cannot be minimized. Fiber is the substance of plant cell walls. All important advisory groups are unanimous in their support of the importance of vegetables, fruits, and whole cereal products for general good health and perhaps also for the prevention of colon cancer. If the latter, colon cancer prevention, indeed proves to be the case, it is possible that it is not the fiber *per se* that protects but it may be one or a combination of many other substances contained in plant cells with fiber that protects. Fiber may be the dietary marker for these other beneficial substances. Or, of course, it may be that fiber itself protects. Be that as it may, we are greatly indebted to Denis Burkitt for instructing us about the hazards associated with the typical western diet and a potential mode for the prevention of cancer. Concerning the possible role of fiber, note again (as indicated in the text) that a dietary change to increased fiber has the potential of reducing colon cancer risk by an astonishing 40 percent (Bingham et al. 2003).

Denis Parsons Burkitt was born 28 February 1911 in Lawnkilla near Enniskillen in Northern Ireland. He obtained his medical degree from Trinity College at Dublin University and studied surgery at Edinburgh. In addition to his election as a Fellow of the Royal Society of London, he was the recipient of the Lasker Award, the Charles Mott General Motors Cancer Award, Bristol-Myers Award for Cancer Research, the Buchanan Medal of the Royal Society and the Bower Award and Prize for Science of the Franklin Institute (among many other prizes, medals, and honorary degrees). Although he died in 1993, it is clear, indeed, that his contributions to cancer research will endure.

8.4b   Correlations between food substances and cancer prevalence: Significance

Before proceeding to other dietary factors that may relate to cancer, consider the following. Epidemiologists have provided much of the insight into the putative role of fiber (and other substances) in protection against cancer. Because colorectal cancer prevalence and dietary fiber consumption differed between African villagers and English boarding school students, Denis Burkitt postulated a protective role for dietary fiber. However, other differences between the populations exist, including affluence, possible racial vulnerability, other dietary factors, occupation, education, geography, and recreational activities.

Fiber is itself a complex mixture of substances difficult to quantify. Furthermore, fiber-containing vegetables and grains contain many substances other than fiber that may act alone or in concert with fiber to reduce risk. Finally, individuals who eat a large quantity and diversity of fruits and vegetables either may not desire or be unable to afford substances related to the etiology of cancer. Minced fatty beef (fast-food hamburgers) and potatoe fragments cooked in hot oil ("French" fries) are less available in a mostly rural country than in metropolitan England or the United States. These cautions are inserted to warn the reader that, thus far, we know very little concerning what it is in a particular food substance that may (or may not) provide protection against cancer.

Consider the remarkable correlation between breast cancer mortality and total dietary fat (see Figure 7–4). The relationship seems compelling. However, the "per capita intake" of a food item, in this instance fat, is calculated by adding the amount of the food item produced to the amount imported less the amounts exported, fed to animals, put to nonfood use, and lost in storage. This quantity is divided by the total population to provide an estimate of the per capita intake. The estimate does not take into account the quantities produced by individuals in gardens or on private farms, waste, and different patterns of consumption by subgroups differing in age and ethnic and economic background (National Research Council 1982). Hence, it is prudent to be circumspect when ascribing a cause/effect relationship of a substance with cancer in populations with differing cancer prevalences.

8.4c   Dietary fat and obesity

A relationship of food intake, dietary fat, and cancer in humans is indicated because individuals who are 25 percent or more overweight were reported to suffer a two-thirds increase in cancer incidence (Doll and Peto 1981) and a more recent study agrees; obese Swedish women were calculated to endure a 37 percent excess incidence in cancer. About two-thirds of Americans over the age of twenty are either overweight or obese (Eyre et al. 2004). Risk elevations of specific cancers, correlated with obesity, included (but are not limited to) malignancies of the small intestine, kidney, endometrium, gall bladder, pancreas, larnyx and connective tissue (Wolk et al.

2001). The report of the National Cancer Policy Board on cancer prevention states that there is convincing evidence of a "large" increase in risk of breast, endometrial, kidney and esophageal cancers with obesity (Curry et al. 2003). Whether it is total dietary fat, a particular kind of fat, cooking of fatty meat at high temperature, total dietary calories, obesity, and/or other factors that predispose to cancer remains to be clarified.

The American Cancer Society Guidelines (Byers et al. 2002) recommend that people limit their intake of high-fat foods. Because limiting intake of saturated fat may be important in cancer prevention, consumers should choose lean meat and fat-free dairy products. (A parenthetical comment is in order here: the most common ingredient in 2% "low fat" milk, other than water, is fat). Vegetable oils, such as olive oil, should be substituted for butter or lard. Baking or broiling instead of frying meat is prudent as well as emphasizing a diversity of vegetables in place of a large serving of meat. One is less likely to consume a huge serving of meat plus potatoes cooked in hot oil when several vegetables are available at the table and are attractively prepared and served.

8.4d   Vitamins and cancer

Vitamin A. Beta-carotene is a plant pigment converted by the gut and liver to vitamin A (retinol). The fat-soluble vitamin is essential for the growth and differentiation of epithelia and bone and it is important in normal reproduction and vision. Foods rich in vitamin A are green, leafy vegetables, yellow vegetables and fruits, as well as milk and liver. A diet rich in beta-carotene, and its related compounds, has been associated with risk reduction for some cancers. This may be due to the consumption of the vegetables and fruits per se, which are known to reduce cancer risk, and not to the beta-carotene that they contain. At this time, vitamin A is not known to reduce cancer risk but consumption of the vitamin seems to be inversely related to lung and possibly other cancers. Be aware that supplements containing large quantities of Vitamin A are toxic (Byers et al. 2002; Curry et al. 2003).

Vitamin C (Ascorbic Acid). The consumption of fruits and vegetables rich in vitamin C has been correlated with reduced risk of oral, larynx, esophagus, stomach, pancreas, lung, and breast neoplasms (National Research Council 1982). However, controversy surrounds the role of vitamin C. For instance, people who have a high intake of vitamin C in their diet also consume generous quantities of fiber, beta-carotene, and other vitamins. As with vitamin A (see preceding discussion), we seek to know if the reduced cancer risk is due to other nutrients in the diet, an interaction of ascorbic acid with the other substances, or to vitamin C itself. Note that vitamin C taken as a dietary supplement has not been demonstrated to reduce cancer risk (Byers et al. 2002).

Vitamin D (Calciferol). Colorectal cancer probably is inversely related to the consumption of the fat-soluble vitamin D. It is obtained in the diet (fish liver oils are an excellent source) or by exposure of the skin to ultraviolet radiation.

Vitamin E (Tocopherol) is not known to reduce cancer risk. However, it is possible that it *may* reduce risk of lung, cervical, colorectal, and prostate cancer (Byers et al. 2002; Curry et al. 2003).

8.4e   Selenium and calcium

Selenium. Epidemiological studies suggest a possible role for selenium in prevention of lung, prostate, and colorectal cancers (Salonen 1986; Curry et al. 2003). Selenium is a semimetallic element which functions as an antioxidant.

The amount of selenium in meat and vegetables varies with the level of that element in animal food, water, and soil. Meat ordinarily contains more selenium than vegetables. Because of the generous use of meat in their diet, most Americans are unlikely to suffer a deficiency of selenium. Selenium is added to fertilizer in Finland because of its putative value as an agent that may reduce cancer prevalence. Note, however, that an excess of selenium is toxic and because of the narrow margin between safe and harmful levels, dietary supplements have not been recommended (Byers et al. 2002).

Calcium. The most abundant metallic element in the body is calcium and adequate amounts (1,000 mg/day for people nineteen through fifty years of age; 1,200 mg/day for people over fifty) of the element can be obtained from dairy products and some green vegetables. Low-fat or fat-free dairy products are the preferred source of calcium. The element has been associated in some studies with a modestly reduced risk for cancer of the colon and rectum (Byers et al. 2002; McCullough et al. 2003; Curry et al. 2003).

8.4f   Non-nutrient organic compounds in food
that may protect against cancer

Chemically diverse, non-nutrient, substances in food may have a role in the chemoprevention of cancer in humans.

Lycopene. Prostate cancer (Chapter 7) is the most common internal malignancy of American males. No certain method of diminishing vulnerability to this nearly ubiquitous neoplasm of older men is known. However, tomatoes and tomato-based products (and watermelons) may provide protection. Risk reduction of about 40 to 50 percent in men who consumed large amounts of tomato products has been reported (Giovannucci 2002; Curry et al. 2003). The protective effect is believed due to the antioxidant lycopene, a carotenoid that gives tomatoes and watermelons their characteristic red color.

Substances found in cruciferous vegetables (currently known as "Brassicacous vegetables"). The consumption of cabbage, broccoli, cauliflower, Brussel sprouts, kale, and turnips is thought to be associated with a reduced incidence of cancer in human populations. Cruciferous vegetable-derived isothiocyanates, metabolized from glucosinolates, and other bioactive substances, protect cells in culture, experimental animals and perhaps humans from carcinogens (Wattenberg 1985; Xiao et al. 2003). The relationship between the isothiocyanates and protection is complex due to several factors which include human polymorphisms in biotransformation enzyme systems that alter protective effects afforded by the vegetables, the manner in which the vegetables are prepared, and even the disparate glucosinolate content of cruciferous vegetables (Lampe and Peterson 2002).

Garlic and other species of Allium. Plants of the genus *Allium* (e.g., garlic, leeks, chives, onions, and shallots) contain organic sulfur compounds, for example, diallyl disulfide, which may have cancer-preventing competence. The American Cancer Society believes there is at present insufficient evidence to support a specific role for these vegetables in cancer prevention (Byers et al. 2002). Nevertheless, an intriguing study from Shanghai, China, the country with the lowest prostate cancer rate in the world (Chapter 7), reports that consumption of *Allium* vegetables was inversely related to the risk of prostate cancer. Who, among the Chinese, was at lowest risk? Those men whose intake exceeds 10 g/day had significantly less prostate cancer than those whose intake was less.

Other phytochemicals. Space precludes discussion of the many other chemicals found in fruits and vegetables that may have salubrious health effects (Park and Pezzuto 2002). Obviously, the salutary effect of these chemicals is obtained by eating the diversity of fruits and vegetables recommended by the American Cancer Society and thus, precise knowledge of what the protective chemicals are is not a prerequisite for a good diet.

Will foods become genetically modified to reduce their health effects? A possibility that the protective effects of fruits and vegetables may be diminished somewhat in the future exists. Many parents are aware that finicky children (and some fastidious adults) disdain the smell and flavor of some plant foods, especially when cooking. Often it is the odor of protective phytochemicals that elicit the negative reaction. Selective breeding and food technology research to eliminate such phytochemicals, with the hope of enhanced acceptance by consumers, is underway. Associated with the flavor "enhancement" may be the loss of some or most of the chemoprotective substances (Barratt-Fornell and Drewnowski 2002).

Alcohol. The only substance thought to cause cancer in humans that has not been shown to cause cancer in animals is alcohol (Tuyns 1990). Hence, if dietary recommendations were made exclusively from animal studies, this section on alcohol would not exist. Further, for many years evidence has suggested that individuals who consume modest quantities of alcohol have a greater life expectancy (probably due to some protection from cardiovascular disease) than those who eschew alcohol consumption entirely (Pearl 1922). Why then does the American Cancer Society (Byers et al. 2002) urge limited consumption of alcohol? "Limited" here means for male drinkers no more than two drinks per day, and for women no more than one. Women at high risk for breast cancer (see Chapter 7) should consider total abstinence.

The reason for the recommendations ensue from convincing studies that report increased risk for cancers of the mouth, pharynx, larynx, esophagus, liver, and breast, and possibly the colon and rectum, with alcohol consumption (Curry et al. 2003). Particularly alarming is the reported synergistic relationship of tobacco and alcohol. Religious groups that do not use alcohol enjoy a low prevalence of breast cancer.

As with so many aspects of cancer biology, we have no simple explanation of how alcohol may cause its reputed pathologic effect. It has been postulated that alcohol stimulates the cytochrome P-450 microsomal enzyme system, which is involved in procarcinogen activation, that alcoholic beverages may contain carcinogen contaminants (some beverages have been reported to contain nitrosamines), and that alcohol induces changes in hormone profiles, resulting in cancer. Alcohol may also have direct cytotoxic effects. Cytotoxicity results in increased mitotic activity (increased cell division) to replace killed cells, resulting in more cells as targets for carcinogenic initiation (Chapter 3). Be that as it may, the multiplicity of purported mechanisms suggests much is yet to be learned. It is perhaps wise to consider the judgment of the International Agency for Research on Cancer (IARC) that alcoholic beverages are carcinogenic to humans and to heed the advice of the American Cancer Society to limit consumption of alcohol (Byers et al. 2002).

8.4g American Cancer Society (2002) Guidelines on Diet, Nutrition, and Cancer Prevention with a note on the efficacy of similar recommendations of the American Institute for Cancer Research

It is not yet possible to ascertain the precise role of specific substances in the diet and their impact on cancer risk. Nevertheless, many studies (some of which were cited here) suggest a consistent relationship between substances in the diet and vulnerability to certain cancers. On the basis of this information, the American Cancer Society has established guidelines (Table 8–1) thought to be prudent and likely to reduce cancer risk. Similar guidelines for cancer prevention were promulgated by the American Institute for Cancer Research (AICR) (World Cancer Research Fund Panel, 1997). Adherence to the recommendations of the AICR does indeed result in a reduced risk for cancer of about one-third among women aged fifty-five to sixty-nine who were followed from 1986 through 1998 (Cerhan et al. 2004). It is not yet known how much risk reduction would follow a lifetime of prudent diet, exercise, and no tobacco.

8.5 Exercise as it relates to cancer

The American Cancer Society recommends physical activity (Table 8–1). Nearly 17,000 Harvard alumni were studied, and not surprisingly regular walking, stair climbing, and sports participation were inversely related to mortality (Paffenbarger et al. 1986). Exercise and a decrease in mortality are related primarily to cardiovascular fitness. However, even though it may take twenty to forty years for cancer to appear because of a particular lifestyle, studies have detected an altered risk associated with exercise. For example, colon cancer was 60 percent to 80 percent higher in sedentary workers of Los Angeles County compared with occupations requiring greater physical activity (Garabrant et al. 1984). Colon cancer among 1.1 million Swedish men was similarly lower in active men (Gerhardsson et al. 1986). Agricultural, forestry, and sawmill workers were at reduced risk for colon cancer when compared with sedentary workers (Fredriksson et al. 1989). Breast and female reproductive system cancers were less common in former athletes compared with nonathletes (Frisch et al. 1987). These studies present compelling evidence that physical activity has a protective role in risk abatement of cancers of the colon and breast (Curry et al. 2003).

How exercise may exert this putative protective effect is not known. Peristalsis is stimulated with exercise, which may result in reduced transit time, thought by Burkitt and others to be important in the genesis of colon cancer (section 7.3e). Perhaps physical activity leads to a healthier diet (also discussed previously) because of appetite stimulation. Furthermore, active people may be less obese, more interested in healthful living, and shun activities that could contribute to certain cancers (Eichner 1987).

8.6 A special note about breast cancer

Risk factors for breast cancer were considered in the previous chapter. An issue that needs to be considered by women who live in "developed" countries is nursing of the newborn. "It is estimated that the cumulative incidence of breast cancer in developed countries would be reduced by more than half, from 6.3 to 2.7 per 100 women by age 70, if women had the average number of births and lifetime duration of breastfeeding that had been prevalent in developing countries until recently. Breast feeding could account for almost two-thirds of this estimated reduction in breast cancer incidence (Collaborative Group on Hormonal Factors in Breast Cancer 2002)." It is, of course, naïve to suggest and impractical to carry out, a recommendation for women to have more children. However, contained in the referenced study is the suggestion that breast feeding has an effect on breast cancer incidence and the take home message is that the longer a woman breast feeds her child, the more the woman is protected.

8.7   Other lifestyle hazards

Probably all human activity has some impact on cancer prevalence. Discussed in the section on breast cancer is the observation that nuns are at a higher risk than other women. With present knowledge, we recognize it is not the religious community that results in the greater risk for breast cancer but rather not having children. Similarly, it is well known that Kaposi's sarcoma (KS) is particularly prevalent among homosexual men. KS is the most common AIDS-associated malignancy. However, it is not being gay per se that causes Kaposi's sarcoma. The cancer occurs in individuals with immune dysfunction, seen among acquired immunodeficiency syndrome (AIDS) patients, gay or not. Present opinion indicates specific viruses and other less known factors, not lifestyle, as the cause of Kaposi's sarcoma (Section 7.5).

8.8   Summary

Lifestyle as it affects cancer prevalence is not a new notion. The former United States Secretary of Health and Human Services who served from 1983 to 1985, Margaret M. Heckler, believed that modification of lifestyle is an enormously rich lode to be mined for cancer risk reduction. Heckler (1984) stated, "Too few Americans realize the simple truth that cancer is usually caused by the way we live, and its risks can be reduced by the choices we make." Heckler estimated that 80 percent of cancers are linked to "lifestyle and environmental factors." That we are not entirely at the mercy of our environment is thus not a new notion.

How many cancers can be prevented by lifestyle modification? The answer is not known. Heckler was correct in noting that about 30 percent of all cancer deaths are due to lung cancer (this is as true in 2006 as it was in 1984; see Chapter 7), and she and her advisers suggested that perhaps as many as 35 percent are due to diet (Doll and Peto 1981; Byers et al. 2002; and Curry et al. 2003). Other environmental factors account for a much smaller percentage of cancer deaths. Heckler did not consider the effects of exercise, but we do and so does the American Cancer Society (Byers et al. 2002) and the American Institute for Cancer Research (Cerhan et al. 2004). What percentage of these deaths could be prevented with education and compliance to simple recommendations remains unknown.

Perhaps more important than speculation about numbers is the reported teaching success for diet modification and efforts to entice smokers to quit. These preliminary successes should serve to motivate educators as to the feasibility of risk reduction and stimulate even greater efforts. Instruction by health professionals is essential. Over 70 percent of the U.S. population sees a physician at least annually, which presents an exceptionally important opportunity to educate many of the millions of smokers. Books on the biology of cancer hopefully will also provide information about cancer prevention and risk reduction – and that is a goal of this chapter.

9

The stem cell basis of cancer treatment: concepts and clinical outcomes

Ralph E. Parchment

9.1 Introduction

Notwithstanding physical removal of malignant tumors by surgery and regional control of malignant tumors by radiation, the concept of cancer as a disease of abnormal stem cell biology leads to three philosophies for treating the disseminated, metastatic disease that most patients will face. Only one of these philosophies has become clinically useful, whereas the others are developing or remain at the conceptual level.

9.1a Therapies remaining at the conceptual level

One treatment to attack this caricature of normal renewing biological tissue is conversion, or perhaps reversion, of the malignant phenotype to a benign state. As studies of the fate of micro-injected cancer cells in the embryo have revealed (detailed in Section 1.15, Chapter 1), it is possible for malignant cells to lose their malignant potential without losing their proliferative potential, such that they contribute to normal formation of chimeric organs in the developing embryo. The development of chimeras containing some functional tissue derived from the implanted malignant cells indicates that the cancer cells reacquire the capacity to properly respond to biological controls on tissue renewal even into adulthood after the embryonic environment disappears. The concept is that once converted back to normalcy by these embryonic control mechanisms, wherein the malignant phenotype is removed or suppressed, cancer-derived normal cells in the adult chimera can no longer cause disease. Therapeutic strategies harnessing this biological control might not be toxic in the adult patient, because they would target malignant, embryonic-like stem cells, converting them into normally functioning adult stem cells. Unfortunately, there are not any approved therapies based on this concept, nor any therapeutic strategies in clinical development. Research continues on the chemical mediators of this phenomenon in the embryo and how to fundamentally eliminate the malignant phenotype without having to destroy the cell itself.

9.1b Therapies being explored clinically: Differentiation therapy and cytostatic therapy

A second treatment philosophy involves slowing the growth rate of malignancies to the point where they would *not* grow fast enough to reach a clinically significant size, or interfere in normal daily activities, during a patient's normal life span. Disease progression is slowed or halted entirely, and quality of life improved, even though cancer cells remain in the body. The proof that an abnormally small fraction of undifferentiated malignant stem cells in tumors terminally differentiate (Chapter 1) suggests it may be possible to increase the proportion of malignant stem cells that participate in this process. Since this would essentially reduce the number of stem cells contributing to tissue renewal, overall tumor growth rate should slow and the time required for tumor burden to reach lethal levels be prolonged. It is also well documented in nonclinical tumor models that individual cancers possess tissue sizer mechanisms which determine the size of a particular tumor when it enters plateau phase growth. Just like the normal renewing tissues that they caricature, as tumor burden increases toward this limit, growth rate will slow, growth fraction decrease, and the proportion of stem cells in the growth fraction decrease (explained in Sec. 9.5). These sizer mechanisms appear to be mediated by systemic factors that circulate in the blood stream. Administering substances that alter the biological properties of a cancer's tissue sizer mechanism to reduce the size at which it enters plateau growth phase would slow tumor growth at much lower tumor burdens than usual.

Vitamin A analogs known as retinoids comprise a class of therapeutic compounds that were originally developed in the clinic as "differentiating agents." In oncology, all-trans-retinoic acid ("ATRA", tretinoin, Vesanoid®) is a highly effective treatment for a specific subtype of leukemia called acute promyelocytic leukemia, which is driven by ectopic expression of the intracellular retinoic acid receptor due to a chromosomal translocation of its gene into a transcriptionally active region of the PML gene. It induces complete remission in patients with this genotype (but not cures), and tretinoin therapy is often cited as an example of differentiation therapy because it induces morphological maturation of APL in patients and in preclinical models. Increased tumor content of differentiated tumor cells has also been observed in nonclinical solid tumor models treated with ATRA (Speers 1982). However, recent experimental evidence in APL using patient specimens and re-evaluating the preclinical models points to induction of a TRAIL apoptosis pathway independent of terminal differentiation in responding leukemia cells (Salih and Kiener 2004), and it is not yet clear what is the relative contribution of apoptosis and differentiation to efficacy. It is conceivable that the apparent increase in the percentage of terminally differentiated cells might be due to selective removal of non-differentiated cells via apoptosis. In this chapter, tretinoin therapy will be considered a cytotoxic therapy that induces tumor cell death because it shows therapeutic features that are commonly associated with chemotherapy: remission followed by relapse, acquired drug resistance, and induction of cell death pathways as a mediator of drug action on the tumor population.

Another theoretical strategy for achieving differentiation therapy is termed "metronomic therapy." The idea is that chronic, low dose chemotherapy induces differentiation in malignant stem cells at exposure levels well below those required to cause apoptosis in the tumor or in any dose-limiting normal tissues. If correct, the idea predicts that metronomic therapy would achieve effective tumor control without the severe toxicity caused by cytotoxicity in normal tissues by regular dose chemotherapy. The idea arose in large part from findings that low dose, chronic therapy with a nucleoside analog named cytarabine ("Ara-C") induced leukemic cell differentiation (Weh, Zschaber, and Hossfeld 1984; Michalewicz, Lotem, and Sachs 1984; Jehn, De Bock, and Haanen 1984, Hoelzer et al. 1984). However, other evidence suggests that metronomic therapy is a form of cytotoxic chemotherapy (Leyden et al. 1984; Perri, Weisdorf, and Oken 1985). In leukemia, drug access to the malignant cells is not limited by solid tissue barriers like membranes to drug diffusion, so lower doses would be expected to have some benefit. In solid tumors, instead of killing malignant cells, the dosing regimen probably kills slowly renewing, normal endothelium in the areas of active angiogenesis in the tumor, cutting off the blood supply to the malignant cells (Bocci, Nicolaou, and Kerbel 2002). Low doses are effective because the endothelium is in direct contact with the circulation, where drug levels are highest, the endothelium does not exhibit the genetic instability that leads to drug resistance in tumors, and long exposure periods assure drug coverage during the long cell cycle. When it reduces nutrient and oxygen delivery to the tumor tissue by killing endothelium, apoptotic and necrotic tumor cell death would also occur over time via these indirect mechanisms. Without convincing evidence to the contrary, this text considers metronomic therapy as low-dose cytotoxic chemotherapy targeting drug action at the endothelium rather than tumor.

In addition to changing the ratio of differentiated, tumor-derived cells to malignant stem cells, it may also be possible to selectively arrest malignant stem and progenitor cells in the G_1 phase of the cell cycle, or induce proliferating tumor cells to exit the cell cycle into G_0 (quiescence). A drug with such a mechanism of action would be called a "cytostatic agent," because it induces stasis – a state of static balance or equilibrium – in the tumor cell population without inducing apoptotic or necrotic cell death. Treatment of metastatic disease with a cytostatic agent would be expected to halt the growth of existing lesions and prevent the appearance of new lesions. In addition, because tumors have an inherent rate of apoptosis due to

The 2001 Nobel Prize in Physiology or Medicine was shared by **Leland Hartwell, R. Timothy Hunt**, and **Sir Paul Nurse** for their seminal research into molecular mechanisms of control of the cell cycle in eukaryotic cells. This research built the current understanding of cell cycle checkpoints and the coordination of cell cycle progression with repair of cellular damage. The enzymes responsible for advancing eukaryotic cells between phases of the cell cycle might become therapeutic targets of true cytostatic agents that one day can be developed to stabilize tumor burden for extended periods of time. Drs. Nurse and Hartwell were awarded General Motors Alfred P. Sloan Prizes for basic cancer research in 1997 and 1991, respectively.

senescence of differentiated cells and death of cells that outgrowth their nutrient supply (Section 1.16, Chapter 1), treatment with cytostatic drugs should block tissue renewal of these dying cells and thereby achieve a slow and gradual shrinkage of tumor dictated by the particular rate of cell loss. Such cytostatic drugs might be particularly good chemopreventive agents to prevent microscopic lesions of malignant tissue from ever reaching clinically detectable size in high risk patients.

Over the last two decades, there has been highly productive and insightful research into the control of cell cycle progression, cell cycle checkpoint control, and enzymes that regulate these processes (see highlight box). These efforts have led to a wealth of new knowledge regarding cell cycle control and its interface with extracellular signals via receptor-mediated, signal transduction pathways involving enzyme kinases and acetylases/deacetylases. Cancer research has revealed that not only are many of these pathways abnormally expressed and/or regulated in malignancies, but also that these abnormalities are critical for sustaining the malignant condition (see Chapter 5). Thus, it is logical to develop drugs for cancer that exert therapeutic control over these processes, and identifying molecular targets which selectively confer a growth advantage upon the malignancy would be predicted to be a promising avenue for new drug development. Small molecules that target molecular engines which drive the growth advantage of malignancies include drugs designed to inhibit specific enzymes, and the last decade has witnessed FDA approval of several molecularly targeted agents that bind to receptor protein kinases like EGFR1 and 2, and *bcr/abl*/PDGFR, to antagonize their functions in signal transduction and stimulating proliferation. These drugs are less toxic than conventional chemotherapy drugs to normal tissue and therefore better tolerated by the patient, because the targeted molecular changes are tumor-specific.

In the current chapter, the question is how to relate the pharmacological action and clinical effectiveness of these new drugs to the stem cell basis of neoplasia. Under controlled in vitro conditions, the EGFR1 inhibitor gefitinib (Iressa®), approved for use in relapsing non-small cell lung cancer, induces G_1 cell cycle arrest in human tumor cell lines without evidence of apoptosis (Giocanti et al. 2004). However, gefitinib exposure must affect regulation of apoptosis in tumor cells because the same study found that the genetic context of the malignant cell dictates how gefitinib alters cancer cell susceptibility to cytotoxic chemotherapy (Knight et al. 2004). Furthermore, in the clinic, gefitinib is active only in a small subset of NSCLC patients (Janne et al. 2004), which is closely linked to the presence of certain mutations in EGRF1 sequence that confer enhanced susceptibility to kinase inhibition (Paez et al. 2004). In vitro, one Leu $\rightarrow$ Arg mutation – L858R – results in a much greater sensitivity of NSCLC cells to gefitinib-induced apoptosis, whereas one NSCLC tumor cell line containing wildtype sequence in the $EGFR1$ was G_1 cell cycle arrested at 50-fold higher concentration and the remaining cells with wild-type $EGFR1$ were not responsive (Tracy et al. 2004). In patients, tumor response to gefitinib is highly associated with these $EGFR1$ mutations (Sordella et al. 2004 Lynch et al. 2004). Although the question remains unresolved, these data point to the possibility that gefitinib achieves complete remission in the small proportion of patients when it acts as a cytotoxic agent to induce apoptosis, partial remission in other patients when it acts as a cytostatic agent to induce cell cycle arrest, and neither effect in the majority of the treated patients. The conclusion that signal transduction inhibitors are cytotoxic rather than cytostatic therapeutics is confirmed by studies of leukemic disease driven by the bcr/abl translocation. A small molecule inhibitor of the tyrosine kinase activity of this protein named imatinib (STI571, Gleevec®) induces cell cycle arrest $plus$ apoptosis only in leukemic cells expressing the chimeric gene product, but does not induce differentiation (Gambacorti-Passerini et al. 1997; Deininger et al. 1997). This drug potently inhibits a second receptor tyrosine kinase named c-kit, mutations in which cause constitutive activation and development of a rare, chemotherapy-resistant malignancy known as gastrointestinal stromal tumor (GIST). Imatinib treatment of GIST tumors achieves a very high response rate while inducing apoptosis (Demetri 2002; Demetri et al. 2002).

Unfortunately, despite the effort of many scientists, it has been difficult to discover chemical agents that are pure "cytostatic" drugs, and the biological consequences of targeted inhibitors of receptor kinases in cancers appear to be generally similar to that of existing cytotoxic drugs. An important difference is that targeted agents to signal transduction pathways which are mutated or overexpressed in malignant cells are not critical targets in normal, healthy tissue, so drug toxicity is much less severe and daily therapy is possible (advantages and disadvantages of these drugs are discussed in Chapter 10). The fact that compounds designed to be cytostatic turn out to induce apoptosis in human tumor cells perhaps should have been anticipated. Many of these drugs interrupt signal transduction pathways for trophic factors of the tumor, and progenitors that are trophic factor-dependent will undergo apoptosis upon withdrawal of the trophic factor action. In addition, cell cycle control checkpoints and apoptosis pathways are interconnected in most cells, such that cells with molecular damage-induced cell cycle abnormalities are eliminated via apoptosis, if the damage cannot be repaired. The conclusion is that development of pure cytostatic drugs, while surely of great potential benefit to patients, has not yet been achieved.

9.1c Eradicating cancer cells – the aim of current cancer therapy

The third treatment philosophy is embodied in the current cancer therapy modalities – surgery, radiotherapy, targeted chemotherapy, and cytotoxic chemotherapy – which aim to destroy the malignancy either by actively removing the tissue (surgery), or by directly killing the malignant cells with radiation or chemical damage (cytotoxic therapies), or by indirectly killing the malignant cells by interfering with critical survival pathways such as trophic factors (targeted therapy). A variety of treatment options have emerged within each modality over the past century, and these options are the clinical armamentarium against cancer and the most effective treatment options for patients. Because they aim to reduce the number of cancer cells in the body (the tumor burden), these therapeutic strategies are "cytoreductive therapies." Although later in the chapter we will see that it is not required for successful clinical outcome, these therapies have been developed with the aim of complete cytoreduction, that is, reducing the tumor burden in the body to zero before it can renew itself and cause relapse. The more realistic therapeutic goal of cytoreductive therapy is the eradication of enough malignant stem cells and their progenitor cell progeny from the body, such that the time required for the malignancy to regrow to clinically detectable relapse exceeds the patient's life expectancy if he/she did not have cancer. However, radiation and cytotoxic chemotherapy inflict damage to proliferating cells not only in the malignancy but also in the healthy cells in normal renewing tissues, thereby inducing apoptotic and/or necrotic cell death that is both desirable (in the tumor) and undesirable (in normal tissues). In this chapter, we will examine how the relative effects of drug toxicity in malignant versus normal tissue, and the rate of recovery from this damage, determine drug selectivity and dictate what is achievable with chemotherapy. Not all chemicals that are cytotoxic to proliferative cells can serve as anticancer drugs; only the chemicals that can selectively damage and kill malignant cells over healthy normal cells can be cancer medicines. Surgery also removes normal tissue, and if the amount of normal tissue that must be compromised to resect the malignant tissue exceeds the reserve capacity of the tissue, then the disease is "inoperable" and cannot be treated surgically.

A few small molecule drugs used in cancer therapy induce apoptosis of malignant cells by indirect mechanisms of action. Instead of directly damaging malignant cells as most cytotoxic drugs do, these few drugs disrupt molecular pathways required for cell survival. Gefitinib and imatinib very likely lead to activation of the apoptosis pathway, as transduction of cell survival signals via the EGFR1 and bcr/abl/c-kit/PDGFR pathways are disrupted, respectively. In contrast to cytotoxic chemotherapy that inflicts damage upon malignant cells, targeted therapies harness control of molecular pathways that regulate apoptosis and induce cell death under conditions where it normally would not occur naturally. Many indirect acting cytotoxic agents are antagonists of trophic factor binding to receptors, and eliminating the sustaining stimulus of these factors leads to removal of the block of apoptosis and atrophy of tumors. Perhaps the best known example of trophic factor blockade is the antisteroid control of estrogen receptor positive breast cancer and androgen-dependent prostate cancer. Since steroid-responsive normal tissues are usually not critical for survival (i.e., breast, prostate, and uterus), some of the well-tolerated drugs in this class that are proven to work against certain cancers are now being used in daily treatment to prevent cancer development in high risk individuals. Chemoprevention trials are starting to evaluate the use of very safe drugs proven to control advanced malignancy to try to affect these same processes under circumstances prior to clinical malignancy, such as the STAR trial comparing the estrogen blockers tamoxifen and raloxifene (http://www.nsabp.pitt.edu/STAR/Index.html). It is important to note, though, that unanticipated toxicities have been encountered in clinical trials of other chemopreventive agents, such as decreased patient survival (Omenn et al. 1996). It is conceivable that these targeted therapies eradicate malignant cells as soon as they arise, so it appears that they prevent cancer when in fact they only prevent growth of microscopic disease to clinically detectable size. A first-in-class proteasome inhibitor named bortezomib (PS-341, Velcade®) that is FDA approved for multiple myeloma therapy induces pleiotropic changes in pathways that regulate apoptosis via differential effects on intracellular degradation of ubiquinated proteins, including a number of ligand receptor kinases and signal transduction enzymes. Another new class of drugs interferes with the trophic factors that promote survival of immature endothelium, block formation of new blood vessels and vessel renewal induced by angiogenesis factors released from tumors, and indirectly kill tumor cells by starving them of essential nutrients. This is similar to the concept of metronomic therapy discussed above, except that inhibition of angiogenesis and vascular destruction were intentional in the design of the drug.

Components of the immune system have been harnessed for use as cytoreductive therapies, and there are examples of immunological drugs that directly induce apoptosis in malignant cells (interferon, interleukin-2) as well as those that indirectly induce apoptosis (monoclonal antibodies). The mechanism of action of these drugs should always be considered complex, because they not only exert direct effects on cancer cells but also can indirectly affect the course of malignancy via modulating effects on cellular immunity, for example, activation or stimulation of LAK, NK, T-cells, and dendritic cells. Significant advances have been made in the past decade to bioengineer monoclonal antibodies with greater potency, safety and pharmaceutical properties to mediate a variety of therapeutic actions. Some monoclonal antibodies interfere with receptor – ligand interactions for trophic factors of tumor cells or their vasculature, such as EGFR1 binding by erbitux (C225), EGFR2 binding by tratuzumab (Herceptin®) or VEGF by bevicizumab (Avastin®) (see highlight box).

There are obviously a large variety of mechanisms of action for cytoreduction, either indirectly or directly eliciting cell death via cellular programs for apoptosis and sometimes necrosis, too. Only the substances that can achieve such tumor cell kill, while sparing enough healthy normal cells in critical tissues for continued survival with good quality of life, ever reach the level of being approved for clinical use. However, therapy with any of these agents follows common biological principles that relate tumor cytoreduction to tissue renewal in malignant and normal tissues.

bevacizumab, trastuzumab, gefitinib, imatinib, bortezomib... Good grief! Who names these drugs? Nonproprietary drug names are selected by the United States Adopted Names (USAN) Council, which is tri-sponsored by the American Medical Association (AMA), the United States Pharmacopeial Convention (USP), and the American Pharmacists Association (APhA). The USAN Council establishes logical nomenclature classifications based on pharmacological and/or chemical relationships (see http://www.ama-assn.org/ama/pub/category/2956.html). Recently approved name stems include the –mab stem for monoclonal antibodies (http://www.ama-assn.org/ama/pub/category/13275.html). Another point of bewilderment is the FDA approval process for new oncology products, which has been nicely summarized in a recent publication by FDA oncologists (see Johnson et al., Journal of Clinical Oncology 2003;21:1404–1411 and Dagher et al. J Natl Cancer Inst 2004;96:1500–1509).

The current chapter connects the concept of malignancy as a caricature of stem cell renewal with these principles of cytoreduction to build an understanding of effectiveness and limitations of therapy. These merits and drawbacks of the particular therapeutic classes, as well as an introduction to their uses and mechanisms of action, will be considered in more detail in Chapter 10.

9.2 Absolute versus fractional cytoreduction

Usually, cancer is first detected in a patient when the tumor burden reaches $10^9 - 10^{10}$ malignant cells, and a tumor burden of 10^{13} is approximately the number of cells that are lethal (Haskell 1990, Haskell et al. 1990). For understanding the principles of cytoreductive therapy, let us use a hypothetical patient with a tumor burden of 10^{12} cancer cells at the time of diagnosis. Immediately this case generates a critical question for planning therapy, defining the course of the disease, and anticipating the future of the patient: where are the 10^{12} cancer cells located? If *all* of the cancer cells are localized to a particular spot of tissue that is operable, the patient will be cured with surgery alone. Since surgery reduces the tumor burden by the absolute number of cells contained within the resectable tissue, tumor burden does not affect the outcome of surgery unless the tumor has become so large and integrated with critical normal tissues that its removal jeopardizes patient survival ("inoperable" tumors). The patient may harbor any number of malignant cells, but will be cured by the surgeon as long as all of the cells can be removed during the surgical procedure. The cytoreduction is definitive because the malignant cells are removed from the body, and the tumor burden removed is simply the absolute number of malignant cells in the resected tissue.

At the other extreme is the patient who harbors numerous malignant deposits throughout the body, which in sum total equal a tumor burden of 10^{12}, yet may be distributed across, for example, 20 metastatic lesions each containing on average

5×10^{10} cancer cells. The malignant cells have infiltrated many tissues, replacing normal tissue parenchyma required for adequate organ function with tumor tissue and perhaps ectopic production of endocrine and paracrine factors. This patient cannot benefit from surgery, because the cells are not localized to just a few particular spots, not all of the cancer cell deposits are large enough to detect with diagnostic procedures, and the tissue loss required to surgically remove 20 metastatic lesions from the body would probably be life-threatening itself. The patient's only recourse is systemic therapy, meaning a therapy that will circulate throughout the body and potentially expose all malignant cells to the treatment – even without knowing where all of the cancer cell clusters are located in the body. Chemotherapeutic agents can be injected into the circulation, or reach it via absorption from the gut after oral administration, where they circulate throughout the body and gain access to the tissue parenchyma via diffusion from capillaries. Hence, cancer drugs provide the systemic therapy that the patient needs. However, in contrast to surgery, each chemotherapy treatment carries a *probability* of encountering a given cancer cell during a susceptible period, and some cancer cells may not be in a state where they are sensitive to the action of the drugs. Therefore, the cytoreduction achieved with chemotherapy is fundamentally and mathematically different from that achieved with surgery, because the former is fractional whereas the latter is absolute. Whereas surgery eradicates the absolute number of cells contained within the resected area, chemotherapy kills a proportion or fraction of the cancer cell burden exposed to drug. This fractional cytoreduction is much more akin to logarithmic reduction each time the tumor burden is exposed to the drug, so it is conventional to quantify cytoreduction using the unit of log-cell kill, where each log-cell kill means a 10-fold reduction in cell number (one logarithm reduction). For example, a 2-log cell kill means that 99 percent of the cells are destroyed (100 reduced to 1). Because it is logarithmic, it is important to note that the absolute amount of tumor cell kill from exposure to a chemotherapeutic drug is mathematically dependent on the initial number of cells. Although a 99 percent log-cell kill rate seems excellent, such a 2-log reduction leaves 10,000 surviving cells in a tumor of 1,000,000 cells, and 10^{10} survivors in the example patient who has a more typical tumor burden of 10^{12} malignant cells. In either case, note that there are residual cancer cells that have survived drug exposure, and multiple drug exposures will be required to eradicate the tumor burden completely, which is why patients must receive multiple cycles (courses) of cytotoxic therapy, or daily exposure to the less toxic targeted agents, to have a chance at cure.

Radiation therapy achieves cytoreduction by using ionizing radiation emitted from external or internal sources of radioisotopes to induce cancer cell death. Radio-therapy possesses some attributes of surgery and some attributes of chemotherapy. Like surgery, radiation therapy is effective when treating cancer that is localized to a particular known or suspected location in the body. However, larger areas can be treated with radiotherapy than surgery because the radiation can be focused on a "tissue field" to treat dispersed cancer cells within a tissue site that is too large to resect and may not be detected by diagnostic procedures. Only a certain volume of tissue can tolerate a certain radiation dose though, before normal tissue damage and loss of function result in toxicity that can be life-threatening. On the other hand, radiotherapy is similar to chemotherapy in the way that the resulting cytoreduction is characterized by log-cell kill rather than the absolute destruction of some number of cells in the irradiated field. Radiation is particularly useful for treating individual lesions, or a cluster of lesions, of cancerous tissue that are inoperable because of location, the critical nature of the involved tissue, or extent of normal tissue involvement (brain, spinal cord, arteries).

The log-cell kill principles of curative cytotoxic therapy were discovered in animal models of chemotherapy (Schabel 1969, 1975; Skipper et al. 1970; Shackney, McCormack, and Cuchural 1978; Norton and Simon 1986; Norton 1979, 1990). These principles show that cure occurs only when the total log-cell kill from all treatment courses has eliminated the last malignant cell, because a single malignant cell has the capacity to regrow into a symptomatic tumor and kill the host (Skipper et al. 1970). Thus, the goal of cytoreductive therapy is often stated as reducing the surviving number of malignant cells to less than one. This principle that cure only occurs after eradication of the last remaining malignant cell is often called the Skipper hypothesis to honor its discoverer (see highlights box).

The concept of renewing tissues as mosaics of discrete and separate populations or phenotypes in a symbiotic relationship, so-called unit characters during tumor progression (Chapter 1), explains why a constant fraction of the tumor cell population is killed by cytotoxic therapy with each cycle. If phenotype A always occupies 90 percent of the malignant tissue at homeostasis, then no matter how many tumor cells are present, a dose of therapy that selectively kills cells of phenotype A will result in one log-cell kill. The surviving cancer cell population will readjust for this loss, and a new patch of phenotype A will develop to occupy 90 percent of the surviving cell population. Other biologic patterns beside mosaics may determine sensitivity to therapy. In addition, mosaic patterns can be further complicated by differences in physiological microenvironment. A cancer cell found in a poorly vascularized

---

Beginning in 1946 at the Southern Research Institute in Alabama, **Howard Skipper** made fundamental discoveries concerning the theoretical basis of cancer chemotherapy, its success and failure. His research established important principles of effective cytoreductive therapy with chemotherapeutic agents, including the concepts of log-cell kill, optimal scheduling of chemotherapy relative to tumor cell kinetics, combination chemotherapy to overcome drug resistance, and curative intent with proper planning of dosage regimens that led to cures of acute leukemia, lymphoma, and solid tumor malignancies. The significance of his work has been recognized by induction into the Alabama State Academy of Honor (1982), and receipt of an Albert Lasker Award for Basic Medical Research (1974) and the Charles F. Kettering General Motors Prize (1982).

area may be quiescent due to low nutrient and oxygen concentrations, insensitive to drug effects or ionizing radiation, and exposed to only low drug levels. Another cancer cell with the same phenotype in a well-vascularized area will be maximally proliferative and oxygenated, highly sensitive to drug action and ionizing radiation, and exposed to the highest drug levels.

There is a very important difference in the cytoreduction and clinical outcome achieved over time between a therapy that kills a constant proportion of cells and one that kills a constant number of cells, which is illustrated in Figures 9–1 and 9–2. This example shows the result of treating three patients, each harboring 10^{12} cancer cells at diagnosis but with very different distributions of these cells in the body. Figure 9–1 presents these outcomes on a linear scale graph where absolute cytoreduction from surgery and from chemotherapy by course is obvious. Cancer cells in the first patient were all localized within an operable tissue site, so resection resulted in an absolute cytoreduction of all the tumor cells to a tumor burden of 0, and the patient was cured by the procedure (solid squares). The second patient had disseminated disease and received chemotherapy every month for seven months that obeyed log-cell kill mathematics (open squares), so each treatment achieved a 2-log reduction in cancer cell burden (99% cell kill) resulting in a reduction in the initial tumor burden of 10^{12} cells to <1 surviving cancer cell upon completion of the seven courses of therapy. After two courses of this therapy, tumor cell burden fell below clinical detection limits of 10^{10} cancer cells and the patient was in remission, even though five more courses of therapy without any clinical evidence of disease were required to achieve a cure. This log-based mathematics is the reason that patients must receive multiple cycles (courses) of cytotoxic therapy to have a chance at cure, even continuing therapy after complete remission has been achieved. The third patient also had disseminated disease and received five treatments but with a theoretical chemotherapy drug that obeys absolute cell kill mathematics, eradicating 200×10^9 cancer cells per course (open circles). The patient was cured after five courses of treatment, having been reduced from a tumor cell burden of 10^{12} to 0, although the patient suffered from clinically detectable disease (>10^{10} tumor cell burden) until the fifth and final drug treatment.

Figure 9–2 is a different mathematical presentation of the same three patients and their clinical outcomes. Tumor burden versus time is now plotted on a log-linear plot, where the important difference between absolute and fractional cytoreduction is easier to understand. The graph indicates an advantage of absolute over fractional cytoreduction with chemotherapy, because five rather than seven courses of therapy are required to reduce tumor burden to <1 surviving cancer cell and cure the patient. However, studies of chemotherapeutic agents and ionizing radiation indicate that log-cell kill principles more accurately describe their anti-tumor effects. Based on a tumor burden of 10^{10} for clinical detection, one can see why effective chemotherapy treatment must continue to be administered while the disease is in remission in order to achieve cure. Most of the courses of chemotherapy are administered to eradicate

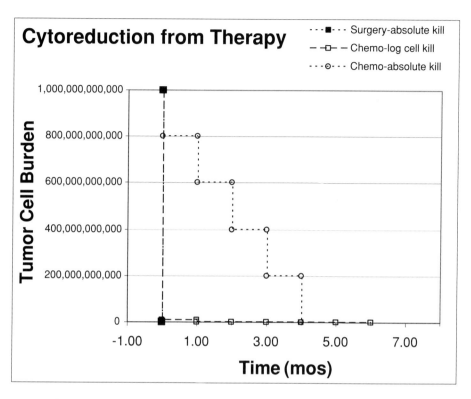

Figure 9–1. The different nature of cytoreduction between absolute cell kill and log-cell kill per cycle of treatment. Modern cytotoxic therapies such as radiotherapy or chemotherapy reduce tumor burden by a fractional amount each cycle, rather than an absolute amount. The proportion of malignant cells killed per cycle is quantified as "log-cell kill." This example shows the result of treating three patients, each harboring 10^{12} cancer cells at diagnosis but with very different distributions of these cells in the body. Cancer cells in the first patient were all localized within an operable tissue site, so resection resulted in an absolute cytoreduction of all the tumor cells to a tumor burden of 0, and the patient was cured by the procedure (solid squares). The second patient, who suffered from disseminated disease, received seven chemotherapy treatments that obeyed log-cell kill mathematics (open squares), so each treatment achieved a 2-log reduction in cancer cell burden (99% cell kill) resulting in a reduction in the initial tumor burden of 10^{12} cells to <1 cell after seven courses of therapy. After two courses of this therapy, tumor cell burden fell below clinical detection limits of 10^{10} cancer cells and the patient was in remission, even though five more courses of therapy without any clinical evidence of disease were required to achieve a cure. The third patient, also with disseminated disease, received five treatments with a theoretical chemotherapy drug that obeys absolute cell kill mathematics, eradicating 200×10^9 cancer cells per course (open circles). The patient was cured after five courses of treatment, having been reduced from a tumor cell burden of 10^{12} to 0, although the patient suffered from clinically detectable disease (>10^{10} tumor cell burden) until the fifth and final drug treatment.

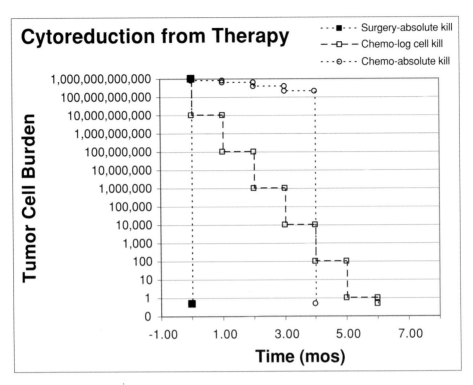

Figure 9–2. This figure presents the same data as Figure 9–1, but now tumor burden over time has been graphed on a semi-log plot, so the differences in tumor burden over time resulting from different strategies for cytoreduction can be seen in greater detail. Clinically detectable tumor burden is 10^{10}. This scale reveals the important mathematical difference between absolute and fractional (log-cell kill) cytoreduction for the clinical course of the treated disease. If the action of chemotherapy drugs followed absolute cytoreduction principles, the patient would continue to experience clinically evident disease ($>10^{10}$ tumor burden) until the fifth and final course of drug therapy, which will reduce the tumor burden to 0 (open circles). However, the activity of chemotherapeutic drugs in clinical use follows log-cell kill principles (open squares), so patients in remission must continue to receive chemotherapy treatment for additional courses to completely eradicate the disease and reach a tumor burden of <1 surviving cell (mathematically, log-reductions never reach "0," they only approach it). In most cases, confounding factors like drug resistance develop in a very small proportion of the tumor burden, so complete cytoreduction is not possible, and a subclinical tumor burden survives to regrow, causing a clinical relapse that is no longer responsive to the original drug therapy.

the subclinical disease that is still present in a patient in remission. The seventh course of drug therapy is required to cure the patient by theoretically eliminating the last cancer cell.

When the 10^{12} malignant cells are concentrated within a particular tissue location, surgery is obviously preferred over cytotoxic therapy, because a single, definitive procedure can eradicate all of the cancer cells and achieve cure. When the same tumor burden is disseminated locally or systemically, and is therefore inoperable, radiotherapy over many weeks or seven courses of toxic chemotherapy ($\sim$6–8 months) are required to eradicate it, assuming that the 2-log cell kill per cycle of therapy can be maintained without the development of drug resistance. If the oncologist is fooled by the remission into thinking that the patient has been cured after fewer than seven cycles and stops chemotherapy, the tumor will relapse to detectable tumor burden, due to the proliferation of the residual tumor burden to replace the lost mass of tumor tissue, as any renewing stem cell system will do. On the other hand if the surgeon is fooled by the inability of diagnostic procedures to detect 1,000 malignant cells that have already metastasized outside the surgical margins before the resection, the patient will be declared "cancer free" after surgery (without clinically detectable disease), but these residual 1,000 cells will eventually regrow and cause a clinical relapse of metastatic disease.

9.3   The meaning of "curing cancer" depends on whom you ask

Although patients conceptualize a cancer cure to mean the complete eradication of cancer cells from their bodies, in fact what they usually mean is that they want to live the rest of their lives without any clinically detectable tumor or, if that is not possible, at least without any symptoms of the disease that affect their daily activities. They want to live the rest of their lives "cancer free." In the patient's experience, any tumor burden between 0–10^{10} is acceptable, because these tumor burdens do not usually cause clinically detectable disease, and cancer would be "cured" if there were therapies that could reduce and maintain these sub-clinical tumor burdens. This understanding of a cancer cure is quite different from the Skipper hypothesis that developed from laboratory studies of chemotherapy, which by definition requires eradication of every viable cancer cell. A third definition of curing cancer, which is the clinical definition, means the continuing absence of any evidence of disease using the best diagnostic and imaging technologies, so it is much closer conceptually to the patient's understanding than the laboratory's understanding of a cancer cure. The purpose of cancer therapy is to achieve a normal life expectancy in the cancer patient with no more than minimum impact of the cancer on daily living.

A major difference between the cancers used in animal models and those that arise in human patients has to be considered when extrapolating the principles of cancer cure to the clinic. Most experimental tumor models in animals have been selected to grow relatively rapidly, so experimental results can be obtained over a period of months. In contrast, most human cancers regrow much more slowly than the tumors used in the animal models, and a clinical case of cancer usually unfolds over a period of years or even decades. The growth rate of residual tumor burden after treatment is an important factor in whether or not a patient will be cured of the cancer. In addition, the life expectancy of the patient relative to the tumor growth rate is also an important determinant of treatment outcome. Consider the following scenarios. It is apparent from Table 9–1 that surgical resection of even the earliest detectable cancers leaves behind some micrometastatic disease in some patients. In a patient

Table 9–1. *Cure rates for early-stage cancers (node negative where indicated) following single-modality surgery. If these diseases were truly localized at time of diagnosis, then cure rates should be 100%. The fact that cure rates are less than 100% indicates that many cases are systemic disease at time of treatment. Surgery is curative in true stage I disease (which is difficult to prove at diagnosis) and in those patients whose life expectancy is less than the time required for the disseminated microscopic disease to grow to clinically detectable levels.*

Malignancy		Long-term survival	References
Breast carcinoma	(node negative)	77–87 percent, 3-year survival	Fisher et al. 1977; Fisher et al. 1985b; Fisher, Redmond, Poisson et al. 1989
Breast carcinoma	(stage I)	98%, 5-year survival	American Cancer Society 2005
Stomach	(node negative)	81–98%, 5-year survival	Haskell, Selch, and Ramming 1990b; Wu et al. 1992
Kidney		60–76%, 5-year survival	Myers, Fehrenbaker, and Kelalis 1968; deKernion and Berry 1980; Neuwirth, Figlin, and deKernion 1990
Ovary (stage I)		96%, 5-year survival	Berek and Hacker 1990
Non-small cell lung		57–86%, 5-year survival	Figlin et al. 1990
Uterus		76–80%, 5-year survival	Barber and Brunschwig 1968; Berman and Berek 1990

where surgery removed the entire tumor burden, a cancer cure has resulted whether the clinical or laboratory definition is used (Figure 9–3, patient 1). In the patients with micrometastatic disease, the residual subclinical tumor burden will regrow at a rate determined by the tumor doubling time and microenvironmental influences, but this regrowth will be irrelevant if the patient dies from other causes before the relapsing tumor reaches clinically detectable tumor burden. By comparison, consider two younger patients who undergo complete surgical resection of clinically evident disease that leaves a tumor burden of 10^3 cells in a micrometastatic deposit (Figure 9–3). The cancer in patient 2 has a fast doubling time that causes a relapse two years after surgery, and after another year of regrowth reaches lethal tumor burden levels three years after surgery. The same amount of residual tumor burden in patient 3 regrows much more slowly, reaching clinical detection after six years and lethal levels after 8.5 years. In both patient 2 and 3, the surgical procedure was equally effective, leaving 10^3 malignant cells that were undetectable outside the resected area at time of surgery. However, the residual tumor in patient 2 regrew with a more rapid doubling time than in patient 3, causing relapse and death to occur much sooner after surgery. Now consider the case that the 10^3 residual tumor burden in patient 3 regrows with a doubling time just slightly slower than the tumor in patient 4, which is on a trajectory to relapse at five and one-half years after surgery. If patient 4 happens to be older and dies from a heart attack five years after surgery, prior to relapse, then the patient, the patient's family, and the doctor would understand this case as a cancer cure. Therefore, patients 3 and 4 survived longer than patient 2, even though surgery was equally effective in all patients, and patients 2 and 3 died of disease, while patient 4 "never relapsed." Both

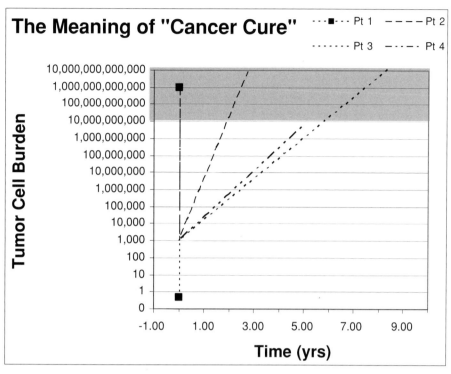

Figure 9–3. The meaning of "curing cancer" depends on the combined influences of the effectiveness of the treatment, the growth rate of residual tumor burden and the life span of the patient. Four patients, each harboring a tumor burden of 10^{12} cancer cells, were treated with surgery. In patient 1, the entire tumor burden was localized within the resected tissue, so the tumor burden was reduced to <1 cell, by the procedure, resulting in a cancer cure that is durable for the patient's life span. In patients 2, 3, and 4, 10^3 malignant cells had spread outside the resected area by the time of surgery. However, the residual tumor in patient 2 regrew very rapidly and caused relapse two years after surgery and death at three years due to lethal tumor burden. The residual tumor in patient 3 regrew much more slowly, reaching clinical detection after 6 years and lethal levels after 8.5 years. The 10^3 residual tumor burden in patient 4 regrows with a doubling time just slightly faster than the tumor in patient 3, on a trajectory to relapse at five and one-half years after surgery. If patient 4 dies from cardiac arrest at five years, prior to detectable relapse, the patient, the patient's family, and the doctor would understand this case as a cancer cure. Therefore, patients 3 and 4 survived longer than patient 2, even though surgery was equally effective in all patients, and patients 2 and 3 died of cancer while patient 4 never relapsed. Both patients 2 and 3 will be cancer fatalities because they lived long enough to die from regrowth of residual disease. The shaded area indicates the tumor burdens typically associated with clinically detectable (10^{10} cells) and clinically fatal disease (10^{13}).

patients 2 and 3 will be cancer fatalities and "surgical failures" because they lived long enough to die from regrowth of residual disease. This example underscores the importance of understanding the term "curing cancer" in the context of the patient's life expectancy. It also should influence concepts for developing effective cancer treatments by considering age of the patient at diagnosis and other factors that influence life span to determine what the therapy will need to achieve to be considered effective at the clinical level. Not all of the malignant cells need to be eradicated from the body, because a patient desires only to never be bothered again by the malignancy.

An interesting point to consider is that chemotherapy or radiation therapy for treating the residual tumor burden outside the resected area in patients 2 and 3 would make a difference in their survival, but treating patient 1 the same way would cause only toxicity without benefit. It is usually quite difficult to distinguish patients with and without micrometastatic disease, so chemo-and/or radiotherapy are often combined with surgery for everyone in a specific prognostic group to make sure that those who will benefit are not under-treated. In essence, the lack of sophisticated diagnostic and prognostic tests results in all patients being treated with combined treatment, including those very few patients who would have been cured by surgery or localized radiotherapy alone. The relatively long time required for repopulation of most tumors after cytoreductive therapy is also the reason why doctors tell patients that they cannot be considered to be cured until they have survived disease-free for five, ten, or even fifteen years, depending on the historical experience of regrowth patterns of particular cancers in patients who were not cured. Before this time, the cancer may be regrowing at tumor burdens that are not clinically detectable.

Log-cell kill principles and the Skipper hypothesis definitely apply to clinical oncology, and the complete eradication of malignant cells is a reasonable goal for patients whose life-expectancy will be long enough for tumor regrowth to clinically detectable relapse to occur. However, these concepts must be tempered by the surrounding clinical circumstances of each case, because the intensity of treatment depends not only on the required log-cell kill to eradicate disease but also on the age of the patient and any other medical conditions that might shorten life expectancy. In summary, patient life expectancy shortened by other causes and medical conditions are variables that simply cannot be factored into the animal models.

9.4   The biological basis of multimodality therapy as optimal cancer treatment

Cancer specialists refer to surgery, radiotherapy, and chemotherapy as therapeutic modalities, and in the earliest days of modern cancer therapy, patients were usually treated with a single modality at a time. Chemotherapy was often reserved for patients failed by surgery and radiotherapy. However, careful biological studies and clinical studies of the natural history of the disease revealed that locally invasive disease (Stage II) at diagnosis is actually disseminated, systemic disease at the microscopic level, which disease will relapse if the patient lives long enough for the microscopic disease to growth to clinically detectable tumor burden. All patients with intermediate stages or high grades must be assumed to have metastasized by time of diagnosis (see Chapter 1, Section 1.4, to understand tumor stage and grade). Even the earliest detectable cancers of very small size (Stage I) are not always curable surgically (Table 9–1). For example, malignant ovarian cells from even grade I stage 1a tumors may be found outside the tumor margin in adjacent tissue, and survival rates are 94 percent to 98 percent at six-year median follow-up rather than 100 percent (Colombo 1994; Young 1994). This clinical experience that almost all patients treated with single modality therapy relapse if they live long enough, the uncertainties mentioned in the preceding section, and the biological understanding of cancer as fundamentally a disease of invasion and metastasis by stem cells combine to necessitate worst-case assumptions about cancer cases. Both local and distant disease must be presumed at diagnosis unless there is overwhelming weight of evidence that the malignancy is truly localized to its site of origin. Sequential therapy with each modality has evolved into a more biologically appropriate strategy that combines proper timing of surgery, radiotherapy, and chemotherapy. Combining these modalities into an integrated treatment strategy is called multimodality therapy, and in most cases it offers a better chance of cure or long-term survival than any single modality therapy. This strategy combines surgery for decreasing the tumor burden, radiotherapy for control of regional invasive and micrometastatic disease and inoperable focal lesions, and systemic chemotherapy for micrometastatic disease in other organs. Curing the cancer patient requires the combined use of effective local, regional, and systemic therapies, which is the rationale behind multimodality therapy. Obviously, systemic drug therapy is the only treatment option if the disease is disseminated at diagnosis, but in these cases surgery and radiotherapy are used for palliation of symptoms or for resolving complications of organ function like obstruction or compression by individual tumor lesions.

Given that in the context of likely micrometastatic disease the purpose of surgery is primarily tumor debulking (reducing tumor burden), restoration of function and relief of symptoms, the effectiveness of radiotherapy and chemotherapy become critical for achieving long-term, disease-free survival (a "cure", if the disease does not growth back during the patient's life span). Consider the example of a patient diagnosed with a 6 cm tumor that is resected successfully with clear margins. This tumor of approximately 100 g contains about 10^{11} cells. However, some regional lymph nodes dissected during the operation were found to contain malignant cells, so the patient is at high risk for relapsing from distant metastases later in life, even though thorough diagnostic imaging procedures have not detected disseminated disease at the present time. Typically, as many as 10^{10} disseminated tumor cells are not clinically detectable at time of diagnosis (Haskell 1990), so the worst-case scenario must be assumed that this many cells exist in distant sites outside the resected tissue. At diagnosis, the patient may harbor a tumor burden of 1.1×10^{11} that is distributed between a primary tumor of approximately 10^{11} cells ($\sim$100 grams) and micrometastatic deposits totalling 10^{10} cells, but any one of which is not clinically detectable ($<10^{10}$). This case can be represented in terms of a tumor burden graph (Figure 9–4). Surgical resection of the primary tumor reduces the total tumor burden to that found in the micrometastatic disease, which if untreated will repopulate to clinically detectable levels and then to lethal levels

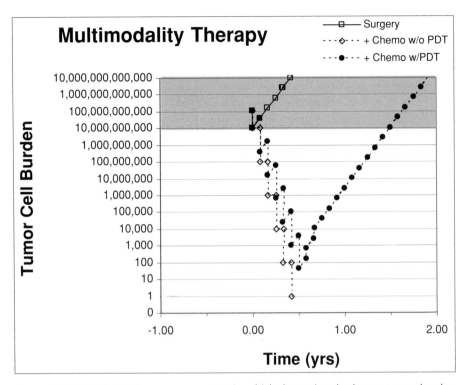

Figure 9–4. This illustration assumes a case in which the patient harbors a tumor burden of 1.1×10^{11} that is distributed between a primary tumor of approximately 10^{11} cells (~100 grams) and micrometastatic deposits containing a total of 10^{10} cells, but any one of which is not clinically detectable ($<10^{10}$). The tumor's population doubling time ("PDT") is two weeks. Surgical resection of the primary tumor reduces the total tumor burden to that found in the micrometastatic disease, which if untreated will repopulate to clinically detectable levels and then to lethal levels (open squares). However, the residual tumor burden is not clinically detectable and therefore cannot be located, so it must be treated with systemic chemotherapy to treat cancer cells that may reside anywhere in the body. Giving chemotherapy every four weeks for seventeen weeks (open diamonds) should reduce this residual tumor burden to <1 tumor cell and cure the patient who was not curable with surgery alone, if each drug treatment achieves a 2-log cell kill and it is assumed that cell proliferation within the residual tumor burden is negligible between cycles of therapy (i.e., PDT essentially 0 once treatment begins). However, this assumption is wrong, because surviving malignant cells do proliferate between drug cycles. If the population doubling time is unchanged by therapy (solid circles), the repopulation between drug treatments prevents a cure from being achieved by even six courses of therapy, although the overall result of adding chemotherapy is a significant increase in life expectancy relative to surgery alone. If drug therapy is not very toxic, most people will choose to try for this increase in disease-free and overall survival. If the toxicity from the chemotherapy is severe, many patients may choose to risk faster tumor progression and a shorter life expectancy with some tumor control from non-toxic therapies instead of compromised daily living activities due to drug toxicity. The shaded area indicates the tumor burdens typically associated with clinically detectable (10^{10} cells) and clinically fatal disease (10^{13}).

(squares). However, the residual tumor burden is not clinically detectable and therefore cannot be located, so it must be treated with systemic chemotherapy to treat cancer cells that may reside anywhere in the body. If chemotherapy achieves a 2-log cell kill per cycle, then five cycles of therapy given every four weeks will reduce the residual tumor burden to <1 viable cancer cell during the fifth cycle, and cure the patient (Figure 9–4 diamonds). However, our experience teaches us that this typical cancer patient is usually not cured and if middle aged or younger has a high probability of recurrence. Why is this? Several factors reduce the effectiveness of chemotherapy, which were not considered in this simple example.

To achieve this cure with a 2 log-cell kill, it was assumed that tumor burden does not increase between cycles of drug therapy. However, this is not the case, for tumors, like any renewing tissue, will try to replace the tissue killed by therapy between therapy cycles. In the example, the tumor's population doubling time ("PDT") is two weeks, and if the PDT is unchanged by therapy, the repopulation between drug treatments prevents a cure from being achieved by six cycles of chemotherapy (Figure 9–4 solid circles). Despite not being curative, the overall result of adding chemotherapy is a significant increase in life expectancy in remission, with good quality of life, relative to surgery alone. If drug therapy is not very toxic, most patients will choose to try for this increase in disease-free and overall survival. If the toxicity from the chemotherapy is severe, many patients may choose to risk faster tumor progression and a shorter life expectancy instead of compromised daily living activities due to drug toxicity.

Tumor repopulation makes it more difficult to cure the residual disease, because the addition of new tumor cells over time increases the tumor log kill that must be achieved by the six cycles of therapy. It takes about seven population doublings of the tumor cell population to replace the 99 percent of cells lost because of the 2-log cell kill of therapy ($2^7 = 128$). For a rapidly doubling tumor with a four-day PDT, the 2-log cell kill achieved by therapy would be replaced in just twenty-eight days (7 doublings of 4 days each). Receiving chemotherapy every four weeks would only maintain the cancer at its post-surgery tumor burden in exchange for drug toxicity, because surviving cancer cells would proliferate and replace the killed ones by the time the next cycle of therapy begins. To achieve a cure of subclinical disease using drug therapy, the patient must be treated with a frequency that is not any slower than the PDT of the tumor. If the tumor doubles every four days, then to be cured in six cycles, the patient must be treated every four days to achieve the 2-log reduction in tumor burden per cycle. There is some evidence that the last one to ten surviving cancer cells are less susceptible than the rest, so the final cycles of therapy might need even closer spacing to increase dose intensity (dose per unit time) and eradicate the last surviving cells (Wilcox 1966, 1970).

The mathematics of the example suggest that the chemotherapy regimen should be modified to involve more frequent drug treatments, so treating every two weeks instead of every four weeks would theoretically result in a curative regimen after six courses (Figure 9–5, open triangles). Although this modification sounds simple,

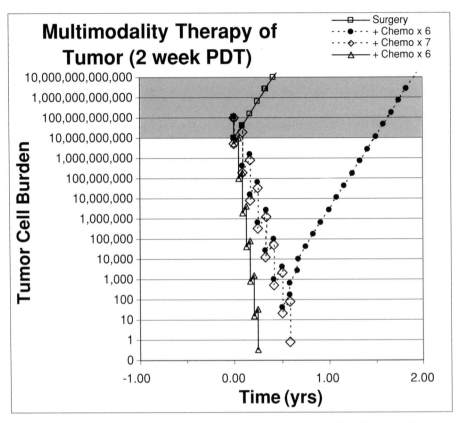

Figure 9–5. Using the same case as in Figure 9–4 and the tumor PDT of two weeks, six cycles of chemotherapy administered every four weeks (closed circles) following surgery would achieve a significant improvement in survival and relapse free survival compared to surgery alone (open squares). The chemotherapy schedule could be modified theoretically in two different ways that could eradicate residual micrometastatic disease after surgery and potentially cure the patient. The first modification involves increasing the duration of chemotherapy treatment to include more courses of therapy. Mathematically, if 2-log cell kill could be achieved by every course of drug treatment, then the residual tumor burden should be eradicated by increasing the length of treatment from six to seven courses of chemotherapy spaced every four weeks (open diamonds). The second modification involves more frequent treatment with chemotherapy, and treating every two weeks instead of every four weeks results in a curative regimen after six courses (open triangles). In practice, neither of these modifications has proven clinically effective. The first modification fails because of the appearance of tumor cell subpopulations that are drug resistant and the development of cumulative dose toxicities that result from total administered dose. The second modification fails because the toxicity of most drugs at full dose intensity does not resolve fast enough to allow dosing every two weeks. Such intensive regimens of cytotoxic therapy cause life-threatening, acute toxicities to gastrointestinal mucosa and bone marrow, which must recover before administering subsequent cycles or be supported by transplantation of normal stem cells to repopulate these tissues. The shaded area indicates the tumor burdens typically associated with clinically detectable (10^{10} cells) and clinically fatal disease (10^{13} cells).

in clinical practice and in animal models, it has not proven to be very effective. Cytotoxic therapy produces dose-limiting toxicities in normal tissues from which the patient must recover before receiving a subsequent course of therapy. Administering additional therapy prior to recovery can exacerbate already suppressed dose-limiting organ function to such an extent that it can never recover. The acute toxicity to the dose-limiting proliferative tissue is generally what limits dose intensity and precludes every two week dosing with many chemotherapeutic agents (but not all). Repopulation must occur to replace the log-cell kill in critical proliferating tissues like bone marrow and gastrointestinal epithelium, which manifest as the acute dose-limiting toxicities, before the next cycle can be tolerated. The clinician uses laboratory tests and physical examination to check that tissue function has recovered before administering the next course of drug therapy. Subsequent courses can be delayed while waiting for evidence of recovery, such as a return of peripheral blood cell counts to adequate levels. It generally takes these tissues about two to three weeks to recover, so chemotherapy with most drugs is scheduled every three to four weeks. A handful of drugs are exceptions, which can be administered more frequently, because recovery from toxicity is rapid (gemcitabine – the active ingredient in the oncology product Gemzar®, and paclitaxel in Taxol®). Many patients can recover from weekly therapy with these few drugs, but even these drugs would likely produce unacceptable, potentially life-threatening toxicity to bone marrow and gastrointestinal epithelium, if administered every four days for six cycles. A few drugs, especially nitrosoureas and mitomycin C that alkylate macromolecules, cause delayed toxicity, or protracted episodes of toxicity, requiring six to eight weeks for recovery. These drugs must be used judiciously in situations where they offer some specific benefit over the others to justify their use; it is possible that much more cytoreduction of tumor burden relative to destruction of normal stem/progenitor cells in marrow and gut must be achieved per cycle of therapy to compensate for longer recovery periods from toxicity, during which tumor repopulation may be more extensive than with the other drugs with more frequent dosing. Scheduling of cytotoxic therapy is a tradeoff between the recovery time of the dose-limiting tissue and the repopulation rate of the malignancy. If repopulation and tissue renewal is faster in the tumor than in the dose-limiting proliferative tissue after each course of drug therapy, then cure with cytotoxic therapy is "clearly impossible" (Schabel 1969). Thus, there are many complications to more frequent administration of anticancer drugs, and the time course of recovery from drug toxicity is too slow for this modification to be effective.

The analysis suggests as second theoretical modification to the chemotherapy regimen to eradicate residual micrometastatic disease after surgery and potentially cure the patient: increasing the duration of chemotherapy treatment to include more courses of therapy (Figure 9–5, open diamonds). If one cannot treat with drug every four to fourteen days, why not continue treatment every three to four weeks for as many cycles as are required to achieve a cure? Mathematically, if 2-log cell kill could be achieved by every course of drug treatment, then the residual tumor burden should be eradicated by increasing the number of courses of chemotherapy spaced every four weeks from six to seven (open diamonds). The reason that this modification does not work in practice is that chronic therapy with cytotoxic modalities, either by drugs or radiation, is limited by cumulative dose-dependent toxicities. These chronic toxicities may occur in different tissues than the acute toxicities, such as nerve or heart instead of bone marrow or gut. Because nonrenewing tissues are being damaged, the toxic effects of the therapy accumulate in the tissue and cause a gradual loss of function. These toxicities are insidious and particularly troublesome clinically, for they are often delayed and are only slowly reversible, or not reversible at all, since there are not any stem cells to produce replacement tissue for the damaged areas. Because these chronic toxicities depend on cumulative dose, they limit the maximum number of treatment cycles that a given patient can tolerate with a particular drug, or with a group of drugs that all cause the same tissue toxicity. Although treatment every twenty-one days can in theory cure the patient in spite of acute, dose-limiting toxicities, in practice chronic toxicities generally limit the maximum number of tolerated cycles to a level below that required for cure. Healthy dose-limiting tissues can only tolerate a limited number of cycles of cytotoxic therapy before their regenerative capacity is impaired or their cumulative damage reduces organ function to critically low levels. Eradication of every malignant cell becomes theoretically impossible, once tumor burden grows larger than can be killed by the maximum tolerated number of cycles of therapy with its characteristic log-cell kill. In addition to considerations of toxicity, this modification also fails because of the presence of tumor cell subpopulations that are drug resistant (see below).

It is noteworthy that curative regimens for childhood cancers like neuroblastoma and hematologic malignancies have been developed that require many months of chemotherapy. Better drug tolerance in young children than in adults seems to contribute to this difference, and the curable childhood tumors have favorable biological factors for responsiveness to chemotherapy. The biological and therapeutic differences in treating adult and childhood malignancies deserves further research attention to find out more about these differences, because these treatment successes in pediatric oncology confirm the clinical value of protracted dosing regimens of chemotherapy drugs for curing cancers.

This understanding of multi-modality therapy according to the Skipper hypothesis of log-cell kill, combined with the biological principles of tissue renewal that apply to malignant tumors as well as normal renewing tissues, point to the clinical requirements for eradicating cancers detected relatively early in their natural history. First, there needs to be surgical debulking of the tumor that leaves only disseminated, microscopic disease ($<10^{10}$ cells); otherwise, the increased tumor burden simply adds to the required number of cycles of therapy to achieve the increased demand for log-cell kill. Second, there must be sufficient log-cell kill to eradicate the tumor burden in about 6 cycles of therapy in spite of the fact that the tumor will repopulate during the typical 21-day recovery period between cycles. Given a more typical doubling time of three weeks for a solid tumor, calculations show that a 2.1 log-cell kill per cycle is required for cure. If these requirements cannot be met, then residual cancer will remain in the body after the six cycles of drug therapy. Whether or not these patients are "cured" is a matter mainly of how long they live relative to the time it will take for the cancer to regrow to clinically detectable levels. A patient with a faster growing tumor will relapse and die sooner than a patient with a slower growing tumor, even though the surgeries and drug-induced log-cell kills were equally successful. If chemotherapy reduces the tumor burden to thirty-two cells and this residuum doubles every three weeks, it will take only 21.5 months (86 weeks) for it to complete the 28.5 doublings required to reach detectable tumor burden, and just another 7.5 months for it to reach a lethal tumor burden of 10^{13} cells (Haskell 1990; Haskell et al. 1990). Although a therapy that reduces 10^{10} tumor cells to just thirty-two sounds highly effective, the patient endured about five months of therapy to experience only twenty-two months of complete remission and twenty-nine months of overall survival. The fact that many patients live longer than three years after multimodality therapy but ultimately relapse means that tumor doubling times during repopulation must be slower than three weeks, or that there is a delay in repopulation. It cannot be due to log-cell kills $> \sim 2.1$, otherwise six cycles would have been curative.

9.5   Biological factors that contribute to treatment success

Understanding the biological factors that contribute to treatment success and failure requires an understanding of tissue renewal itself. As explained in Chapter 1, there are two cell types with different roles in tissue renewal, whether malignant or normal (Pierce and Speers 1988; Lehman et al. 1974). Progenitors (also called transient amplifying cells) are cells that divide rapidly during a relatively short period of time to generate tissue mass, but then terminally differentiate after a finite number of population doublings into the mature cells of the tissue. Stem cells divide rapidly and indefinitely, but only periodically, in order to produce progenitor cells or to self-replicate to replenish their own numbers. When not proliferating, they remain in a nonproliferative state that is relatively resistant to cytotoxic chemo- and radiotherapy. When tissue mass decreases through cell loss, the stem cell content of the growth fraction increases rapidly and transiently to produce cells that restore the tissue to its steady-state size (Pierce and Speers 1988; Al-Dewachi et al. 1975, 1977, 1979; 1980, Bromley et al. 1996; Hendry et al. 1982; Levine et al. 1994, 1995; Potten et al. 1984, 1994b, 1995; Potten and Chadwick 1994). Tumors also appear to have a maximum size that is tightly regulated through homeostatic mechanisms: when tumor burden is lost, proliferation in the remaining tumor burden increases, usually via increasing growth fraction, in order to reestablish homeostatic tumor mass (Laster et al. 1969; Schabel 1969, 1975, 1976; Dorr and Fritz 1980; Vaughan, Karp, and Burke 1984; Fisher et al. 1989a, 1989b; Cameron, Hardman, and Skehan 1990; Shackney et al. 1978; Wilcox 1966, 1970). This is the process called repopulation, and when stem cells transiently enter the growth fraction, they become more sensitive to chemo- and radiotherapy.

In modern multimodality therapy, surgery is often used to reduce solid tumor burden to the point that repopulation is stimulated and the remaining tumor is sensitized to chemotherapy. This use of cytoreductive surgery without intent to cure is called "debulking," and the use of properly timed radiotherapy and/or chemotherapy after surgery to kill malignant cells during repopulation is called "adjuvant therapy" (Fisher, Gunduz, and Saffer 1983; Wittes 1986; Forbes 1990). This approach to chemotherapy and radiotherapy is far superior clinically to waiting to administer post-surgical chemo- or radiotherapy at the time of relapse, and it has become the dominant concept in multi-modality cancer treatment. The superior efficacy of adjuvant therapy compared to therapy at relapse is evident in animal models, in which the most sensitive tumors were the micrometastatic deposits that were below clinically detectable size (Schabel 1969; Wilcox 1966, 1970; Laster et al. 1969; Fisher et al. 1983, 1989a; 1989b; Shackney et al. 1978). There is experimental evidence that chemotherapy is curative only when preceded by surgical debulking (see Shapiro and Fugmann 1957).

To maximize the exposure of tumor stem cells to drug or radiation, expose tumor during repopulation, and decrease acute toxicity, a dose of cytotoxic therapy is split into fractions and administered over time (usually on different days), a strategy called fractionation. With fractionated dosing, the malignant cells that were out of cycle during the first dose fraction may be at a sensitive point during the second or third. A very few cancer drugs circulate in the body for weeks at a time after administration, and these therapeutics provide systemic drug exposure that is analogous to fractionated dosing of drugs that are eliminated from the body within hours of dosing. Several monoclonal antibody products have been approved by the FDA in recent years that have been humanized by genetic engineering so as not to be recognized as foreign substances. As any antibody does, these drugs circulate for weeks to months in the plasma and lymph of the body. The long duration of exposure following a single dose achieves the same type of constant exposure as one would achieve by daily dosing of a drug eliminated from the body every twelve to twenty-four hours.

Dose fractionation originated in the finding that radiation doses which sterilize rams are equally effective but better tolerated by the overlying normal skin when fractionated (Regaud and Ferroux 1927) and has developed into an effective strategy for both radiotherapy and chemotherapy (see highlight box). Dose fractionation allows differences in growth fraction or cell cycle time between the target tissue and the dose-limiting normal tissue to be exploited. Fractionation prevents toxicity that is caused by exceeding threshold concentrations of the drug or threshold doses of radiation, which overwhelm mechanisms for preventing and repairing damage in normal tissue, but still delivers the same total dose of radiation or drug to the malignant cells. As an example, hyperfractionated radiotherapy (exposure 2 to 3 times per day, each below the maximum tolerated dose for the CNS) delivers 6,141 cGy instead of the 5,800 cGy dose in 200 cGy fractions (Batzdorf, Black, and Selch 1990). Unfortunately, fractionation cannot be used to protect very slowly renewing tissues such as kidney and liver from toxicity, in which cellular proliferation is slower than in most cancers.

Some tissues may be treated with fractionated adjuvant therapy because of a high likelihood, based on clinical experience, that they contain metastatic cells, even though cancer is not yet detectable. This type of adjuvant therapy, based on the predictable nature of the natural history of cancers, is called "prophylactic therapy." This is an unfortunate term for this type of adjuvant therapy because it wrongly implies that metastasis can be prevented by treating the tissue with radiotherapy or chemotherapy before any cancer cells have arrived. Instead, its mechanism must involve the log-cell kill of micrometastatic disease that is already present in the tissue of most patients but has not yet grown to sufficient size to be clinically detectable. Radiotherapy of the brain, or direct injection of drug into the brain, is often part of multimodality therapy for leukemia and small cell lung carcinoma to "prevent" brain metastases, which previous clinical experience shows are often the first evidence of relapse and sometimes the sole cause of treatment failure (Holmes, Livingston, and Turrisi 1993; Schiffer 1993; Trigg 1993).

Early detected tumors have probably just completed a period of maximum growth rate called log-phase growth, and the growth rate is just beginning to decelerate as tumor burden approaches its homeostatic size (Dorr and Fritz 1980). Just like the normal renewing tissues that they caricature, as tumor burden increases toward this limit, growth rate will slow, growth fraction decrease, the proportion of stem cells in the growth fraction decrease, and the tumor will become refractory to chemo- and radiotherapy ("refractory" means reversible resistance as physiologic conditions change, distinct from "resistant," which is a cellular property that is irreversible).

In 1998, **H. Rodney Withers** received the Charles F. Kettering General Motors Prize for his important contributions to the conceptual development and practical use of "hyperfractionation" in radiation therapy. This strategy divides the total delivered radiation dose to the malignant tissue into a large number of parcels that are administered over closely spaced time intervals, allowing higher doses of radiation to be delivered to tumors while killing fewer healthy cells. His early research investigated normal tissue reactions to radiation, tumor cell division late in a course of treatment, and the rate of cell growth in tumors, which ultimately led to his hyperfractionation and accelerated treatment advances to administer radiation over a shorter period of time. These methods have improved cure and response rates and decreased the incidence of side effects from radiation, particularly in patients with head and neck cancer.

(see http://www.cancer.mednet.ucla.edu/newsmedia/news/pr061098.html).

It is easy to understand why chemotherapy and radiotherapy are most effective against small tumor burdens (Schabel 1969, Shackney et al. 1978, Wilcox 1966, 1970; Wilcox et al. 1965; Laster et al. 1969). Not only is a lower log-cell kill required for complete cytoreduction, but also a larger growth fraction containing cycling stem cells maximizes sensitivity to cytotoxic therapies. In patients with advanced stages of cancer, the tumors exist in so-called plateau (stationary) phase growth, in which growth fraction is low, consisting of a few progenitor cells and rarely stem cells. Consequently, these tumors are refractory to cytotoxic therapy. However, a sudden reduction in tumor burden will trigger repopulation, increasing the growth fraction by activating stem cells as the tumor attempts to restore homeostatic mass. The growth fraction increases, the stem cell contribution to the growth fraction increases, and the tumor becomes *transiently* more sensitive to cytotoxic therapy (Schabel 1969; Shackney et al. 1978; Laster et al. 1969; Vaughan et al. 1984; Fisher et al. 1983, 1989a, 1989b, 1993; Cameron et al. 1990; Wilcox et al. 1965; Gunduz, Fisher, and Saffer 1979).

Repopulation can be induced in advanced tumors not only by surgical debulking, but also by treatment with some cytotoxic drugs or radiation. In animal tumor models, cyclophosphamide induces a large log-cell kill in slowly growing malignancies and sensitizes remaining tumor cells to cell cycle – specific drugs to which they were originally refractory (Schabel 1969; Schabel et al. 1965). This activity may explain why cyclophosphamide is one of the most active drugs against human and animal solid tumors (Griswold et al. 1968). This phenomenon, which occurs in leukemia as well, is important for determining the schedule for chemotherapy cycles (Vaughan et al. 1984). Repopulation can also be induced in micrometastatic deposits by surgical debulking of the primary tumor, possibly a distant growth control mechanism via humoral factors (Fisher et al. 1989a, 1989b).

The fact that gastrointestinal epithelium and blood cells reappear following denuding and bone marrow suppression caused by radio- or cytotoxic chemotherapy indicates that stem cells in these tissues are more resistant than the progenitors to cytotoxic therapy. The relative resistance of the stem cell pool is due to several characteristics. Stem cells can express mechanisms of drug resistance not expressed in progenitors (Drenou et al. 1993), and most of them are quiescent at any one time. However, stem cells can be killed by cytotoxic therapy if treatment is timed to expose the stem cells while they are transiently in the cell cycle, such as during repopulation or after growth factor stimulation (Molineux et al. 1994). Malignant stem cells would behave similarly, and this may be an important factor in the success of adjuvant therapy. Normal stem cells may be less sensitive than their malignant counterparts to cytotoxic damage because of properly functioning cell cycle check point controls that prevent apoptosis and promote repair of damaged cells. Therapeutic selectivity for the malignant cells may be created by genetic abnormalities only in malignant cells that compromise cellular repair and the coordination between inhibition of cell cycle progression and the presence of macromolecular damage (Chapters 4 and 5). Despite normal progenitors being more sensitive, causing the patient to experience severe toxicity from each cycle of cytotoxic therapy that induces just small log-cell kills in the malignant progenitor pool, the periodic success stories of chemo- and radiotherapy might be due to an underlying difference in stem cell susceptibility that favors eradication of the tumor tissue. This would be an interesting area to investigate once methods are available to isolate, quantify and cultivate stem cells from renewable tissues.

9.6   Biological factors that contribute to treatment failure

Despite several biological features of malignant stem cells that may lead to a greater susceptibility than their normal counterparts to radiation and drug-induced damage, there seem to be many more aspects of malignant tissue renewal that confer survival advantages over normal tissue during chemo- and radiotherapy. The average-sized human produces about 40 kg of intestinal epithelium each year, but it is rare indeed for a tumor to produce 40 kg of new tumor tissue in its entire lifetime. Many patients are diagnosed with 10^{11} to 10^{12} tumor cells (100 to 1000g of tumor, about $\frac{1}{5} - 2$ pounds). If the log-cell kills in the malignant and normal cell populations are equal, then the tumor should be eradicated by fewer cycles of therapy than the normal tissue because normal cells far outnumber tumor cells. Because clinical experience shows this is not the case, there must be mechanisms that skew cytoreduction toward the 40 kg of intestine and away from the 100 g of tumor. These mechanisms are the subject of this section.

To eradicate a cancer (or normal renewing tissue), the stem cells that can repopulate all of the cellular elements of the tissue must be killed. Only 1 in 500 cells are estimated to be stem cells in the normal mucosa (Schmidt, Winton, and Ponder 1988; Winton and Ponder 1990), yet cloning experiments indicate that up to 25 percent of the tumor cells in a malignancy may be malignant stem cells (Pierce and Speers 1988). This exaggerated capability for tissue renewal in the tumor translates into as much as a 125-fold advantage in stem cell numbers and reduces the target cell differential between normal and malignant tissue from 400 (40 vs. 0.1 kg) to about 3. Thus, eradicating a clinically advanced tumor requires an intensity of therapy that nearly eradicates the intestinal epithelium.

Additional physiologic variables further skew log-cell kill toward normal tissues and against malignant tissue. The growth fraction of a tissue is the proportion of proliferation-competent cells that are progressing through the cell cycle at any given time (Mendelsohn 1960; Mendelsohn et al. 1960). Only the cells in the growth fraction are sensitive to cytotoxic therapies; quiescent cells will not be sensitive unless they enter the growth fraction during repopulation before the cellular damage has been repaired. Unfortunately, the growth fraction is usually lower in the malignant than the dose-limiting normal tissue. Slowly growing solid tumors have growth fractions that average less than 10 percent, although there may be large interpatient variability, but even the high end of this range is significantly below the 30 percent to 60 percent growth fraction in gut epithelium or hematopoietic tissue of the bone marrow (Al-Dewachi et al. 1974, 1975, 1979, 1980; Skipper and Perry 1970; Charbit, Malaise, and Tubiana 1971; Malaise, Chavaudra, and Tubiana 1973, Malaise et al. 1974; Wright et al. 1973; Rijke, Plaisier, and Langendoen

1979; Smith and Jarvis 1980; Smith, Jarvis, and King 1980). Because only cycling cells are affected by cytotoxic therapy, the three- to six-fold or greater difference in growth fraction neutralizes the remaining three-fold advantage that the normal tissue seemed to have over the cancer based on stem cell content (see preceding paragraph).

Although high growth fraction confers sensitivity to chemo- and radiotherapy, it does not confer curability. For example, higher grades of ovarian cancer have a higher mitotic index, yet a worse chance of cure (Berek and Hacker 1990), and radio-resistant astrocytomas exhibit the highest mitotic index (Batzdorf et al. 1990). Likewise, it is an important clinical conclusion that sensitivity does not correlate with curability (Haskell 1990; Bingham 1978; Fisher et al. 1993). This is why complete remission is necessary but not sufficient for cure of the cancer patient. This is also why oncology drugs are approved by the FDA based on hard evidence of benefit (efficacy) at tolerated doses, such as increased overall survival or quality of life. The identity of the proliferative cells in the growth fraction may be very important. The growth fraction of curable tumors (testicular cancer, neuroblastoma, childhood leukemias) might include all of the malignant stem cells, whereas the growth fraction of incurable yet highly chemosensitive cancers like small cell lung carcinoma may contain exclusively progenitors and few, if any, stem cells. Because cytotoxic therapy kills all cycling cells and the progenitors make the largest contributions to tissue mass, all of these tumors shrink dramatically in response to cytotoxic therapy as cycling progenitors are destroyed by therapy. However, it may be only in small cell carcinoma that the stem cell population has not been affected and relapse is unavoidable. Perhaps the most difficult tumors to treat with cytotoxic therapy are those with rapid cell cycle times but small growth fractions. The high mitotic index may exist in a very small growth fraction, and this fraction is so small that its eradication does not affect tumor mass and it can be quickly replaced by repopulation from quiescent stem cells.

The cancer has yet another advantage over the normal renewing tissue: poorly functional vasculature. Many physiological phenomena contribute to poor tumor perfusion, which is an important reason for the lower growth fractions in tumors than in normal counterpart tissues. Tumor cells far away from the blood supply become dormant, and enter the G_0 phase of the cell cycle, in which they are resistant to cytotoxic therapy but from which they can become mitotically active again (Schabel et al. 1965; Pittillo, Schabel, and Skipper 1970). This means that the most drug-refractory cells (because they are not progressing through the cell cycle) are exposed to the lowest levels of anticancer drugs and radiation damage (due to hypoxia). To grow (and metastasize), tumors must stimulate development of new vasculature. Adequate and pervasive blood flow throughout the tumor is essential for delivery of sufficient molecular oxygen for effective radiotherapy, and delivery of sufficient drug levels for effective chemotherapy. Tumor growth may exceed the rate of formation of new blood vessels, so malignant cells may lay far from a blood supply. The blood vessels in tumors are also very dynamic, opening and closing in seemingly random fashion in response to paracrine factors produced by the tumor cells (Chaplin and Trotter 1990). Thus, entire regions of the cancer may transiently lose blood perfusion, although the duration is brief and the cells remain in the growth fraction. But a transient loss of perfusion during a cycle of therapy will diminish the log-cell kill in this region of the tumor, either by diminishing drug delivery or decreasing oxygen delivery which is required for radiotherapy with ionizing radiation. Because this physiologic response does not occur in the vasculature of normal tissues, it further diminishes the effectiveness of the therapy (Simpson-Herren and Noker 1990). Note that this physiologic mechanism is non-specific – conferring resistance to all drugs, and in many cases radiation too, because the critical drugs (and oxygen) cannot get to the malignant cells. Recent work indicates that, unlike normal blood vessels, tumor blood vessels are chaotic, irregular, and leaky, which leads to uneven delivery of nutrients and therapeutic agents to the tumor (Jain 2002). Furthermore, tumor vasculature is dilated to vessel diameters much larger than normal and contains sluggish flow, and the dilated vessels are too large to support rapid, efficient diffusion of oxygen, nutrients, and drugs into the tumor tissue, limiting the effectiveness of radio- and chemotherapy. Even sophisticated biotechnology approaches to cancer therapy will not work if the therapy cannot be delivered to the tumor or diffuse from the bloodstream to the malignant cells.

One would predict that drugs that could normalize the size and permeability properties of these small tumor vessels would normalize their delivery properties toward those of capillaries in healthy tissue. Preclinical studies with a monoclonal antibody drug named bevacizumab (Avastin®) that blocks the action of the tumor-derived angiogenesis factor VEGF has shown that treatment of human tumor xenografts in mice led to reduced tumor vessel permeability and caused vascular regression. Suppression of angiogenesis induces tumor vessels to acquire more normal diameters, blood flow, and perfusion properties, and the reduced vascular permeability, resulting from inhibition of VEGF, led to increased delivery of oxygen and therapeutic agents to tumors. Anti-VEGF therapy did not decrease the effectiveness of radiation or chemotherapies in the preclinical models, and these studies disproved the idea that bevacizumab would decrease effectiveness of cytotoxic drugs by decreasing vascular delivery to the tumor. These findings were confirmed in a clinical study of colon carcinoma patients given a single dose of bevacizumab (Willett et al. 2004). This research explained clinical trial results that some investigators feared would be the opposite: adding bevacizumab to cytotoxic chemotherapy regimens actually increased efficacy (Kabbinavar et al. 2003). Combining bevacizumab with radiation therapy should also increase efficacy, as a result of more efficient oxygen delivery to the malignant tissue.

Another problem of delivery of the therapeutic agents occurs when cancers metastasize into, or arise within, tissues which have blood-tissue barriers formed by endothelial cells with specialized tight junctions that exclude noxious chemicals. One such barrier, called the blood-brain barrier, not only protects brain tissue from drug exposure but also can inhibit drug penetration from the blood into a tumor in the brain. Only a few drugs can penetrate these barriers, such as nitrosoureas, and these drugs are often used to treat CNS tumors (Batzdorf et al. 1990). Some cancers are not sensitive to nitrosoureas and must be treated by direct injection of drug into the cerebrospinal fluid. But only a few drugs are suitable for intrathecal injection; many are not because of acute or delayed neurotoxicity. A similar endothelial barrier for chemotherapeutics exists in the testes (Smith and Haskell 1990). Radiotherapy is often the only treatment for cancers behind these pharmacologic barriers, especially when surgical resection is risky.

Another cellular variable that reduces the effectiveness of many cytotoxic therapies is the slower than normal cell cycle time of most malignant cells, which is consistent with the protracted natural history of most cancers (11 years to reach 10^{10} cells, a 10 g tumor [(Haskell 1990)]). Slowed cell cycle times decrease the therapeutic index of cell cycle phase–specific chemotherapeutics and radiotherapy by decreasing the probability that a cycling cell will be exposed during the sensitive phase of the cell cycle (Skipper, Schabel, and Wilcox 1967). For example, a six-hour infusion of an S phase-specific drug might provide eight hours of systemic exposure – six hours of infusion plus two hours circulating in the blood before being cleared from the body via excretion and metabolism. This exposure time represents 50 percent of the sixteen-hour cell cycle time of rapidly dividing progenitors in gastrointestinal mucosa and bone marrow, but only 10 percent of the eighty-hour cell cycle time in an adenocarcinoma (Al-Dewachi et al. 1974, 1975, 1979, 1980; Skipper and Perry 1970; Charbit et al. 1971; Malaise et al. 1973; Wright et al. 1973; Rijke et al. 1979; Smith and Jarvis 1980). The more rapidly dividing cell will more likely be exposed during S phase, and the log-cell kill from the six-hour infusion will be greater in the rapidly dividing normal tissues than in the tumor. A continuous infusion of drug for 38 hours would be needed to achieve the same log-cell kill in the adenocarcinoma that the 6-hour infusion achieves in the normal tissues, and an infusion of seventy-eight hours would be required to expose every adenocarcinoma cell in the growth fraction during S phase. Continuous infusions of many days' duration with some drugs have been tolerated by patients without irreversible toxicity, but other drugs are extremely toxic to normal tissues when administered on these schedules, probably because the time frame is long enough to kill progenitors and stimulate stem cell repopulation during drug exposure.

Drug resistance can also be due to cellular features that are restricted to specific subpopulations of the tumor. Genetic instability and phenotypic selection generate drug-resistant cells in tumors even before they have been exposed to drug – the Goldie-Coldman hypothesis (Goldie and Coldman 1979). These cellular phenotypes of drug resistance likely arise in untreated tumors via the same genetic mechanisms responsible for the heterogeneity of other phenotypic traits (Berek and Hacker 1990), and therefore cellular drug resistance is fundamentally an extension of the unit character concept of tumor progression espoused by Foulds (Foulds 1965). Every cycle of chemo-or radiotherapy kills nonresistant cells, including cells in small unit characters that have yet to express a resistant phenotype. But resistant cells survive the initial exposure, and by proliferating to replace the cell mass lost to drug sensitivity come to represent an ever greater proportion of the tumor with each cycle of therapy. Each successive cycle of therapy will be less effective until at some point resistant disease will develop that is no longer responsive to the treatment (Figure 9–6). This is the most likely reason why cycles of drug treatment become ineffective after a certain number of courses even though cumulative dose toxicity has not appeared. It also explains why relapsing cancers often fail to respond a second time to drugs that initially induced complete remission. It is also possible that additional mutations that affect drug sensitivity are caused by exposure to the cytotoxic therapy, since many of these drugs induce genetic damage in the exposed cells. To cure tumors under these conditions, theoretical principles indicate that subsequent courses of therapy need to be of increasing intensity (Wilcox 1966, 1970), but patients are most likely unable to tolerate increased dose intensity because of normal tissue damage from prior drug exposures. It is interesting that some solid tumors are still highly curable with chemotherapy, and it is an interesting question why the Goldie-Coldman hypothesis does not apply in those cases.

Some cellular resistance is specific for one particular drug or drug class. For example, resistance to anti-pyrimidines and anti-folates is due to overexpression and mutation of specific target enzymes in biochemical pathways of nucleic acid synthesis or specific transport proteins in the plasma membrane (Allegra 1990; Grem 1990). In these cases, resistance to the anti-folate drugs will not cause cross-resistance to the anti-pyrimidine drugs. Mutations and other modifications in a specific molecular target like topoisomerase II can cause sweeping resistance to all drugs that act on this target, and increased capacity for DNA repair can confer cross-resistance to many alkylating agents. For example, a nuclear repair enzyme called alkyl transferase removes O-alkyl groups from guanine bases in DNA caused by alkylating drugs and some carcinogens. So-called mer[-] cells lacking this enzyme are extremely sensitive to alkylating agents, whereas high levels of this enzyme confer resistance to many alkylating agents, not just the drug which originally selected the mer[+] clones during chemotherapy (Ikenaga 1994).

However, malignant cells can also express a multidrug-resistant phenotype. When this happens, development of resistance to one drug brings cross resistance to other drugs with different pharmacological mechanisms of action, which have not yet been used against the particular cancer (Ueda et al. 1986; van der Bliek and Borst 1989; Chabner 1990; Chin, Pastan, and Gottesman 1993). Glutathione transferases that rapidly conjugate electrophilic drugs (quinones, alkylating agents) and clear them from the cell may be induced by one drug but then act on others. Perhaps the most thoroughly studied cellular mechanism of multidrug resistance involves a family of proteins encoded by the mdr genes ("multidrug resistance" genes), the best known of which is the P-glycoprotein ("P170") (Ueda et al. 1986; van der Bliek and Borst 1989; Chabner 1990; Chin et al. 1993; Thorgeirsson, Gant, and Silverman 1994; Gottesman et al. 1995). These plasma membrane proteins utilize energy in the form of ATP to pump drug molecules out of the cancer cell (Cornwell 1987). This

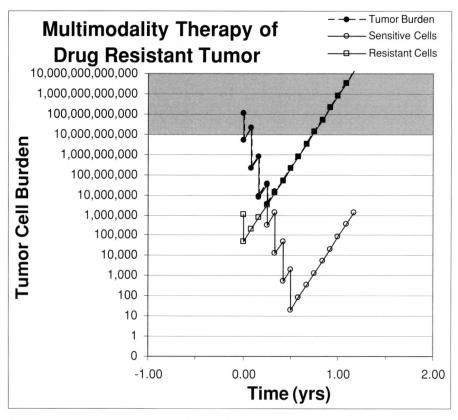

Figure 9–6. Drug-resistant tumor cells are a frequently encountered problem of cytotoxic drug therapy caused by the overgrowth of tumor cell subpopulations that have adapted to drug exposure. Six cycles of adjuvant therapy that achieves a 2-log cell kill, administered as frequently as the doubling time of the tumor, should eradicate metastatic disease after surgery (Section 9.4). However, the Coldman-Goldie hypothesis predicts that genetic instability creates drug-resistant cells with a frequency as high as 10^{-5}, so micrometastatic disease remaining after surgery could contain 10^5 cells resistant to single agent chemotherapy even before the first cycle is administered. Although initial surgery removes the same proportion of drug-sensitive and drug-resistant tumor cells, the resistant cells survive every cycle of chemotherapy, growing at the population doubling time until they dominate the cancer, and cause a relapse which will not respond to a therapy that was effective in the patient in the past, or any other therapy that is cross-resistant with the selective pressure. The figure shows the change in tumor burden over time resulting from the combined contributions of sensitive tumor cell populations as well as a drug resistant population at an initial frequency of 10^{-5}. This example assumes the same PDT of fourteen days for resistant and nonresistant cells. The presence of the drug resistant subpopulation causes lethal tumor burdens to be reached much earlier than would otherwise have been the case. This phenomenon can be averted in some patients by using combination chemotherapy against micrometastatic disease, because the probability that a cell is resistant to three drugs from non-cross-resistant drug families (10^{-15}) is much smaller than the probability of resistance to one drug (10^{-5}). The shaded area indicates the tumor burdens typically associated with clinically detectable (10^{10} cells) and clinically fatal disease (10^{13}).

mechanism causes the mdr phenotype to create cross resistance to many important drugs that act on diverse molecular targets, ranging from topoisomerases to microtubules to mitochondrial function. The mdr phenotype causes much wider drug resistance patterns than alterations in molecular targets of drug action: for example, the mdr phenotype confers cross resistance not only to many topoisomerase II inhibitors but also to vinca alkaloids that affect microtubule function (Chabner 1990). Tumors that arise in tissues with physiological functions dependent on excretion or secretion (kidney, large bowel, liver, pancreas, and adrenal gland) express mdr proteins as part of the normal differentiation program (Chin et al. 1993). The mdr system may protect stem cells throughout the body from damaging dietary and environmental chemicals, and protect those same stem cells from chemotherapy damage when they transform into malignant caricatures. Both immature leukemic cells and their normal hematopoietic stem cell counterparts express P170 (Drenou et al. 1993). P170 can also be expressed in cancers deriving from tissues that do not express the protein during normal cellular renewal and differentiation, and such ectopic expression of the mdr phenotype has been found in breast cancer (Chin et al. 1993). Overexpression of P170 in human tumor cells is most often due to transcriptional enhancement, which is different from the molecular mechanism of overexpression in rodent tumors (Thorgeirsson et al. 1994; Gottesman et al. 1995; Glazer and Rohlff 1994). It should be noted that plasma membrane transporters responsible for the mdr phenotype are just one family of proteins that remove drug substances from cancer cells, and other transporter systems have been discovered that recognize anticancer drugs as noxious environmental chemicals and pump them out of the cytoplasm.

Mdr protein inhibitors and other strategies to reverse cellular drug resistance have been developed, but the biochemical modulation is not tumor specific and the result is generally increased drug potency; that is, less drug is required for a certain level of biologic effect both in the tumor and in the dose-limiting normal tissue. Thus, there is not any gain in therapeutic index or effectiveness because there is not any selective enhancement of drug toxicity in the tumor. In addition, some modulators of drug resistance exhibit pharmacologic effects alone and are therefore not clinically useful (Sikic 1993; Patel and Rothenberg 1994), but these prototype modulators of resistance can nevertheless provide structural leads for developing pure antagonists of drug resistance that are pharmacologically inert (Hait and Aftab 1992). A quite different clinical approach to overcoming drug resistance is to genetically modify normal stem cells in the dose-limiting tissue (usually bone marrow) with the genes responsible for cellular drug resistance in order to increase drug tolerance and dose (Hesdorffer et al. 1994; Ward et al. 1994). However, the increase in dose gained with this approach is modest because higher doses are limited by more serious toxicity in different tissues that are more difficult to modify by gene therapy.

The Goldie-Coldman and Foulds hypotheses suggest that drug-resistant subpopulations arise spontaneously via genetic instability and natural selection prior to exposure to any chemotherapy, and continue to arise during therapy as a result of positive selection of resistant clones (Norton 1979, 1990; Norton and Simon 1986; Skipper, Schabel, and Lloyd 1978). Goldie and Coldman (Goldie and Coldman 1979) estimated that drug resistance occurs at a mutation frequency of about 10^{-5}. This phenomenon may seem to present overwhelming odds against successful treatment of cancer. However, the hypotheses are related to two very important principles of modern chemotherapy. First, a fixed frequency of drug resistance (mutation rate) means that the probability of encountering drug resistant cells increases as tumor burden increases. In a tumor of 10^{12} cells, one would expect 10^7 resistant cells to any drug one might choose for therapy, as long as mutation can lead to drug resistance. In a residual tumor burden of 10^4 cells following surgical debulking, one would be less likely to encounter cellular drug resistance (only 0.1 resistant cell expected on average, or maybe one of ten tumors if the tumors were identical biologically). Thus, the optimal way to combat drug resistance is to administer chemotherapy when tumor burden is minimal – another reason why adjuvant therapy may be more effective than chemotherapy at relapse. A second critical discovery about drug therapy is the much greater effectiveness of chemotherapy when using multiple drugs in combination rather than single agent therapy. The more useful alternative in the clinic to combat the problem of drug resistance is to treat cancers with a combination of drugs which do not share cross resistance. Combination chemotherapy uses simultaneous or closely timed exposures to several drugs that do not share mechanistic or structural similarities, and therefore should not exhibit cross resistance, to minimize the impact of de novo cellular drug resistance. By combining drugs which require different mechanisms of resistance, one is increasing the odds that every malignant cell of the tumor will be sensitive to at least one drug to which it is exposed (Schabel et al. 1980). Assuming drug resistance occurs at a frequency of about 10^{-5} (Goldie and Coldman 1979), a tumor of 10^{12} cells would be expected to contain 10^7 resistant cells to any single drug selected. However, assuming independent genetic events and lack of cross-resistance, cells resistant to two drugs would be present with a frequency of 10^{-10} ($10^{-5} \times 10^{-5}$). Although 10^7 tumor cells would be present that are resistant to therapy with a single drug, only 100 cells would be present that are resistant to both drug mechanisms. If three unrelated drugs are used, the frequency would be 10^{-15}, implying it is highly unlikely that any tumor contains such a phenotype. Even the largest nonlethal tumor burden of 10^{13} cells would have only a small probability of containing cells resistant to all three drugs, and it is interesting to note that combination regimens developed clinically usually contain three to four drugs.

The key to successful combination chemotherapy is selecting drugs that are affected by different mechanisms of resistance and therefore do not exhibit any cross resistance. For example, a combination of even ten drugs would be ineffective against a mdr$^+$ tumor if all ten were substrates for the P170 pump. It is also important that the different mechanisms do not interfere with each other, so called drug antagonism, but should show additive log-cell kill, if not synergy. Combinations of drugs with different cell cycle phase specificities provide the additional advantage that tumor cells are more likely exposed to at least one cytotoxic compound during a sensitive phase of the cell cycle. If the drugs are carefully selected not only according to complementary mechanism of action but also to complementary dose-limiting toxicities, an increase in log-cell kill over monodrug therapy is usually achieved without an increase in overall toxicity (Chabner 1990). In fact, the severe and serious side effects of combination chemotherapy are generally predictable from the toxicities of the individual agents used alone (Smith and Haskell 1990). It is also interesting to note that there are rare tumor types that respond to drug monotherapy and combination therapy equally, so adding more drugs to the regimen increases toxicity but not efficacy – to the detriment of the patient (Haskell 1990; Haskell et al. 1990; Morton, Cochran, and Lazar 1990).

Drug resistance can occur at the cellular and physiological levels. Drug resistance resulting from molecular characteristics of the malignant cell (like mdr) is a very different problem from drug resistance resulting from physiologic characteristics of the tumor, such as fluctuations in blood flow or growth fraction. The first mechanism will not reverse during the course of treatment and therefore represents permanent drug resistance of the tumor. As the results of the bevacizumab studies prove, the second mechanism can reverse during therapy as physiologic conditions in the tumor change, and therefore physiological mechanisms create drug-refractory tumors that are not necessarily drug resistant. Under this mechanism, cancer cells can be killed by the drug therapy if only the drug could be delivered to the tumor tissue during times of functioning vasculature, or when the growth fraction is actively cycling and susceptible to drug action. Cellular mechanisms of drug resistance can be studied in the laboratory (Greenberger, Cohen, and Horwitz 1994), but physiologic drug resistance is difficult to model in vitro.

9.7 Treatment of intermediate-stage breast cancer as a clinical science success story

Modern multimodality treatment of breast cancer reflects the newer understanding of the biologic basis of cancer, and in this section, we examine how three stages of breast cancer are treated and why. In selecting which modalities to use for a particular case, it is important to remember the strengths and weaknesses of each. Surgery, most effective for bulky and/or localized disease, is local therapy and used whenever a large proportion of the tumor burden can be resected or needs to be resected for immediate relief of life-threatening conditions. In contrast to fractional log-cell kill with radio-and chemotherapy, the 100 percent cell kill with surgery makes it illogical in most cases to treat localized disease without using surgery. Regional therapy with radiation is used to control disseminated cancer confined to a radiation field of the body and is most effective against low tumor burden found in microscopic disease. If metastases are present or micrometastases are likely based on the natural history of the disease in previous patients, then systemic therapy is required, and chemotherapy with a combination of non-cross-resistant drugs offers a chance of curing disseminated disease. However, its systemic nature also exposes all of the healthy tissues of the body, and systemic therapy goes hand in hand with systemic toxicity, which is more severe than the local toxicity of radiotherapy or surgery. The benefits of these therapies must be weighed against the toxicities and risks, and the decision to use them is on a case-by-case basis. Clinical experience with a wide variety of cancers indicates that single modality therapy is rarely curative. Multimodality therapy, which combines local, regional, and systemic therapy, is emerging as the best treatment strategy. Multimodality therapy protocols include treatment with surgery and/or radiotherapy initially to dramatically decrease the local or regional tumor burden and provide quick relief of symptoms, and carefully timed chemotherapy with drug combinations that exhibit different toxicity profiles and not cross resistance is then continued for an extended period of time to treat the micrometastatic residual disease.

A patient presenting with stage I breast cancer has the best prognosis (Figure 9–7a). The small tumor can be completely resected with clear margins and with no sign of cancer in local lymph nodes (Figure 9–7d). Greater than 90 percent of true stage I patients will be cured surgically, and single modality radiotherapy and chemotherapy are not generally used in place of surgery because of the chance that log-cell kill will not be sufficient for cure and a residuum will lead to relapse (Figure 9–7g and j). In node-negative patients, radical resection with regional lymph node removal (radical mastectomy) does not provide better long-term survival or lower distant relapse rates than conservative resections that remove only enough nodes for pathologic sampling and spare as much of the healthy breast tissue as possible (lumpectomy) (Haskell et al. 1990). Reconstructive surgery, using prostheses or implants, helps compensate for deformities resulting from cancer surgery. Proper surgical planning and technique includes a conscious effort to minimize morbidity and disfigurement and retain tissues that will be necessary for reconstructive surgery of good cosmetic quality. There is a trend toward combining the resection of the tumor with reconstruction during a single operation, which has been shown to reduce psychological morbidity in breast cancer patients (Dean, Chetty, and Forrest 1983). The low probability of local or distant micrometastatic disease in stage I patients means that the risks and toxicities of radio- and chemotherapy outweigh the benefits, and routine follow-ups are generally indicated without any postsurgical treatment. The fact that patients with node-negative, early-stage breast cancer treated only with surgery have less than 100 percent chance of disease-free survival provides strong evidence for the idea that cancers can disseminate prior to reaching clinically detectable size. Although adjuvant chemotherapy would be effective treatment for the micrometastatic disease responsible for the distant relapse (Figure 9–7m), exposing all of these patients to toxic drug therapy when the overwhelming majority will not benefit is considered unethical. Diagnostic tests are urgently needed that can accurately stratify the high-risk patients in this good prognosis

STAGE OF DISEASE
(● = Clinically detectable mass;   • = microscopic lession)

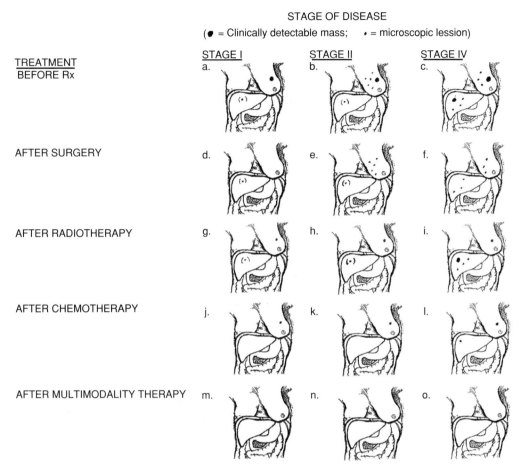

Figure 9–7. The strengths and weaknesses of each modality are illustrated using breast cancer as an example. (a,b,c) Stage I, II, and IV indicate three stages of disease that are possible at diagnosis; only stage IV has clinically detectable metastatic lesions, but micrometastatic disease might be present in some patients staged to I and II disease. (d,e,f) Surgery removes the obvious tumor lesions but cannot be used to eradicate locally invasive or systemic disease. Radiotherapy (g,h,i) and chemotherapy (j,k,l) are best at eradicating microscopic disease that is localized to a region or is systemic/unknown locations, respectively. The cytotoxic modalities work poorly against bulky disease (note partial responses of primary tumors and gross metastatic lesions in g–l). When combined in an integrated approach to treatment, localized bulky disease is treated surgically, regional disease after tissue-sparing surgery with radiotherapy, and micrometastatic disease with properly timed adjuvant chemotherapy (m,n,o). A minority of patients with stage II disease may not have disseminated cancer and therefore don't benefit from chemotherapy, but it is difficult to identify these patients with current prognostic markers, so all patients are treated. Conversely, the majority of patients with stage I disease are cured surgically, so adjuvant chemotherapy is not usually given to these patients, even though it is recognized that a few do have undetectable systemic disease at time of diagnosis.

group. Oncogenes linked to unusually aggressive behavior of breast cancer cells (Chapter 5) may provide a prognostic marker for stratification of stage I patients.

A patient presenting with stage II breast cancer has a much greater chance of local and distant micrometastatic disease that is not clinically detectable except as an incidental finding during biopsy or resection (Figure 9–7b). Although these cancers appear to be local-regional disease that extend no farther than regional lymph nodes, radical mastectomy does not provide any better long-term survival than lumpectomy. In a direct comparison, lumpectomy or total mastectomy produced the same rates of survival, disease-free survival (lack of relapse), and distant disease-free survival (lack of distant metastases), regardless of age, tumor size, and number of positive nodes (Fisher et al. 1989, 1993). The only clinical difference between the two techniques is a higher incidence of local relapse following lumpectomy that can be controlled with radiotherapy and does not affect survival rates (Fischer et al. 1993). Optimal treatment is currently a multimodality approach and includes lumpectomy as the first choice to remove the primary tumor (Figure 9–7e), regional radiotherapy to control local relapse (Figure 9–7h), and adjuvant chemotherapy to eradicate the micro-metastatic disease responsible for distant relapse and shortened survival (Figure 9–7k). The multimodality approach has a much higher chance of curing the patient than each modality used alone (Figure 9–7n) because single modality therapy will likely leave residual disease no matter which one is selected (Figures 9–7e, h, and k). Note that each individual patient is treated based on the clinical evidence in the average patient, and tests for high-risk cancers might also benefit this patient group by identifying the minority of patients who will never relapse with distant disease and do not require systemic chemotherapy.

Some patients present with stage IV breast cancer with clinically detectable metastatic disease, which is usually an early manifestation of extensive, noncurable disease (Figure 9–7c). Surgery is used sometimes to debulk the tumor burden (Figure 9–7f), unless there are so many detectable lesions that removing a few will not significantly decrease the tumor burden. Radiotherapy and chemotherapy alone will achieve only partial remission because of the large tumor burden and the unfavorable cytokinetics (Figures 9–7i and l). In this case, the purpose of multimodality therapy becomes prolongation of life and ultimately palliation – alleviation of pain and suffering – and stabilizing a patient's condition while the disease completes its natural course, for example, removing intestinal obstruction and isolated brain lesions (Batzdorf et al. 1990). Even with highly effective multimodality therapy, usually the best that can be achieved is cancer control because a relatively large tumor burden remains after maximum tolerated doses of cytotoxic therapy (Figure 9–7o), and drug resistant subpopulations are likely that will regrow to relapse from this residuum.

Targeted therapies are gaining importance in the management of breast cancer. A monoclonal antibody called trastuzumab (Herceptin®) has been approved for treatment of advanced breast cancer that overexpresses the *her-2/neu* oncogene (EGFR2) (Birner et al. 2001). Trastuzumab is being used in combination with taxanes (paclitaxel – Taxol® and docetaxel – Taxotere®) as first-line therapy in this group of patients to improve overall survival relative to taxane-based chemotherapy alone. Its role in adjuvant chemotherapy for Stage II disease of *her-2/neu* overexpressing tumors is being explored in clinical trials to evaluate its safety and effectiveness. The therapy is only approved for use in patients with documented overexpression of the target; it is contraindicated in patients with low or no expression of *her-2/neu*, because it is ineffective against cells that are not dependent on dominantly acting receptor activity and causes only toxicity in these patients (cardiotoxicity). Overexpression of EGFR2 is associated with a much more aggressive disease. The development of a targeted therapy specific for this molecule is clearly clinically beneficial, but only 25 percent of advanced breast cancer cases that overexpress EGFR2 respond to trastuzumab therapy (Simon et al. 2001). This drug will be described in more detail in Chapter 10 as an important example of targeted therapy for growth factor pathways.

9.8 Summary

In conclusion, four modalities are available to treat cancer: surgery, radiotherapy, cytotoxic chemotherapy, and targeted chemotherapy. With the exception of skin cancer, surgical cure rates for early detected cancers (Table 9–1) indicate that cancer is generally a disseminated disease at the time of diagnosis, and single modality surgery is curative only in patients with life expectancies shorter than the time required for residual micrometastatic disease to grow to detectable levels. These data have changed the purpose of cancer surgery from "complete resection" of all involved tissue to control of local disease, reduction of bulky tumor burden and life-threatening lesions, and optimal timing of adjuvant therapy, all while sparing normal tissue. Radio- and chemotherapy are indicated if there is a likelihood of local-regional dissemination and systemic metastases, respectively, even if there is no evidence of disease from standard diagnostic tests. The dose of adjuvant therapy is fractionated over time (Wittes 1986; Forbes 1990; Dutreix et al. 1971) to expose the remaining malignant cells several times during repopulation, their most susceptible time, and to attempt to expose the entire growth fraction of the tumor during a sensitive phase of the cell cycle. If adjuvant therapy is delayed until the disseminated disease becomes symptomatic or clinically detectable, then one has lost the window of opportunity to treat the cancer with radio- and chemotherapy under optimal conditions: lower tumor burden, better tumor perfusion, larger growth fraction that likely contains stem cells, lower numbers of drug and radiation resistant cells, and fewer dormant or quiescent cells. Combination chemotherapy has a better chance than single drug therapy because of the likelihood of killing preexisting drug-resistant cells. Cures with adjuvant therapy are possible as long as the healthy dose-limiting tissues can tolerate the required number of cycles. The slow doubling times of most solid tumors necessitate many years of disease-free survival, for example ten to fifteen years for breast cancer patients, to be certain of cure (Haskell et al. 1990; Mueller and Jeffries 1975).

Early detection and diagnosis increase the probability of cure with multimodality therapy by decreasing the probability of dissemination, minimizing required log-cell kill, decreasing the likelihood of resistance to cytotoxic therapy, and increasing the probability of optimal cell cycle kinetics for log-cell kill. The emphasis on early detection of cancer from warning signs like unusual bleeding, a change in bowel habits, a persistent cough, or an unusual or changing mole can make an important contribution to cancer cure. It cannot be overemphasized that the chances of curing cancer are much higher when two events coincide: the patient goes to a physician early during the course of the disease when symptoms are just appearing and the physician quickly recognizes the symptoms, establishes the diagnosis, and begins treatment. Delays by either person make cure ever less likely. Many late-stage patients cannot tolerate surgical debulking because of poor overall health caused by the cancer and must receive chemotherapy or radiotherapy under highly unfavorable biologic conditions that limit cytoreductive therapy to palliative purposes (Charlson 1985).

Cytotoxic therapy is ineffective when cancers repopulate faster than dose-limiting normal tissues, when cancers have small but active growth fractions, or when the log-cell kill is significantly greater in normal tissue than the cancer. Unfortunately, one or more of these appears to be the case in most malignancies. This is not to say some cancers cannot be cured; in fact, some of the less common ones are cured in a very high number of cases. Methotrexate or etoposide monotherapy cures 95 percent of cases of stage I choriocarcinoma (Berkowitz and Goldstein 1990). Adjuvant chemotherapy after surgical debulking is given with curative intent in patients with advanced seminoma of the testis (Smith and Haskell 1990). Nonseminomatous tumors of the testis are also chemosensitive, allowing the "luxury" of delaying chemotherapy after surgery until relapse is detected, thereby sparing surgically cured patients the drug toxicity (Smith and Haskell 1990). Neuroblastoma, Ewing's sarcoma, retinoblastoma, and acute lymphocytic leukemia in children and Hodgkin's lymphoma in adults are additional examples of malignancies treated with the intention of cure. The biologic basis for the curability of these cancers has received little attention. The vast majority of mechanistic research on cancer pharmacology is dedicated to answering why malignant cells are resistant to cytotoxic agents. Perhaps a more important question is why cytotoxic therapy works so well against a few types of cancers. Curing cancer is obviously a tremendously complicated goal. The development of curative strategies for several cancers, many of them childhood malignancies, is a tribute to slow but steady progress. It is hoped that insights from these curable malignancies will lead to better approaches for treating the more common, incurable cancers.

10

Oncology: The difficult task of eradicating caricatures of normal tissue renewal in the human patient

Ralph E. Parchment

In the previous chapter, the biology of malignancies and their caricature of normal tissue renewal suggested important principles that guide the development and use of treatment strategies. The current chapter presents how the treatment principles have been reduced to practice and how treatment principles relate to biological principles governing malignant growth. The chapter is organized around the four conventional treatment modalities available today: surgery, radiotherapy, cytotoxic chemotherapy and targeted therapy. (The term "conventional" refers to therapy accepted as the best available standard treatment.) Because of their different strategies, each modality is associated with specific risks and side effects, and this chapter builds a scientific understanding of the modalities' approved uses, successes, limitations, and toxicities. As explained in the preceding chapter, the goal of using the modalities is cytoreduction – hopefully the complete eradication of all cancer cells from the body. If not possible, then the goal becomes reducing the number of cancer cells in the body to the point that the time required for malignant stem cells to replace them is longer than the patient's life, giving rise to a cure. If sufficient cytoreduction is not achievable, then relapse ensues at some point in the future that depends on the amount of surviving malignant tissue and its rate of repopulation. The younger the patient at time of diagnosis, the more effective the treatment must be at eradicating malignant cells. Typically this translates into more aggressive therapy in younger patients, but as it turns out, younger

Table 10–1. *Rates of surgical cure by mastectomy of diagnosed non-metastatic breast cancer grouped by size and nodal status*

Tumor stage*	Tumor diameter	10-Year death rate from cancer
T1N0M0 (Stage I)	≤ 1.0 cm	7 %
T1N0M0 (Stage I)	1.1 – 2.0 cm	18 %
T1N1M0, 1 node (Stage II)	Not stratified	22 %
T1N1M0, ≥ 2 nodes (Stage II)	Not stratified	30 %

*Note: Tumor stage is denoted using the "TNM Classification System," where "T" indicates tumor stage, "N" indicates number of positive lymph nodes, and "M" indicates status of metastatic disease. From Rosem et al (1989).

patients typically are the best at tissue repair and can recover from damage caused by the more aggressive therapy, whereas older patients cannot recover as quickly.

Specific cancers and their treatments have been selected to use in this chapter to illustrate principles. Cancers of interest to you might not be mentioned because they were not the best examples to use for explaining the scientific topics. For detailed information about treatment of specific cancers, see Cancer Treatment (Haskell and Berek 2001), Cancer: Principles and Practice of Oncology (de Vita, Hellman, and Rosenberg 2004), Cancer Medicine (Kufe, Pollock, Weichselbaum, Bast, and Gansler 2003), and Medical Oncology: Basic Principles and Clinical Management of Cancer (Calabresi and Schein 1993). Likewise, this chapter is not intended to be a compendium of drug targets and mechanisms. Rather, drug examples are selected to illustrate the pharmacologic principles of cancer treatment. For detailed information about anticancer drugs and their use, the reader is referred to current handbooks and texts about cancer chemotherapy, Concise Clinical Pharmacology: Cancer Therapeutics (Hohl 2005) and Handbook of AntiCancer Pharmacokinetics and Pharmacodynamics (Figg and McLeod 2004).

10.1 Surgical oncology

Surgical procedures are used to physically remove malignant tissue. If surgery is the only treatment, the assumptions are made that the location of all malignant tissue is known, it can all be removed, and undetectable micrometastatic disease is not left behind, or if some is left behind, it will not re-grow to clinically detectable levels within the life expectancy of the patient. There is no more effective modality for treating localized disease than surgery, as long as the lesion can be localized and safely resected. Diagnosis at the earliest possible stage minimizes the probability of local and distant spread and increases the probability of localized disease, resulting in a higher probability of surgical cure (Table 10–1).

The toxicity (morbidity) and the risk of death (mortality) of tumor resections are determined in large part by the location of the tumor, the percentage of healthy tissue removed during the operation, and the amount of lost normal tissue that can be replaced by the body. When the resected tumor is small compared to the size of the involved tissue, or in a noncritical site, surgical morbidity is lower. For example, surgical removal of a localized squamous carcinoma of the skin has low morbidity and mortality because a small proportion of skin is removed, the surrounding normal tissue can regrow and organ function can be preserved, and the tumor is easily accessible. Resection of a brain stem tumor of the same size is a higher risk procedure in a tissue that cannot regrow, and morbidity can be much higher. Cancer surgery carries the risk associated with any general surgical procedure: dangers of anesthesia, loss of hemostasis, and infection. As with all major surgeries, risks increase with age: for example, the mortality rate of surgery for colorectal cancer is 15% higher in octogenarians than those less than 70 years of age (Haskell et al. 1990). However, the experience of the surgical team with the procedure can also be an important factor (Garnick 1994).

Even very large tumors can be successfully resected. However, larger tumors are more likely to be associated with tumor extensions into critical or fragile normal tissue, greater risk of lethal hemorrhage during resection, and suppressed immune function that increases the risk of infection. For example, the mortality rate of optimal cytoreductive surgery for ovarian cancer is 3% (Berek and Hacker 1990), but it rises to 16% in patients with advanced ovarian cancer when the operation includes relief of bowel obstruction, exploratory surgery, or extensive tumor resections (Pecorelli 1994).

Adjacent healthy tissue is also usually removed to provide a surgical "margin" between the diseased and healthy tissues that for a cure pathological examination must prove to be free of malignant cells. Local lymph nodes may also be removed ("lymph node dissection") and analyzed as "sentinels" for locally disseminated disease from which to derive a probability of systemic disease (Chapter 1). The detection of occult disease in surgical margins and in the draining lymph nodes is important information for staging the patient and for determining the prognosis relative to the patient's age and overall health. It is still an inexact clinical science though, even with the most sophisticated detection technology. In many cases, clinical experience indicates that the cancer has very likely spread to other sites in the body by the time of surgery, and even though not detectable at the time of the procedure, this micrometastatic disease will cause a relapse despite successful surgery. For example, in early stage melanoma, 15% of patients who do not have any detectable melanoma cells in sentinel lymph nodes and who are also negative for local spread of disease using polymerase chain reaction (SLN- PCR-) at a median followup of 42 months from diagnosis nevertheless experience disease relapse by median 62 months followup, and 67% of these patients relapse with systemic disease that indicates invasive primary lesions were present at time of resection (Kammula et al. 2004). In some cases, inaccessible anatomic sites or extensive intermingling of tumor and critical normal tissues results in an "unresectable tumor."

In most cases, the location of residual tumor, clinical experience with previous case histories and sensitive diagnostic tools will indicate a high probability of relapse after surgery, unless nonsurgical treatments are added to the management of the patient's condition. Serial case studies of even the earliest diagnosed tumors treated with surgery alone indicate that many patients have occult, micrometastatic disease by the time of tumor detection and surgical removal (Tables 9–1 and 10–1). The high likelihood of micro-metastatic spread by the time of diagnosis, combined with a high probability that the patient will live long enough after surgery for micro metastatic disease to grow to a clinically detectable tumor burden and relapse, have caused the evolution of treatment strategy toward multimodality therapy (recall Figure 9–7). These advances in the biologic and clinical understanding of cancer biology from its beginnings in surgical oncology (Halsted 1894–1895) have modified thinking about the purpose of cancer surgery, and surgical techniques have been modified to spare tissue and preserve function (Baum 1976, Charlson 1985, Fisher et al. 1977, 1985a, 1985b, 1985c, 1993, Haskell et al. 1990). As the understanding of cancer as an intrinsically metastatic disease even at diagnosis in many patients evolved, radical resection and exenteration (complete removal of a block or region of the body) were phased out, because they cause significant morbidity and mortality without any improvement in cure rate. Tissue-sparing procedures have been developed for breast cancer (Fisher et al. 1993, Fisher et al. 1985c, Margolese et al. 1987), ovarian carcinoma (Colombo 1994), sarcoma of the limb (Morton et al. 1993), and prostate carcinoma (Jewett 1970, Walsh et al. 1983, Walsh et al. 1990). In lung cancer especially, surgical access of the diseased areas is possible via the airways using endoscopic procedures, and tumor tissue on the walls or in the lumen of the airways can be treated very specifically with a new technique called photodynamic therapy (PDT). In this procedure, the patient is dosed systemically with a drug called a photosensitizer, which becomes cytotoxic when exposed to certain wavelengths of light, and then the surgeon can specifically irradiate only those tissue areas that contain malignant tissue using laser light of the correct wavelength from a flexible fiberoptic source. PDT achieves specificity for the malignant tissue at two levels: only malignant tissue along the airways is exposed to the activating laser light, and some photosensitizers selectively accumulate in the malignant tissue compared to surrounding normal epithelium. PDT is very useful for relieving airway obstructions and controlling bleeding into the airways due to bronchogenic carcinomas, and further developments in this emerging field are expected. Conservative surgical strategies play an important role in increasing the quality of life of the patient by preventing the malformations and dysfunction that result from radical surgery, with equally good or even improved treatment outcome (see highlight box).

Surgical oncology has played an important role in the management of cancer, from the time when it was the only possible treatment option to the current day as part of multi-modality strategies. Several surgeons have been recognized for their roles in adapting surgical procedures, or developing new surgical procedures, in response to basic discoveries about the bilogy of cancer. These surgical advances include early procedures for complete resection of breast cancer by **William Halsted** at Johns Hopkins Hospital in the late 1800s, to the development of tissue sparing procedures for definitive treatment of breast and prostate cancer that maintain structural and functional integrity of surrounding tissue and organs by **Bernard Fisher** and **Patrick Walsh**, respectively. In 1993, Dr. Fisher shared the Charles F. Kettering Prize with Dr. Gianni Bonadonna for developing a combined conservative surgical and chemotherapeutic regimen that was a more effective treatment for breast cancer, and their research contributed much of what is known about optimally treating breast cancer (see Section 9.7). In 1996, Dr. Walsh shared the same prize with Dr. Malcolm Bagshaw for contributions to the treatment of prostate cancer with retention of urinary continence and sexual potency.

10.2   Radiation oncology

Radiation therapy ("radiotherapy") for cancer originated in the finding that x-rays sterilize rams by killing the proliferating germ cells in the testes that maintain spermatogenesis (Regaud 1930, Regaud and Ferroux 1927). The historical understanding of cancer as a disease of overly rapid cellular proliferation made it logical to treat cancer patients with x-rays, and the initial tumor responses encouraged the development of this treatment (Coutard 1932). During radiotherapy, malignant cells are exposed to ionizing radiation from either an external or implanted radiation source, and the resulting damage causes the death of the cell when it tries to divide, which leads to a gradual reduction in tumor mass – hence the descriptive name cytotoxic therapy or cytoreductive therapy. Radiotherapy is regional therapy; the radiation is focused like a beam of light on the treated area called a radiation field, but cancer cells that reside outside of the irradiated area will not be damaged. Radiotherapy is also very useful in control of localized disease when surgical resection is not prudent because of potential complications from tumor involvement of major blood vessels or other critical tissue sites.

Radiation is most toxic to proliferating cells, and higher doses are required to kill cells that are capable of proliferation but are not actively dividing (quiescent cells) at the time of exposure. Mammalian cells are most sensitive to radiation-induced damage in the late G2 and M phases of the cell cycle. A radiation dose that kills a cell in G2/M will usually not kill the same cell in the G1 or S phase, or in a quiescence state, because there is time for the cellular damage to be repaired before progressing through G2/M (Kaplan 1981, Parker 1990, Weichselbaum et al. 1993). Cellular damage produced by radiotherapy is an indirect result of ionization of chemicals in the cell to very reactive compounds. Oxygen is the predominant electron capturer in cells, and cytotoxicity is due primarily to damage caused by oxygen free radicals like hydrogen peroxide (H_2O_2), superoxide anion (O_2-), and hydroxyl radicals ($OH\bullet$). In fact, in the absence of oxygen or at very low concentrations of oxygen (hypoxia) such as exist in tumors, mammalian cells become two- to four-fold more resistant to radiation toxicity (Batzdorf et al. 1990, Kaplan 1981, Parker 1990, Weichselbaum et al. 1993). Therefore, poor blood perfusion of the tumor or a poorly developed microvasculature in the tumor, either of which causes poor oxygenation, leads to decreased effectiveness of radiotherapy.

Chemicals that can substitute for molecular oxygen have been developed as radiosensitizers for radiotherapy of poorly oxygenated tumors. However, it may not be possible to deliver blood-borne radiosensitizers to the tumor if oxygen itself cannot be delivered. More recently, targeted therapies that inhibit the action of angiogenesis factors have been found to normalize tumor vasculature and improve oxygen perfusion of tumor tissue (Jain 2002, Willett et al. 2004). Scheduling administration of angiogenesis inhibitors so that vascular normalization occurs at the time of radiotherapy should improve the effectiveness of radiation therapy.

Because malignant cells and their normal counterparts use the same basic mechanisms for cell division, it is not surprising that radiotherapy kills tumor cells as well as rapidly dividing normal cells within renewing tissues. Thus, radiation treatments cause cytoreduction not only in tumor tissue but also in healthy, normal tissues (toxicity) that lie within the radiation field. Death of rapidly dividing normal cells in renewing tissues causes the acute side effects (toxicities rapid in onset) of radiotherapy, and cells that divide as fast or faster than cancer cells are highly sensitive: hair follicles, gastrointestinal epithelium including the oral cavity, and hematopoietic tissue (blood cell–producing cells) in the bone marrow. Death of the dividing cells in these normal tissues produces the side effects that people associate with cancer therapy: infection and hemorrhage (bone marrow toxicity), diarrhea (gut mucosa toxicity), mucositis (oral mucosa toxicity), and hair loss (hair follicle toxicity). For example, abdominal radiation for metastatic ovarian cancer also exposes the intestine, and causes nausea, vomiting, or anorexia in 75% of the patients (Berek and Hacker 1990). In contrast, most of the bones that contain the hematopoietic marrow lie outside of this radiation field, so decreased peripheral blood counts of neutrophils (neutropenia) or platelets (thrombocytopenia) occur in only 10% of treated patients (Berek and Hacker 1990).

The dose of radiotherapy administered to the treatment field can also be limited by chronic (irreversible) side effects in nonproliferative tissues or in slowly renewing tissues like liver and kidney. Clinicians view severe chronic toxicities as more serious side effects than acute toxicities, which can be reversed if the patient can be supported medically during the recovery period. Chronic toxicities are more insidious in that many months or years may pass before they manifest clinically, more time is required for recovery, and they may be irreversible. There is also a general difficulty in the clinical management of patients with these side effects, and recovery may take weeks to months to reverse toxicity in tissues with protracted tissue renewal. Damage to nonproliferative, terminally differentiated tissues like peripheral nerve and heart may be irreversible. These chronic toxicities most often limit the total cumulative dose that can be administered. A relatively slow rate of repair of sublethal damage in these tissues may cause the delayed nature of these toxicities that depend on cumulative total dose because repair from the prior exposures has not been completed by the time of the subsequent exposure. The tissues susceptible to chronic toxicity are shielded during radiotherapy whenever possible. For example, shielding is used during abdominal radiotherapy for ovarian cancer to limit the total radiation dose to the kidney and liver.

The radiation dose might be limited by the severity of an acute toxicity to a rapidly renewing tissue or by the severity of an irreversible toxicity that appears several months after treatment is completed. Usually one healthy organ or tissue in the treatment field is more sensitive to radiation damage than the others, and the patient cannot be exposed to higher levels of radiation than this dose-limiting tissue can safely tolerate (Grever and Grieshaber 1993). As a general rule, the dose given per unit time, called dose intensity, is limited by acute toxicity usually to rapidly renewing tissues, whereas cumulative dose is limited by irreversible toxicity or delayed toxicity to non-renewing tissues. Dose-limiting toxicity (DLT) in the dose-limiting tissue limits the radiation exposure a patient can safely receive. Too high a dose poses risks of protracted or permanent organ damage, significant morbidity, and risk of death. The maximum tolerated dose (MTD) is defined as the dose that does not produce life-threatening, irreversible toxicity in the dose-limiting tissue in most patients. Therefore, this is the most commonly used dose for treatment.

For any therapy, the term "efficacy" denotes its effectiveness. The term "therapeutic index" is the ratio between the MTD and the efficacious dose, in other words how much differential sensitivity exists between the target tissue (the cancer) and the dose-limiting tissue. For cytotoxic therapy, the therapeutic index is usually low, which means the radiation dose that causes tumor regression and that which causes DLT usually differ by very little. Thus, radiotherapy has little margin for error in dose calculation. Recall from Chapter 9 that the radiation dose is usually fractionated, or even hyperfractionated, so that daily or hourly portions of the total dose remain below threshold levels for normal tissues, but deliver a large total dose to the malignant tissue, which is defective in its ability to repair and respond to subcellular damage.

In recent years, several new therapeutic approaches have developed in the field of radiation oncology to destroy tumor mass by achieving extreme changes in temperature, which are particularly useful in destroying inoperable lesions without exposure to ionizing radiation. The techniques combine the use of specialized needles that serve as sources of radiofrequency radiation or ultra-cold freezing materials and image guided placement of the needles into the tumor. A technique called radiofrequency ablation ("RFA") generates heat within the tumor via focused delivery of microwaves (low frequency radiation) from specialized needles that are implanted into the treatment field, similar to the heat generated from water-containing materials in a microwave oven. A second relatively new radiologic technique called cryotherapy is finding an important niche among treatment options for localized tumor lesions, especially those that are not resectable because of proximity to critical tissue sites like major vessels. Specially designed metal needles are inserted into the center of the malignant mass, and then super-cooled to withdraw heat radially from the surrounding tissue. Since water expands when freezing into ice, the tissue freezing effect kills the cancer cells within a certain distance of the super-cooled probe. The morbidity of this procedure is also considerably reduced relative to surgical resection.

10.3   Chemotherapy

10.3a   Directly cytotoxic chemotherapy drugs

Like radiotherapy, cytotoxic chemotherapy is designed to kill proliferating cells, so it is another type of cytoreductive therapy. Cytotoxic chemotherapy works best against cells actively progressing through the cell cycle, and these drugs are generally less effective against the same cells in a quiescent state. Like radiotherapy, chemotherapy eradicates dividing cells in renewing tissues, and the anticancer drugs act as cytoreductants in both tumor tissue (efficacy) and in the gastrointestinal mucosa, bone marrow, hair follicles, and germ cells (toxicity). The most sensitive normal tissue in which life-threatening toxicity occurs is again called the dose-limiting tissue, and the dose just below that which causes life-threatening toxicity is called the maximum tolerated dose. Usually the therapeutic index of chemotherapy (MTD to efficacious dose ratio) is quite small, and patients are dosed with the drugs to toxic levels in order to achieve maximum clinical benefit. Chemotherapeutics are administered by injection directly into the bloodstream, or are absorbed into the blood from the gut after oral dosing, where they circulate. In contrast to surgery and radiotherapy, which are local and regional therapies, respectively, chemotherapy is a form of systemic therapy because the dose is distributed throughout the body. Although this means toxicity will be more extensive, systemic therapy is the only conventional modality that potentially can treat every malignant cell of a metastatic cancer, even in micrometastatic disease with unknown locations of tumor foci. However, tumor foci with poor perfusion or poorly functioning capillaries may be exposed to suboptimal drug levels because "drug delivery" (diffusion from blood stream into tumor tissue) is very inefficient. Increasing vascular delivery of chemotherapy may be one reason why newer drugs that target angiogenesis factors paradoxically improve effectiveness of chemotherapy (Jain 2002, Willett et al. 2004).

10.3b   Reactive chemicals as cytotoxic anticancer drugs

Much of the current pharmacopoeia of cytotoxic chemotherapeutic agents for cancer arose from pursuit of two separate lines of reasoning. One line of reasoning was similar to that of radiotherapy and came from observed effects of reactive chemicals on rapidly proliferating tissues of the body. The bone marrow suppression and lymph node atrophy that followed combat exposure to nitrogen and sulfur mustard gases during World War I suggested a trial of these toxicants against malignancies of the bone marrow and lymphoid tissues, i.e., leukemia and lymphoma (Gilman and Philips 1946, Goodman et al. 1946). The clinical effectiveness and manageable toxicities of these compounds in prolonging survival of cancer patients led to a scientific effort to discover improved cytotoxic drugs for each type of cancer (Noble et al. 1958). Initially, this effort explored alkylating agents that were chemical relatives of the original mustard gases used in the war, but later expanded into a diversity of chemical structures that directly alkylated macromolecules in exposed cells, or released alkylating chemicals after spontaneous decomposition or activation by enzymatic action (e.g., thiotepa, procarbazine, dacarbazine, cyclophosphamide, ifosfamide). Serendipity also played an important role: the lead compound (cisplatin) in one of the most active families of alkylating agents for solid tumors (analogs carboplatin and oxaliplatin) was discovered as a contaminant in electrode baths that killed actively dividing bacteria (Rosenberg et al. 1965, Rosenberg et al. 1969). New and improved alkylating agents continue to be developed, with two new ones introduced to the market in 1999 (temolozomide as Temodar®) and 2002 (oxaliplatin as Eloxatin®). Some common alkylating agents are listed in Table 10–2.

Pharmacological levels of corticosteroids produce lymphocytopenia and thymic atrophy via activation of the apoptosis pathways in normal lymphocytes, and analogs which do not exhibit mineralocorticoid activity (prednisone, prednisolone, dexamethasone) have been developed to treat lymphocytic leukemia and lymphoma. These compounds induce apoptosis of lymphoma and lymphocytic leukemia with a T-cell phenotype at particular stages of differentiation by binding to an intracellular glucocorticoid receptor (Robertson et al. 1978, Wyllie and Morris 1982). Even as a single agent, prednisone induces complete remissions in 70% of children with acute lymphoblastic leukemia (Berenson and Gale 1990). These drugs are also useful in oncology for controlling inflammatory reactions caused by tumor or its treatment, such as intracranial pressure from brain tumors (Fosså et al. 1990).

10.3c   Selective cytotoxicity as a screening tool to discover more cytotoxic drugs

Once alkylating agents were found to have therapeutic potential in treating cancer, screening programs were established to identify other cytotoxic chemicals – either natural or synthetic or semi-synthetic analogs of active natural products – that exhibit selective cytotoxicity to malignant tumor cells over the normal, usually

Table 10–2. *FDA approved cytotoxic chemotherapeutic agents. Bold font identifies drugs approved between 1995–2004, and the recent approvals indicate an important role for, and continuing interest in, cytotoxic drugs. Italics identify natural products, or drugs developed from natural product leads.*

Alkylating agents or prodrugs	Non-alkylating cytotoxics
Altretamine	**Arsenic trioxide (Trisenox)**
Busulfan and **Busulflex**	*Bacillus Calmette-Guerin (BCG)*
Carboplatin	*Bleomycin*
Carmustine	**Bortezomib (Velcade)**
Carmustine wafer (Gliadel)	*Dactinomycin*
Chlorambucil	*Daunorubicin*
Cisplatin	***Daunorubicin liposomal (DaunoXome)***
Cyclophosphamide	***Docetaxel (Taxotere)***
Dacarbazine	*Doxorubicin*
Ifosfamide	***Doxorubicin liposomal (Doxil)***
Lomustine	***Epirubicin (Ellence)***
Melphalan	*Etoposide*
Mitomycin C	Hydroxyurea
Mustargen	***Idarubicin***
Oxaliplatin (Eloxatin)	***Irinotecan (Camptosar)***
Procarbazine	***Mitoxantrone***
Temozolomide (Temodar)	***Paclitaxel (Taxol)***
Thiotepa	*Streptozotocin*
	Teniposide
	Topotecan (Hycamtin)
	Valrubicin (Valstar)
	Vinblastine
	Vincristine
	Vinorelbine (Navelbine)

proliferative cell types that limit the dose that can be administered to patients. These screening programs often evaluated a large number of organic and aqueous extracts from plants, marine organisms and microbes (and their fermentation broths) in models of leukemia that would grow in mice, so as to identify extracts from which chemicals could be purified that were drug candidates because they exerted selective cytotoxicity toward malignant tissue over normal tissue, i.e., they possessed favorable therapeutic index where efficacy exceeded toxicity in vivo. A variety of compounds have been approved for clinical use based on this paradigm (Table 10–2), including some of the most active drugs in current use like taxanes (paclitaxel, docetaxel), anthracyclines (doxorubicin, daunorubicin), mitoxantrone, bleomycin, epipodophyllotoxins (etoposide, teniposide), vinca alkaloids (vincristine, vinblastine, vindesine, vinorelbine), and camptothecins (topotecan, irinotecan).

New cytotoxic mechanisms continue to be discovered and validated as legitimate therapeutic targets for human cancers, and the field of cytotoxic drug development remains active. A novel proteasome inhibitor – bortezomib (PS-341, Velcade®) – was approved in 2003 for third line therapy in multiple myeloma patients in whom two previous chemo regimens have failed and disease is progressing on the second one. This drug represented a first-in-class therapeutic agent that inhibits an ATP-dependent multicatalytic protease that plays a crucial role in the ubiquitin protein degradation pathway of proteins involved in cell cycle control and tumor growth (Adams et al. 1999, Grisham et al. 1999, King et al. 1996, Teicher et al. 1999). Development of the optimal dosage regimen in adult patients depended on titration of dose to a particular level of proteasome inhibition and duration of inhibition. Interestingly, 1.04 mg/m^2/dose given twice weekly for 4 weeks with two weeks rest achieved 70% inhibition of proteasome activity and caused dose-limiting thrombocytopenia, hyponatremia and kalemia, and fatigue/malaise at higher doses. When the dose schedule was changed to twice weekly for two weeks followed by two weeks rest, the maximum tolerated dose increased to 1.56 but the dose-limiting toxicities inexplicably changed to diarrhea and sensory neuropathy without bone marrow toxicity. When given twice weekly for two weeks followed by one-week rest in a pediatric population, only bone marrow toxicity was observed (Blaney et al. 2004).

Since these agents were discovered by screening programs against disease models, understanding their mechanism of action was not as important as the fact that they exhibited a favorable therapeutic index (ratio of toxic dose to efficacious dose) that qualified them as drugs. However, this is not to say that cytotoxic drugs do not act specifically upon certain molecular targets, and over the years, the molecular pharmacology of some of these compounds has been elucidated. A few of them appear to have very specific molecular targets that they act upon to cause cytotoxicity and eventually cell cycle arrest and apoptosis. For example, the vinca alkaloids appear to act solely via disassembly of microtubules by binding to tubulin. In contrast, other natural products show more promiscuous mechanisms of action at multiple targets, some of which are well understood but others are not. For example, the taxanes stabilize microtubules and convert them from dynamic functional machinery that moves molecules inside the cell into stationary cabling, but may exert other molecular effects in certain target cells. Anthracyclines may function simultaneously to inhibit topoisomerase II, produce free radicals and interfere with plasma membrane function, but the relative contribution of these activities to efficacy and toxicity in different tissues is the subject of ongoing research. Some of these cytotoxic natural products are produced by the source organism as a form of chemical defense against predators or parasites, and it is not surprising that they exert a number of molecular actions inside mammalian cells, which may all contribute to tissue selectivity and overwhelm the impaired capability of malignant cells to repair cellular damage.

10.3d Indirect tumor cytotoxicity by nutrient deprivation ("antimetabolite therapy")

A second line of scientific reasoning that has led to the development of a large number of useful anticancer drugs was based on the assumption that cancer cell proliferation could be halted by eliminating its supply of critical metabolic building blocks for DNA synthesis and other biochemical processes required for cell proliferation. By inhibiting proliferation of the tumor tissue, tumor regression would ensue due to the unchanged intrinsic rate of cell death from senescence of terminally differentiated or damaged cells. This strategy is known as antimetabolite therapy because the early chemotherapeutic agents targeted key biochemical steps for intracellular synthesis of intermediates for macromolecular synthesis. The success of these nucleoside analogs as anticancer agents verified the clinical value of this therapeutic strategy. For the past 35 years, it has been known that tumors require new blood vessel formation for continued growth and metastasis (Folkman 1971, Folkman et al. 1971). Otherwise, they outgrow their nutrient and oxygen supply and atrophy. The ultimate extension of the concept of antimetabolite therapy would be the elimination of the tumor's blood supply to deprive the tumor of all its essential nutrients. The treated tumor either outgrows its blood supply and atrophies, or it growth arrests to match the available level of nutrients. In 2004, a new therapeutic class of cancer drug that was intended to prevent develop of new tumor vasculature received FDA approval for treating colon cancer.

1. Depriving tumors of thymidine – the specific DNA building block. In the 1940s, under the direction of George Hitchings, the Wellcome Research Laboratories initiated a developmental therapeutics program in nucleoside analogs that led eventually to his receipt of a Nobel Prize in 1988 (see highlight box). These chemicals are man-made mimics of naturally occurring purine and pyrimidine building blocks used in the synthesis of the nucleic acids. Although the chemical modifications introduced into the pharmacologically active analogs were small enough to maintain their function as enzyme substrates in the salvage pathways of nucleosides, the structural changes lead to incorrect nucleic acid structure and function and eventually to cell death. A large number of analogs were synthesized and explored for biological

---

In 1988, **George H. Hitchings Jr and Gertude B. Elion** shared the Nobel Prize in Physiology or Medicine with Sir James W. Black "for their discoveries of important principles for drug treatment" in cancer. While at the Wellcome Research Laboratories in the 1940s, their scientific research into rational drug design for targeted theory based on chemical modifications to purine and pyrimidine structures led to the development of a class of therapeutics called antimetabolites, many of which are useful in treating solid tumor and hematological malignancies. (see http://nobelprize.org/medicine/laureates/1988/index.html).

---

activity. In 1947, scientists at Memorial Sloan Kettering made the key finding with a Wellcome compound synthesized by Gertude Elion, 2,6-diaminopurine. This purine analog was effective in vivo against a mouse model of sarcoma, and two purine analogs, 6-mercaptopurine and 6-thioguanine, were highly active clinically

Table 10–3. *FDA approved antimetabolite drugs that block tumor utilization of vital nutrients required to maintain cell proliferation.*

Class of antimetabolite	Used to treat solid tumors	Used to treat hematologic cancers
Purine analogs		Cladribine
		Fludarabine
		Mercaptopurine
		Pentostatin
		Thioguanine
Pyrimidine analogs	Capecitabine	Cytarabine
	5-Fluorouracil	Cytarabine liposomal
	Floxuridine	
	Gemcitabine	
Antifolate antimetabolites	Methotrexate	Methotrexate
	Pemetrexed	

in pediatric and adult leukemia, respectively. She shared the 1988 Nobel Prize with Hitchings for her work. These compounds were referred to as "antimetabolites," because they seemed to interfere with the function of essential nutrients for growth like sulfonamides do in bacteria regarding *p*-aminobenzoic acid utilization. Other lines of research led Hitchings to discover selective, species-specific inhibitors of the folate cofactor required by the enzyme dihydrofolate reductase (DHFR), which generates tetrahydrofolate required for conversion of uridine monophosphate into thymidine monophosphate. Since thymidine nucleotides are required for DNA synthesis, inhibitors of DHFR were expected to block formation of this critical building block, inhibit DNA synthesis and indirectly block cell replication. The DHFR inhibitor methotrexate has been an important folate analog for treating several types of cancers, and in 2004 a new antifolate drug called pemetrexed (Alimta®), which inhibits DHFR as well as some other folate dependent enzymes, was FDA approved for mesothelioma treatment as part of a multi-drug regimen. Another drug named 5-fluorouracil (5FU) inhibits another target in the pathway known as thymidylate synthase, which utilizes the reduced folate from DHFR to synthesize thymidine from uridine. In colon cancer chemotherapy, 5FU is administered with reduced folate called leucovorin, which increases 5FU binding and inactivation of thymidylate synthase. Since that time, several antimetabolite drugs have been synthesized (Table 10–3) and been found to be clinically active against certain cancers. Many more have been synthesized and found to be ineffective against clinical cancer. In addition to depriving malignant cells of thymidine or other metabolites, nucleoside analogs are designed to terminate DNA and/or RNA synthesis.

Recent pharmaceutical efforts have led to the clinical introduction of new nucleoside analogs with enhanced tissue selectivity for malignant tissue. The tissue selectivity of 5FU therapy has been improved by creating a prodrug of 5FU called capecitabine (Xeloda®), which is approved for second line therapy of breast cancer after anthracycline failure or anthracycline/taxane failure. This drug is designed to be activated in a series of enzymatic reactions beginning in the liver after absorption from the lumen of the gut following oral dosing, and ending with release of the active 5FU species by an enzyme that exists at much higher content in malignant tissue than normal tissue. This strategy minimizes exposure of normal tissues in the body to 5FU, which is lower than in infusional 5FU therapy, and maximizes the tumor delivery of 5FU. A cytosine analog known as cytarabine is now available in a liposome-encapsulated product called DepoCyt®, which embodies the idea of maximizing delivery to malignant tissues and minimizing systemic exposure of normal tissues, relative to giving the naked drug directly.

Interestingly, some of these drugs are much more active against hematological malignancies than solid tumors, whereas others are more active against solid tumors than leukemia and lymphoma. Regardless, seminal research by Hitchings and Elion established the proof of principle that depriving malignancies of essential nutrients could lead to significant clinical benefit, which led to the development of a number of effective anticancer drugs.

2. Antiangiogenesis therapy – depriving tumor of all nutrients and oxygen? For solid tumors to grow from microscopic size into clinically detectable disease, they must establish a vigorous blood supply that can match tissue perfusion requirements as the tumor enlarges (Fidler and Ellis 1994, Fidler et al. 1998, Folkman 1971, Folkman et al. 1971). Otherwise, tumor growth would be nutrient and/or oxygen limited, and tumor size at homeostasis, where tumor cell proliferation and cell death from senescence are balanced, would be very small and not clinically significant. Tumor cells can secrete any of a number of cytokines that act as chemotactic factors on endothelial cell (Fidler et al. 1998, Kim et al. 2001, Kuniyasu et al. 2003, Yano et al. 2000), including vascular endothelial growth factors (VEGFs), fibroblast growth factors (FGFs), and interleukin-8 and -15 (IL-8, IL-15). These factors stimulate new blood vessel formation by inducing budding and branching of endothelium in nearby established vessels, a process termed angiogenesis, and undifferentiated endothelial precursors migrate along the chemotactic pathway into the tumor, where they probably anastomose with differentiated tumor cells (Folberg et al. 2000, Hendrix et al. 1998, Hess et al. 2001, Maniotis et al. 1999, Van Der Schaft et al. 2004), and provide access to additional blood flow.

Angiogenesis is an attractive target for developing new anticancer therapies. Since the endothelium is directly exposed to the blood stream, there should be no obstacles effective drug levels to the areas of active angiogenesis following IV to delivering injection. In contrast, drugs targeting the tumor tissue itself must diffuse out of the capillaries, pass through several basement membrane barriers and connective tissue, and diffuse into the tumor tissue against oncotic pressure to be active. Because anti-angiogenesis therapy targeted a normal cell type in the endothelial cell that should be genetically and phenotypically stable, it was thought that the Goldie-Coldman hypothesis of drug resistance would not apply to anti-angiogenesis therapy, unlike cytotoxic chemotherapeutic agents targeting the tumor itself. Finally, it was expected that clinical toxicity would be mild, with the exception of possible worsening of wound healing, including recovery from cancer surgery and biopsies, because it naturally involves angiogenesis.

Initial clinical trials evaluated naturally occurring antagonists of angiogenesis, which were proteins that did not have suitable pharmaceutical properties, and failed to demonstrate any therapeutic response or clinical benefit from therapy. Several companies developed small molecule inhibitors of receptors for cytokines that stimulate angiogenesis. However, at the current time, there is only one FDA approved product for inhibiting angiogenesis named bevacizumab (Avastin®), which is a humanized IgG1 monoclonal antibody that contains a murine complementarity-determining region (CDR) that binds VEGF and neutralizes biological activity by preventing its binding to two VEGF cell surface receptors named Flt-1 and KDR. This drug was approved in 2004 for use in combination with infusional 5FU + leucovorin (IFL) regimens for first-line therapy for colorectal carcinoma ("first-line" means the first chemotherapy regimen a patient receives). As expected, serious, but uncommon, side-effects of Avastin include formation of holes in the colon (gastrointestinal perforation) generally requiring surgery to repair to avoid intra-abdominal infections, impaired wound healing, and bleeding from the lungs or internally.

It is paradoxical that the addition of an inhibitor of new blood vessel formation could improve the outcome of chemotherapy, since one might expect that the chemotherapeutic drugs might not work as well if their delivery to the tumor was interrupted by inhibiting blood vessel formation. However, unlike normal blood vessels, tumor blood vessels cannot deliver drugs or nutrients very efficiently to the tissue, because they are leaky and engorged with stagnant blood flow (Jain 2002, Willett et al. 2004). In preclinical studies, treatment of human tumor xenografts with an anti-VEGF monoclonal antibody reduced tumor vessel permeability and caused vascular regression. The reduced vascular permeability led to *increased* delivery of oxygen and therapeutic agents to tumors. In human patients, a single infusion of bevacizumab decreased tumor perfusion, vascular volume, and interstitial fluid pressure. The data indicated that VEGF blockade has a direct and rapid antivascular effect in human tumors. Given the fact that bevacizumab improves the delivery and effectiveness of chemotherapeutic agents, one would expect that it would provide similar improvement in radiotherapy as a result of improved tumor oxygenation.

Antiangiogenic therapy seems to meet high expectations regarding its safety and relatively uncommon serious toxicities. This safety profile makes bevacizumab a candidate for combination with other chemotherapy agents to explore clinical effectiveness when added to other regimens and in other malignancies. However, the Goldie-Coldman hypothesis appears to still apply – not because endothelial genotype is unstable, but because of the biological adaptability of the tumor to anti-VEGF therapy. There are a number of soluble factors in addition to VEGF that stimulate angiogenesis (Fidler et al. 1998, Kim et al. 2001, Kuniyasu et al. 2003, Yano et al. 2000), providing the tumor with options for an "angiogenic swap" from

VEGF to some other angiogenic factor and thereby becoming resistant to beva-cizumab. The change from VEGF to another angiogeneic factor can result from the tumor's response to nutrient deprivation (Bobrovnikova-Marjon et al. 2004, Marjon et al. 2004). Bevacizumab would no longer be effective if tumor angiogen-esis became, e.g., FGF- or IL-8 dependent, rather than VEGF-dependent, because the antibody specifically binds and neutralizes VEGF. It remains to be seen if this phenomenon causes drug resistance problems clinically. To exploit the stable phe-notype and genotype of normal angiogenic endothelium and avoid development of drug resistance, new therapeutics will be needed that selectively inhibit convergence points in signal transduction pathways for multiple angiogenic factors secreted by tumors.

Note that the clinical effectiveness of bevacizumab in improving combination chemotherapy actually is evidence that disproves the original rationale of antian-giogenic therapy. The original idea was that inhibiting angiogenesis would decrease nutrient and oxygen availability to the tumor, resulting in growth stasis and a slow atrophy of tumor burden toward remission levels at a rate determined by the inherent apoptosis rate of a starving tumor. Instead of specifically inhibiting tumor access to a single critical nutrient for a required process for cell proliferation like the antimetabolite drugs do, eradicating new vasculature would remove access to all nutrients and oxygen and jeopardize tumor survival, not just proliferation. How-ever, the clinical evidence, supported by laboratory studies (Jain 2002, Willett et al. 2004), suggests that antiangiogenic therapy improves tumor delivery of small drug molecules, so it seems very likely that it improves nutrient and oxygen delivery as well. This is the opposite of the expected and planned biological effect when these agents were first developed as "rational" new drugs. The erroneous initial concept of the biological action of these drugs may explain why clinical trials of previous angiogenesis inhibitors tested alone, and not in combination with chemotherapeutic agents or radiotherapy, failed to demonstrate any clinical efficacy. The lack of clinical effectiveness of these new drugs tested as single agents is easily understandable, and the biology of bevacizumab therapy suggests that many of the previous angiogenesis inhibitors should be revaluated clinically in combination with chemotherapeutic drugs. This is especially important for developing angiogenic inhibitors for differ-ent molecular targets that provide alternative therapies to bevacizumab if tumors swap VEGF for some other angiogenic mechanism.

3. Indirect tumor cytotoxicity by depriving tumors of trophic factors. As described in detail in Chapter 1, tissue renewal is maintained by the balance between several subpopulations of cells: stem cells, progenitors committed to expansion and differ-entiation, and terminally differentiated cells which perform the specialized function of the tissue (Figure 10–1). Tissue mass can be increased or decreased by altering the number of differentiated cells produced by each progenitor and/or by altering the number of progenitors which produce differentiated progeny. Although there are exaggerated stem cell numbers and activity in malignancies (Pierce and Speers 1988),

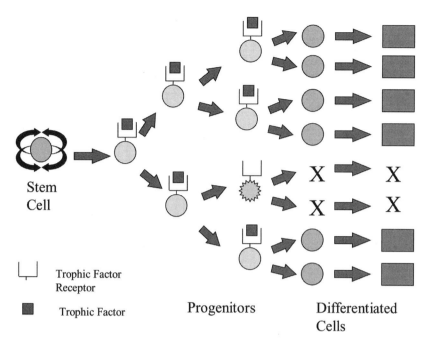

**Stem
Cell**

Trophic Factor
Receptor

Trophic Factor

Progenitors

**Differentiated
Cells**

Figure 10–1. Schematic representation of the stem cell model of tissue renewal. At homeostasis, tissue renewal is the result of balance between production of new differentiated cells and the loss of older ones through sloughing or apoptosis (cell death). New specialized cells (rectangles) are produced via the differentiation of immature precursor cells (densely stippled circles), which in turn have been produced via the finite proliferation and maturation of progenitor cells (lightly stippled circles). Although usually quiescent, stem cells can enter the cell cycle periodically to produce committed progenitor cells or replenish the stem cell population as needed. Malignancy is a caricature (exaggeration) of normal tissue renewal in that cell production and differentiation become disconnected and cell production and loss are not in balance (see Chapter 1). Progenitors seem to be biologically programmed to undergo apoptosis, unless trophic factors are present to block apoptosis and sustain proliferation. In the absence of trophic factor stimulation, the progenitors undergo apoptosis before tissue expansion. After a finite number of cell divisions, the replicating progenitor pool terminally differentiates into the non-proliferative, functional cells of the particular organ.

both normal and malignant tissue have mechanisms which maintain homeostasis and stimulate repopulation after loss of differentiated tissue mass (Al-Dewachi et al. 1980, Cameron et al. 1990, Fisher et al. 1989a, 1989b, Mayo et al. 1972, Potten et al. 1994, Potten and Chadwick 1994, Schabel 1975).

An important biological principle is the control of tissue mass by two opposing mechanisms: trophic factor stimulation of cell accumulation and increased tissue mass versus inhibition of cell accumulation by negative regulators (called chalones in the older literature). Trophic factors prevent apoptotic cell death in progenitors and often in terminally differentiated cells as well which results in an increase in tissue mass (Figure 10–1). When the progenitors cannot survive, new cell production is lost, senescent and sloughed terminally differentiated cells cannot be replaced, and the tissue atrophies. Inhibition of apoptosis is a universal function of trophic factors from all renewing tissues: hemato- and lymphopoietic (Abbrecht and Littell 1972, Borthwick et al. 1996, Gombert et al. 1996, Hassan and Zander 1996, Jelkmann and Metzen 1996), endometrial (Hopwood and Levison 1976, Sandow et al. 1979), breast (Anderson et al. 1982, Ferguson and Anderson 1981, Ferguson and Anderson 1981), and prostate (Kerr and Searle 1973, Sandford et al. 1984, Van Steenbrugge et al. 1984).

Trophic factors can be endocrine in nature (humoral) or paracrine (locally produced and secreted). For example, soluble factors are present in blood plasma after cytoreduction of normal and tumor tissue that initiate and then halt cell division in the affected tissue (Fisher et al. 1989a, 1989b, Potten et al. 1994, Potten and Chadwick 1994). Some of these factors have been isolated and identified, but it seems likely that many more remain to be characterized. The trophic factors include sex steroids, produced by the gonads under pituitary control and adrenal gland, and several cytokines produced by specialized cells in the kidney and lymphoid system. Paracrine trophic factors of tumors also include peptide growth factors like EGFs, PDGFs, FGFs, and TGFs. Trophic factor over-expression or ectopic expression, and constitutive receptor activation for these pathways, drive survival of malignant progenitors and growth of many malignancies. New therapeutic agents that interfere with signal transduction from these peptide growth factors by blocking ligand-receptor interactions, receptor tyrosine kinase response to ligand binding, or downstream signal transduction targets have started to appear in the US pharmaceutical market.

This therapeutic approach for blocking the action of trophic factors is known by many names: endocrine therapy (more appropriately anti-endocrine therapy), hormonal therapy, hormone ablation therapy, or anti-hormonal therapy. Recently, the term "targeted therapy" has entered the pharmacological literature to mean therapies that target specific molecules that are driving malignant growth and cell survival. Although many targeted agents are inhibitors of the action of paracrine trophic factors in tumors, many other targeted therapies including antimetabolite therapy and antiproteasome chemotherapy are not directed at trophic factor mechanisms. The recently approved therapies that target EGFR1, EGFR2 (Her-2/neu), bcr/abl and c-kit will be presented as drugs that interfere with trophic factor signaling. They are examples not only of "targeted therapies", but also of therapy by trophic factor withdrawal because they interfere with peptide growth factor receptor signaling.

Some trophic factors are beneficial because they act on normal tissue that is damaged by cancer therapy. Some trophic factors for the blood-forming progenitors in the bone marrow are used clinically to minimize therapy-induced toxicity and avoid delays in scheduled chemotherapy (Table 10–4, Section 10.3e). The trophic factors that cause enlargement of tissue mass by increasing cell size (hypertrophy) rather than cell number probably do not play a role in balancing cell production and cell loss in renewing tissues or in tumors, and have not been discussed.

Table 10–4. *FDA approved trophic factors that are used to stimulate recovery of peripheral blood cell counts via inhibition of apoptosis in the hematopoietic progenitor populations of the bone marrow. These recombinant factors are used in some patients to shorten the time required for recovery from drug-induced bone marrow toxicity and maintain the optimal chemotherapy scheduling.*

Stimulation of red blood cell levels	Stimulation of white blood cell levels	Stimulation of platelet levels
darbepoetin-alfa epoetin alfa	filgrastim pegfilgrastim sargramostim	oprelvekin

4. Antagonism of trophic factors causes atrophy and decreases tissue mass in responsive normal and malignant tissues. Cancers from tissues that depend on trophic factors often retain the factor-dependence of their tissue of origin. Therapies that block (antagonize) the biological activity of trophic factors cause atrophy of both the malignancy and its normal counterpart (Figure 10–2). These treatments can induce tumor shrinkage and provide years of control of malignant disease. Because of the tissue-specificity of the trophic factor, the toxicity of anti-hormone therapy occurs mostly in the normal tissue counterparts that express the same receptor and response mechanisms, and clinical toxicity is usually mild compared to toxicity from conventional chemotherapy.

There are at least three requirements for effective control of cancer with anti-endocrine therapy. First, the cancer must originate from a tissue that is not critical for life, or must have acquired ectopic expression of the trophic factor response pathway, where ectopic means abnormal expression for its tissue of origin. This requirement is responsible for the mild toxicity associated with therapy. For example, inhibiting the trophic factor for prostate epithelium and causing atrophy (involution) of both prostate cancer and the normal prostate does not cause life-threatening toxicity, whereas inhibiting the trophic factor for white blood cell production would quickly lead to a life-threatening neutropenia. Second, the malignancy must retain responsiveness and dependency to the trophic factor, including receptors, a signal transduction mechanism, and the hormone-response genes. Tumors from trophic factor-dependent tissues can evolve into factor-independent phenotypes that are refractory to anti-hormonal therapy. This evolution is probably the result of genetic instability of the Goldie-Coldman type (Goldie and Coldman 1979) and positive selection of subpopulations that gain a growth advantage when trophic factor action is blocked (Isaacs and Coffey 1981, Isaacs et al. 1982). For some reason, some malignant stem cells eventually adapt to trophic factor deprivation and may at any time begin to produce progenitors which no longer require trophic factor for expansion (Isaacs and Coffey 1981). The biological explanation for the appearance of "hormone-resistant" ("hormone independent") malignancy remains an enigma,

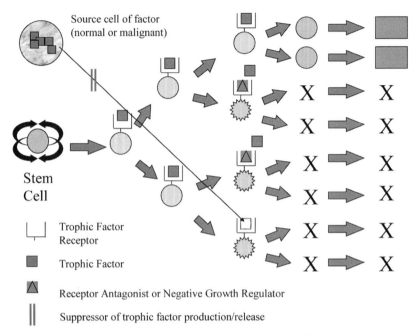

Source cell of factor
(normal or malignant)

Stem
Cell

⊔ Trophic Factor
   Receptor

■ Trophic Factor

△ Receptor Antagonist or Negative Growth Regulator

‖ Suppressor of trophic factor production/release

Figure 10–2. Trophic factors (dark squares) that serve as targets for cancer therapy are endocrine or paracrine mediators that block apoptosis in the progenitor population of a renewing tissue. They may interact with cell surface or intracellular receptors to promote progenitor survival so that differentiated cells can be generated. Progenitors that fail to receive adequate exposure to the specific trophic factor die via apoptosis (indicated by the abnormally shaped cells) and therefore do not produce any differentiated cells. Homeostasis usually occurs at a level of trophic factor that cannot block apoptosis of all of the progenitors. This creates a reserve for rapidly increasing cell production if required and a mechanism for physiologic response to the environment. Drugs that block trophic factor binding to its receptor (light triangles), or that interfere with the supply of the trophic factor (double line) will increase the number of progenitor cells that die via apoptosis (indicated by the abnormally shaped cells), substantially diminish new tissue production, and result in reduced tissue mass over time. Note that stem cells do not die as a result of antagonizing trophic factor activity, although there is some evidence that stem cell renewal and their production of progenitors may be slowed by trophic factor withdrawal.

because hormone-independence never occurs in normal tissue deprived of trophic factor. Third, anti-hormonal therapy is most effective when the trophic factor is tissue specific. For example, insulin and growth hormone, which have systemic effects on many target tissues, would not be suitable targets for anti-endocrine chemotherapy, because hormone withdrawal would cause serious complications in many organ systems. The less the tissue specificity of the targeted trophic factor, the more complicated the side effects, and the less effective the biological therapy.

In contrast to progenitor cells, repopulation of atrophied tissue in response to a new source of trophic factor proves that stem cells survive for long time periods without trophic factors. For example, androgen withdrawal causes involution of the prostate, but the secretory epithelium rapidly repopulates with androgen replacement therapy even after a dormancy of many months (Huggins and Hodges 1941, Isaacs and Coffey 1981, Isaacs et al. 1982, Kerr and Searle 1973, Leav et al. 1978). This cycle of re-growth and atrophy can be repeated many times (Sandford et al. 1984). The human hematopoietic stem cell can survive in the absence of human- and myeloid-specific trophic factors in an immune deficient mouse, without producing any surviving progenitors, yet it will produce human leukocytes when human trophic factors like GM-CSF and IL-3 are administered to the animal (Lapidot et al. 1992, Murray et al. 1995, Murray et al. 1994). However, an ability to survive in the absence of trophic factor does not mean that the stem cell won't respond to the factor. In fact, trophic factors may accelerate repopulation of the tissue by stimulating stem cell proliferation or increasing growth fraction. Therefore, therapies which block trophic factor activity should only be expected to achieve tumor control initially via large reductions in tumor burden following apoptosis of progenitors and then subsequently via slowed stem cell activity. Interfering with trophic factor stimulation should not be expected to result in the complete eradication of the malignant tissue, because the cancerous stem cell would be expected to survive just like its normal counterpart.

Even though they do not eliminate the malignant stem cell, anti-endocrine therapies are quite useful clinically. Anti-hormonal therapies can control disease for years despite likely survival of the malignant stem cell pool, and theoretically "cure" patients whose life-span due to other medical conditions is shorter than the time required for evolution to a hormone-independent state. A reduction in tumor burden with mild, non-life-threatening toxicity may be sufficient in many of these patients to reach their life expectancy with minimal interruption in quality of life. The anti-endocrine therapy may also slow repopulation by the stem cell, so continuing to take the anti-endocrine chemotherapeutic even after the patient enters remission (maintenance therapy) offers a theoretical advantage. However, chronic trophic factor deprivation might also accelerate the development of trophic factor-independence via biological selection of stem clones with a growth advantage under these conditions, but there are a variety of mechanistic approaches for disrupting endocrine pathways driving tumor growth. Switching from one therapeutic class of anti-endocrine therapy to a second one that circumvents the resistance mechanism of the first may achieve additional months-to-years of tumor control. The important issues of intermittent versus chronic anti-endocrine therapy, and switching between therapeutic strategies to maintain endocrine disruption, should be resolved soon by on-going clinical trials.

Trophic factor withdrawal can be accomplished surgically by removing the organ which produces the trophic factor, such as testes or ovaries, as long as the organ is not critical for survival. In the US, surgical castration is not used as commonly as "medical castration," which uses medicines to disrupt production or action of endocrine factors. Even with surgical resection of hormone-producing tissues, anti-endocrine therapeutics are usually required to block any possible effects of the lower yet constant levels of hormone secreted by other tissues which are not resected because of their important role in regulating other physiological processes (adrenal, liver, adipose).

5. Antagonism of steroid trophic factors. In the reproductive system, steroid hormones serve as trophic factors. In the female, progesterone and estrogens cause increases in cell number and maintain differentiated epithelium in breast and uterus (Going et al. 1988, Helgason et al. 1982, Horwitz et al. 1978, Potten et al. 1988, Sandow et al. 1979). In the male, the androgenic steroids maintain the cellularity and specialized differentiation of prostatic epithelium. "Androgen ablation therapy" which means eradicating the supply of testosterone and/or dihydrotestosterone, or antagonizing their activity, results in rapid apoptosis of prostatic epithelium and atrophy of the prostate gland (Huggins and Hodges 1941, Isaacs and Coffey 1981, Isaacs et al. 1982, Kerr and Searle 1973, Sandford et al. 1984, Van Steenbrugge et al. 1984). However, the stem cells of the epithelium survive androgen ablation and repopulate the epithelium in response to androgen replacement (Leav et al. 1978, Sandford et al. 1984). This knowledge has been used to develop anti-endocrine therapies for cancers originating in these tissues.

6. Carcinoma of the prostate and androgen ablation. Charles Huggins was the first to show that removal of the testes (orchiectomy) in prostatic carcinoma patients induced clinical regressions with high frequency (Huggins and Hodges 1941), a discovery for which he was awarded the Nobel Prize in Medicine in 1966 (see highlight box). Today anti-endocrine therapy plays a major role in the treatment of prostatic carcinoma (Neuwirth and De Kermion 1990). Building upon the pioneering work of Huggins, modern androgen ablation can be achieved pharmacologically as well as surgically in the patient with prostatic carcinoma. Most of the testosterone is made in the testes by Leydig cells, although a small amount is also produced in the adrenal glands. When androgen levels are low, the hypothalamus secretes pulses of luteinizing hormone-releasing hormone/gonadotropin-releasing hormone (LHRH/GnRH) and this in turn stimulates release of luteinizing hormone (LH) and follicle-stimulating hormone (FSH) from the pituitary gland that co-stimulate androgen synthesis by the Leydig cells. Therefore, anti-androgen therapy can be accomplished in several ways: surgical removal of the androgen source (testes) with orchiectomy; inhibiting the response of the pituitary to low androgen levels using chronic exposure to estrogen, diethylstilbestrol (DES), or LHRH/GnRH agonists like Leuprolide (Barradell and Faulds 1994, Clarke et al. 1990, Santen et al. 1986); inhibiting the action of testosterone on the target organs using cyproterone acetate (Barradell and Faulds 1994) or the androgen receptor antagonists flutamide, bicalutamide and nilutamide (Klein 1996); or in emergency situations, using ketaconazole, a non-specific inhibitor of the cytochrome P450 that produces testosterone (Clarke et al. 1990, English et al. 1986).

In 1996, **Dr. Charles Brenton Huggins** shared the Nobel Prize in Physiology or Medicine (with famed tumor virologist Dr. Peyton Rous) "for his discoveries concerning hormonal treatment of prostatic cancer." While at the University of Chicago in the 1930s, he found that following orchiectomy (removal of the testicles – the major tissue source of androgens), the prostate gland atrophies and secretions stop, a condition reversed by administering testosterone. Similar atrophy of prostate cancer cells during androgen withdrawal forms the basis of treatment for many patients with this disease today (see http://nobelprize.org/medicine/laureates/1966/huggins-bio.html).

In most cases in the US, androgen ablation is achieved by "medical castration," in which medicines are used to disrupt the androgen factor pathway that promotes prostate cancer growth, rather than by surgical castration. In rare cases, the source tissue of low androgen levels, the adrenal glands, might by surgically removed.

Most prostate cancers at early stages of their natural history are composed primarily of androgen-dependent progenitors and differentiated epithelium, and androgen-independent stem cells are at most only a minor component (Isaacs et al. 1978, Isaacs et al. 1982, Smolev et al. 1977, Smolev et al. 1977). Thus, androgen ablation therapy causes clinical regression and the atrophy of the tumor tissue occurs quickly enough to provide prompt relief from bone-pain in metastatic disease (Neuwirth and De Kermion 1990). Androgen ablation therapy relieves symptoms without life-threatening toxicities, and remission may be of sufficient duration to achieve normal life expectancy in older patients.

However, just as normal prostatic epithelium stem cells survive androgen ablation (Leav et al. 1978, Sandford et al. 1984), malignant stem cells also survive deprivation of androgens (Isaacs and Coffey 1981, Isaacs et al. 1982). But, in contrast to normal stem cells, these malignant stem cells will adapt and begin to repopulate the tumor despite the lower exposure to testosterone, leading eventually to relapse while bring treated with anti-androgen therapy. As expected in relapsed cancer after orchiectomy, LHRH/GnRH agonists show no activity (Neuwirth and De Kermion 1990), but therapy can be switched to a different anti-androgen strategy to interfere with trophic factor activity at a different site of action in the physiological pathway, and circumvent the adaptive response of the malignant prostatic tissue to the low androgen exposure. Patients with progressive disease after hormonal therapy respond to additional anti-hormonal therapies, and clinical trials are addressing which treatment, treatment combinations and treatment sequences are optimal. There is a need for an larger number of non-cross-resistant therapeutic options to allow for more longer-term tumor control, as early detection strategies lead to diagnosis of younger patients.

Repopulation of the prostatic carcinoma and the presence of differentiated cells under conditions in which the normal prostate remains atrophied (Kraljic et al. 1994) lead one to try to explain the difference between stem cells and progenitor cells from normal versus malignant prostatic epithelium. Androgen hypersensitivity might develop via a Goldie-Coleman-type mechanism or via environmental pressure, mutation, and natural selection. This hypothesis proposes that the relapsing cells are surviving and/or proliferating in response to very low but detectable blood levels of androgen that persist after androgen ablation, perhaps via a mutated androgen receptor that binds testosterone with greater affinity. However, there is legitimate disagreement about the importance of these low levels of circulating androgens in these patients and the mechanisms underlying conversion to apparent androgen independency.

Androgen suppression strategies continue to be the mainstay of treatment for prostate cancer that cannot be cured by surgical resection or local radiotherapy alone. Recent efforts in developmental therapeutics have produced new products that increase patient convenience and compliance by requiring only infrequent, injectable dosing every one to four months. More convenient depot forms are made from pharmaceutical formulations that slowly release the drugs over time from a deposit in the muscle or under the skin, such as injectable suspensions of GnRH agonists abarelix (Plenaxis®), leuprolide acetate (Eligard) subcutaneous injection, and Lupron depot.

7. Carcinoma of the breast and estrogen ablation. Estrogen is a steroid which acts as a trophic factor for the epithelium in the glands of the breast (Anderson et al. 1989, Anderson et al. 1982, Ferguson and Anderson 1981, Ferguson and Anderson 1981, Going et al. 1988, Potten et al. 1988), so anti-endocrine therapy plays an important role in breast cancer therapy (Goldhirsch and Gelber 1996). Anti-estrogen therapy in pre-menopausal women is more complex than androgen ablation therapy in men because estrogen and progesterone levels fluctuate during the menstrual cycle (Ferguson and Anderson 1981, Going et al. 1988) and estrogen levels drop dramatically at menopause (Anderson et al. 1982, Potten et al. 1988). Similar to anti-androgen therapy for prostatic carcinoma, anti-estrogen therapy for breast carcinoma does not eliminate the malignant stem cell but rather slows its proliferation and promotes apoptosis of progenitors and their progeny. As expected with this mechanism, relapse during estrogen-ablation therapy can be due to estrogen-independent cells (Fisher et al. 1993, Reddel et al. 1988), although relapsing tumors may still respond to anti-estrogen therapy administered at higher doses or using anti-estrogens of different mechanism of action (Manni and Arafah 1981, Muss et al. 1987).

Estrogen-ablation in pre-menopausal women is achieved by removing the primary tissue source of estrogen (ovaries) in a procedure called an oophorectomy, because the high levels make pharmacological blockage of estrogen action hard to achieve. In post-menopausal women, estrogen production by the ovaries is very low, and there is no need to remove them. The possible trophic activity of the low estrogen levels (from diet or the adrenal gland) in the absence of ovarian production can be blocked pharmacologically with chronic tamoxifen therapy (Vogel 1996), which is a selective estrogen receptor modulator (SERM) rather than a pure estrogen antagonist. In other words, the modulation of estrogen receptor activity depends on the tissue-specific context in which it is found. It does antagonize estrogen binding to its intracellular receptor in breast tissue, but it functions like an estrogen agonist in the endometrium of the uterus, where it causes cancer in some treated patients (Anthony et al. 2001, Parczyk and Schneider 1996). Infrequently, adrenal production of estrogen may be high enough to warrant treatment with aminoglutethimide to inhibit the enzymes that produce estrogen, but the greater toxicity of aminoglutethimide therapy relegates it to the role of salvage therapy after tamoxifen failure (Fisher et al. 1993).

Tamoxifen has been the treatment of choice when the tumor is expected to be estrogen-responsive in a low estrogen environment, when the toxicity of cytotoxic chemotherapy is not tolerable, or when the age or health of the patient indicates only a short period of tumor control will be required (Fisher et al. 1993, Jordan 1988). Clinical cure is more likely in patients with slower growing tumors or in older patients with shorter life-expectancies. As predicted from the mechanism, maintenance therapy (constant exposure to tamoxifen) is more effective at controlling tumor regrowth than brief exposure (Fisher et al. 1993). Adjuvant therapy with tamoxifen after surgery is effective in pre-menopausal women with node-negative disease and in post-menopausal patients with estrogen receptor-positive tumors (Haskell et al. 1990). Five years of tamoxifen are superior to 1–3 years of treatment (Fisher et al. 1993). Randomized trials mostly support the use of concurrent tamoxifen plus standard adjuvant chemotherapy over tamoxifen alone in post-menopausal women with node-positive, estrogen receptor-positive early stage breast cancer. In 2000, the National Institutes of Health convened a consensus conference on Adjuvant Therapy for Breast Cancer (see http://consensus.nih.gov), which resulted in the following conclusions about the role of hormonal therapy in adjuvant therapy of breast cancer: adjuvant hormonal therapy is recommended for all women with tumors that contain estrogen receptors, regardless of age, menopausal status, tumor size, or axillary lymph node involvement with the exception of pre-menopausal women with tumors <1 cm in size; tamoxifen therapy should be limited to 5 years; and tamoxifen may be combined with combination chemotherapy in pre- and post-menopausal women upto age 70 years (insufficient data for conclusions in the older patient population). Although tamoxifen is employed as an estrogen antagonist, under some conditions tamoxifen demonstrates estrogenic activity (Boccardo et al. 1981, Fornander et al. 1991, Horwitz et al. 1978, Noguchi et al. 1988). This ambivalent activity may contribute to slow repopulation in breast carcinoma that leads to relapse and to an increased risk of uterine cancer in patients treated with tamoxifen (Fornander et al. 1991, Jordan et al. 1991, Spinelli et al. 1991). Tamoxifen is extensively metabolized by the liver to other chemical species, which might contribute to its confusing array of biological activities (Campen et al. 1985, Katzenellenbogen et al. 1984, Loser et al. 1985, Loser et al. 1985, Lyman and Jordan 1985, Mccague et al. 1990, Murphy et al. 1990). It is known that there are substantial species differences in metabolism and the pharmacological activity of the metabolites (Jordan and Robinson 1987, Phillips et al. 1996).

New SERMs with improved pharmaceutical properties and/or patient safety have been recently approved by the FDA for treating breast cancer. These drugs inhibit estrogen binding to its receptor and inhibit estrogen-dependent receptor signal transduction. Toremifene may cause similar menopausal systems to tamoxifen and maybe endometrial hyperplasia, but it is associated with lower risk of cardiovascular side effects than tamoxifen (Hayes et al. 1995). Raloxifene, approved for treating osteoporosis, is being evaluated clinically as a possible replacement for tamoxifen for controlling breast cancer in post-menopausal women with lower risk of uterine cancers and for chemoprevention of breast cancer in high risk subjects.

SERMs cause a brief tumor flare (a transient "worsening" of disease) 1–2 weeks after therapy is initiated, so they can only be recommended for patients who can wait the 6–8 weeks required to ascertain the drug-responsiveness of the tumor (Fisher et al. 1993). In addition to the tumor flare, tamoxifen benefits must be weighed against other toxicities, including a menopause-like condition, increased risk of uterine cancer, and increased risks of blood clots (Fisher et al. 1993). Single agent SERM therapy is definitely contraindicated in patients with rapidly relapsing breast cancers ("aggressive disease") or liver metastases, because these patients need a rapid regression of the lesion (Haskell et al. 1990) and because estrogen receptors are less likely in metastases than in primary lesions (Fisher et al. 1993). Recent evidence from clinical trials in patients progressing after tamoxifen therapy has shown that some of these newer anti-estrogen agents remain clinically effective against tamoxifen-refractory disease. Not only is it an exciting finding that drug cross-resistance is not broad enough to involve all classes of anti-estrogen drugs, but this finding focuses studies of mechanisms of drug resistance onto molecular changes which are specific for particular SERMs, such as receptor binding specificity and conformation changes after binding the drugs, rather than complete loss of the estrogen response mechanisms.

Because of the potential therapeutic disadvantages of the weak estrogenic activity of SERMs, second generation, pure estrogen receptor antagonists without any estrogenic activity have been developed (Osborne et al. 1995). Fulvestrant (Faslodex®) is a new type of endocrine agent, which is an estrogen receptor antagonist without any agonist effects that down-regulates the receptor, probably by inducing its degradation. Such a "pure" estrogen receptor antagonist should show equivalent anti-tumor effects to tamoxifen, but without the side effects caused by receptor modulation by the SERMs. It is approved for ER+ metastatic breast cancer in post-menopausal patients failed by prior antiestrogen therapy, and shows much reduced risk of thromboembolic events over SERMs.

In addition to blocking the action of low level estrogens in the post-menopausal patient or the patient after oophorectomy with receptor antagonists, another strategy to accomplish the same therapeutic goal is to shut off the low level production of estrogen using inhibitors of the enzyme that produces it, named aromatase. Aromatase inhibitors specifically target the enzyme that synthesizes estrogen from androgens, a reaction that lies "downstream" of mineralo- and corticosteroid synthesis, and in post-menopausal women, blocks the low level production by adipose, muscle, skin and breast tissue itself. Exemestane (Aromasin®) is a steroid analog that irreversibly inhibits aromatase, while anastrozole (Arimidex®) and letrozole (Femara®) are non-steroidal, reversible enzyme inhibitors. Exemestane, letrozole and anastrozole are approved for second-line treatment of advanced breast cancer after tamoxifen failure. Anastrozole and letrazole are approved for advanced breast cancer in post-menopausal patients with ER+ or unknown ER status; anastrozole is also approved for adjuvant chemotherapy in post-menopausal patients with ER+ tumors. Like some of the newer anti-androgen products for prostate cancer, letrozole is available as an injectable form that produces an intramuscular depot that releases active drug over a 1-month period. In pre-menopausal women, aromatase inhibition results in stimulation of the pituitary-gonadal axis, and increased enzyme levels in the ovary. Consistent with the value of the strategy of switching therapeutics to target a new and independent point of control in a trophic factor pathway once tumor adapts to inhibition from the first, letrozole and anastrozole are effective second-line endocrine therapies after tamoxifen (Goss et al. 2003) with similar time-to-progression (Rose 2003). The long-term results of on-going clinical trials that compare these drugs in second-line therapy after tamoxifen failure, that evaluate the significance of drug sequencing and switching prior to completion of optimal tamoxifen therapy, and that explore their combinations are eagerly awaited (ATAC Trialists' Group 2005).

In theory, all breast cancers containing the estrogen receptor ("ER+" tumors) should respond to anti-estrogen therapy, but only 50–60% of ER+ tumors respond regardless of whether anti-estrogen therapy is surgical or pharmacological (Osborne et al. 1980). It is poorly understood why 40–50% of ER+ tumors fail to respond, but this could be due to signal transduction defects in pre-menopausal women (Fisher et al. 1993). Since estrogen binding to the estrogen receptor regulates expression of progesterone receptor (Horwitz et al. 1978), dysfunctions in estrogen signal transduction can be detected in women with adequate estrogen levels by an absence of progesterone receptors ("PR−") in ER+ tumor cells. This hypothesis is consistent with, but does not completely explain, response rates of only 74% in ER+/PR+ tumors (Fisher et al. 1993, Osborne et al. 1980). Another confusing result is why 10% of the ER- PR- tumors respond to tamoxifen therapy (Fisher et al. 1993). It is possible that assay cutoffs used by clinical laboratories to define "receptor negative" may be misleading (Encarnacion et al. 1993, Fisher et al. 1993).

Preventing and then managing post-menopausal breast cancer is a complex decision making process to reach a conclusion about the optimal use of available therapeutic options and their combinations. Some of these options and possible clinical outcomes are represented in Figure 10–3 using the concepts of tumor burden and population doubling time that were introduced in the previous chapter, and assuming that repopulation rate is independent of initial tumor burden. The complexity of choosing any one of several possible therapeutic interventions is indicated by the comparable clinical outcomes resulting from the degree of cytoreduction

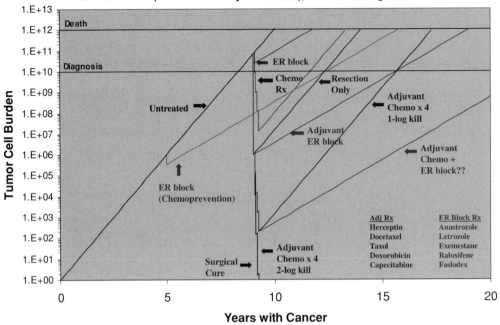

Figure 10–3. Preventing and then managing post-menopausal breast cancer involves a complex decision making process about optimal use of available therapeutic options and their combinations. This figure presents outcomes of possible treatment scenarios using the log-cell kill, tumor burden and tumor growth rate concepts from Chapter 9. The reference point is the growth curve labeled "Untreated," which shows the increase in tumor burden to detectable and lethal levels. Anti-estrogen therapies could be used to delay the growth of tumor in high risk patients (chemoprevention), treat stage 4 disease (ER block), or prolong surgical remission of stage 2–3 disease (adjuvant ER block). Antiestrogens appear to be able to slow tumor growth in concert with chemotherapy, but withdrawal could also kill the same target cells that chemotherapy affects (adjuvant chemo + ER block??). The mathematics of treatment outcome shows that ER block that slows the re-growth of micrometastatic tumor after surgery could improve survival more than adjuvant chemotherapy alone at 1-log cell kill, even though relapse would become clinically detectable at the same time after surgery.

combined with the rate of re-population of residual disease. Note that the regrowth curves reflect slower tumor growth when treatment includes estrogen blockade, reflecting removal of trophic effects of estrogen. Given the parameters of the example, 4 cycles of adjuvant chemotherapy will cure, if 2-log kill can be achieved per cycle and drug resistance does not develop, but the curve showing the consequences of a 1-log cell kill per cycle is the more typical clinical result. Adding anti-estrogen therapy not only causes an initial reduction in tumor mass as trophic factor-dependent tumor cells die, but also a slowed tumor growth rate due to diminished capacity of the tumor to replace lost tissue in the low estrogen environment. Effective adjuvant chemotherapy does add significant years to survival, but adjuvant anti-estrogen therapy may achieve similar tumor control and overall survival in post-menopausal patients, without the toxic side effects of chemotherapy. However, the relapsing tumor will likely grow faster after chemotherapy than while still on anti-estrogen therapy (compare the years required for the relapsing tumor to progress from clinically detectable to lethal tumor burden in patients treated with chemotherapy alone or anti-estrogen receptor therapy). As indicated by the Figure, the difficult choices and trade-offs in selecting therapy in the context of life expectancy and overall health, discussed in detail in Chapter 9, confront most breast cancer patients.

8. Carcinoma of the uterus. Estrogen and progesterone also affect the proliferative status and cellularity of the uterine epithelium during the menstrual cycle (Hopwood and Levison 1976, Sandow et al. 1979), so anti-endocrine therapy has been tested against carcinoma of the endometrium of the uterus (Grenman et al. 1988, Grenman et al. 1988, Gronroos et al. 1987). Stage III/IV endometrioid cancer that is ER+ PR+ responds to the progesterone antagonist, medroxyprogesterone (Rendina et al. 1982). Other anti-progestins like megastrol acetate and 17-alpha-hydroxyprogesterone are indicated in recurrent or metastatic endometrial carcinoma, and achieve a higher response rate than cytotoxic chemotherapy (Berman and Berek 1990).

9. Antagonism of cytokine trophic factors. In addition to steroids, there are many polypeptide growth factors that function as trophic factors in hematological and solid tissues to block progenitor cell apoptosis and promote cell proliferation and tissue expansion. In epithelial tissues, EGF, IGF and TGFs are common, whereas FGF and PDGF are common in mesenchymal tissue, and CSFs and interleukins are common in lymphohematopoietic tissues. All of these factors stimulate target cells by first binding to extracellular receptors, because being water soluble peptides cannot pass through the cell membrane. Upon ligand binding, the receptor utilizes one of a number of possible signal transduction mechanisms for propagating the signal to the immediate intracellular mediator, such as tyrosine phosphorylation, serine/threonine phosphorylation, activation of PKC, etc. The key to designing effective therapeutics designed to block the action of peptide trophic factors is identifying which ones are critical determinants of malignant cell growth and survival. As with steroid blockade, inhibiting the action of these trophic factors can result in major tumor shrinkage and long-term patient survival, but they would not be expected to achieve cures because malignant stem cells likely do not require these factors for survival or self-renewal.

10. Interferon-alpha: a protein antagonist of trophic factors in the immune system. Interferon-α is FDA-approved for the treatment of hairy-cell and chronic myelogenous leukemia as well as melanoma and Kaposi's sarcoma (Kim 1996). Although its clinical activity is well established (Bouroncle 1994, Gutterman 1994, Vedantham et al. 1992), the mechanism responsible for this activity has been actively debated. On the one hand, interferon-alpha may be clinically effective because it functions as a trophic factor for the immune system, reversing immunosuppression caused by the leukemia and enhancing the anti-tumor response of the immune system. On the other hand, the activity of interferon-alpha against hairy cell leukemia may result from a direct anti-proliferative effect upon the malignant cells (Vedantham et al. 1992). This conclusion is consistent with the finding that co-administration of prednisone, which suppresses the immune system, reduces the severity of side effects of interferon-alpha therapy, but not its efficacy (Fosså et al. 1990). This direct effect on tumor does not appear to be cytotoxicity because the tumor burden declines more slowly than with cytoreductive chemotherapy. Although unrelated to tumor regression, the immune system stimulation is an added benefit for the patient, since opportunistic infections are the leading mortality-associated complication of hairy cell leukemia (Vedantham et al. 1992).

B-cell growth factor and tumor necrosis factor (TNF) are trophic factors for hairy-cell leukemia, as is interleukin-2 (IL-2) after the cells respond to B-cell growth factor. Interferon-alpha binds to a cell surface receptor, which leads to a diminished proliferative response to these trophic factors (Vedantham et al. 1992). Interferon-alpha inhibits IL-2 responsiveness by decreasing the number of cell surface IL-2 receptors; however, inhibition of TNF responsiveness is independent of the number of cell surface TNF receptors (Vedantham et al. 1992). Interferon also stimulates the hairy cells to differentiate to a mature phenotype which is less responsive to growth factors. Clinical benefit is derived from the slowed natural course of disease progression as a result of reduced proliferative responses to these trophic factors (slower rate of increase in tumor burden). A soluble form of the interferon-alpha receptor is found at elevated levels in body fluids of patients with hairy cell leukemia, and pharmacological doses of interferon-alpha may be required to overcome the neutralizing effects of this factor (Novick et al. 1992).

Resistance to interferon-alpha may occur in some cases of hairy cell leukemia. In general, resistance is associated with hairy cells at differentiation stages slightly before or after the stage of B-cell differentiation which responds to BCGF, TNF, and IL-2, and in fact, resistant cells generally do not express cell surface IL-2 receptors (Vedantham et al. 1992). These resistant cells might be at a stage of differentiation which utilizes other trophic factors which have receptors that are not regulated by interferon-alpha. Moreover, the presence of a cell surface receptor for a negative regulator does not necessarily imply responsiveness to that factor. For example, chronic lymphocytic leukemia cells also express surface receptors for interferon-alpha, but they lack the intracellular pathway that inhibits response to trophic factors after receptor binding of interferon (Vedantham et al. 1992). Molecular defects explain other cases of interferon resistance: for example blocked mRNA translation of cell surface receptors for interferon-alpha (Platanias et al. 1992) or

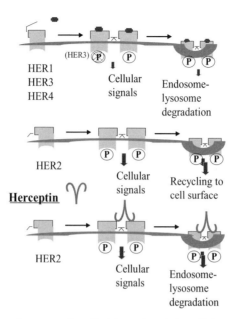

Figure 10–4. The cell surface receptors that transduce intracellular growth signals upon binding ligands of the EGF family constitute a multi-member family including EGFR1 (HER1/erbB-1), EGFR2 (HER2/neu/erbB-2), EGFR3 and EGFR4. (top) Two EGFR1, 3 and 4 receptors dimerize upon ligand binding and phosphorylate opposing molecules, activating the tyrosine kinase that tranduces the signal inside cell. As with most dimerized cell surface proteins, the dimers are internalized into the endo-lysosomal system, where the ligand is degraded and the receptors are either degraded or recycled to the plasma membrane. (middle) EGFR2 (Her2/neu/c-erbB-2) is a truncated receptor form that does not extracellular ligand bind but forms hetero- or homo-dimers that activate the kinase domain and transduce signal in the absence of ligand. Its constitutive activity increases as membrane receptor density of the EGFR family increases. (bottom) Herceptin (trastuzumab) most likely works by forcing receptor dimerization and internalization without kinase activation, effectively removing Her2 as a constitutive signaling system.

in an experimental system blocked signal transduction from normal cell surface receptors (Ozes et al. 1993).

11. Antagonism of EGFR-dependent trophic factors with trastuzumab, cetuximab, erlotinib and gefitinib. The cell surface receptors that transduce intracellular growth signals upon binding EGF-family ligands constitute a multi-member family including EGFR1 (HER1/erbB-1), EGFR2 (HER2/neu/erbB-2), EGFR3 and EGFR4 (Figure 10–4). EGFR2 is a truncated receptor form that does not possess an extracellular ligand binding domain. As a result, when EGFR2 is over-expressed or when present within a membrane environment containing high levels of another EGFR family member, it readily forms receptor dimers that cross-activate the opposing receptor's tyrosine kinase activity, leading to constitutive signal transduction independent of ligand. HER2+ over-expressing tumors are therefore more aggressive, and this receptor is a logical target for development of therapeutic agents (Slamon and Pegram 2001). The first FDA approved therapy that targets EGFR2 is a "humanized" monoclonal antibody against an extracellular domain of HER2/neu (erbB-2) called trastuzumab (Herceptin®). When tested as a single agent in previously untreated patients ("first-line therapy") with advanced breast cancer, this drug achieved a 34–35% response rate in patients whose tumors over-expressed HER2 as measured by either of two laboratory techniques, but no more than a 7% response rate in patients whose tumors expressed lower levels of HER2 (Vogel et al. 2002). When added to standard paclitaxel-containing chemotherapy regimens for HER2-overexpressing metastatic breast cancer in first-line therapy, trastuzumab increased response rate (50–80% v. 25–40%), time to tumor progression (7.4 vs 4.6 mos) and overall survival (25 vs 20 mos) compared to chemotherapy alone (Slamon et al. 2001). The requirement for high level HER2 expression for observing efficacy is consistent with the hypothesis behind the development of the drug that constitutive activation of EGFR2 and tumor proliferation require high probability of homo- or hetero-dimerization in the membrane due to over–expression of the protein. Most likely, trastuzumab's mechanism of action involves the formation of a dimer of two EGFR2 proteins via divalent binding by the antibody (Roskoski Jr 2004). Unlike natural receptor dimerization, the antibody does not induce an active conformation in the receptor and the opposing kinase domains of the two receptor molecules are not activated. Like binding of many divalent antibodies to cell surface proteins, dimerization induces endocytosis of EGFR2, which at the least minimizes its time in the plasma membrane as it circulates through the endosomal pathway and may also induce degradation of receptor protein in the lysosome.

Surprisingly, trastuzumab is cardiotoxic, causing ventricular dysfunction and congestive heart failure. It doubled the risk of serious cardiomyopathy of doxorubicin-cyclophosphamide therapy (8% to 16%), and is contraindicated in combination with anthracyclines like doxorubicin, a family of cytotoxic drugs that cause cardiomyopathy. It increases the cardiac risk of paclitaxel therapy from 1% to 13%, but this rate of toxicity is no worse than standard therapy with the cardiotoxic anthracyclines like doxorubicin. Recent studies have found low level expression of HER2 in myocardium from just 5% of patients (Fuchs et al. 2003), so the mechanism for this cardiotoxicity is not yet established.

Although EGFR2 is expressed in colon carcinoma and to a lesser extent in rectal cancers, levels of expression of EGFR2 are similar in malignant tissue and surrounding normal mucosa, and HER2 expression is not a prognostic marker in this disease (Mckay 2002a, 2002b). These findings suggest that trastuzumab would not be effective against colorectal carcinoma. In contrast, EGFR1 is expressed in about 80% of colorectal tumors progressing in patients being treated with chemotherapy (Cunningham et al. 2004). Cetuximab (Erbitux®) is a recombinant human-mouse chimeric monoclonal antibody (mouse Fv region + human IgG1 heavy and kappa light constant regions) that binds specifically to EGFR1 on normal and malignant cells, inhibits ligand binding, and blocks phosphorylation and activation of receptor kinases. It is FDA approved as a single agent therapy for EGFR1 (HER1) expressing metastatic colorectal carcinoma unresponsive to irinotecan-based therapy (Cunningham et al. 2004). It will find utility in combination with irinotecan and likely oxaliplatin to treat EGFR1 expressing metastatic colorectal cancer refractory to irinotecan-based therapy, because the addition of the antibody to the chemotherapy appears to reverse the resistance of the tumor to irinotecan cytotoxicity in some patients (Cunningham et al. 2004, Iqbal and Lenz 2004). A recent trial found that cetuximab is clinically effective for colorectal cancer refractory to 5-FU and irinotecan, which are the best chemotherapy regimens for this disease, and therefore provides an important therapeutic option for third line therapy where none existed before (Saltz et al. 2004). The clinical toxicity of cetuximab is consistent with the known biological function of EGFR1 in maintenance of epithelial integrity: dermatologic toxicity (acneform rash), asthenia, malaise and abdominal pain. Scientifically, it is important to note that the clinical trials did not find a relation between response and the percent positive cells or intensity of staining, and clinical trials in patients with tumors that do not express EGFR1 are in progress (Iqbal and Lenz 2004). This is an important issue for further clinical exploration of its use to treat carcinoma derived from other tissue sites. Additionally, there is not good correspondence between EGFR1 expression in the primary tumor and its expression in metastatic sites in colorectal cancer (Scartozzi et al. 2004), and this anatomic heterogeneity could explain some of the confusing clinical findings regarding EGFR1 expression in the primary tumor biopsy and the clinical effectiveness of the agent against advanced disease.

Gefitinib (Iressa®) represents a second, completely different therapeutic strategy for targeting EGFR1 function, which is approved for use in relapsing non-small cell lung cancer. In contrast to antibodies which circulate in extracellular fluids, gefitinib is a small molecule that can enter the cell and inhibit the intracellular kinase domain of the EGFR1, blocking signal transduction. Initially, gefitinib was thought to be a pure cytostatic agent, because in vitro treatment of human tumor cell lines resulted in G1 cell cycle arrest without apoptosis (Giocanti et al. 2004). However, in the clinic, gefitinib is active only in a small subset of NSCLC patients (Janne et al. 2004), which is closely linked to the presence of certain mutations in EGRF1 sequence that confer enhanced susceptibility to kinase inhibition (Paez et al. 2004). In vitro, one Leu→Arg mutation (L858R) results in a much greater sensitivity of NSCLC cells to gefitinib-induced apoptosis than when the wildtype gene sequence is present. Only one NSCLC tumor cell line containing wildtype sequence in the EGFR1 responded to gefitinib by arresting in the G1 phase of the cell cycle, but it required exposure to 50-fold higher drug concentration. The remaining cell lines with wildtype EGFR1 were not responsive to gefitinib (Tracy et al. 2004). In patients, tumor response to gefitinib is highly associated with EGFR1 mutations (Lynch et al. 2004, Sordella et al. 2004), suggesting that the patients who respond to gefitinib therapy harbor tumors with mutated EGFR1 sequences that cause dominant, trophic-like effects to drive tumor proliferation, and gefitinib achieves a block in this trophic factor pathway and therefore apoptosis by blocking signal transduction These data point to the possibility that gefitinib achieves response and remission in the small proportion of patients with mutations where it acts as a cytotoxic to induce apoptosis, but will be ineffective in the majority of the treated patients. In fact, a trial of gefitinib in an unselected patient population with Stage III non-small cell lung cancer was terminated early due to lack of efficacy (Tamura and Fukuoka 2005). In November 2004, another small molecule inhibitor of EGFR named erlotinib (Tarceva®) was approved by the FDA for second-line treatment of locally advanced or metastatic non-small cell lung cancer (see http://www.fda.gov/bbs/topics/news/2004/new01139.html).

12. Antagonism of c-Kit and c-abl-dependent trophic factors using imatinib. A specific t(9:22) chromosomal translocation responsible for chronic myelogenous leukemia involves the translocation of a portion of the c-abl gene, including the kinase domain, into the break-point cluster region (bcr) to produce the Philadelphia chromosome so famously characteristic of this disease. The constitutive tyrosine kinase activity of the fusion protein is the driving force of the cell proliferation of CML, so compounds that inhibit the abl kinase activity have been developed as potential drugs for this disease. A small molecule inhibitor of the bcr/abl tyrosine kinase activity named imatinib (STI571, Gleevec®) induces cell cycle arrest plus apoptosis only in leukemic cells expressing the chimeric gene product, but does not induce differentiation (Deininger et al. 1997, Gambacorti-Passerini et al. 1997). This drug received accelerated approval from the FDA for this use based on the surrogate endpoint of remission rate, so the agency will require mature survival, efficacy and safety data to prove clinical benefit.

Although this drug was developed as an inhibitor of bcr/abl kinase activity, it also inhibits receptors for two other peptide growth factors: PDGFR for PDGF and c-kit proto-oncogene for stem cell factor (SCF). Constitutive activity in c-kit, possibly mutation related, causes development of a rare, chemotherapy-resistant malignancy known as gastrointestinal stromal tumor (GIST). GISTs arise from interstitial cells of Cajal, a network of innervated cells that coordinate peristalsis in the intestine. Imatinib therapy achieves a very high response rate in GIST patients and induces apoptosis in the tumor (Demetri 2002, Demetri et al. 2002), so it is FDA approved for treating metastatic or unresectable malignant gastrointestinal stromal tumors. Mis-sense mutations in the c-kit gene tightly associated with the emergence of imatinib resistance in 12 patients with GISTs that initially responded to therapy (Chen et al. 2004), another example of the Goldie-Coldman hypothesis. Imatinib is typically administered to patients whose tumor biopsy is positive for the surface marker CD117, which is the epitope contained in the c-kit receptor. Caution has been urged in the prognostic value of CD117 positivity, because a recent study of c-kit "negative" GISTs based on CD117 evaluations found that a high proportion contain imatinib-sensitive c-kit or PDGFR-A mutations (Medeiros et al. 2004).

10.3e Trophic factor therapy to treat hematologic side-effects of chemotherapy

A frequent dose-limiting toxicity of cytotoxic chemotherapy is suppression of new blood cell production by the bone marrow, leading to dose reductions or treatment delays that compromise the efficacy of the treatment (see Chapter 9). Suppression of blood cell formation results from the direct cytotoxic action of many anticancer drugs on the hematopoietic progenitor cells that proliferate and then differentiate into myeloid cell types (Parchment 1998, Parchment et al. 1998). Trophic factors for the erythrocyte, neutrophil and platelet lineages are used clinically to accelerate recovery from anemia, neutropenia and thrombocytopenia, respectively (Table 10–4) (Anaissie et al. 1996, Vadhan-Raj 1996, Vadhan-Raj et al. 1992). The trophic factor for red blood cell precursors in the bone marrow is called erythropoietin (Epo), which is synthesized and secreted into the blood by interstitial cells in the kidney cortex (Koury et al. 1988a, 1988b, Lacombe et al. 1988a, 1988b). Epo production and secretion into the blood is regulated by blood oxygen tension (Abbrecht and Littell 1972, Schuster et al. 1987). An initial recombinant form of Epo called Epogen was approved for the treatment of anemia associated with renal failure (Epo deficiency) (Jacobs et al. 1985, Recny et al. 1987) and epoetin alfa (Procrit®) for chemotherapy-induced anemia in nonmyeloid malignancies. These factors are administered by subcutaneous injection weekly or 3x per week, similar to insulin injections. A derivative of Epo called darbepoetin-alfa (Aransep®) is a recombinant protein that contains two more N-glycosylation sites than Epo, resulting in the incorporation of 5 N-linked oligosaccharide chains rather than 3. The increased glycosylation slows the elimination of the drug from the body, so it is easier for patients to take weekly, compared to the 3x per week with Epoetin-alfa.

Three trophic factors are used clinically to treat neutropenia. Since neutropenic complications often delay subsequent chemotherapy cycles or result in dose reduction to remain on the schedule of the dosage regimen (Link et al. 2001, Lyman et al. 2003), it is a major contributing factor to suboptimal chemotherapy. G-CSF (granulocyte-colony stimulating factor) is available commercially in recombinant form as Neupogen, which is used to stimulate neutrophil recovery after cytotoxic chemotherapy. A new derivative if G-CSF called pegfilgrastim (Neulasta) is a pegylated form of G-CSF (derivatized with polyethylene glycol) to reduce its elimination rate from the body, so a single fixed dose per chemotherapy cycle shortens the duration of neutropenia and reduces the incidence of febrile neutropenia (Green et al. 2003, Holmes et al. 2002, Holmes et al. 2002). The opportunity for patients to self-administer a single subcutaneous dose once per 3–4 week period, rather than once per day, represents an advance for patient convenience and compliance. A recombinant form of GM-CSF (granulocyte/macrophage-colony stimulating factor) known as sargramostim (Leukine®) is used to accelerate repopulation of myelopoietic tissue in the bone marrow after transplantation. It is also used to accelerate neutrophil recovery and decrease risk of serious infection and infection related deaths in older patients following induction chemotherapy for AML. It must be given daily by subcutaneous injection.

Thrombocytopenia often accompanies neutropenia and is also serious toxicity. When platelet levels in the blood are too low, the patient is at risk of hemorrhage. The FDA-approved trophic factor for stimulating new platelet production by the bone marrow is named oprelvekin (Neumega). Neumega is indicated for the prevention of severe thrombocytopenia and the reduction of the need for platelet transfusions following myelosuppressive chemotherapy in adult patients with non-myeloid malignancies who are at high risk of severe thrombocytopenia. This drug is the recombinant form of the protein interleukin eleven (IL-11), which directly stimulates the proliferation of hematopoietic stem cells and megakaryocyte progenitor cells in the bone marrow, as well as inducing megakaryocyte maturation, which together increase platelet production and release into the blood. IL-11 is a member of a family of human growth factors which includes granulocyte colony-stimulating factor (G-CSF), but each member of the family exerts trophic effects on just particular cell lineages. It is self-administered by subcutaneous injection, but its short half-life in the body of ~7 hours requires daily injections for 10–21 days.

10.3f Therapy that exploits differentiation processes in malignancies

When first discovered and proven in models of teratocarcinoma and squamous cell carcinoma of the skin (Chapter 1), the concept that undifferentiated stem cells could give rise to the terminally differentiated daughter cells in tumors was a foreign one and competed with the concept of tumors as the result of blocked differentiation. In addition to the scientific proofs that cancer is a caricature of tissue renewal - imbalanced proliferation and differentiation, clinical proof exists in the successful use of this concept to develop effective therapeutics. Unfortunately, no one has yet figured out how to restore the normal balance between proliferation, differentiation, and apoptosis in tumors, such that tissue renewal is no longer exaggerated. But, there are currently four therapeutic approaches that utilize the fact that some daughter cells from malignant stem cell renewal are terminally differentiated.

1. Retinoic acids for acute promyelocytic leukemia and cutaneous malignancies. One of these therapeutic approaches based on tumor differentiation processes is treating cancers with substances that induce cell differentiation. Naturally occurring retinoids like retinoic acid are inducers of differentiation in the developing embryo and influence tissue renewal in epithelium systems in vivo. Retinoids are structural relatives of vitamin-A (all-*trans*-retinol) that exert biological effects via binding to two families of intracellular retinoic acid receptors known as RAR and RXR, which differ in affinity for different retinoids and in their formation of heterodimers with structurally similar receptors for other ligands. RXRs can form heterodimers with various receptor partners such as retinoic acid receptors (RARs), vitamin D receptor, thyroid receptor, and peroxisome proliferator activator receptors (PPARs).

All-*trans*- retinoic acid is the active ingredient in a cancer drug called tretinoin (Vesanoid®) used for remission induction in patients with acute promyelocytic leukemia (APL), French-American-British (FAB) classification M3 (including the M3 variant), characterized by the presence of the t(15;17) translocation. This specific translocation defines this form of leukemia, and results in a PML/RARalpha gene translocation that causes ectopic expression of the RAR (Daenen et al. 1986, Dong et al. 1993, Eardley et al. 1994, Fenaux 1993, Fenaux et al. 1993, Flynn et al. 1983, Gallagher et al. 1989, Huang et al. 1988, Ikeda et al. 1994, Ikeda et al. 1994, Wu et al. 1993). The activity of retinoic acid in APL is impressive, considering its use in relapsed or chemoresistant APL (Liang et al. 1993, Tallman 1994). Differentiation therapy could maintain smaller tumor burden by promoting differentiation of progenitors after fewer rounds of cell division. The increased abundance of differentiated tissue in the tumor might result in signals to the stem cell compartment that homeostasis has been reached, resulting in some tumor control via the slowing of stem cell repopulation. However, these therapies should not be expected to eradicate the malignant stem cell and therefore cannot be used for making a patient "tumor-free". As with all biological therapy, resistance can develop and the duration of the remission is relatively short (Gallagher et al. 1989, Ikeda et al. 1994, Ikeda et al. 1994), perhaps caused by genetic alteration in the RAR (Trayner and Farzaneh 1993), or altered intracellular drug metabolism.

Tretinoin is not a cytolytic agent but instead induces cytodifferentiation and decreased proliferation of APL cells in culture and in vivo. Tretinoin induces maturation of acute promyelocytic leukemia (APL) cells in culture. In APL patients, tretinoin treatment produces an initial maturation of the primitive promyelocytes derived from the leukemic clone, followed by a repopulation of the bone marrow and peripheral blood by normal, polyclonal hematopoietic cells in patients achieving complete remission. The complete mechanism of differentiation induction is not understood, but may involve the secretion of soluble factors that induce terminal differentiation of the leukemic cells as observed in preclinical models (Falk and Sachs 1980, Liebermann et al. 1982). By coupling trophism and differentiation, progenitors divide a finite number of times before differentiation and tissue expansion is automatically limited (Figure 10–5). Malignancies characteristically have a defect in this coupling of progenitor expansion and differentiation (Lotem and Sachs 1982, Lotem and Sachs 1992, Symonds and Sachs 1982, Symonds and Sachs 1982, Symonds and Sachs 1982), which probably results in too many rounds of progenitor cell division (infinite number?) before differentiation begins. Malignancies that result from this loss of coupling may respond to exogenously supplied differentiating factor that by-passes the defect, or perhaps pharmacological levels of the factor if uncoupling is due to decreased sensitivity. Accordingly, replacement therapy with the differentiating factor or biological therapy that by-passes the defect can be used to achieve long-term control of malignancy, and this approach

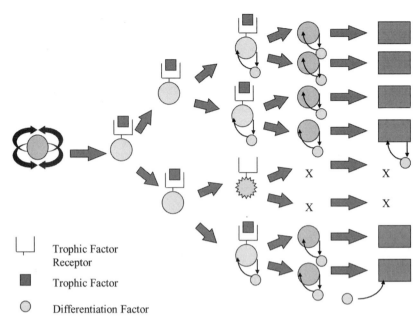

Figure 10–5. In normal stem cell systems, there is precise coupling of progenitor proliferation (requiring presence of trophic factor, dark squares) and terminal differentiation into functional cell types restricted to the tissue of origin of the stem cell. Based on studies with leukemia, the mechanism that couples differentiation to proliferation likely involves trophic factor-induced secretion of soluble factors that induce terminal differentiation (light circles). By coupling trophism and differentiation, progenitors divide a finite number of times before differentiation – automatically limiting tissue expansion. Malignancies characteristically have a defect in the coupling of progenitor expansion and differentiation. When production of the differentiating factor is not properly triggered, or when cells are de-sensitized to its effect, progenitors continue to proliferate well beyond the normal number of cell divisions (perhaps "indefinitely"). However, progenitors in these malignancies may still respond to exogenously supplied differentiating factors that circumvent the defect, especially at the higher pharmacological levels achieved by cancer treatment.

to cancer therapy is called differentiation therapy (Pierce and Speers 1988, Reiss et al. 1986).

The RXR receptor family is also a useful therapeutic target for treating cancers. Bexarotene (Targretin®) is a member of a subclass of retinoids that selectively activate retinoid X receptors (RXRs) that is used to treat cutaneous manifestations of cutaneous T-cell lymphoma in patients who are refractory to at least one prior systemic therapy. These retinoid receptors have biologic activity distinct from that of RARs. Bexarotene selectively binds and activates retinoid X receptor subtypes (RXRa, RXRb, RXRg). Once activated, these receptors function as transcription factors that regulate the expression of genes that control cellular differentiation and proliferation. Bexarotene inhibits the growth in vitro of tumor cell lines of hematopoietic and squamous cell origin, and induces tumor regression in vivo in animal tumor models. The exact mechanism of action of bexarotene in the treatment of cutaneous T-cell lymphoma (CTCL) is unknown.

Azacitidine (Vidaza®) is a nucleoside analog of cytosine and therefore probably functions partly as an antimetabolite and cytotoxic drug. However, it also shows a very interesting property of reducing methylation of genes throughout the genome. DNA methylation is a common mechanism of "silencing" gene expression, including suppressor genes that regulate coupled proliferation and differentiation. In addition to being a cytotoxic agent that like most other nucleosides causes bone marrow suppression as its major toxicity, azacitidine is also a drug known as a hypomethylating or "demethylating" agent that can unsilence these and other genes to restore control of the cell cycle. Drug concentrations that maximally inhibit DNA methylation do not inhibit DNA synthesis in S-phase. Azacitidine is approved for treating a group of hematological disorders known as myelodysplastic syndromes (MDS) caused by a problem in the blood-forming cells of bone marrow. It is especially effective against one subgroup in MDS known as chronic myelomonocytic leukemia (CMMoL).

2. Cell surface differentiation markers as targets for therapy. Daughter cells of stem cells that are differentiating according to the program appropriate for their tissue of origin express specific proteins, which are characteristic for the phenotype and function of the appropriate differentiated tissue. Some of these proteins expressed as part of the differentiation program are plasma membrane proteins that are accessible to antibodies in the extracellular fluids. Other cell surface proteins may be tissue specific receptors for polypeptide ligands that cannot readily pass across the plasma membrane and require cell surface receptors for signal transduction. A second approach to treating cancer based on differentiation principles is to direct the binding of cytotoxic drugs to these differentiation markers on malignant cells. The most useful markers for effective therapy would be those expressed early in the differentiation program on committed, but immature progenitors, because the therapy would eradicate the cell type responsible for expansion of the tumor burden. However, stem cells do not express differentiation markers and, therefore, cannot be direct targets of this type of therapy, so cure in the sense of eradicating the malignant stem cells is likely impossible. Nevertheless, if stem cell renewal is slow enough, biotherapy can shrink the tumor to a non-symptomatic size and maintain this reduced tumor burden for decades or longer, which can equal the life-span of older patients. The cell surface target will also be expressed on normal cellular counterparts of the malignant cells at the same stages of differentiation, so these healthy cells will also be killed by drug therapy. Just like with all cancer therapies, targeted cytotoxicity to cell surface markers of differentiation will be effective if the loss of normal cells is restricted to tissues not critical for survival, or if recovery from cytoreduction in normal tissues is faster than recovery by the malignant tissue. Clearly, these biological therapies are cytoreductive in nature, even if they are targeted to a particular cell surface protein.

Alemtuzumab (Campath®) is a recombinant DNA-derived humanized monoclonal antibody (Campath-1H) that is directed against a cell surface glycoprotein containing the epitope named CD52. The Campath-1H antibody is an IgG1 kappa with human variable framework and constant regions, and complementarity-determining regions from a murine (rat) monoclonal antibody (Campath-1G). Alemtuzumab binds to CD52, a non-modulating antigen that is present on the surface of essentially all B and T lymphocytes, a majority of monocytes, macrophages, and NK cells, and a subpopulation of granulocytes. It is used to treat B-cell chronic lymphocytic leukemia (B-CLL) in patients who have been treated with alkylating agents and who have failed fludarabine therapy. It is also expressed on a subpopulation of CD34+ bone marrow stem cells, explaining why it causes myelosuppression like many cytotoxic chemotherapeutic drugs. After binding to CD52, the antibody likely mediates antibody-dependent cellular cytotoxicity (ADCC) of leukemic cells by immune cells (Golay et al. 2004, Stanglmaier et al. 2004).

Rituximab (Rituxan®) was approved by the FDA for treating relapsed or refractory, B-cell, low-grade or follicular non-Hodgkin's lymphoma (NHL) that expresses the surface protein containing the differentiation marker epitope CD20, which is expressed on both normal and malignant B lymphocytes. It can be used as single agent therapy or in combination with chemotherapy. The IgG1 kappa monoclonal antibody is a chimera of mouse sequence in the variable region and human sequence in the constant regions. Its mechanism of action is most likely also ADCC mediated by immune cells (Stanglmaier et al. 2004). Since CD20 is a normal differentiation marker of the B-lymphocyte lineage, the toxicity of rituximab is predictably lymphocytopenia due to B-cell destruction. B-cell levels rapidly decline after the first dose of drug and stay depressed for 6 months following completion of therapy, but there does not seem to be any increased risk of infection during this time (McLaughlin et al. 1998).

3. Cell surface differentiation markers as targets for delivering cytotoxic therapy. A third approach to treat cancer based on differentiation principles is to use cell surface markers expressed during a differentiation program as binding sites for antibodies, or for ligands if the cell surface protein is a receptor, which are conjugated to cytotoxic materials that will kill malignant cells expressing the epitope or receptor. If the antibody or ligand is conjugated to a radioisotope, then even malignant stem cells that do not express differentiation markers may be killed by the radiation exposure if they lie in close proximity to maturing cells expressing the cell surface markers and binding the radiation source. This is an important mechanistic distinction between this type of directed, cytotoxic therapy (Figure 10–6) and the binding of antibodies to stimulate cell destruction by the immune system (previous section). Antibodies or ligands conjugated to toxic chemicals can also kill surrounding cells that do not express the marker via a

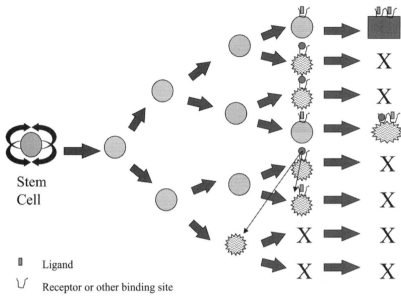

▌ Ligand

ʊ Receptor or other binding site

● Receptor-binding Ligand with Toxin or Radionuclide

Figure 10–6. Malignant tissue contains a proportion of terminally differentiated cells, albeit a much lower proportion of the total cellularity than is found in the normal counterpart tissue. This concrete expression of the caricature concept can be exploited for therapy. Since the differentiation program leads to expression of cell surface receptors for specific ligands or cell surface markers, recombinant ligands or monoclonal antibodies coupled to cytotoxic chemicals or high energy radioactive nuclides can be directed to these targets to deliver the cytotoxic material (small circles), thereby killing them (abnormally shaped cells). Radiation from high energy nuclides may also damage cells in the vicinity of binding (indicated by the two arrows), even though they do not express the cell surface target or do not bind the cytotoxic conjugate, which provides a means to kill malignant stem cells that by definition have not committed to differentiation. Furthermore, a phenomenon called the by-stander effect may lead to the death of neighboring, undifferentiated cells (target-negative) when differentiated cells die as a result of exposure to the cytotoxic payload on the ligand or antibody. Of course, normal tissues that express the same target will also be damaged by the therapy. This strategy would be expected to work poorly against anaplastic tumors that express little if any differentiated features that can function as targets.

phenomenon called the "by-stander effect." This poorly understood biological consequence of some therapies means that the destruction of one cell leads to cytotoxicity in its neighboring cells, probably via exposure to the cytotoxic agent released by the dying cell as it lyses or undergoes phagocytosis by adjacent cells, or perhaps by release of physiological chemicals that become toxic when outside the cell membrane.

Gemtuzumab ozogamicin (Mylotarg®) is a recombinant humanized IgG4 antibody composed of human constant region and framework regions but mouse complementarity-determining regions. The antibody molecule is conjugated to the cytotoxic antitumor antibiotic, which is a minor groove selective agent that causes DNA double strand breaks and cell death (Maiese et al. 1989, Uesugi and Sugiura 1992). Approximately 50% of the antibodies are linked to 4–6 moles calicheamicin per mole, while the remaining 50% is not linked to calicheamicin. Although if administered as a cytotoxic drug, this chemical does not have an adequate safety profile, its conjugation to the antibody directed against a marker on malignant cells limits exposure of healthy tissue and increases its therapeutic index (toxic dose to efficacious dose ratio). Gemtuzumab binds to the CD33 antigen found on the surface of leukemic blasts and normal progenitors committed to myelomonocytic differentiation, but not normal hematopoietic stem cells. After binding of the divalent antibody to two CD33 surface antigens, the complex is internalized into the endolysosomal system of the target cell, where calicheamicin is released and gains access to DNA. The cytotoxic effect of gemtuzumab on adult leukemic bone marrow cells and normal myeloid precursors leads to efficacy and toxicity (severe myelosuppression), respectively. The myelosuppression is reversible because pluripotent hematopoietic stem cells are spared. Since the CD33 antigen is expressed on the surface of leukemic blasts in more than 80% of patients with acute myeloid leukemia (AML), gemtuzumab is approved for treating CD33+ acute myeloid leukemia in patients in first relapse who are not candidates for cytotoxic chemotherapy.

Ibritumomab tiuxetan (Zevalin®) is a conjugate between a monoclonal antibody and a heavy metal chelator called tiuxetan, which provides high affinity binding of metal radioisotopes indium-111 or yttrium-90. Ibritumomab is a mouse IgG1 kappa monoclonal antibody against the CD20 antigen - human B-lymphocyte-restricted differentiation antigen or Bp35 (Einfeld et al. 1988, Valentine et al. 1989). The CD20 antigen is a differentiation marker expressed on normal pre-B and mature B lymphocytes in bone marrow, lymph nodes, spleen and lymphoid nodules, and on > 90% of B-cell non-Hodgkin's lymphomas (NHL) (Anderson et al. 1984, Tedder et al. 1985). The CD20 antigen does not internalize upon antibody binding (Press et al. 1987), so it remains bound to the cell surface where the radioisotope can access the tiuxetan and be chelated. Based on the biology of CD20 expression, it is approved for treating relapsed or refractory low-grade, follicular, or transformed B-cell non-Hodgkin's lymphoma (NHL), including patients refractory to rituximab therapy. It appears to kill malignant and normal B-cells via two independent mechanisms. Ibritumomab, like rituximab, induces apoptosis in CD20+ B-cells (Press et al. 1987), but unlike rituximab, the antibody can at the same time chelate a metal radioisotope which provides localized exposure to ionizing radiation (beta emission). The therapy is fundamentally a form of radiotherapy that has been restricted to B-cells by antibody binding. Any normal cells in range of the radioisotope emission will also be damaged by the free radical formation.

Tositumomab (Bexxar®) is another antibody designed to deliver a radiation dose to follicular NHL using the cell surface CD20 epitope to target antibody binding to malignant as well as normal B-cells. The treatment principles are the same, but instead of chelating a radioactive metal atom, tositumomab has been labeled with radioactive iodine-131 as the radiation source. This drug is approved for second line treatment of patients with CD20 positive, follicular, non-Hodgkin's lymphoma, whose disease is refractory to Rituximab and has relapsed following chemotherapy.

There is one clinically approved drug product that exploits the specificity and affinity of ligand binding to a cell surface receptor, instead of antibody recognition, to deliver a cytotoxic dose specifically to cells expressing the receptor. Denileukin diftitox (Ontak®) is a recombinant protein composed of the amino acid sequences for diphtheria toxin fragments A and B (Met 1-Thr 387) linked to the sequence for interleukin-2 (IL-2; Ala 1-Thr 133). It is a fusion protein designed to direct the diphtheria toxin into cells expressing the IL-2 receptor. At the doses used clinically, the drug binds only to the high affinity IL2 receptor (CD25/CD122/CD132) expressed on activated T and B lymphocytes, activated macrophages, and certain leukemias and lymphomas including cutaneous T-cell lymphoma (CTCL). After binding, the ligand-receptor complex likely internalizes, where proteolytic degradation releases the diphtheria toxin into the cell to inhibit protein synthesis and induce apoptosis. It does not bind to low (CD25) or intermediate (CD122/CD132) affinity receptors. The drug is approved for treatment of patients with persistent or recurrent cutaneous T-cell lymphoma that are CD25+.

4. Exploiting unique and tissue-specific biochemistry of differentiation programs. Although not usually categorized as "differentiation therapy," it is possible to take advantage of the biological function of the differentiated cells in neuroendocrine tumors by administering a radioactive precursor compound that is concentrated to high levels in the malignant tissue and the corresponding normal tissue. The high levels of radioactivity that accumulate in the differentiated tumor cells not only expose these cells that contain it but also neighboring cells, even though the neighbors might be undifferentiated and do not accumulate the drug. For example, I-131 is used to treat tumor residuum and metastases from thyroid carcinoma (Dworkin et al. 1995, Sweeney and Johnston 1995). The normal function of the thyroid epithelium is to incorporate iodine into precursors of thyroid hormone, so I-131 can be delivered with a high degree of tissue selectivity to cancerous thyroid tissue as long as there is some degree of thyroid differentiation in the tumor. Likewise, I-131 labeled iodo-benzylguanidine is used to treat metastatic neuroendocrine tumors (Gelfand 1993, Sloan et al. 1996).

10.4 Pharmacological issues arising from tumor biology

As explained the previous chapter, the Goldie-Coldman hypothesis has significant implications for the chemotherapy of cancer, and experience with targeted agents to growth factor receptors and even to endothelium suggests that tumor heterogeneity and selection are facts that must be addressed by cancer therapy. As applied to drug therapy, the hypothesis predicts that the probability of finding pre-existing cancer cells resistant to multiple cytotoxic chemotherapy agents of different mechanism of action and structure is much lower than the probability of resistance to a single drug. With the exception of a few rare diseases that are curable with single agent drug therapy (like gestational trophoblastic neoplasia using methotrexate), modern chemotherapy regimens are combination regimens. The improved effectiveness of the regimens compared to single agent therapy confirms the clinical significance of the hypothesis. The student should be aware of some notable examples of combination regimens, some of which combine not only non-cross-resistant drugs but also different therapeutic strategies: irinotecan+5FU and 5FU+oxaliplatin +/− bevacizumab in colorectal carcinoma; multiple combination regimens +/− trastuzumab for first-line therapy of metastatic breast cancer; and pemetrexed+cisplatin better than cisplatin alone for malignant pleural mesothelioma. There are also drug regimens to which have been added nutrients or metabolites to reduce toxicity without reducing drug action on the target: tegafur with uracil; 5FU with leucovorin, which also increases inhibition of thymidylate synthase by 5FU; cisplatin with amifostine to reduce cumulative renal toxicity and limit treatment delays (Kemp et al. 1996); and pemetrexed with folic acid plus B12 to reduce bone marrow and GI toxicity.

The scheduling of the multiple cycles of therapy is also important for maximum effectiveness, and for minimizing toxicity so that therapy can be continued for as long as possible. For many drugs, cytotoxicity requires exposure during a sensitive phase of the cell cycle, caused either by phase-specific presence of the target or susceptibility of the cell to target inhibition. These drugs are called cell cycle phase–specific drugs, and to achieve exposure of every target cell during this specific phase, prolonged exposures to these drugs are required, leading to development of the fractionated dosing schemes administered over multiple days. Improper scheduling of drug dosing usually leads to a dramatic loss of effectiveness; hence, these drugs are also called "schedule-dependent drugs" (Berenson and Gale 1990, Dorr and Fritz 1980). Relatively few schedule-dependent drugs are selectively toxic to cells in G0, G1, or G2. Although it is usually considered an inhibitor of mitosis, paclitaxel (Taxol®) is most toxic to cells making the G0:G1 transition (Donaldson et al. 1994). During the G1:S transition and in early S phase, camptothecins (topotecan, Hycamtin®; irinotecan, Camptosar®) inhibit topoisomerase I, an enzyme that relieves tension in DNA (Slichenmyer et al. 1994). Bleomycin causes fragmentation of DNA into small pieces and inhibits G2:M progression. Because exposed cells accumulate at the G2/M boundary, bleomycin has been used to synchronize tumor cells for subsequent exposure to other cytotoxic therapies (Chabner 1990). Since S-phase is the cell cycle phase during which DNA is synthesized, there are many S phase–specific drugs, including most of the large number of antimetabolites that are patterned after the endogenous purine and pyrimidine building blocks in DNA and RNA. Inhibition of DNA synthesis can be accomplished with 5-fluorouracil or floxuridine, which inhibit thymidylate synthase, and methotrexate to inhibit DHFR production of reduced folate required by thymidylate synthase (Allegra 1990, Dexter et al. 1973, Nesnow et al. 1973, Tone and Heidelberger 1973). The combination of methotrexate and 5-fluorouracil is sometimes used to increase the effect (Grem 1990). Tumors that rely heavily on the "salvage pathway" that phosphorylates thymine and bypasses the need for de novo synthesis will be resistant (Rabinowitz and Wilhite 1969). Some S phase–specific nucleoside analogues act as competitive inhibitors of nucleoside kinases, and some of these have additional mechanisms, such as terminating DNA chain elongation. Examples of such nucleoside analogues are ara-C, 6-mercaptopurine, and 2-chlorodeoxyadenosine, which are highly active against leukemia (Calabresi and Parks Jr. 1980, Carson et al. 1980, Carson et al. 1980, Chabner 1990, Elion and Hitchings 1965, Mccormack and Johns 1990, Skipper et al. 1967). During the S:G2 transition, etoposide (VP-16) inhibits topoisomerase II, which contributes to organization of DNA topology for mitosis (Bender et al. 1990). The M phase–specific cytotoxics disrupt microtubule function. Vincristine and vinblastine, extracted from the periwinkle plant, depolymerize microtubules (Johnson et al. 1963, Johnson et al. 1960, Noble et al. 1958). Taxanes, like paclitaxel originally extracted from bark and needles of yew trees, hyper-stabilize the microtubules, making them rigid and static rather than dynamic (Bender et al. 1990, Fuchs and Johnson 1978, Wani et al. 1971).

A second class of drugs causes the death of cells that have been exposed during any phase of the cell cycle. Because cytotoxicity does not require drug exposure during a particular phase of the cycle, they are called schedule-independent drugs, and they are administered to patients by a variety of schedules from bolus to long IV infusions. Several of these drugs are mainstays of modern cancer chemotherapy - doxorubicin, the nitrosoureas, cisplatin, and cyclophosphamide, while others are less frequently used - melphalan (L-PAM) and chlorambucil. These drugs cause many of the same side effects as the schedule-dependent drugs: bone marrow suppression, mucositis, and hair loss. Some of them also cause cumulative dose-dependent toxicities to non- or slowly proliferating normal tissues like kidney, liver, nervous system, and heart. For example, doxorubicin is limited to a cumulative total dose of about 550 mg per m^2 of body surface area because of the risk of congestive heart failure (Myers and Chabner 1990).

Why are these drugs schedule independent even though they are preferentially toxic to dividing cells? As alkylating agents or their prodrugs, they covalently modify proteins and nucleic acids and cause potentially long-lasting chemical modifications within the cell (Calabresi and Welch 1962, Johnston et al. 1963, Schabel et al. 1963). Cytotoxicity results when exposed cells attempt cell division before repairing the critical damage, and cell cycle dysregulation and stress response genes activate the apoptosis pathway (Chapter 5). Second, schedule independency could be due to more than one drug action, multiple drug targets, and a high probability that at least one critical drug target is expressed during every phase of the cell cycle. For example, cisplatin is most toxic to cells in early G1, but it is cytotoxic throughout the cell cycle (Donaldson et al. 1994, Rosenberg et al. 1965, Rosenberg et al. 1969). Doxorubicin interacts directly with topoisomerase II and the plasma membrane, intercalates into DNA, and generates free radicals (Burke et al. 1987, Gamba-Vitalo et al. 1987, Lane et al. 1987, Myers and Chabner 1990). Multiple mechanisms of action may explain why these drugs are active in such a high percentage of patients and also in such a broad range of diseases: multiple mechanisms of drug resistance would have to be present for a cancer cell to be completely resistant to their effects.

The optimal schedule for administering chemotherapy is also determined in large part by the time required for normal tissues to recover from drug toxicity. Most drug-induced toxicities require 2–3 weeks for complete recovery, leading to the common clinic visit schedule for most cancer patients of therapy every 3–4 weeks with weekly checkups for monitoring toxicity. Some drugs (alkylating agents like nitrosoureas, mitomycin C) cause delayed or protracted toxicity and patients usually require 6–8 week recovery periods before the next cycle of therapy can be tolerated. On the other hand, a few drugs (gemcitabine and paclitaxel) cause rapid-onset and rapidly reversible toxicities, so they are exceptions to this general rule and can be administered more frequently for several weeks and even months. Changing the treatment schedule can also alter the dose-limiting toxicity of certain agents and increase long-term tolerability. For example, lowering the dose level of paclitaxel allowed it to be administered weekly for a median of 19 weeks as second-line therapy for metastatic breast cancer, but changed the most common serious toxicity of treatment from neurotoxicity to neutropenia (Lombardi et al. 2004).

A final determining factor of the dosage regimen is the pharmacokinetic profile of the drug. The field of pharmacokinetics is a specialty of pharmacology concerned with what the body does to the drug, specifically how drug concentration in the body changes over time. Concentration over time is determined by the balance between the in-rate from intravenous, subcutaneous or oral dosing, and the elimination rate by excretion and metabolism. Unless protected from degradation (pegfilgrastim, darbe-poetin), therapeutic proteins are the most unstable drugs in the blood and must be administered frequently for several days to stimulate blood cell production in the bone marrow (epoetin, sargramostim, denileukin, diftitox). Small molecule drugs are somewhat intermediate in their residence time in the body, and so are usually administered once per day over a period of a few days. Enzyme inhibitors may be administered for many days in a row, especially if they achieve small log-cell kills per dose and are available in an oral formulation that is easy to administer outside of a clinical setting. Small molecule drugs encapsulated into liposome particles or polymers (Doxil, Caelyx, DaunoXome, DepoCyt) that are protected from uptake by spleen and liver can circulate for days and require convenient monthly dosing. Antibodies naturally circulate throughout the body for weeks in blood plasma and lymph. Antibody-based drugs exhibit similar characteristics, so most of them are administered initially in a large "loading dose" to rapidly achieve therapeutic levels, which are then maintained in the plasma by intermittent "maintenance doses". Bevacizumab and gemtuzumab have elimination half lives of ~10–20 days and are administered every 2 weeks, while cetuximab's half-life of ~100 hours (Baselga et al. 2000, Thomas and Grandis 2004) necessitates weekly maintenance doses after an initial IV dose. A few antibody products that circulate with long half lives (alemtuzumab half-life 12–21 days; (Rebello et al. 2001) still require frequent dosing to accumulate sufficient drug in the body while avoiding high peak plasma levels that cause acute toxicities.

10.5 Unknowns, the future, and the emergence of molecular oncology

As described in the preceding sections, there are many therapeutic strategies being used to treat cancer patients. The effectiveness of all of them is determined by biological principles of tissue differentiation and renewal in tumors, and pharmacological principles of optimal scheduling, toxicity and efficacy. Furthermore, some of the strategies directly and intentionally exploit the malignant caricature of tissue renewal for therapeutic gain. Therapy to block trophic factor action teaches us that trophic factors are required for progenitor survival and optimal growth of stem cells, whether malignant or normal, but as predicted by biology is not curative unless tumor control exceeds the patient's life span. Since stem cells survive trophic factor withdrawal, attempts at curative therapy in younger patients will require the use of cytotoxic chemotherapy with its small, but concrete chance of killing stem cells using properly timed drug therapy during repopulation when they are susceptible.

There is a large gap in our knowledge about why different drugs with similar mechanisms of action cause different toxicities in different organs and tissues. There is also a poor understanding of the mechanistic basis for selective toxicity of drugs in malignant cells over their normal counterparts, which distinguishes the useful drugs from the many chemical toxicants that have no value as anticancer agents. One interesting example is the comparative toxicology of two vincas: the dose-limiting toxicity of vincristine is neurotoxicity, but the closely related analogue vinblastine is myelosuppressive (Dorr and Fritz 1980). The alkylating agent cisplatin is much more toxic to proximal tubule function in the kidney than to bone marrow function, while one analogue (carboplatin) shows the reverse toxicity profile and another analog (oxaliplatin) primarily causes neurotoxicity (Reed and Kohn 1990). Cisplatin is effective primarily against cancers of the upper GI tract and head and neck, but oxaliplatin (Eloxatin®) is currently used for treating lower GI cancers in combination with infusional 5FU/LV regimens (FOLFOX"). A few drugs are not myelosuppressive in most patients at standard doses: bleomycin, streptozotocin, floxuridine, vincristine, and cisplatin (Dorr and Fritz 1980). Because they do not cause any bone marrow toxicity, they are often included in combination chemotherapy protocols to achieve additional antitumor effect without additional suppression of white blood cell and platelet production.

Regarding the scientific basis for selective drug action on malignant cells compared to normal counterparts, methotrexate-sensitive leukemic cells accumulate intracellular drug faster than their resistant counterparts (Whitehead et al. 1987), and polyglutamation of methotrexate that traps the drug inside the cell is less efficient in hematopoietic progenitors then leukemic cells (Fabre et al. 1984, Koizumi et al. 1985). This differential handling of methotrexate is also the basis for the success of "leucovorin rescue," which is the planned administration of this folate analogue following high-dose methotrexate therapy. It decreases bone marrow but not tumor log-cell kill from methotrexate exposure, because the slower metabolism of methotrexate in normal progenitors provides time for leucovorin to compete with methotrexate, whereas faster metabolism in the leukemic cells result in such high accumulation of polyglutamated methotrexate before leucovorin is administered that it cannot compete (Allegra 1990).

Some of the drugs described in this chapter are the results of stellar advances in the understanding of the molecular drivers of the malignant process and the development of inhibitors of these molecular targets. However, practical problems with molecular targeted oncology are foreseeable on the horizon, having to do with the funding of new drug development and the financing of patient care. If future molecular therapies are effective in just a small subset of patients with a certain diagnosis, such as gefitinib against tumors with mutated EGFR1 (Janne et al. 2004, Paez et al. 2004), then the high selectivity of the drug for its molecular target by definition means that is will only be useful in a small minority of patients whose tumors contain the molecular lesion. Yet, these drugs are some of the most expensive drugs to develop, and the overall cost of developing any drug is rising. How will society support the cost of this kind of drug development, where market revenue from sales will not make the drugs profitable, unless they are priced very expensively to raise this level of revenue. But, the high cost of drugs is a high profile political and medical-economic issue.

These drugs against molecular targets that may likely be mutated in some tumors have taught an important lesson for the design of clinical studies to prove their clinical effectiveness: patients participating in the clinical trials must be selected for the expression of the target of the drug because natural incidence of the molecular lesion may be too low to demonstrate an overall clinical benefit in an unselected patient group.

On the positive side, molecular oncology provides new tools for genetic testing for identifying the subsets of patients, who are particularly susceptible to certain drugs' toxicities, usually because they cannot metabolize them fast enough to avoid severe toxicity. Some of these tests are already being used to identify DPD-deficient patients, who cannot efficiently eliminate 5FU and essentially over-dose despite receiving standard doses (Harris et al. 1991, Meinsma et al. 1995, Morrison et al. 1997), and identify patients with an inherited disorder of bilirubin metabolism (Gilbert's syndrome) or other defect in the activity of hepatic uridine diphosphate glucuronosyltransferases (UGT1A1) that leads to very slow elimination of irinotecan and severe toxicity (Iyer et al. 1998, Wasserman et al. 1997).

Description of Selected Tumors

G. Barry Pierce and Ivan Damjanov

A.1 Adenocarcinoma of the breast

Adenocarcinoma of the breast usually presents as a small painless mass (Figures A–1a,b). The disease rarely occurs in the second and third decades but then increases in incidence through and past the menopause. Because of the seriousness and great range in age incidence of the disease, any mass in the breast must be viewed as cancer until proven otherwise (see letter in Introduction).

The stem cell population of the breast that will give rise to the milk-producing acini are located in the small ductules and persist in the fibrotic stroma of the senile breast. These stem cells are the target in carcinogenesis (Figures A–2 a,b,c). The majority of adenocarcinomas of the breast arise in these same small ductules. They may present in several ways: the vast majority develop as small solitary painless nodules, but some within the ductules may present with a bloody discharge from the nipple as the nodule develops. As the adenocarcinomas grow, they infiltrate surrounding tissues and eventually become fixed to the deep tissues and to the skin. This may cause dimpling of the skin or, if it invades the nipple, retraction of the nipple (Figure A–1a). With further growth and development it may spread first to the lymph nodes in the axilla, and shortly thereafter to the lungs, pleura, and to the bones of the spine and pelvis.

The initial diagnosis is usually made by fine needle aspiration biopsy or a biopsy with a thicker needle through which a core of tissue can be removed under radiologic guidance, essential for localizing small mammographically detected lesions. If the lesion is malignant the patient will typically undergo further surgery, including complete resection of the tumor ("lumpectomy") usually accompanied by axillary lymph node dissection. In the recent past the treatment of choice for adenocarcinoma of the breast has been radical mastectomy. The breast, the muscles underlying the breast, and the contents of the armpit were removed in an attempt to extirpate all tumor cells. Lumpectomy followed by irradiation and chemotherapy is as effective as radical mastectomy in terms of cure and has completely replaced radical mastectomy, which is performed only exceptionally. If distant metastases are

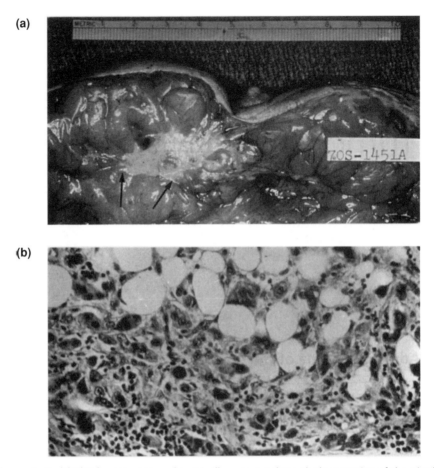

Figure A–1. (a) This breast, removed surgically, was cut through the margins of the nipple and through the adenocarcinoma. The tumor is several centimeters in diameter with gray streaks of invading cancer radiating into the fat (arrows). Invasion to underlying muscle and nipple caused retraction of the nipple. This is a bad prognostic sign. (b) Anaplastic adenocarcinoma of the breast. Huge, irregularly shaped nuclei of cells infiltrating fat and cords of anaplastic cells are the clue to the diagnosis.

present, the surgical resection will be followed by radiation therapy and adjuvant chemotherapy.

The prognosis of breast carcinoma depends primarily on the clinical stage assigned to the tumor at the time of diagnosis. Stage 1 cancers include tumors that measure 2.5 cm or less and have no metastases. Surgical removal of the tumor is associated with an 85 percent, five-year survival rate. Compare these excellent therapeutic results with the survival of less than 10 percent of patients who have stage IV tumors. Obviously, breast cancers are curable only in early stages and thus early detection by self examination and mammography are crucial factors in the combat against this neoplastic disease.

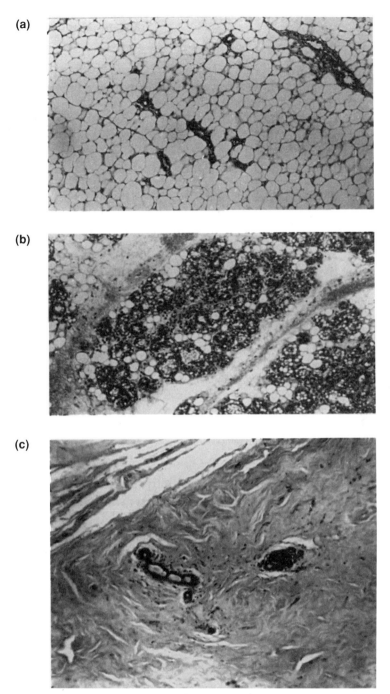

Figure A–2. (a) Photomicrograph of breast of virgin mouse. Note all of the epithelial elements of the breast are included in the primitive tubules embedded in the fat that characterizes the breast at this stage. (b) Photomicrograph of mouse breast in mid-pregnancy. The stem cells in the primitive tubules have multiplied in response to the hormones of pregnancy and become hyperplastic. The epithelial cells have displaced fat cells. Glandular acini are forming. (c) Photomicrograph of breast of old mouse. As in the virgin state there are only a few ducts present, but unlike the virgin state these are embedded in dense scar tissue. The epithelial cells (stem cells) in these ducts must be the target in carcinogenesis because there are no other cells present.

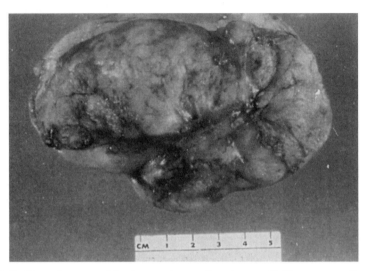

Figure A–3. Fibroadenoma of the human breast, female age 30. The mass, unlike adenocarcinoma, which is stone hard on palpation, is rubbery in consistency. This is such a subjective distinction that an extirpational biopsy is peformed. The mass is encapsulated and was shelled out of the organ with blunt dissection. The smooth capsule suggests invasion has not occurred; however, final diagnosis is made microscopically. The patient is cured by extirpation of the mass that was easily shelled out of the breast. It is benign and the patient has been cured by simple resection.

We should finish our discussion of breast cancer on a positive note. Not all breast tumors are malignant, and in young women most of the masses in the breast parenchyma are actually benign. Figure A–3 illustrates such a benign tumor, a fibroadenoma, removed from the breast of a 30-year-old woman. The lesion is encapsulated, and could be easily shelled out during surgery. Histologically, it was composed of benign, well-differentiated ducts and stroma, with no evidence of anaplasia. Simple removal results in cure.

A.2   Adenocarcinoma of the prostate

Adenocarcinoma of the prostate is the most common malignant tumor in males, and the third most common cause of cancer related death in men. The American Cancer Society estimates that over 200,000 men will be found to have this cancer every year, and approximately 30,000 men will die. Typically, it is a tumor of elderly males, and is quite uncommon in men who are younger than fifty years of age.

Prostatic carcinoma originates most often in the peripheral parts of the gland. Histologically, it is usually composed of small cells forming tubules and glandlike structures arranged in a back to back pattern with no stroma separating them (Figure A–4). Similar cells are found in perineural lymphatics and in tumor invading adjacent organs such as the seminal vesicle. Metastases are most often found in regional lymph nodes, but also in the bones and other distant organs, such as the

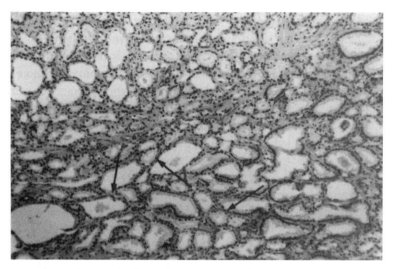

Figure A–4. Photomicrograph of an adenocarcinoma of the prostate gland taken at very low power. Evidence of lobular pattern is still present, but the epithelial stromal relationships are abnormal. No stroma separates acini (arrows). Adenocarcinoma of the prostate is usually not very anaplastic, and the diagnosis is made at low magnification based on the abnormal epithelial stromal relationships.

liver or the lungs. The metastases in the bones are often osteoblastic, that is, elicit the formation of new bone.

Prostate carcinoma is often associated with an increased prostate specific antigen (PSA), which can be measured in the serum and used diagnostically. Unfortunately, serum PSA levels are within normal limits in a significant number of patients who have prostatic carcinoma. Furthermore, PSA in the serum may be elevated in patients with prostatic inflammation but who have no cancer. These false positive and false negative results limit the specificity of the PSA test, which, if combined with biopsy is, nevertheless, still the best approach to diagnosing prostatic carcinoma in early stages of the disease.

The prognosis of prostatic carcinoma depends on the stage and the grade of the tumor. Tumor grading is based on histologic examination of prostate tumor tissue according to the system developed by Gleason. Tumors of higher Gleason grade have a worse prognosis than tumors that have a lower Gleason grade, but in all instances the staging is of far greater prognostic significance than grading. Tumors limited to the prostate have a five year survival rate of 75 percent, but those that have spread beyond the prostate have only a 35 percent chance to survive five years.

A.3   Adenocarcinoma of the colon

Adenocarcinoma of the colon is a common disease of individuals fifty-five to seventy years of age. In men, the lesion usually occurs in the sigmoid colon or rectum and at

percent of them are within reach of the examining finger. A routine physical
...dividual over the age of fifty is incomplete if rectal examination has not
been done. The lesion invades circumferentially and results in an ulcerated napkin
ring–like constriction of the colon that causes bowel obstruction and bleeding
(Figures 1–7 and 1–8). In women the adenocarcinomas are more common in the
caecum and ascending colon where large fungating lesions occur that ulcerate and
bleed, but do not cause bowel obstruction. This is because there is less constriction
of the colon and the fecal stream in the right colon is liquid, whereas that in the left
is solid. Bowel obstruction is a surgical emergency, and a person with a complete
obstruction will die in about a week if untreated. In addition to obstruction or
bleeding, a multiplicity of symptoms, many nonspecific, can herald the disease. For
example, anemia and colicky (spastic) abdominal pain can be symptoms.

Etiology of adenocarcinoma of the colon remains poorly understood, despite
many epidemiologic studies on potential carcinogens in food. Today, it is known
that tumors develop sequentially from benign tubular adenoma (Figure 1–11) and
villous adenomas as well as in microscopic focal intramucosal neoplastic changes.
The sequence of events leading to the formation of adenocarcinoma includes activation of several oncogens. In approximately 10 percent to 15 percent of cases
the tumors are related to a hereditary predisposition including conditions such as
the autosomal dominant familial adenomatous polyposis coli and hereditary nonpolyposis colorectal cancer (HNPCC), also known as the Lynch syndrome (Lynch
and de la Chapelle 2003). Ulcerative colitis and Crohn's disease, two chronic inflammatory bowel diseases, predispose to adenocarcinoma. Yet, in most patients with
adenocarcinoma of the colon it is not possible to identify any predisposing conditions or definitive risk factors.

Early diagnosis of colon cancer is the best approach in the fight against this
neoplastic disease. The early diagnosis is based on colonoscopic examination of
the large intestine, which should be accompanied by a biopsy of all polyps and
other suspicious lesions. These biopsies should be examined histologically by a
pathologist. If the malignancy is limited to a pedunculated tubular adenoma, and
the malignant cells do not invade the stalk, a "snare-polypectomy" performed during
colonoscopy is curative. If the malignant cells invade the stalk or even deeper, a
loop of the intestine containing the tumor must be resected, and the local lymph
nodes dissected. Multiple polyps, polyps larger than 2 cm and those that have
a villous structure and a broad base deserve special attention, because they are
much more prone to malignant transformation than the small solitary pedunculated
tubular adenomas. It has been estimated that 4 percent of tubular adenomas undergo
malignant transformation, where the incidence of malignancy is ten times higher
in villous adenomas. Carcinoembryonic antigen (CEA) and glycoprotein antigen
is synthesized by the tumor cells and can be assayed in the serum. It is particularly
useful in detecting recurrences as metastases grow and produce the antigen.

Carcinomas of the large intestine are almost all histologically classified as adenocarcinomas. The histologic grading of these tumors is of limited clinical significance,

and the prognosis depends primarily on the tumor stage at the time
Tumors can be staged according to the international TNM system, w.
porates the basic diagnostic tenets established by Dukes some seventy years ago.
As in other sites, tumors resected in early stages of development have a very good
prognosis, whereas those that have already spread beyond the large intestinal wall
have a much worse prognosis.

Adenocarcinomas of the colon produce carcinoembryonic antigen (CEA), a glycoprotein that appears in the serum of colon cancer patients. Unfortunately, elevated
levels of CEA can be found in patients who have adenocarcinomas in other sites,
as well. Furthermore, elevated serum CEA may be found in patients who have
chronic ulcerative colitis, and even in heavy smokers. Thus, CEA cannot be used
as a screening test for colon carcinoma. Nevertheless, the test used during the post-
operative period is useful for estimating the completeness of the tumor removal or
detecting recurrences and new metastatic tumors in a patient who was treated for
colon cancer.

A.4 Squamous cell carcinoma

Squamous cell carcinomas develop wherever squamous epithelium occurs normally,
such as the skin, mouth, cervix, vagina, or esophagus. It may develop from foci of
squamous metaplasia that may occur in the respiratory tract, renal pelvis, or bladder.

Squamous cell carcinomas develop most commonly in skin, bronchus, and cervix.
Even though all squamous cell carcinomas are histologically composed of the same
cells, each of these tumors has its own clinical features. Those in the skin are usually
removed before they metastasize, but given time they can spread to regional lymph
nodes and later to distant sites. Skin tumors ulcerate early (Figure A–5). In the
bronchus they metastasize early to regional lymph nodes and distant sites. Those in
the cervix invade locally in an extremely aggressive fashion, and usually destroy the
adjacent ureters, rendering the host incapable of excreting liquid wastes, and cause
death from uremia. For reasons unknown, they rarely metastasize to distant organs.

Histologically all squamous cell carcinomas vary from well-differentiated, to
highly anaplastic poorly differentiated tumors. Well-differentiated cancers consist
of tumor islands which show gradual differentiation toward the central parts, which
often contain numerous squamous pearls (areas of fully differentiated squamous
epithelial cells arranged into concentric whorls) (Figure A–6). In anaplastic tumors,
the cells form widely invasive strands of cells that show very little squamous differentiation. Obviously, the well differentiated tumors grow slower than the anaplastic
ones. Nevertheless, it is not known why tumors of the same histologic appearance
behave differently one from another in different locales. On average, cure rates are
about 10 percent for squamous cell carcinoma of bronchus, 50 percent for cervix,
and 100 percent for skin. Squamous cell carcinomas of the mucous membranes
such as the mouth, vulva, glans penis, bladder, and bronchus have a much worse

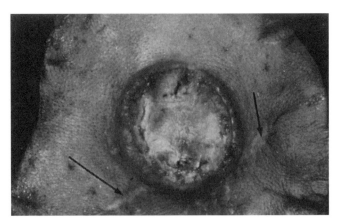

Figure A–5. Photograph of squamous cell carcinoma of the skin. The lesion measures about 1.5 to 2 cm in diameter and has a raised, rolled edge (typical of malignant ulcers). This edge is a reflection of invasion and proliferation of the tumor cells beneath the skin. The base of the ulcer is crusted and composed of necrotic cancer and stromal cells. Note the surgical scar at the arrows. This lesion is a recurrence of an incompletely removed carcinoma. It is such inadequately treated lesions that are prone to metastasize because the surgery opens lymphatics and small venules, chief avenues for metastasis.

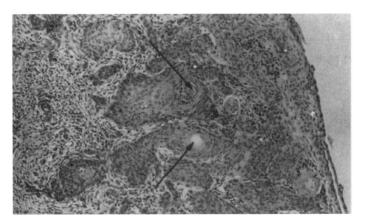

Figure A–6. Photomicrograph of squamous cell carcinoma of skin illustrating plugs and columns of anaplastic cancer cells invading stroma. Note the squamous pearls (arrows), good examples of squamous differentiation.

prognosis than those of the skin, presumably because they are much more advanced when diagnosed. Squamous cell carcinoma of the lung is discussed with other lung cancers in section A.7.

Studies of squamous cell carcinoma of the cervix are illustrative of the pathogenesis of squamous cell carcinoma in general, so cervical carcinoma is discussed here. Squamous cell carcinoma of the cervix is most common in women 45 years of age or older. It develops from carcinoma in situ at the junction of squamous and glandular epithelium of the cervix (Figures A–7 and A–8). The lesion is first evident

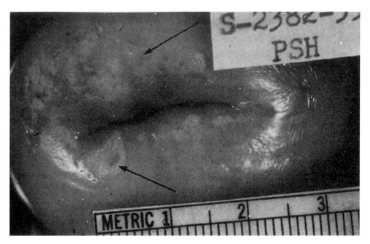

Figure A–7. Photograph of cervix with leukoplakia (white patch, arrows). This white patch is in reality a carcinoma in situ of the cervix, but leukoplakia could be dysplasia, inflammation, or an invasive carcinoma. The term is merely descriptive.

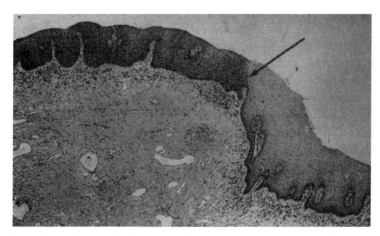

Figure A–8. Photomicrograph of carcinoma in situ. Note the sharp line of demarcation (arrow) between the carcinoma in situ (to the left) and normal squamous epithelium (on the right). Note the progressive differentiation of squamous cells in the normal epithelium, from the dark-stained basal (stem cell layer) to the terminally differentiated cells at the surface. This differentiation is lacking in the carcinoma in situ, and the surface cells are as undifferentiated as the basal layer of cells. Notice there is no evidence of invasion.

as a white patch, and with progression it becomes invasive and ulcerates, but seldom metastasizes to distant organs. Invasive squamous cell carcinoma occurs about ten years later than carcinoma in situ.

Invasive carcinoma, carcinoma in situ, or inflammation may all present as leukoplakia, which means "white patch." Therefore, presence of a white patch on examination does not necessarily mean a diagnosis of cancer. The diagnosis of cancer is made using the Papanicolaou technique (Pap smear). In this procedure, the cervix is scraped and the surface cells are smeared on a slide and examined microscopically.

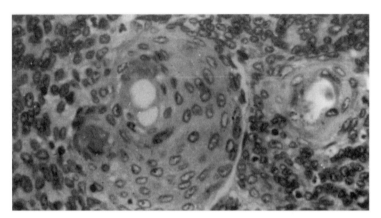

Figure A–9. Photomicrograph of squamous cell carcinoma of the skin of a rat to illustrate the masses of dark-stained anaplastic tumor cells surrounding islands of well-differentiated squamous cells with keratotic centers. These are referred to as pearls.

The surface cells of a carcinoma in situ are anaplastic and appear so on the slide (Figure A–8).

Histopathologically, carcinoma in situ is recognized by the fact that the neoplastic cells occupy the entire thickness of the epithelium. Carcinoma in situ develops in the basal layer and the transformed cells, instead of differentiating in orderly successive steps until senescence occurs and the dead cells are sloughed from the surface of the mucous membrane, form layer after layer of cells closely resembling the dark-staining basal cells. Thus, the full thickness of the epithelium shows no evidence of maturation (Figure A–8). Why the lesions do not invade is not known, but through the use of the Papanicolaou technique, carcinomas in situ are easily identified and resected. As a consequence, fewer cases of invasive carcinoma develop now than prior to the use of this simple technique. Once the squamous cell carcinoma develops it will invade the stroma and extend into adjacent organs (Color Figure 4). In this respect carcinoma of the cervix behaves similar to carcinomas in other sites such as the esophagus, skin, or the mouth.

The cell biology of squamous cell carcinomas was clarified in a study using rats carrying transplantable squamous cell carcinomas of the skin (Figure A–9). The relationship of the anaplastic tumor cells and the apparently differentiated ones in the squamous pearls was studied. Rats were injected with ^{3}H-thymidine to identify tumor cells systesizing DNA. Autoradiography was performed on the tissues at 1 hour and at 96 hours after administration of the thymidine. The autoradiography showed that labeled nuclei were found only in the anaplastic cancer cells at the first time point (Figure A–10), indicating that the pearls were composed of postmitotic differentiated cells and the proliferative pool of the tumor lay in the anaplastic cells. Samples taken at 96 hours showed the presence of labeled cells not only in the anaplastic tumor but in the pearls, indicating that the anaplastic cells labeled in the first hour migrated to or invaded the pearls as well. Autoradiography with electron microscopes showed that the anaplastic cells lacked the features of squamous

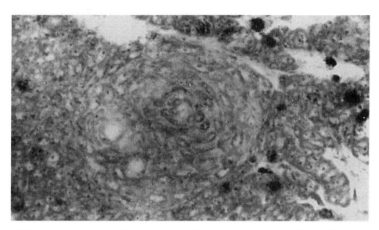

Figure A–10. Autoradiogram of squamous cell carcinoma of a rat injected 1 hour previously with tritiated thymidine. The tumor was fixed, sectioned, and the sections were covered with photographic emulsion and stored in the dark. The photographic emulsion was exposed by decay of the tritiated thymidine at sites of its incorporation into DNA (see clusters of black dots over anaplastic cancer cells primarily in right side of the illustration). Note that in this illustration, as well as throughout the carcinoma, no black dots were seen over pearls. This indicates that the pearl cells did not synthesize DNA. They did not synthesize DNA because they were differentiated, postmitotic and benign squamous cells. DNA synthesis was observed only in the undifferentiated, mitotically active, carcinoma cells in this 1 hour preparation.

epithelium and that the labeled cells of the pearls were well-differentiated squamous cells with typical membrane and cytoplasmic characteristics. It was concluded that the anaplastic cells had migrated into the pearls and had differentiated. When the squamous pearls were dissected from the tumor and transplanted to histocompatible hosts, they failed to produce tumors, whereas equivalent amounts of anaplastic cells formed tumor on subcutaneous transplantation. It was thus concluded that in the process of spontaneous differentiation, anaplastic cancer cells gave rise to benign squamous cells no longer dangerous to the host (Pierce and Wallace 1971).

A.5 Teratocarcinomas

An emphasis disproportionate to the incidence or clinical importance is given to these tumors because of the impact their study has made on oncology. Teratomas, literally benign tumors resembling monsters or malformed babies (terata), commonly occur in the ovaries. They may contain almost any tissue found in the body, and recognizable organs, such as fingers, eyes, or jaws with teeth, may be present in them. Teratocarcinomas, like teratomas, contain tissues representing each of the three embryonic germ layers (endoderm, mesoderm, and ectoderm). As the term implies, teratocarcinomas are malignant. For reasons unknown they are more commonly found in the testes than in the ovary (Figure A–11). They contain

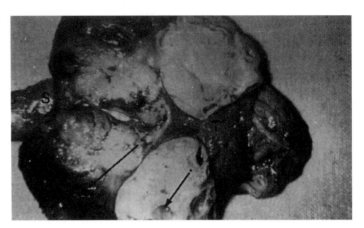

Figure A–11. Photograph of a human testis cut in half and opened like the wings of a butterfly. The spermatic cord is at S, normal testis T, and numerous nodules of teratocarcinoma; note epithelium-lined cysts (short arrow) and choriocarcinoma (long arrow).

embryonal carcinoma (Figure A–12a), an exceedingly anaplastic carcinoma that is interspersed with the other tissue of the tumors (Figure A–12b) (Pierce et al. 1978). Embryonal carcinoma cells resemble the inner cell mass cells of the blastocyst (Chapter 1).

Teratocarcinomas usually present as a painless enlargement of the testis in the third decade of life. They are so lethal that survival is measured in two years; the patient is either dead or cured two years after diagnosis and treatment. Most of the tumors have metastasized at the time of diagnosis to the periaortic lymph nodes and to the liver, lungs, and other organs. Teratocarcinomas and pure embryonal carcinomas are extremely sensitive to chemotherapy; cure can be achieved in over 80 percent of cases.

With the exception of two strains of inbred mice, teratocarcinomas are ultrarare in lower animals. Strains 129 and LT have high incidences of testicular teratocarcinoma and ovarian teratocarcinomas, respectively. They have proved to be useful models of the human tumors (Pierce, Shikes, and Fink 1978). As in humans, the differentiated tissues are well formed in the murine tumors but haphazardly arranged, so it is not uncommon to find islands of brain, gut, muscle, bone, and gland all mixed up together (Figure A–12b–d) with embryonal carcinoma (Figure A–12a). Organization of tissues is seen in the tumors more often than can be accounted for by chance (Figure A–12c). In other words, gastrointestinal glands may be surrounded by a double layer of smooth muscle, or masses of cartilage may be ossifying and marrow may form in the developing bone. A tooth may lie in juxtaposition to membrane bone (Figure A–12d). Structures named embryoid bodies, because they resemble early preimplantation mouse embryos, are frequently found in them, particularly in the ascites form (Figure A–12e). These structures have proved useful in studies of the tissue interactions in teratocarcinomas.

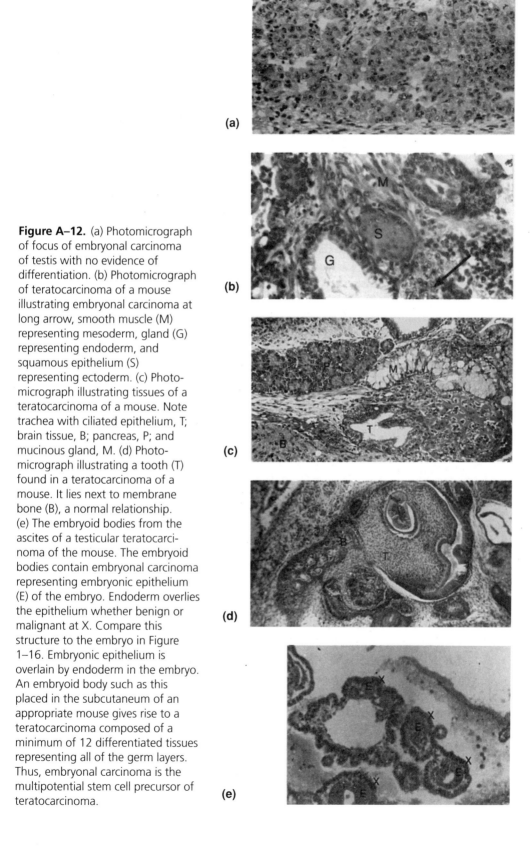

Figure A–12. (a) Photomicrograph of focus of embryonal carcinoma of testis with no evidence of differentiation. (b) Photomicrograph of teratocarcinoma of a mouse illustrating embryonal carcinoma at long arrow, smooth muscle (M) representing mesoderm, gland (G) representing endoderm, and squamous epithelium (S) representing ectoderm. (c) Photomicrograph illustrating tissues of a teratocarcinoma of a mouse. Note trachea with ciliated epithelium, T; brain tissue, B; pancreas, P; and mucinous gland, M. (d) Photomicrograph illustrating a tooth (T) found in a teratocarcinoma of a mouse. It lies next to membrane bone (B), a normal relationship. (e) The embryoid bodies from the ascites of a testicular teratocarcinoma of the mouse. The embryoid bodies contain embryonal carcinoma representing embryonic epithelium (E) of the embryo. Endoderm overlies the epithelium whether benign or malignant at X. Compare this structure to the embryo in Figure 1–16. Embryonic epithelium is overlain by endoderm in the embryo. An embryoid body such as this placed in the subcutaneum of an appropriate mouse gives rise to a teratocarcinoma composed of a minimum of 12 differentiated tissues representing all of the germ layers. Thus, embryonal carcinoma is the multipotential stem cell precursor of teratocarcinoma.

Dixon and Moore (1952) proposed that embryonal carcinoma cells were multipotent cells that differentiated into the somatic tissues of the teratocarcinomas. This was proved correct when embryonal carcinoma cells were shown to differentiate into the three germ layers, which in turn differentiated into the somatic tissues characteristic of the tumors (Pierce and Dixon 1959). These tissues were benign and of no danger to the host (Pierce, Dixon, and Verney 1960). These observations were confirmed by cloning embryonal carcinoma in vivo. Single embryonal carcinomas transplanted into the intraperitoneum of mice gave rise to teratocarcinomas containing the twelve or more tissues characteristic of the tumors plus embryonal carcinoma (Kleinsmith and Pierce 1964). Because the histiotypic potential of embryonal carcinoma so resembled that of the inner cell mass of the blastocyst, which is the only other tissue capable of giving rise to the three germ layers, it was postulated that embryonal carcinoma was the neoplastic equivalent of inner cell mass cells.

There are three principal sources of embryonal carcinoma cells: primordial germ cells, inner cell mass cells, and transplantation of preimplantation embryos. Stevens (1967) proved that the spontaneous teratocarcinomas of strain 129 mice originated from primordial germ cells. On the twelfth day of gestational age, he transplanted genital ridges into the testes of adult animals. If the genital ridges contained germ cells, they produced islands of embryonal carcinoma cells visible seven days after transplantation. Tumors several centimeters in diameter were formed in the ensuing two weeks that contained the multiplicity of differentiated tissues characteristic of teratocarcinoma. If the genital ridge did not contain germ cells (Stevens transplanted genital ridges from animals with a congenital absence of germ cells when in the homozygous state), teratocarcinomas did not occur. The primordial germ cells may be construed as the stem cells of the species, and when the fine structure of the primordial germ cells was compared to that of the embryonal carcinoma cells to which they gave rise, they proved to be equally undifferentiated (Pierce, Stevens, and Nakane 1967). This eliminated the need for the concept of dedifferentiation as a mechanism in the genesis of these tumors and led to the idea that tumors in general probably originate from the stem cells in any differentiated tissue. Preimplantation mouse embryos of appropriate strain transplanted into the testes of adult mice also give rise to teratocarcinoma. The most widely used teratocarcinoma cell line, OTT6050, originated in such a manner (Stevens 1970).

Strains of low-grade embryonal carcinoma cells, which are called ES cells (embryonic stem cells), have been developed by culturing inner cell mass cells in vitro where they become spontaneously transformed (Evans and Kaufman 1981; Martin 1981). This confirmed the idea that embryonal carcinoma cells are the neoplastic equivalent of the inner cell mass. ES cells form beautiful teratocarcinomas in vivo, differentiate well in vitro, and are less progressed than any of the transplantable spontaneous teratocarcinomas.

Another important observation was made by Stevens (1967) in the conduct of the transplant experiments that give rise to teratocarcinomas. As many as eleven or more islands of embryonal carcinoma were found in individual testicular tubules by the seventh day after transplantation of genital ridges. It is widely accepted that most tumors are monoclonal in origin. To rationalize these observations and ideas, it is clear that the monoclonality of tumors must result from selection of the fittest clone. Thus, in the embryonal carcinomas, one of the eleven clones would dominate, be selected, and the resultant tumor would be monoclonal. The significance of the observation is that selection of cells is an important process in the latent period during carcinogenesis and later during progression of the established tumors.

Brinster (1974) was the first of many who have shown that embryonal carcinoma cells injected into the blastocysts of mice are regulated by the blastocyst to behave as normal inner cell mass cells, take part in embryonic development, eventuating chimeric animals. The animals are chimeric because their tissues are composed of embryo-derived cells and embryonal carcinoma-derived cells. A majority of chimeras, in which aneuploid embryonal carcinoma cells were used, were chimeric in only a few tissues. If an animal was chimeric at all, it was in coat color and then possibly in some of the internal organs. These animals were almost always sterile (Mintz and Illmensee 1975). In contrast, chimeras produced by injection of ES cells into blastocysts give a much greater percentage of chimeras, many of which are fertile. These are used in the construction of transgenic mice (Brinster et al. 1984).

The technique of mutating ES cells or transfecting them with genes of various kinds provides a powerful tool for studying molecular events in genetics. For example, if gene X is transfected into ES cells and a fertile male chimera is produced with them, in one backcross a pure inbred line carrying gene X will be produced. This is of great value in modeling a variety of genetic diseases.

Finally, how does one rationalize teratocarcinoma with other carcinomas and sarcomas? It can be stated unequivocally that all of the lessons learned from the study of teratocarcinoma extrapolates to all other kinds of cancer. The phenotypes of tumors like those of their normal counterparts differ only in the potency of their stem cells. Embryonal carcinoma cells have the multipotency of inner cell mass cells. Multipotent tumors differ from unipotent ones, such as squamous cell carcinomas, in that the potential of the stem cell originating the squamous cell carcinoma has been reduced during development to a single tissue. Carcinogenesis does not alter the histiotypic potential of the responding stem cell – it adds the malignant phenotype to it.

Why some embryonal carcinomas differentiate into somatic tissues and others do not is not known. Retinoic acid, applied to cells that normally do not differentiate in vitro or in vivo, causes them to differentiate (Strickland, Smith, and Marotti 1980). Therefore, they have the potential to differentiate, but for reasons unknown do not express it. Apparently, environment controls the expression of the determined state. These observations form the basis for the chemical induction of differentiation as a form of therapy for cancer.

A.6   Liver cell carcinoma

Liver cell carcinoma is rare in North America, but it is common in parts of Subsaharan Africa and South-East Asia. In North America, individuals with this disease often have cirrhosis of the liver and the clue to hepatoma is an unexplained deterioration in the individual's condition as a complication of cirrhosis. Weight loss, a right upper quadrant mass, or abdominal pain may be present. The role of cirrhosis as an etiologic agent is unexplained, although the multiple nodules of regenerating liver in the cirrhotic condition represent an ever-present hyperplasia, which in turn could present more cellular targets for the carcinogenic event. Macronodular cirrhosis is a hepatitis-related condition and indicates a role for hepatitis B virus and hepatitis C virus in the etiology of hepatocellular carcinoma.

Aflatoxin, produced by the mold *Aspergillus flavus*, is a potent liver carcinogen in the Third World, where food is often contaminated by the carcinogen. Butter yellow (P-dimethyl aminoazobenzene) causes liver cell carcinoma without cirrhosis, indicating that the cirrhotic process itself is not essential for the production of the disease.

The tumors may be single or multifocal in origin, usually well demarcated from the surrounding normal liver tissue, and have either of two histologic patterns. One is composed of well-differentiated-appearing hepatocytes (parenchymal liver cells) arranged in trabeculae (fibrous cords of connective tissue) separated by sinusoids (minute blood vessels). Many of the cells are functional because some of the nodules are bile stained (the bile canaliculae of the tumor do not connect with the normal bile ducts) and contain liver enzymes that can be demonstrated histochemically. The second pattern is anaplastic adenocarcinoma, which arises in small bile ducts. These bile duct carcinomas are rarely associated with cirrhosis and very rarely associated with parasitic infections, such as liver flukes. The tumors metastasize widely, especially to the lung.

Because the vascular drainage of the entire gastrointestinal tract passes through the liver, it is not surprising that the liver is a common site for metastatic tumors, which are much more numerous than primary ones (Figure A–13). The liver is the primary organ of detoxification of xenobiotics. In this process, poisons are oxidized to inactive forms and solubilized so they can be excreted in either the bile or the urine. Procarcinogens can be metabolized to highly active carcinogens by these processes, and not only liver cells, but those of the bladder, are exposed to the carcinogens if they are excreted in the urine. We supposedly live in a sea of synthetic chemical carcinogens that require metabolic activation, and it is interesting that the liver does not have a much higher incidence of primary carcinomas than other organs.

A.7   Lung cancer

Lung cancer is the most common malignant tumor of internal organs in North America. More than 30 percent of men and 27 percent of women dying of cancer will die of lung cancer. The death rate grimly portrays our inability to cure people

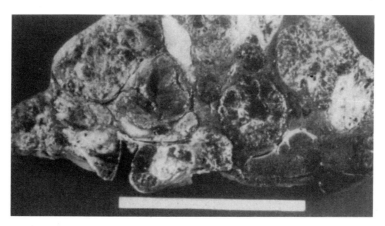

Figure A–13. Photograph of metastatic embryonal carcinoma to the liver. Metastatic carcinoma is the most common tumor of the liver in North America. It is easily distinguished from the normal liver parenchyma, and even in a case such as this with massive replacement of liver by embryonal carcinoma, death does not result from liver failure, but rather from cachexia and infection.

with the disease and stresses the need for prevention. When one thinks of epidemic, one thinks of infectious diseases, but in the case of lung cancer, the epidemic is caused by inhaled pollutants. The most important of these is cigarette smoke. Numerous studies have implicated cigarette smoking in the dramatic increase in incidence of lung cancer from 1930 to 1990. Compounding the effect is carcinogenic synergism that occurs when multiple carcinogens are applied simultaneously. Inhalation of asbestos by a smoker is a case in point. As a rule, if a smoker has a ten-fold increased risk of lung cancer and an asbestos worker has a ten-fold increased risk, then the cigarette smoker exposed to asbestos has a hundred-fold increased risk (Selikoff et al. 1980). These numbers are approximations, but sufficiently accurate to make the point that if no one smoked, the risk for lung cancer would be less than 5 percent of the current incidence. For the first time in the past ten years, education has resulted in a decline in the epidemic in the younger age groups.

Other carcinogenic agents are of lesser importance but still deserve to be mentioned. Among others they include exposure to irradiation (the uranium miners of Colorado), and heavy metals such as chromates and nickel as industrial hazards.

The airway has been designed to clean, moisten, and warm inhaled air. In addition, it has a mechanism to lubricate and sweep particulate matter and fluid from the smaller to larger airways (the cough sensitive portion of the lung) (Figure 1–1) where the contaminants can be expelled by coughing. Anything that interferes with ciliary movement or secretion of mucus perturbs the housecleaning mechanism of the lung and results in chronic inflammation. Cigarette smoking bombards the bifurcations of the bronchi with hot dry gases, particulates, and carcinogens. This leads to squamous metaplasia, which in turn is believed to be the site of origin of squamous cell carcinoma of the bronchus. To give a measure of the hazards, a smoker is at high risk with a 20-pack-year smoking history ("pack years" equal the packs smoked per day times the number of years of smoking). The cancer risk varies directly with the number of cigarettes smoked, but with at least the square of the duration of smoking. Thus, longtime smokers do not protect themselves by merely cutting back on the number of cigarettes smoked per day. To accomplish this goal, smoking must be stopped.

Lung cancers can be histologically classified as adenocarcinomas, squamous cell carcinomas, small cell carcinomas and large cell undifferentiated carcinomas (Color Plates 5 to 11). For clinical purposes, it is common to divide these tumors into two major groups: small cell carcinomas and non-small cell carcinomas. Small cell carcinomas differ from others in that these tumors are usually inoperable but respond, at least temporarily, to chemotherapy. Non-small cell carcinomas are still treated by surgery, which is usually accompanied by radiation therapy and adjuvant chemotherapy.

Prognosis of lung cancers is very poor, and the five-year survival rate is in the range of 15 percent to 25 percent. Peripheral adenocarcinomas found in women who are non-smokers (Color Plates 6 and 7) have a somewhat better prognosis, especially if discovered in early stages before the tumor has metastasized to lymph nodes. The overall survival of patients with small cell carcinoma, which earlier was approximately six months from the time of diagnosis, has been prolonged with chemotherapy to approximately forty months, but nevertheless few if any of these patients are completely cured.

Primary lung cancers must be distinguished from metastatic carcinomas. Since the lungs are full of blood, hematogenous metastases to the lungs are common. Typically they present in the form of multiple nodules (Color Plate 11). On X-ray examination of the thorax such nodules are described by radiologists as "cannon-ball" or "shotgun-lesions."

A.8   Malignant melanoma

Melanomas develop most commonly in the skin of the lower extremities or other sunbathed areas, but they can occur wherever pigmented cells are found in the body (mucous membranes, eye, central nervous system). About one-third of the tumors develop from preexisting moles or "beauty spots" (pigmented nevi).

The clinical course varies from extremely malignant lesions that invade and metastasize and cause death in a matter of months to indolent ones that pursue a protracted course over years. The best prognostic indicator is the stage of the disease at the time of diagnosis. This is determined by microscopic examination of the lesion and actual measurement of the depth of penetration of malignant cells into the dermis. Melanomas that have not invaded beyond the superficial dermis are cured at the 100 percent level by widespread excision. When the lesion has penetrated into the subcutaneous tissue, the cure rate is less than 15 percent. This

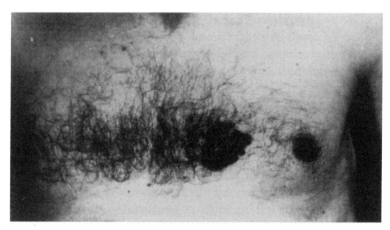

Figure A–14. Gross photograph of malignant melanoma of the skin of the chest of a man, illustrating various degrees of pigmentation and an irregular margin.

dramatic change in prognosis occurs when the tumor has penetrated to a depth of only 1.5 mm.

Malignant lesions are usually discovered accidentally. Change in color, shape, size, or consistency of a preexisting nevus are considered danger signs. Invasive melanomas are almost always inflamed. Therefore, itching or mild inflammatory change in a mole must also be construed as a danger sign (Figure A–14).

Three kinds of melanoma are described, and although none is as important as the stage of the disease at the time of diagnosis, they are of some interest. The first is melanoma in situ that develops in a Hutchinson's freckle. Hutchinson's freckles are brownish, slightly raised spots about 3 cm in diameter on exposed surfaces of elderly patients. Over prolonged periods of time these freckles develop melanoma in situ and eventually invasive melanoma (Figure A–14). Like other carcinomas in situ, melanoma in situ is cured by local removal. Superficial spreading melanomas are the most common and are slightly raised, variably colored lesions that may show evidence of growth in one area and regression in another. The actively growing areas are light brown, whereas the regressing areas may be blue or inflamed. Melanoma cells are found at the epidermal-dermal junction and spread throughout the epidermis. Nodular melanoma carries the worst prognosis and has a tendency to infiltrate into the subcutaneous tissues, resulting in a more advanced stage at the time of diagnosis.

Metastasis of malignant melanoma may occur early and extensively via the lymph nodes and blood vessels, with widespread involvement of many organs. Often the metastases are amelanotic and rapidly growing. Under very rare circumstances, metastases can be dormant and the patients survive asymptomatically for many years after the removal of the primary tumor, only to have metastases eventually grow in the liver, for example. The reasons for this dormancy are not known.

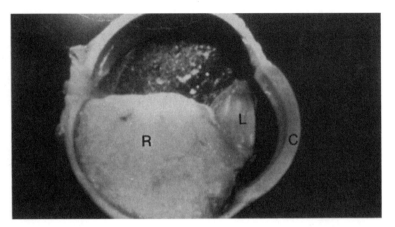

Figure A–15. Photograph of bisected eye with retinoblastoma (R) pushing the retina forward into the posterior chamber of the eye. The lens is at (L), the cornea at (C).

A.9   Retinoblastoma

Retinoblastomas are malignant eye tumors of infancy and childhood. Tumors develop from fetal retinal neuroblasts that have become malignant due to a deletion of a tumor suppressor gene known as Rb-1. Retinoblastomas occur in most instances at random without any obviously identifiable risk factors, but a minority of cases are familial. In spontaneous cases the tumors are usually unilateral, but in those that are familial they may be bilateral. The tumors occur in the posterior chamber of the eye and push the retina forward into the vitreous (Figure A–15). They are composed of masses of small darkly staining, rapidly proliferating retinoblastoma cells that differentiate into rosettes typical of developing normal neuroblasts. The tumors metastasize, but cure can be achieved with appropriate therapy in a large number of cases.

A.10   Neuroblastoma

Neuroblastoma, as the name implies, is derived from neuroblasts in the autonomic nervous system, in the adrenal medulla or the central nervous system of children. Most commonly it occurs in the adrenal medulla (Figure A–16). The tumors are composed of masses of rapidly growing, highly invasive, dark-staining cells that may form rosettes resembling primitive neuroblasts of the embryo (Figure A–17a). They invade and destroy the adrenal gland. The liver and other contiguous organs may be invaded directly. Distant metastases occur early and involve the lymph nodes, lung, liver, and bones, especially the skull.

In the embryo, neuroblasts differentiate into adult nerve cells termed ganglion cells. In some tumors the neuroblastoma cells also differentiate into the same

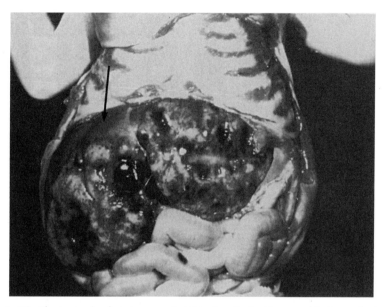

Figure A–16. Photograph of abdomen of child who died with neuroblastoma that developed in the right adrenal gland. Note massive tumor that has infiltrated the liver (arrow). The absence of fat indicates cachexia.

terminally differentiated ganglion cells in which case the tumors are called ganglioneuroblastoma (Figure A–17b). It is apparent from studies of neuroblastoma in vitro and in vivo that this tumor is a caricature of the process of development of nerve cells.

Neuroblastoma may undergo spontaneous regression, in which case all of the neuroblastoma cells either differentiate into ganglion cells or disappear. Spontaneous regression is rare to the point that it is a medical curiosity. If infants are to be saved from this disease, early and effective therapy must be undertaken. Interestingly, very young children with the disease have a better prognosis than older children.

Much experimentation has been done on spontaneous and induced differentiation of neuroblastoma cells.

A.11   Wilms tumor (nephroblastoma)

Wilms tumor is a rare, rapidly growing tumor of infants or children under the age of five years. Hematuria is rare because the renal pelvis (the expanded proximal end of the ureter that receives urine) is not involved. It usually presents as a large abdominal mass that invades locally and metastasizes widely. It is briefly discussed here not because of its clinical importance, but because of its potential for new types of therapy.

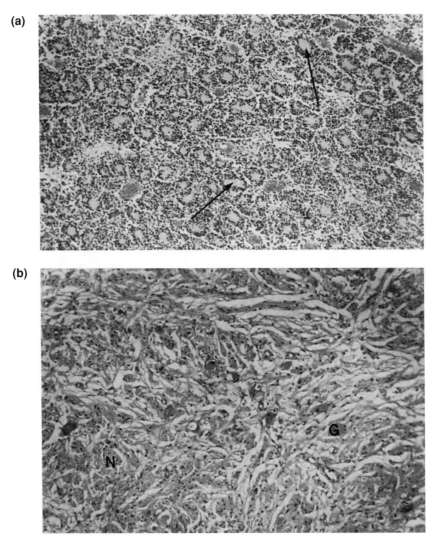

Figure A–17. (a) Photomicrograph of a neuroblastoma of the adrenal gland illustrating the abortive differentiation and organization of primitive neuroblasts into structures resembling embryonic spinal cords (arrows). It is of interest that these cells palisaded around central accumulations of nerve fibers, migrated to the center of the embryonic organ, divided, and then migrated back out, and if the environment was appropriate would differentiate into mature ganglion cells. (b) Part of ganglioneuroma to show tangles of nerve filaments (N) and ganglion cells (G) derived from tissue illustrated in Figure A–17a.

The tumor is composed of masses of undifferentiated spindle cells and epithelial elements arranged in tubules and even in structures resembling primitive glomeruli (small structures in the malpighian body of the kidney made up of capillaries). Occasionally muscle or cartilage is present as an integral part of the tumor.

The kidney is derived from the mesoderm, which under the influence of the ureteric bud differentiates into the tubules and glomeruli of the kidney. Because the stem cells of the tumor are multipotential, the normal counterpart must be the multipotent mesenchymal cells capable of forming the adult kidney. Nephrogenesis is still in progress at birth. Therapy includes surgical removal of the tumor with irradiation and chemotherapy. The tumors are responsive and significant cures can now be achieved.

This is an ideal tumor on which to test the idea that the appropriate embryonic field should regulate its closely related cancer.

A.12 Sarcomas

Sarcomas are malignant tumors that originate in mesenchymal tissues such as muscle, bone, fibrous connective tissue, cartilage, or fat. (Color Plates 12 and 13). Although they can and do develop in the connective tissue stroma of epithelial organs, they are a much less common malignancy than the carcinomas of those organs. To emphasize the point, a liposarcoma can occur in the breast but it is much less common than an adenocarcinoma of the breast. Sarcomas developing in the stroma of deep-seated organs such as stomach or colon are difficult to distinguish preoperatively from their respective carcinomas and are usually discovered at surgery.

The mesenchymal tissues originate from multipotent mesodermal precursors of the embryo. Similar stem cells are most likely present in all tissues. This potential is parceled out to produce fibroblasts, lipoblasts, chondroblasts, osteoblasts, and hematoblasts, which in turn develop into the mature tissues characteristic of their lineages. Whether or not the potentials of cells of these lineages are irreversibly reduced, as appears to be the case with most epithelial cells on differentiation, is not known. The fully differentiated fibrocyte on injury synthesizes DNA, divides, and aids in repair. Thus, terminal differentiation as it occurs in many epithelial tissues may not occur in the mesenchymal ones.

Sarcomas in general are soft fleshy tumors with a pseudocapsule often penetrated by invading sarcoma cells. Thus, wide excision of the lesions is required at the time of first therapy to avoid local recurrence. The sarcoma cells are able to excite a stroma that often has thin-walled vascular spaces, which are easily penetrated by invading cells, resulting in widespread metastases. Hemorrhage can result from these sinusoidal capillaries.

Histologically, sarcomas are classified into several groups: malignant fibrous histiocytoma, fibrosarcoma, liposarcoma, chondrosarcoma, leiomyosarcoma, osteosarcoma, synovial cell sarcoma, Ewing sarcoma, etc. The nomenclature used by pathologist and clinicians is often confusing, and often reflects our inadequate understanding of the basic biology and histogenesis of these tumors. Many clinical studies have been conducted, but the progress has been slow: the results of surgical treatment combined with chemotherapy are, in most instances, quite disappointing.

(a)

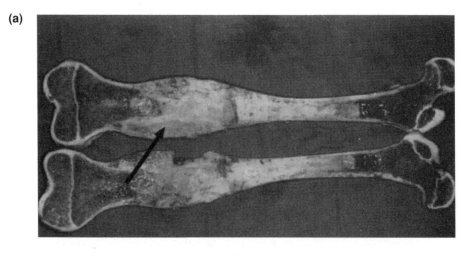

(b)

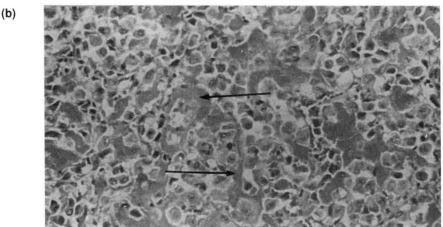

Figure A–18. (a) Osteosarcoma in the shaft of a femur of a young person. The marrow has been invaded, the cortex thinned, and the tumor is growing in the soft tissues outside of the periosteum (arrow). (b) Photomicrograph of an osteogenic sarcoma that could be subclassified as an osteoblastic tumor because masses of osteoid (bone matrix, arrows) are interspersed among myriads of anaplastic sarcoma cells.

The best results have been recorded with some sarcomas of childhood, most notably rhabdomyosarcomas (striated muscle tumors), Ewing sarcoma, and osteosarcomas (Figure A–18). Sarcomas of adults, with the exception of some well-differentiated chondrosarcomas and liposarcomas, cannot be adequately resected surgically, and are generally resistant to chemotherapy.

A.13   Lymphoma and leukemia

Lymphomas are neoplasms of lymphoid cells typically originating in lymph nodes or other lymphocyte-containing organs such as spleen, bone marrow, or thymus.

Lymphomas fall into two major groups: Hodgkin lymphomas and non-Hodgkin lymphomas. Non-Hodgkin lymphomas are classified according to their cell of origin as B cell lymphomas or T cell lymphomas. Neoplasms of histiocytic and other support and antigen-presenting cells are included clinically together with lymphomas.

Major advances in the classification of lymphomas have been made by applying modern techniques of laboratory medicine, such as immunocytochemistry, flow cytometry, cytogenetics and molecular biology. Using immunocytochemical techniques, scientists have discovered that normal lymphocytes can be classified as T or B cells. T and B cells, like their malignant equivalents, express lineage-specific clusters of designation (CD) on their surface. Using monoclonal antibodies to these CDs (more than 150 of which have been identified so far) it is possible to state whether the lymphoma cells are of T or B cell origin. Immunocytochemistry applied to cell suspensions, which are allowed to flow through a flowcytometer permits the separation of malignant T and B cells, which may then be studied with molecular procedures. It is also worth notice that the maturation of normal lymphocytes is characterized by stage-specific activation of genes, resulting in the expression of specific antigens on the cell membrane of these cells. These antigens are expressed in a permanent form on some malignant cells, allowing the scientist to determine the specific stage of development at which the malignant transformation of lymphoid cells has taken place.

Cytogenetic and molecular biology studies have further validated the concepts based on immunohistochemistry and have shown that specific subtypes of lymphoma have specific genotypic profiles. These express unique oncogenes or have specific chromosomal (karyotypic) changes. Current chemotherapy of lymphomas depends on a precise identification of the malignant cells, and modern therapy is unimaginable without cell biology and molecular biology data.

The classification of lymphomas is based on immunocytochemical identification of neoplastic cells as belonging to either B or T cell lineage. Subsequent subtyping is based on the immunocytochemical of flowcytometric demonstration of early or late differentiation antigens on the surface of malignant cells. Thus, a lymphoblastic large cell lymphoma will express early differentiation antigens, whereas a lymphoma composed of small well-differentiated lymphocytes will express antigens typical of fully differentiated lymphocytes. Hodgkin disease is thought to be a B cell disease, but its precise classification is still defying scientific understanding. Histiocytic tumors involve lymph nodes and are clinically grouped with lymphomas, but genetically and phenotypically they are neither T nor B cell lymphomas.

Malignant lymphoid cells can enter the blood and this will result in leukemia, a condition literally meaning "white blood." It is worth remembering that lymphocytes are just one of several white blood cells in the normal blood. The bone marrow precursors of other white blood cells, that is, neutrophils, eosinophils, basophils, and monocytes, collectively known as myeloid cells, can also become malignant and give rise to leukemia. Like lymphomas, leukemias are currently studied by immunocytochemistry, and modern techniques of cytogenetics and molecular biology. The discoveries made about the nature of genetic changes in various leukemias are too numerous to list and thus, let us mention only the first major cytogenetic break-through – the discovery of the Philadelphia chromosome (Ph), found in 90 percent of chronic myelogenous leukemia (for details see Chapter 4).

Leukemias are clinically classified as either acute or chronic. Previously, before the era of modern chemotherapy, acute leukemia was lethal within three to six months, whereas the patients who had the chronic form lived one or more years after diagnosis. Depending on the nature of malignant cells that spill over into the blood one can classify leukemia as lymphoid or myeloid. Accordingly, four major forms of leukemia are recognized: acute lymphoblastic leukemia (ALL), chronic lymphocytic leukemia (CLL), acute myelogenous leukemia (AML) and chronic myeloid leukemia (CML). Each of these leukemias has distinct clinical and bio-logic features. For example, ALL is predominantly found in children, whereas CLL is a disease of older men and women. ALL responds well to modern chemother-apy, and this previously lethal disease can be cured today in more than 60 percent of cases. On the other hand, CLL responds poorly to chemotherapy, but patients live with their disease five, six, and even ten years. As mentioned previously, ALL and CLL may be associated with lymphoma and are often accompanied by lymph node enlargement. AML and CML involve primarily the bone marrow, but can colonize other organs, such as the spleen and liver and cause their enlargement. As the malignant cells proliferate, the normal hematopoietic elements are displaced from the marrow by leukemia cells. As a result, the final reserves for producing functional hematopoietic elements are lost, and the patient presents with the pic-ture of bone marrow failure. These signs and symptoms will include the inabil-ity to mount effective defense mechanisms against bacterial, fungal, or parasitic infections, and inability to stop capillary bleeding as a result of displacement of megakaryocytes (petechial hemorrhages – small purplish hemorrhagic spots – can occur in the skin and mucous membranes, with bleeding around the teeth, for example). As the erythroid elements are displaced, anemia results. Finally, leukemia cells will grow in the stroma of any organ, and although it is difficult to pinpoint symptoms directed at the particular organs, the organs are all enlarged, can give a sense of heaviness, and the tremendous number of proliferating leukemic cells can result in cachexia.

Glossary

This glossary is *not* a dictionary. Listed terms have been defined primarily in the context of how they are used in this text. It is provided as a convenience to students and others who read this book. A medical dictionary obviously is of greater value in that it provides several definitions of terms when appropriate, etymology and pronunciation.

acute toxicity   A treatment-related toxicity that occurs during therapy or shortly after, and usually reverses if treatment is interrupted

adduct   Covalent product formed from the reaction of DNA with an activated carcinogen.

adenocarcinoma   Malignant tumor composed of glandular or ductal epithelial cells; most often found in the large intestine, uterus, stomach, pancreas, and lungs.

adenoma   Benign tumor composed of glandular or ductal epithelium; most often found in the large intestine in form of so-called tubular adenoma.

adhesion plaque   Fibroblasts adhere to a substratum at centers known as adhesion plaques.

adipose   Refers to fat or to fatty cells or tissues.

adjuvant therapy   The additional therapy that is timed after a definitive procudure to remove a tumor in order to eradicate disease that is not clinically detectable but is very likely present, based on the natural history of similar grade and stage of disease in previous patients.

Adriamycin   Trademark name for doxorubicin (q.v.).

agonist   Pharmaceutical agent that stimulates a specific physiologic response.

alkylating agent   Highly reactive cytotoxic chemical that transfer alkyl (straight or branched chain hydrocarbons) groups to nucleotides forming DNA adducts (q.v.). Cells that are rapidly dividing are most vulnerable to the toxic effects of these agents.

aminoglutethamide   Inhibitor of adrenocortical synthesis of hormones such as estrogen.

amitrole   Toxic (carcinogenic) herbicide not used on edible plants.

anaplasia   Lack of differentiation or abnormal differentiation of malignant neoplastic cells, usually associated with rapid tumor growth and poor clinical prognosis.

anaplastic   Adjective describing tumor cells that show very little or no differentiation.

androgen   Substance with the potential for inducing masculine characteristics.

anemia   Reduction in the number of red blood cells.

aneuploid   Abnormal chromosome number.

angiogenesis   The biological process of establishing new blood vessels. In embryonic development angiogenesis is normal and essential; however in tumors, it is permissive for further neoplastic growth.

angiogenesis inhibitors   Drugs designed to inhibit any of several biological processes required for establishing new blood vessels.

antagonist   A chemical that blocks the action of an active drug without itself causing any physiological response.

antiangiogenesis therapy   Treatment designed to interfere with the establishment of new blood vessels, such as those that supply tumors with essential nutrients for continued growth.

antigenic   Adjective describing a substance having the competence to induce the formation of a specific antibody.

antimetabolite   Drug that interferes with the availability or utilization of biochemicals essential for cell survival or proliferation.

apoptosis   Programmed cell death; a mode of biologically regulating cell numbers.

ara-C   Chemotherapeutic agent, also known as cytarabine, which inhibits DNA synthesis by the inhibition of DNA polymerase.

aralkylating agent   Cytotoxic substance that transfers aryl (aromatic hydrocarbons) groups to DNA.

astrocytoma   The most common brain tumor, composed of glial cells known as astrocytes.

ataxia telangiectasia   Hereditary disorder characterized by lack of muscular coordination, immunodeficiency, inadequate DNA repair, and an increased prevalence of malignancies.

atrophy   Loss of size in a tissue or organ; shrinkage and death of tissue due to deprivation of a trophic factor.

autonomous   State of a tumor when it no longer responds to its usual control signals.

bacillus Calmette-Guérin (BCG)   Vaccine produced from a particular strain of *Mycobacterium bovis*. BCG is also known as *Bacille Calmette-Guérin*.

B-cell growth factor   Substance derived from T-cells that stimulates growth of B-cells in vitro.

benign   Not malignant. Benign tumors do not invade, are not metastatic, and have a limited growth potential, thus having a clinically favorable prognosis.

bioassay Biologic assay for determination of drug or chemical effect on living organisms or living cells or tissues.

blastoma Tumor comprised of cells, many of which appear less than fully differentiated, and thus resemble cells of an embryo.

bleomycin Chemotherapeutic agent that functions by binding to and cutting DNA with resultant inhibition of DNA synthesis. Bleomycin is obtained from the fungus-like bacterium *Streptomyces verticillus*.

Bloom syndrome Genetic disorder characterized by dwarfism, redness of the face, chromosomal abnormalities, and an elevated prevalence of leukemia.

bolus A drug ready to swallow.

Burkitt lymphoma Most common childhood malignant tumor of central Africa composed of neoplastic B lymphocytes that carry a specific chromosomal translocation. The malignancy displays a dysregulation of the *myc* oncogene.

cachexia Weak and wasted appearance of terminally ill cancer patients.

camptothecins Chemotherapeutic alkaloids from the Chinese woody plant *Camptotheca acuminata* (family Nyssaceae) thought to inhibit topoisomerase I.

cancer A general term for malignant tumors derived from the Latin term *cancer*, meaning crab. The term refers to the infiltrative growth of tumors.

carboplatin Chemotherapeutic agent similar in its action to cisplatin (q.v.).

carcinogen Chemical substance, virus, or radiation that produces or incites cancer.

carcinogenesis Production of cancer.

carcinoma Malignant tumor of epithelial origin. It can occur in several forms, such as squamous cell carcinoma, transitional cell carcinoma, basal cell carcinoma, adenocarcinoma, etc. Most often found on the skin, but also in the gastrointestinal tract, respiratory tract, female genital tract and endocrine glands.

carcinoma in situ Malignant tumor identified in an early stage before it has demonstrated its competence to transgress the basement membrane and become invasive. The tumor is confined to the epithelial cells of its origin.

caricature An exaggeration of normal features.

carotenoid Fat-soluble yellow or red pigments; examples include beta-carotene, which can be converted to vitamin A in the body, and the red pigment lycopene, found in tomatoes.

castration Removal of the testes or ovaries.

cathepsin B Hydrolytic enzyme that digests proteins and peptides.

C. elegans Scientific shorthand for *Caenorhabditis elegans*, a small free-living roundworm that is of particular interest to developmental biologists who wish to trace the history of every single cell.

cell cycle A regulated series of stages (G1-S-G2-M) resulting in cell growth and cell division. The stages, G1 for Gap 1, S for synthesis, G2 for Gap 2, and M for Mitosis are repeated (cycled) as long as the cells are dividing.

cellular heterogeneity   A malignant tumor is generally thought to be composed of a diversity of cells with differing growth rates, gene products, and vulnerability to chemotherapeutic agents – this diversity is known as cellular heterogeneity.

chemotherapy   Treatment of cancer with chemical agents (drugs).

chlorambucil   Alkylating chemotherapeutic agent of a class of drugs known as mustards.

chlorodeoxyadenosine (2-chlorodeoxyadenosine)   A chemotherapeutic agent that inhibits DNA synthesis and DNA repair.

chondrosarcoma   Malignant tumor composed of cartilaginous cells often found in bones.

choriocarcinoma   Malignant tumor composed of placental trophoblast cells.

chromosome banding   Differential staining along the length of a chromosome that permits positive identification of each individual chromosome as well as the identification of chromosomal aberrations such as deletions or translocations.

chronic lymphatic leukemia   Chronic leukemia of the lymphocytic type, usually with abnormal proliferation of small lymphocytes. This leukemia is rare in young people.

chronic myelogenous leukemia   Chronic leukemia, characterized by differentiated cells of the myeloid granulocytic type, almost always associated with the Philadelphia chromosome (q.v.). Also known as chronic myeloid leukemia.

chronic toxicity   a therapy-related toxicity that occurs long after completion of a treatment, or as a result of the total dose of treatment administered over a long period of time.

cisplatin   Chemotherapeutic agent containing platinum used in the treatment of solid cancers (e.g., testicular and ovarian cancers).

clonal expansion   Production of a population of cells descended from a single cell.

collagenase   Generic term for one of several enzymes with the competence to degrade collagen. Collagens are specific fibrous proteins associated with connective tissue, skin, tendon, bone, cornea, cartilage, and basement membranes.

colostomy   Opening of the colon (large intestine) to the exterior of the body formed as the result of a surgical procedure.

corticosteroids   Hormones elaborated by the adrenal cortex; the two groups are glucocorticoids (q.v.), which influence the metabolism of carbohydrates, fats, and proteins, and mineralocorticoids (q.v.), which affect electrolytes and water balance.

crocidolite   Blue asbestos associated with lung cancer and mesotheliomas.

cross-resistance   The development of resistance to one treatment because of the biochemical response or cellular selection due to prior exposure to a different treatment.

cruciferous vegetables   Plants of a family that include broccoli, Brussels sprouts, cabbage, cauliflower, kale, mustards, radishes, and turnips.

Cushing syndrome   Overproduction of glucocorticoids by the adrenal cortex.

cyclophosphamide Alkylating chemotherapeutic agent included within the class of drugs known as mustards.

cyproterone acetate Synthetic hormone with antiandrogen properties.

cystoscope Instrument for the visual examination of the interior of the bladder.

cytochrome P-450 Intracellular protein, similar to hemoglobin, that occurs in most cells and is involved in oxygenation of many reactions; P refers to "pigment" and "450" refers to the 450 nm absorption maximum in a spectrophotometer.

cytoreductive surgery Surgical procedure that results in diminished tumor mass; also known as debulking.

cytoreductive therapy Therapy that results in diminished tumor mass.

cytotoxic Poisonous to cells; used in chemotherapy to describe a drug or agent poisonous to cancer cells.

cytotoxic T-cells Lymphocytes that function in graft rejection and the killing of virus-infected cells and tumor cells.

debulking Elimination of much of a tumor to enhance subsequent therapy because of reduced tumor load (see cytoreductive therapy).

dedifferentiation Loss of mature or organ specific properties that the cells have acquired upon differentiation and maturation.

deletion Loss of genetic material from a chromosome.

desmoplasia Proliferation of connective tissue stroma within a malignant tumor. It accounts for the "rock-hard" consistency of many carcinomas, most notably breast carcinoma and prostatic carcinoma.

determination Commitment of cells to a particular differentiation pathway. Determined cells are not visibly differentiated.

dexamethasone Synthetic glucocorticoid (q.v.) drug.

dicotyledon Plant having two seed leaves (cotyledons); the group includes most trees that lose their leaves and most herbs and shrubs.

diethylstilbestrol (DES) Synthetic estrogen that may be used for treating the symptoms of prostate cancer.

differentiation Acquisition of the mature phenotype of cells by a process of selective gene activation resulting in specialized tissue or organ specific function and features. It takes place normally during embryonic development but also to a limited extent in many tumors.

differentiation therapy Cancer therapy with agents that induce expression of mature (and benign) characteristics.

dimer The association of two molecules, either homodimer (two identical molecules) or heterodimer (two different molecules).

diverticular disease Sacs or pouches of the large intestine as a result of weaknesses in the muscular wall of that organ.

DNA methylation Transfer of methyl groups to DNA. Extensively methylated DNA is transcriptionally silent; an important mode of gene regulation and repression.

dormancy   State of a malignancy in which component cells do not proliferate but remain quiescent and clinically unapparent.

dosage regimen   The dose and dose frequency (schedule) of a drug therapy.

dose-limiting toxicity (DLT)   The adverse effect(s) of drug therapy or radiotherapy that is so severe and dangerous for the patient that the dose cannot be increased further.

doxorubicin   Chemotherapeutic agent of a class of compounds known as anthracyclines derived from *Streptomyces peucetius* that interfere with DNA synthesis.

drug delivery   The transfer of a drug from the site where it enters the body to its target in the tumor.

duplication   Genetic material that is repeated in a chromosome.

dysplasia   Abnormalities in cancer cell morphology resulting from abnormal differentiation or maturation of tissues.

E-cadherin   A specific calcium-dependent epithelial cell adhesion molecule.

ectopic   a normal physiological process that takes place in the wrong anatomic site or at the wrong developmental time.

endocrine therapy   Blocks action of tumor trophic factors leading to death of tumor progenitors but not tumor stem cells; also known as endocrine ablation, hormonal therapy, antihormonal therapy, antiendocrine therapy.

endogenous   Arising from within the body.

endothelium   Cells that line the inside of blood vessels and form capillaries; specialized endothelium, such as that found at the blood brain barrier, protects the central nervous system from toxic chemicals.

epidemiology   Study of factors that influence the prevalence and distribution of disease.

epigenetic   Differentiation and morphology due to selective gene activation, expression or repression; epigenetic changes occur without altered DNA sequences. Because changed genetic activity is not due to structural alteration of genes, it is potentially reversible.

epithelial cell   Cells that cover and line the internal surfaces of the body. Epithelial cells line the hollow organs of the respiratory, alimentary, and urogenital tracts. The parenchyma of all internal organs except the heart and the brain are epithelial.

epitopes   Antigenic determinants.

Epogen   Trademark name of epoetin alpha. Epoetin alpha is human erythropoietin produced by recombinant DNA technology and is used to stimulate red blood cell production.

Epstein-Barr herpesvirus (EBV)   Specific herpesvirus that has been linked with Burkitt lymphoma (q.v.). EBV is also associated with infectious mononucleosis and nasopharyngeal carcinoma.

estrogens   Female sex hormones produced by the ovary responsible for secondary sexual characteristics and monthly changes in the lining of the vagina and uterus.

etiology   The study of the causes of diseases.

etoposide Chemotherapeutic agent that is a semisynthetic derivative of podophyllotoxin (obtained from the mayapple plant, *Podophyllum peltatum*). The drug functions by affecting the topologic state of DNA.

euploid Normal chromosome number for a species or an exact multiple of the haploid number. The haploid number for humans is 23; thus, any exact multiple of 23 is euploid.

extravasation Exit of cells or fluid from a capillary or lymph vessel. Extravasation of cancer cells is a prerequisite for metastasis.

familial adenomatous polyposis Genetic condition that leads to the production of thousands of small benign tumors (polyps) in the colon which have a high potential to become malignant.

Fanconi syndrome Rare genetic disease with a deficiency in blood cells, inadequate bone marrow development, brown facial spots, multiple muscular and skeletal abnormalities, and genital and urinary malformations.

fecal occult blood test Assay to reveal blood in the feces which may result from colon cancer.

fibroblast Connective tissue cell producing collagen and other extracellular matrix components of the connective tissue.

fibrosarcoma Malignant tumor of soft tissues composed of fibroblasts.

floxuridine Chemotherapeutic agent that is metabolically altered to an intermediate with the same properties as 5-fluorouracil (q.v.); a drug that inhibits DNA synthesis.

fluorouracil (5-fluorouracil, 5-FU) Chemotherapeutic agent that inhibits DNA synthesis.

flutamide Synthetic antiandrogen used in prostate cancer treatment.

fractionation The division of a treatment dose into small increments that are administered with only short intervals of separation, in contrast to giving the total dose at one time.

free radical Highly reactive constituent (component) of a molecule with an unpaired electron.

genomics The study of the entire genome and its expression in normal development and cancer; with sequencing of the entire genome of a species, microarray analysis (q.v.) permits characterization of expression levels of many thousands of genes simultaneously.

germline Refers to sperm or eggs and the cells which give rise to them, not somatic (body) cells. **germline mutations** occur in the germline and are transmitted to progeny in contrast to somatic mutations (q.v.) which are not transmitted to progeny.

Giemsa banding Chromosome banding (q.v.), with a specific mixture of stains.

glioma Tumor of the brain or spinal cord composed of glial cells (non-neural supporting cells of the central nervous system).

glucocorticoid Adrenal cortical hormones involved in the metabolism of carbohydrates; see corticosteroids.

GM-CSF   Granulocyte macrophage colony stimulating factor.

granulomatous reaction   Response to chronic inflammation that is characterized by the presence of granulomas. Granulomas are usually composed of aggregations of mononuclear inflammatory cells.

growth fraction   The proportion of proliferation-competent cells in a renewing tissue that are progressing through the cell cycle at any given time.

hairy cell leukemia   A chronic leukemia. The blood of these patients has many large mononuclear cells with cytoplasmic projections that give the "hairy" cell appearance.

hemangioblastoma   Nonmalignant tumor of proliferating blood vessel cells and angioblasts found in the brain, spinal cord, or retina.

hemostasis   Stoppage or blockage of blood flow.

hepatoma   Malignancy of the liver, also known as hepatocellular carcinoma.

histiocytoma   A tumor containing macrophages.

Hodgkin's disease   Malignant neoplasm involving the lymph nodes.

homeostasis   The condition of dynamic balance between opposing forces and mechanisms.

homeothermic   Refers to an animal that maintains constant body temperature despite changes in the environment. The animals are referred to as warm blooded, in contrast to cold-blooded animals.

homogenate   Tissue minced into extremely fine subcellular particles.

homologue   The other chromosome of a pair; only the y chromosome lacks a homologue in a metaphase spread of chromosomes.

17-alpha hydroxyprogesterone   Intermediate compound in the formation of male and female hormones.

hyperchromatic   A term used in histology to describe increased staining of cells.

hyperplasia   Abnormal increase in the number of normal cells in a tissue.

hypophysectomy   Surgical removal of the hypophysis (pituitary gland).

hypoxia   Diminished supply of oxygen to cells and tissues.

I^{131}   Radioactive iodine.

initiation   Initial preneoplastic change in the genetic material of cells caused by exposure to a chemical carcinogen. Overt cancer occurs with subsequent exposure of the initiated cells to the same or another carcinogen.

in situ   Localized, in position. A term usually used for early carcinoma that has not breached the basement membrane and is limited to the epithelium.

interferon   Member of a group of glycoproteins with antiviral activity. Interferon alpha, an interferon produced by leukocyte cultures, is used in the treatment of hairy cell leukemia.

interleukin   Interleukins are nonantibody proteins; interleukin-2 (IL-2) is produced by T-cells and stimulates specific T-cell replication; it is used in the treatment of solid tumors. Interleukin-3 (IL-3) is produced by T-cells and stimulates production of bone marrow blood cells and lymphoid stem cells.

intrathecal injection   Injection of a drug under the sheath that covers the spinal cord.

intravasation   Entrance of cancer cells into capillary or lymph vessels.

invasion   Movement of cancer cells into a contiguous tissue.

in vitro   Term used to indicate that a cell manipulation is outside the body. Originally, the manipulations were performed in glass dishes (hence, "in vitro") but now most studies are carried out in plastic laboratory dishes.

in vivo   In the living body or organism. An experimental procedure using an intact live animal.

inversion   Chromosomal abnormality formed by a double break with subsequent reuniting of the broken parts with the sequence of genes reversed in their linear order in relation to the remainder of the chromosome. Inversions can be detected by chromosomal banding (q.v.).

ionizing radiation   Energy transmitted by electromagnetic waves with the potential of causing ionization.

Kaposi sarcoma   Malignant tumor composed of endothelial cells typically found in the skin, but also in internal organs. It has been linked to infection with human herpesvirus type 8 (HHV-8) and is found with increased frequency in male homosexuals infected with the human immunodeficiency virus (HIV).

karyotype   Photograph of metaphase chromosomes arranged by pairs in descending order of size.

keratin pearl   A spherical mass of cells comprised of concentric layers containing keratin, found in some squamous cell carcinomas.

ketaconazole   Antifungal drug that in high doses reduces testosterone production.

latency   Period of time between the application of carcinogen and the appearance of cancer.

leiomyoma   Benign tumor composed of well differentiated smooth muscle cells frequently found in the uterus.

leiomyosarcoma   Malignancy of neoplastic smooth muscle cells and their precursors. Smooth muscle is composed of involuntary muscle cells (e.g., the muscle cells of the uterus and the intestines).

leukemia   Neoplastic disease of hematopoietic precursor cells in the bone marrow or the lymph nodes characterized by the appearance of malignant white cells or their precursors in the blood. It may involve either the myeloid or lymphoid cell lineages. It occurs in several distinct clinical forms such as acute myelogenous leukemia, chronic myelocytic leukemia, acute lymphoblastic leukemia, chronic lymphocytic leukemia, hair cell leukemia, etc. Lymphocytic leukemias are related to lymphomas.

Leukine   Trademark name of sargramostim, which is a granulocyte/macrophage colony stimulating factor (GM-CSF) produced with recombinant DNA technology. GM-CSF enhances blood cell production in the bone marrow.

leukocytosis   Increase in the number of leukocytes (white blood cells) in the blood.

leuprolide Synthetic gonadotropin-releasing hormone used to relieve symptoms of prostate cancer.

ligand A chemical substance that binds with high affinity and specificity to a receptor that converts binding into a biochemical signal.

lignin Polysaccharide component of plant cell wall.

lipoma Benign tumor composed of well-differentiated fat cells (lipocytes).

liposarcoma A malignant tumor derived from fat precursor (lipoblastic) cells.

log-cell kill The mathematics of fractional reductions in cell number, where the number of cells killed by therapy depends on the number of cells being treated but the proportion of cells killed is constant and independent of initial cell number. Each 10-fold reduction in viable cell number equals 1 log-cell kill.

Lucké renal adenocarcinoma Kidney cancer that is the most common malignancy of the North American Leopard Frog, *Rana pipiens*.

lumpectomy Surgical procedure, also known as limited or localized resection, during which only the tumor and a small amount of surrounding normal tissue are removed from the breast. Contrast with radical mastectomy (q.v.)

luteinizing hormone Anterior pituitary hormone important to corpus luteum formation in the female and testicular interstitial cell development in the male.

luxury molecules Gene products that are specific to specialized cells.

lymph node dissection The surgical removal of lymph nodes that drain a tissue area containing cancer.

lymphocyte Nongranular mononuclear white blood cells. There are two groups which include B- and T-lymphocytes. T-lymphocytes are involved in cellular immunity.

lymphocytopenia Decrease in the number of lymphocytes in the blood.

lymphokine-activated killer cells (LAK cells) Cytotoxic T-cells and natural killer cells activated by culture in interleukin-2 (q.v.).

lymphoma Malignant tumor of lymphoid tissues, most often involving the lymph nodes, spleen, thymus, and the bone marrow. Malignant lymphoid cells may spill into the blood and give rise to acute or chronic lymphocytic (lymphoblastic) leukemia.

lysosome Cellular structure that contains hydrolytic enzymes.

malignant An adjective, which when modifying "tumor" or "neoplasm," indicates a tendency to grow relentlessly, metastasize, and kill the host.

malignant conversion Changes to a normal cell that result in a cancer cell via initiation, promotion, and progression.

malignant melanoma See melanoma.

mammogram X-ray of the breast.

mastectomy Surgical removal of the breast. If the surgeon removes, in addition to the entire breast, the axillary lymph nodes and chest wall tissue the operation is called radical mastectomy.

maximum tolerated dose (MTD) The highest dose of a drug or radiation that does not cause life-threatening or irreversible toxicity.

medroxyprogesterone   Progestational agent, that is, a substance that stimulates the endometrium as does progesterone.

megestrol acetate   Synthetic drug with an effect similar to progesterone.

melanocyte   Pigmented skin cell with the competence to produce melanin.

melanoma   Pigmented malignant tumor, the most lethal of skin cancers, arising from melanocytes (pigment cells).

melphalan   Alkylating chemotherapeutic agent, also known as L-PAM (L-phenylalanine mustard), and a member of the class of drugs known as mustards.

meningioma   Benign tumor of the membranes that surround the brain and spinal cord.

menopause   Event that marks the permanent termination of menstruation; pre-menopausal refers to the time before menopause and postmenopausal refers to the time after menopause.

mercaptopurine   (6-mercaptopurine) Chemotherapeutic agent that inhibits purine synthesis, which in turn blocks DNA synthesis.

mesenchymal   Adjective describing cells and extracellular matrix of mesodermal origin surrounding the epithelial components of the tissues and organs.

mesothelioma   Malignant tumor arising from the mesothelial lining of the pleural, peritoneal, and pericardial cavities often associated with exposure to asbestos.

metaplasia   Change in cell type of a mature cell to another cell type that is not found at that site, e.g. replacement of esophageal squamous epithelial cells by intestinal-like mucus producing cells.

metastasis   Cancer growth at sites anatomically distant from the original tumor.

methotrexate   Chemotherapeutic agent, a folic acid antagonist, that inhibits synthesis of DNA by interfering with the enzyme dihydrofolate.

microarray analysis   A system of characterizing the expression of hundreds to thousands of genes arrayed on a DNA chip. Specific sequences of DNA molecules ("probes") are attached to a substrate in a specific order. Fluorescent labelled mRNA from cells or tissues is hybridized to the DNA probes of the substrate. Expression of specific genes can then be characterized quantitatively and qualitatively by their location on the DNA chip in what is known as a "gene expression profile."

micro-metastatic disease   Cancer that is growing in tissues and organs outside of the site of origin of the primary tumor, but has not grown large enough to be detectable with clinical diagnostic techniques.

microsome   Subcellular fragment of the endoplasmic reticulum formed after disruption of the cell.

mineralocorticoid   Hormones of the adrenal cortex important in retention of water and sodium with reduction of potassium; see corticosteroids.

minute   Adjective to describe an abnormally small chromosome.

mitogen   Something that will provoke cell proliferation.

modality   A treatment concept. Surgery, radiotherapy, cytotoxic chemotherapy, and targeted therapy are four modalities available in 2005.

monoclonal   Population of cells derived from a single cell.

monoclonal antibody   An immunoglobulin protein encoded by a single rearranged Ig gene that binds to a single antigen epitope.

morbidity   A disease state; also denotes the frequency of complications of treatment.

mortality   Death; also denotes fatal outcome of a disease or death rate from a disease.

mucosa   Inner lining of hollow organs, composed of a surface epithelium, basement membrane, and underlying connective tissue (lamina propria).

mucositis   Infection or inflammation of a mucosa (q.v.).

multimodality therapy   Therapy involving more than one therapeutic agent; the clinical strategy of combining the treatment modalities (surgery, radiotherapy, cytotoxic chemotherapy, targeted therapy) in a way consistent with its biological effects on the cancer for the optimal benefit to the patient.

mummichog   Common name of the fish *Fundulus heteroclitus.*

myeloid   Refers to bone marrow.

myeloma   Tumor of bone marrow.

myelosuppressive   Inhibition of blood- and platelet-forming cells in bone marrow by a drug or other toxic agent.

natural killer cells   Large lymphocytes produced in the bone marrow, distinct from T-cells, which kill virus-infected and tumor cells.

necrosis   Cell death.

neoplasm   Synonym for tumor; abnormal continuously proliferating growth serving no useful function to the host.

neoplastic   Adjective describing abnormal growth of tissue. If the growth is invasive with the potential for metastasis, it is cancerous.

Neupogen   Trade name for filgrastim, a granulocyte colony stimulating factor (G-CSF), produced by recombinant DNA technology, used to enhance neutrophil production.

neuroblastoma   Malignant tumor composed of neuroblasts, i.e., immature precursors of nerve cells, most often found in the adrenal, peripheral ganglia and in the brain.

neutropenia   Reduction in the number of neutrophilic leukocytes.

nevi   Plural of nevus, congenital mark or blemish, a mole.

nitrosoureas   Class of alkylating chemotherapeutic agents that inhibits DNA synthesis.

nonrenewing tissue   A tissue which cannot replace damaged or lost differentiated cells.

nucleosome   Chromatin is composed of DNA and protein; nucleosomes are bead-like packages of chromatin arranged on a "string."

nulliparity   Condition of having borne no children.

oligonucleotide   Polymer of 20 or fewer nucleotides.

oncogene   Gene with the competence, under specific circumstances, to induce normal cells to form cancer cells.

oncogenic   Having the capacity or competence to cause cancer.

orchiectomy   Male castration; removal of the testes.

osteoma   Benign tumor composed of well-differentiated bone cells (osteocytes).

osteosarcoma   Malignant tumor composed of neoplastic bone cells and their precursors, also known as osteogenic sarcoma.

P-450 cytochrome   Oxygenating enzyme found in most tissues important in the detoxification of many drugs.

palliation   Treatment that provides relief but no cure. Cancer therapies not intended to achieve cure or remission but rather to alleviate or minimize pain and suffering.

papilloma   Benign epithelial tumor with the appearance of fingerlike protrusions from the surface of the epithelium.

Pap test/smear   Method of staining shed (exfoliated) vaginal cells that are examined microscopically to detect early cancer; the test, named for Dr. George Nicolas Papanicolaou, is used as a screening method for early diagnosis of carcinoma of the female genital tract.

paraneoplastic syndrome   Symptoms caused by a cancer that cannot be attributed directly to the mass of the primary tumor or of its metastases, but rather indirectly with remote effects by a substance(s) produced by the cancer; typically symptoms include various endocrine, immunologic, neurologic, hematologic complications of the neoplastic disease.

parenchymal   Refers to the functional cell types of an organ as distinguished from connective tissue stroma.

pathogenesis   A sequence of events leading to a disease.

pathology   Study of diseases.

pedigree   A record of a person's ancestors used in studies of cancer inheritance. The term is derived from the French *pied de grue*, a "crane's foot," by analogy of the crane's foot with a stemma.

pellagra   Disease characterized by dermatitis, diarrhea, and dementia caused by a dietary deficiency of niacin.

pharmacokinetics   Concerns the fate of drugs in the body over a period of time including absorption, organ and tissue distribution, metabolism and excretion; a study of how drug concentration changes over time.

pharmacology   The study of drug action and the fate of drugs in the body.

phenobarbitol   Long-acting sedative that may be used as a promoter to enhance carcinogenicity.

pheochromocytoma   Usually benign neoplasm of the adrenal medulla that may produce symptoms caused by overproduction of epinephrine.

Philadelphia chromosome   Minute (q.v.) chromosome associated with chronic myelogenous leukemia; the chromosome was jointly described by a professor from the University of Pennsylvania, Philadelphia, and an investigator at the Fox Chase Cancer Center, Philadelphia, hence its name "Philadelphia chromosome."

phosphorylation   Process of incorporation of a phosphate group into an organic molecule.

pleomorphic   Characterized by the presence of cells with various shapes and forms; pleomorphism is a feature of malignant tumors.

pluripotent   Possessing differentiative potential for more than one kind of cell type.

polarity   Property of having an axis with different properties at the extremes; epithelial cells are said to possess polarity when their basal portions clearly differ from their apical portion; loss of polarity is a characteristic of anaplastic (q.v.) cells.

polysaccharides   Carbohydrates that yield more than two molecules of simple sugar upon hydrolysis; includes but not limited to starch, glycogen, and cellulose.

polysome   Synonym for polyribosome; a cell structure involved in peptide synthesis.

prednisone   Synthetic drug derived from cortisone used to reduce inflammation and to suppress the immune system.

prodrug   Drug inactive in the administered form but, after metabolic conversion within the body, becomes a pharmacologic active drug.

progenitors (transient amplifying cells)   The daughter cells of stem cells which divide rapidly for a finite number of population doublings. In the presence of trophic factors, cell proliferation is permanently stopped and cell maturation is induced allowing for the function of the tissue of origin.

progesterone   Ovarian and placental hormone responsible for preparation of uterine lining for implantation of the early embryo.

prognosis   Prediction of the likely, or expected, outcome of a disease.

progression   Progressive acquisition of greater malignancy with increased growth rate, invasiveness, and metastasis; said to be preceded by initiation and promotion (q.v).

promoting agent   Co-carcinogen that activates carcinogenesis after initiation usually by a different chemical carcinogen.

promotion   Expression of the malignant potential of initiated cells after exposure to the same or to a different chemical carcinogen.

prophylactic   Treatment taken to prevent disease.

prophylactic therapy   The administration of a treatment to a tissue or organ to kill any micrometastatic disease before it can grow to clinically detectable size. Chemoprevention differs from prophylactic therapy in preventing the appearance of malignant cells during oncogenesis.

proto-oncogene   Gene which, when activated by mutation or other change (i.e., becomes an oncogene), has the potential for causing a normal cell to become malignant. Normal proto-oncogenes are thought to function in normal growth and differentiation.

pyelonephritis   Bacterial infection of the kidney.

pyknosis   Condensation (shrinking) of the cell nucleus that occurs in dead and dying cells.

radiation   Energy transmitted by means of electromagnetic waves or by a stream of particles such as electrons, protons, neutrons, or alpha particles.

radical mastectomy   Removal of the breast including underlying pectoral muscles, lymph nodes of the axilla, and associated skin and subcutaneous tissue. Contrast with lumpectomy (q.v.).

radiosensitizer   Agent that augments efficacy of radiotherapy.

radiotherapy   Treatment of disease with radiation.

refractory   Reversible resistance to a therapy, i.e., existing in a state not responsive to therapy such as would happen if a sensitive population of tumor cells lost blood perfusion during administration of a chemotherapy drug due to a dynamic occlusion of its feeding vessel.

renewing tissue   A tissue that can replace the loss of its differentiated, functional cell types via stem cell activity and progenitor expansion and coupled maturation.

resection   Surgical removal of a tissue mass or organ.

resectoscope   Optical instrument provided with a wire loop for the surgical removal of tissue from the bladder, urethra, or prostate.

residuum   Remainder of cells or tissues after initial treatment.

resistance   Permanent changes in a cell's genotype or phenotype that allows it to survive a particular type of cytotoxic insult.

response elements   DNA sequences, near genes that they regulate, that bind specific proteins which then stimulate transcription.

reticuloendothelial cells   Any of a group of cells of the reticuloendothelial system with the competence to ingest foreign particulate matter (phagocytosis) located at many sites in the body.

retinoblastoma   Pediatric eye malignancy composed or retinal cell malignant precursors; linked to a loss of the tumor suppressor gene Rb-1. It may be either sporadic or hereditary.

retinoic acid   Form of vitamin A. The all-trans retinoic acid (ATRA) isomer is also known as tretinoin.

retinoid   Vitamin A or similar type compound.

retrovirus   Small RNA virus whose genome serves as a template for production of cell DNA via reverse transcriptase. The new DNA is integrated into the host cell. Most retroviruses are believed to be oncogenic.

rhabdomyosarcoma   Tumor composed of malignant precursors of striated muscle cells.

Rous sarcoma virus   Retrovirus (q.v.) of a chicken. It was the first virus shown to cause a malignancy; named for Nobel laureate Francis Peyton Rous.

sarcoma   Malignant tumor of connective tissues; occurs in several forms such as fibrosarcoma, leiomyosarcoma, rhabdomyosarcoma, osteosarcoma etc.

schedule dependency   The dependency of a drug's efficacy and toxicity upon the frequency of its administration.

secondary tumors   Metastases.

seminoma   Malignant germ cell tumor of the testis composed of cells resembling fetal germ cells or precursors of sperm (spermatogonia).

senescence   Aging, the process of growing old. The term applies not only to individuals but to their constituent cells.

sentinel lymph nodes   Lymph nodes that drain a tissue area containing cancer that can be removed surgically and analyzed for disseminating tumor cells to determine the prognosis.

septicemia   Infection of the blood with disease-causing organisms or their toxic substances; also known as blood poisoning.

sigmoidoscopy   Visual examination of the large intestine with an illuminated tube-like optical instrument inserted into the anus.

signal transduction   The conversion of one biochemical signal into a different biochemical signal, e.g., ligand binding to the extracellular domain of a receptor that stimulates intracellular phosphorylation by the receptor kinase.

Skipper hypothesis   The premise that a cancer cure requires the eradication of the last remaining cancer cell. While true in animal models of cancer, this hypothesis may not apply to human patients, where tumor reduction need only be sufficient to prevent regrowth of the cancer to clinically detectable levels during the patient's life span.

somatic   Body; used to describe nongametic cells; in contrast to germline cells.

somatic mutations   Genetic alterations not transmitted to progeny.

spontaneous regression   Return to a nonmalignant condition either by death of tumor cells or by differentiation to a benign or nontumorous state without intentional external influence.

squamous cell carcinoma   Malignant tumor composed of neoplastic squamous epithelial cells, most often found on the skin, but also in the mouth, cervix of the uterus and bronchi. It is the second most common skin cancer that occurs at about half the frequency as basal cell carcinoma.

stem cell   Mother cell with the capacity to give rise to stem cells identical to itself as well as to cells that differentiate into other types.

stroma   Connective tissue elements of a tissue or organ surrounding the tissue/organ specific parenchymal cells. Stroma is also found in tumors, in which it forms the non-neoplastic framework supporting the growth of neoplastic cells.

subcutaneous   Underneath the skin, between the dermis and underlying muscle tissue.

suppressor   Gene whose loss of activity leads to cancer.

synovial cell sarcoma   Malignant soft tissue tumor composed of undifferentiated connective tissue cells. The name, which is a misnomer, was coined by pathologists who thought that the tumor originates from the synovial lining of the joints.

systemic   Something that circulates in the bloodstream, exposing all tissues of the body.

tamoxifen   Drug with antiestrogen activity.

targeted therapy   Drugs that are directed to specific molecules that drive a malignant process and preferably do not exist, or at least are not critical, in normal tissues.

Taxol   Trademark name for paclitaxel, a chemotherapeutic drug that affects polymerization and stability of intracellular microtubules. The agent was originally derived from the Pacific yew tree (*Taxus brevifolia*).

telomerase   Enzyme that adds telomere sequences to the ends of chromosomes; young cells and cancer cells have longer telomeres than do aged cells; length of telomeres is correlated with cell division competence.

teratocarcinoma   Germ cell tumor composed of developmentally pluripotent malignant stem cells, equivalent to embryonal carcinoma, and differentiated somatic tissues derived from the stem cells. It occurs most often in the testis and the ovary but may be found in extragonadal locations such as mediastinum and retroperitoneum. The latter tumors originate most likely from germ cells that were misplaced during fetal development.

teratoma   Benign tumor of germ cell origin composed of haphazardly arranged mature tissues derived from all three germ layers, i.e., ectoderm, mesoderm and endoderm; most often found in the ovaries.

terminal differentiation   The appearance of a cellular phenotype(s) responsible for tissue function coupled to the cessation of proliferation.

ternary complex   Comprised of three chemical components.

testosterone   Male hormone produced by the testes.

therapeutic index   The ratio between the maximum tolerated dose (MTD) and the efficacious dose. An effective therapy will have a therapeutic index >1, i.e., the MTD is a higher dose than that required for efficacy against the disease.

thrombocytopenia   Diminished platelets in the blood.

tissue renewal   Continuous replacement of skin, the lining of the gut, and blood cells that would otherwise be lost due to wear and senescence.

tocopherol   Any of several substances having vitamin E activity.

transfected cell   Transfer and incorporation of DNA from one cell type to another.

transformation   Change that a normal cell undergoes as it becomes malignant.

translocation   Movement of genetic material from one site in a chromosome to another site in the same or another chromosome.

tremolite   White or gray mineral of the amphibole group.

trophic   Adjective referring to nutrition.

trophic factor   Diffusable molecules, made either locally (paracrine/autocrine) or distantly (systemic), which increase the mass of the target tissue by blocking the apoptosis program in tissue progenitors, allowing them to increase the number of viable differentiated cells.

trophoblast   Extraembryonic cells of the mammalian embryo that attach the blastocyst (the young embryo) to the uterine wall. Trophoblast cells are essential for the nutrition of the embryo.

tubulin   The constituent molecule of microtubules; microtubules form the mitotic spindle as well as a significant portion of the cytoplasmic cytoskeleton.

tumor   Synonym for neoplasia; abnormal new growth of cells that may be either benign (q.v.) or malignant (q.v.).

tumor burden   The number of cancer cells in the body at any given time whether localized within a single tumor or disseminated within metastases.

tumor doubling time   The amount of time required for a tumor to double its cellularity and mass. Doubling time is not usually equal to cell cycle time because tumor size is the net result of cell gain from proliferation and cell loss from sloughing and senescence.

tumor grade   Degree of differentiation as well as growth rate, as indicated by mitotic frequency.

tumorigenesis   Formation of tumors, used as a synonym for carcinogenesis (q.v.).

tumor-infiltrating lymphocytes (TIL cells)   Lymphoid cells that invade solid malignant tumors. TIL cells can be grown in interleukin-2, and some TIL cells have competence to lyse specific tumor cells.

tumor markers   Molecular products of cancer found in serum of tumor-bearing hosts. They may be useful in diagnosis, prognosis, and assessment of therapeutic effect and possible recurrence.

tumor necrosis factor (TNF)   Substance released by macrophages that causes, among many effects, the stimulation of T-cell production and cell death in some tumors.

typhus   Group of severe bacterial diseases caused by rickettsiae.

tyrosine kinase   Enzyme(s) that catalyze phosphorylation of the amino acid tyrosine in target proteins.

ultraviolet radiation   Radiation with wavelengths from 200 and 400 nm, between the visible violet and X-ray wavelengths. UV between 290 and 320 nm (ultraviolet B) causes sunburn and DNA damage leading to skin cancer.

urogenital ridge   Mesodermal area in the celomic cavity of the early embryo that will give rise to both the kidney and the gonads.

vinblastine   One of a class of chemotherapeutic agents known as the vinca alkaloids (q.v.).

vinca alkaloids   Cytotoxic chemotherapeutic agents, derived from the Madagascar periwinkle, *Catharanthus roseus*, that target microtubules ("*Vinca*" is an older generic name for *Catharanthus*).

vincristine   One of a class of chemotherapeutic agents known as the vinca alkaloids (q.v.).

von Recklinghausen's disease Synonym for neurofibromatosis, an autosomal recessive hereditary tumor syndrome, characterized by the formation of numerous peripheral nerve sheath tumors (neurofibromas and neurilemmomas) on the skin and in internal organs.

Wilms' tumor Synonym for nephroblastoma, a pediatric kidney cancer, pathogenetically linked to a loss of the tumor suppressor gene (WT-1). It is familial in a minority of cases.

xenobiotics Chemical substances foreign to the organism.

xenograft The growth of a foreign tissue in an immunocompromised host, such as growing human tumors in nu/nu (nude) mice or SCID mice that cannot reject the foreign tissue.

xeroderma pigmentosum Hereditary disease, caused by inadequate DNA repair, characterized by extreme sensitivity to the sun with multiple skin problems including malignancies.

zinc finger Many DNA-binding proteins, including transcription factor IIIA, are associated with zinc. The protein with the zinc forms a loop structure known as a zinc finger.

References

Abbrecht PH and Littell JK. 1972. Plasma erythropoietin in men and mice during acclimatization to different altitude. *J Appl Physiol* 32:54–8.

Adachi M, Cossman J, Longo D, Croce CM, and Tsujimoto Y. 1989. Variant translocation of the bcl-2 gene to immunoglobulin lambda light chain gene in chronic lymphocytic leukemia. *Proc Natl Acad Sci USA* 86:2771–4.

Adams J, Palombella VJ, Sausville EA, Johnson J, Destree A, Lazarus DD, Maas J, Pien CS, Prakash S, and Elliott PJ. 1999. Proteasome inhibitors: A novel class of potent and effective antitumor agents. *Cancer Res* 59:2615–22.

Adams JM. 2003. Ways of dying: multiple pathways to apoptosis. *Genes Dev* 17:2481–95.

Adams JM, Harris AW, Pinkert CA, Corcoran LM, Alexander WS, Cory S, Palmiter RD, and Brinster RL. 1985. The c-*myc* oncogene driven by immunoglobulin enhancers induces lymphoid malignancy in transgenic mice. *Nature* 318:533–8.

Adams MD, Celniker SE, Holt RA, Evans CA, Gocayne JD, Amanatides PG, Scherer SE, Li PW, Hoskins RA, Galle RF et al. 2000. The genome sequence of Drosophila melanogaster. *Science* 287:2185–95.

Adler I. 1912. *Primary Malignant Growths of the Lungs and Bronchi.* New York: Longmans, Green.

Adnane J, Jackson RJ, Nicosia SV, Cantor AB, Pledger WJ, and Sebti SM. 2000. Loss of p21WAF1/CIP1 accelerates Ras oncogenesis in a transgenic/knockout mammary cancer model. *Oncogene* 19:5338–47.

Advisory Committee to the Surgeon General. 1964. *Smoking and Health.* US Public Health Service Publication 1103, Washington, DC: US Dept Health, Education and Welfare.

Alberti L, Carniti C, Miranda C, Roccato E, Pierotti MA. 2003. RET and NTRK1 proto-oncogenes in human diseases. *J Cell Physiol* 195:168–86.

Al-Dewachi HS, Appleton DR, Watson AJ, and Wright NA. 1979. Variation in the cell cycle time in the crypts of Lieberkuhn of the mouse. *Virchows Arch B Cell Pathol Incl Mol Pathol* 31: 37–44.

Al-Dewachi HS, Wright NA, Appleton DR, and Watson AJ. 1974. The cell cycle time in the rat jejunal mucosa. *Cell Tissue Kinet* 7:587–94.

1975. Cell population kinetics in the mouse jejunal crypt. *Virchows Arch B Cell Pathol* 18:225–42.

1977. The effect of a single injection of hydroxyurea on cell population kinetics in the small bowel mucosa of the rat. *Cell Tissue Kinet* 10:203–13.

1980. The effect of a single injection of cytosine arabinoside on cell population kinetics in the mouse jejunal crypt. *Virchows Arch B Cell Pathol Incl Mol Pathol* 34:299–309.

Alexander WS, Adams JM, and Cory S. 1989. Oncogene cooperation in lymphocyte transformation: malignant conversion of E mu-myc transgenic pre-B cells in vitro is enhanced by v-H-ras or v-raf but not v-abl. *Mol Cell Biol* 9:67–73.

Allegra CJ. 1990. Antifolates. In: BA Chabner and JM Collins (eds.). *Cancer Chemotherapy*, pp. 110–53. Philadelphia, PA: JB Lippincott.

Almoguera C, Shibata D, Forrester K, Martin J, Arnheim N, and Perucho M. 1988. Most human carcinomas of the exocrine pancreas contain mutant c-K-*ras* genes. *Cell* 53:549–54.

Al-Mulla F, Going JJ, Sowden ET, Winter A, Pickford IR, and Birnie GD. 1998. Heterogeneity of mutant versus wild-type Ki-ras in primary and metastatic colorectal carcinomas, and association of codon-12 valine with early mortality. *J Pathol* 185:130–8.

Alpert ME, Hutt MS, and Davidson CS. 1968. Hepatoma in Uganda A study in geographic pathology. *Lancet* I:1265–7.

Alpha Tocopherol, Beta Carotene Cancer Prevention Study Group. 1994. The effect of vitamin E and beta carotene on the incidence of lung cancer and other cancers in male smokers. *N Engl J Med* 330:1029–35.

Alt JR, Gladden AB, and Diehl JA. 2002. p21(Cip1) Promotes cyclin D1 nuclear accumulation via direct inhibition of nuclear export. *J Biol Chem* 277:8517–23.

Alvarez OA, Carmichael DF, and Declerck YA. 1990. Inhibition of collagenolytic activity and metastasis by a recombinant human tissue inhibitor of metalloproteinases. *J Natl Cancer Inst* 82:589–95.

American Cancer Society 1996 Advisory Committee on Diet, Nutrition, and Cancer Prevention. 1996. Guidelines on diet, nutrition, and cancer prevention: Reducing the risk of cancer with healthy food choices and physical activity. *CA Cancer J Clin* 46:325–41.

American Cancer Society. 2005. *Cancer Facts & Figures 2005*. http://www.cancerorg/statistics.

Ames BN and Gold LS. 2000. Paracelsus to parascience: the environmental cancer distraction. *Mutat Res* 447:3–13.

Ames BN, Magaw R, Gold LS. 1987. Ranking possible carcinogenic hazards. *Science* 236:271–80.

Anaissie EJ, Vartivarian S, Bodey GP, Legrand C, Kantarjian H, Abi-Said D, Karl C, and Vadhan-Raj S. 1996. Randomized comparison between antibiotics alone and antibiotics plus granulocyte-macrophage colony-stimulating factor (escherichia coli-derived in cancer patients with fever and neutropenia. *Am J Med* 100:17–23.

Anderson KC, Bates MP, Slaughenhoupt BL, Pinkus GS, Schlossman SF, and Nadler LM. 1984. Expression of human b cell-associated antigens on leukemias and lymphomas: A model of human b cell differentiation. *Blood* 63:1424–33.

Anderson T. 1994. Palaeopathology: More than just dry bones. *Proc R Coll Physicians Edinb* 24:554–80.

Anderson TJ, Battersby S, King RJB, McPherson K, and Going JJ. 1989. Oral contraceptive use influences resting breast proliferation. *Human Pathology* 20:1139–44.

Anderson TJ, Ferguson DJ, and Raab GM. 1982. Cell turnover in the "resting" human breast: Influence of parity, contraceptive pill, age and laterality. *Br J Cancer* 46:376–82.

Andrews PW. 2002. From teratocarcinomas to embryonic stem cells. *Philos Trans R Soc Lond B Biol Sci* 357:405–17.

Anonymous. 1994. A 44 year campaign, but the epidemic is growing. *BMJ* 309: Number 6960, unnumbered contents page (8 October 1994).

Anthony M, Williams JK, and Dunn BK. 2001. What would be the properties of an ideal SERM? *Annals of the New York Academy of Sciences* 949:261–78.

Aoki T, Takeda S, Yanagisawa A, Kato Y, Ajioka Y, Watanabe H, Kudo S, and Nakamura Y. 1994. APC and p53 mutations in de novo colorectal adenocarcinomas. *Hum Mutat* 3:342–6.

ATAC Trialists' Group. 2005. Results of the ATAC (arimidex, tamoxifen, alone or in combination) trial after completion of 5 years' adjuvant treatment for breast cancer. *Lancet* 365:60–2.

Aragane Y, Kulms D, Metze D, Wilkes G, Poppelmann B, Luger TA, and Schwarz T. 1998. Ultraviolet light induces apoptosis via direct activation of CD95 (Fas/APO-1) independently of its ligand CD95L. *J Cell Biol* 140:171–82.

Armstrong BK. 1984. Melanoma of the skin. *Br Med Bull* 40:346–50.

———. 1988. Epidemiology of malignant melanoma: Intermittent or total accumulated exposure to the sun? *J Dermatol Surg Oncol* 14:835–49.

Armstrong PB. 1984. Invasiveness of non-malignant cells. In *Invasion, Experimental and Clinical Implications*, MM Mareel and KC Calman, eds, pp. 126–167. Oxford, UK: Oxford University Press.

Ashkenazi A. 2002. Targeting death and decoy receptors of the tumour-necrosis factor superfamily. *Nat Rev Cancer* 2:420–30.

Ashworth TR. 1869. A case of cancer in which cells similar to those in the tumours were seen in the blood after death. *Aust Med J* 14:146–7.

Assoian RK and Zhu X. 1997. Cell anchorage and the cytoskeleton as partners in growth factor dependent cell cycle progression. *Curr Opin Cell Biol* 9:93–8.

Aszterbaum M, Epstein J, Oro A, Douglas V, LeBoit PE, Scott MP, and Epstein EH, Jr. 1999. Ultraviolet and ionizing radiation enhance the growth of BCCs and trichoblastomas in patched heterozygous knockout mice. *Nat Med* 5:1285–91.

Ault JG, Cole RW, Jensen CG, Jensen LC, Bachert LA, and Rieder CL. 1995. Behavior of crocidolite asbestos during mitosis in living vertebrate lung epithelial cells. *Cancer Res* 55:792–8.

Bach PB, Kattan MW, Thornquist MD, Kris MG, Tate RC, Barnett MJ, Hsieh LJ, and Begg CB. 2003. Variations in lung cancer risk among smokers. *J Natl Cancer Inst* 95:470–8.

Bader JP. 1972. Temperature-dependent transformation of cells infected with a mutant of Bryan Rous sarcoma virus. *J Virol* 10:267–76.

Bagshawe KD. 1992. Choriocarcinoma: A model for tumour markers. *Acta Oncol* 31:99–106.

Baker SJ, Markowitz S, Fearon ER, Willson JK, and Vogelstein B. 1990. Suppression of human colorectal carcinoma cell growth by wild-type p53. *Science* 249:912–15.

Baldus CD, Tanner SM, Ruppert AS, Whitman SP, Archer KJ, Marcucci G, Caligiuri MA, Carroll AJ, Vardiman JW, Powell BL et al. 2003. BAALC expression predicts clinical outcome of de vovo acute myeloid leukemia patients with normal cytogenetics: a Cancer and Leukemia Group B study. *Blood* 102:1613–8.

Bale AE and Brown SJ. 2001. Etiology of cancer: cancer genetics. In: VT DeVita, S Hellman, and SA Rosenberg (eds.). *Cancer, Principles and Practice of Oncology*, 6th ed., pp. 207–17. Philadelphia, PA: Lippincott Williams & Wilkins.

Balendran A, Casamayor A, Deak M, Paterson A, Gaffney P, Currie R, Downes CP, and Alessi DR. 1999. PDK1 acquires PDK2 activity in the presence of a synthetic peptide derived from the carboxyl terminus of PRK2. *Curr Biol* 9:393–404.

Barber HR K and Brunschwig A. 1968. Treatment and results of recurrent cancer of corpus uteri in patients receiving anterior and total pelvic exenteration. *Cancer* 22:949–55.

Bargmann CI, Hung MC, and Weinberg RA. 1986. Multiple independent activations of the *neu* oncogene by a point mutation altering the transmembrane domain of p185. *Cell* 45:649–57.

Barker N, Morin PJ, and Clevers H. 2000. The Yin-Yang of TCF/beta-catenin signaling. *Adv Cancer Res* 77:1–24.

Barradell LB and Faulds D. 1994. Cyproterone A review of its pharmacology and therapeutic efficacy in prostate cancer. *Drugs Aging* 5:59–80.

Barratt-Fornell A, Drewnowski A. 2002. The taste of health: nature's bitter gifts. *Nutr Today* 37: 144–50.

Bar-Sagi D and Feramisco JR. 1985. Micro-injection of the *ras* oncogene protein into PC12 cells induces morphological differentiation. *Cell* 42:841–8.

Bartsch H, Ohshima H, Pignatelli B, and Calmels S. 1992. Endogenously formed N-nitroso compounds and nitrosating agents in human cancer etiology. *Pharmacogenetics* 2:272–7.

Baselga J, Pfister D, Cooper MR, Cohen R, Burtness B, Bos M, D'Andrea G, Seidman A, Norton L, Gunnett K et al. 2000. Phase I studies of anti-epidermal growth factor receptor chimeric antibody c225 alone and in combination with cisplatin. *J Clin Oncol* 18:904–14.

Baskaran R, Dahmus ME, and Wang JYJ. 1993. Tyrosine phosphorylation of mammalian RNA polymerase II carboxy-terminal domain. *Proc Natl Acad Sci USA* 90:11167–71.

Bates CM, Kharzai S, Erwin T, Rossant J, and Parada LF. 2000. Role of N-myc in the developing mouse kidney. *Dev Biol* 222:317–25.

Batzdorf U, Black KL, and Selch MT. 1990. Neoplasms of the nervous system. In: CM Haskell (ed.). *Cancer Treatment*, 3rd ed., pp. 436–68. Philadelphia, PA: WB Saunders.

Batzer AG, Blaikie P, Nelson K, Schlessinger J, and Margolis B. 1995. The phosphotyrosine interaction domain of Shc binds an LXNPXY motif on the epidermal growth factor receptor. *Mol Cell Biol* 15:4403–9.

Baum M. 1976. The curability of breast cancer. *Br Med J* 1:439–42.

Baumann PC and Harshbarger JC. 1995. Decline in liver neoplasms in wild brown bullhead catfish after coking plant closes and environmental PAHs plummet. *Environ Health Perspect* 103: 168–70.

Baxevanis CN, Sotiropoulou PA, Sotiriadou NN, and Papamichail M. 2004. Immunobiology of HER-2/ *neu* oncoprotein and its potential application in cancer immunotherapy. *Cancer Immunol Immunother* 53:166–75.

Bayreuther K. 1960. Chromosomes in primary neoplastic growth. *Nature* 186:6–9.

Beckman KB and Ames BN. 1997. Oxidative decay of DNA. *J Biol Chem* 272:19633–6.

Beirut LJ, Dinwiddie SH, Begleiter H, Crowe RR, Hesselbrock V, Nurnberger JI Jr, Porjesz B, Schuckit MA, and Reich T. 1998. Familial transmission of substance dependence: alcohol, marijuana, cocaine, and habitual smoking: a report from the Collaborative Study on the Genetics of Alcoholism. *Arch Gen Psychiatry* 55:982–8.

Bell B. 1794. *A Treatise on the Hydrocele or Sarcocele, or Cancer and Other Disease of the Testes.* Edinburgh.

Bellacosa A, Testa JR, Staal SP, and Tsichlis PN. 1991. A retroviral oncogene, akt, encoding a serine-threonine kinase containing an SH2-like region. *Science* 254:274–7.

Bender RA, Hamel E, and Hande KR. 1990. Plant alkaloids. In: BA Chabner and JM Collins (eds.). *Cancer Chemotherapy*, pp. 253–75. Philadelphia, PA: JB Lippincott.

Beral V, Peterman TA, Berkelman RL, and Jaffe HW. 1990. Kaposi's sarcoma among persons with AIDS: a sexually transmitted infection? *Lancet* 335:123–8.

Berchuck A, Kamel A, Whitaker R, Kerns B, Olt G, Kinney R, Soper JT, Dodge R, Clarke-Pearson DL, Marks P et al. 1990. Overexpression of HER-2/neu is associated with poor survival in advanced epithelial ovarian cancer. *Cancer Res* 50:4087–91.

Berek JS and Hacker NF. 1990. Ovary and fallopian tubes. In: CM Haskell (ed.). *Cancer Treatment,* 3rd ed., pp. 295–325. Philadelphia, PA: WB Saunders.

Berenblum I and Shubik P. 1947. The role of croton oil application associated with a single painting of a carcinogen in tumour induction of the mouse's skin. *Br J Cancer* 1:379–83.

Berenson JR and Gale RP. 1990. Acute lymphoblastic leukemia. In: CM Haskell (ed.). *Cancer Treatment,* 3rd ed., pp. 606–20. Philadelphia, PA: WB Saunders.

Berkowitz RS and Goldstein DP. 1990. Gestational trophoblastic neoplasia. In: CM Haskell (ed.). *Cancer Treatment,* 3rd ed., pp. 366–72. Philadelphia, PA: WB Saunders.

Berman ML and Berek JS. 1990. Uterine corpus. In: CM Haskell (ed.). *Cancer Treatment,* 3rd ed, pp. 338–51. Philadelphia, PA: WB Saunders.

Beukers R and Berends W. 1960. Isolation and identification of the irradiation product of thymine. *Biochim Biophys Acta* 41:550–1.

Beutler B and Cerani A. 1986. Cachectic and tumor necrosis factors as two sides of the same biological coin. *Nature* 320:584–8.

Bevona C and Sober AJ. 2002. Melanoma incidence trends. *Dermatol Clin* 20:589–95.

Biggs J, Hersperger E, Steeg PS, Liotta LA, and Shearn A. 1990. A Drosophila gene that is homologous to a mammalian gene associated with tumor metastasis codes for a nucleoside diphosphate kinase. *Cell* 63:933–40.

Biggs PM. 2001. The history and biology of Mareks disease virus. *Curr Top Microbiol Immunol* 255:1–24.

Bingham CA. 1978. The cell cycle and cancer chemotherapy. *Am J Nurs* 78:1201–5.

Bingham SA, Day NE, Luben R, Ferrari P, Slimani N, Norat T, Clavel-Chapelon F, Kesse E, Nieters A, Boeing H et al. 2003. Dietary fibre in food and protection against colorectal cancer in the European Prospective Investigation into cancer and nutrition (EPIC): an observational study. *Lancet* 361:1496–1501.

Binns AN. 2002. T-DNA of Agrobacterium tumefaciens: 25 years and counting. *Trends Plant Sci* 7:231–3.

Birner P, Oberhuber G, Stani J, Reithofer C, Samonigg H, Hausmaninger H, Kubista E, Kwasny W, Kandioler-Eckersberger D, Gnant M et al. 2001. Evaluation of the United States Food and Drug Administration-approved scoring and test system of HER-2 protein expression in breast cancer. *Clin Cancer Res* 7:1669–75.

Bister K and Jansen HW. 1986. Oncogenes in retroviruses and cells: Biochemistry and molecular genetics. *Adv Cancer Res* 47:99–188.

Bittner JJ. 1936. Some possible effects of nursing on the mammary gland tumor incidence in mice. *Science* 84:162.

1937. Mammary tumors in mice in relation to nursing. *Am J Cancer* 30:530–8.

Bjorge JD, Jakymiw A and Fujita DJ. 2000. Selected glimpses into the activation and function of Src kinase. *Oncogene* 19:5620–35.

Bjorkoy G, Overvatn A, Diaz-Meco MT, Moscat J, and Johansen T. 1995. Evidence for a bifurcation of the mitogenic signaling pathway activated by Ras and phosphatidylcholine-hydrolyzing phospholipase C. *J Biol Chem* 270:21299–21306.

Blain SW, Montalvo E, and Massague J. 1997. Differential interaction of the cyclin-dependent kinase (Cdk) inhibitor p27Kip1 with cyclin A-Cdk2 and cyclin D2-Cdk4. *J Biol Chem* 272:25863–72.

Blakeslee AF. 1934. New Jimson weeds from old chromosomes. *J Hered* 25:80–108.

Blaney SM, Bernstein M, Neville K, Ginsberg J, Kitchen B, Horton T, Berg SL, Krailo M, and Adamson PC. 2004. Phase I study of the proteasome inhibitor bortezomib in pediatric patients with refractory solid tumors: A children's oncology group study (advl0015). *J Clin Oncol* 22:4752–7.

Blasco MA, Lee HW, Hande MP, Samper E, Lansdorp PM, DePinho RA, and Greider CW. 1997. Telomere shortening and tumor formation by mouse cells lacking telomerase RNA. *Cell* 91: 25–34.

Blezinger P, Wang J, Gondo M, Quezada A, Mehrens D, French M, Singhal A, Sullivan S, Rolland A, Ralston R, and Min W. 1999. Systemic inhibition of tumor growth and tumor metastases by intramuscular administration of the endostatin gene. *Nat Biotechnol* 17:343–8.

Blount WP. 1961. Turkey "X" disease. *J Br Turkey Fed* 9:522–8.

Blume-Jensen P and Hunter T. 2001. Oncogenic kinase signalling. *Nature* 411:355–65.

Bobrovnikova-Marjon EV, Marjon PL, Barbash O, Vander Jagt DL, and Abcouwer SF. 2004. Expression of angiogenic factors vascular endothelial growth factor and interleukin-8/cxcl8 is highly responsive to ambient glutamine availability: Role of nuclear factor-kappab and activating protein-1. *Cancer Res* 64:4858–69.

Boccardo F, Bruzzi P, Rubagotti A, Nicolo GU, and Rosso R. 1981. Estrogen-like action of tamoxifen on vaginal epithelium in breast cancer patients. *Oncology* 38:281–5.

Bocci G, Nicolaou KC, and Kerbel RS. 2002. Protracted low-dose effects on human endothelial cell proliferation and survival in vitro reveal a selective antiangiogenic window for various chemotherapeutic drugs. *Cancer Res* 62:6938–43.

Bodmer WF, Bailey CJ, Bodmer J, Bussey HJ, Ellis A, Gorman P, Lucibello FC, Murday VA, Rider SH, Scambler P et al. 1987. Localization of the gene for familial adenomatous polyposis on chromosome 5. *Nature* 328:614.

Bodnar AG, Ouellette M, Frolkis M, Holt SE, Chiu CP, Morin GB, Harley CB, Shay JW, Lichtsteiner S, and Wright WE. 1998. Extension of life-span by introduction of telomerase into normal human cells. *Science* 279:349–52.

Boffetta P. 2002. Involuntary smoking and lung cancer. *Scand J Work Environ Health* 28, Suppl 2:30–40.

Bogenrieder T and Herlyn M. 2003. Axis of evil: molecular mechanisms of cancer metastasis. *Oncogene* 22:6524–36.

Boice JD Jr, Greene MH, Killen JY, Ellenberg SS, Keehn RJ, McFadden E, Chen TT, and Fraumeni JF Jr. 1983. Leukemia and preleukemia after adjuvant treatment of gastrointestinal cancer with semustine methyl-CCNU. *N Engl J Med* 309:1079–84.

Borthwick NJ, Bofill M, Hassan I, Panayiotidis P, Janossy G, Salmon M, and Akbar AN. 1996. Factors that influence activated CD8+ T-cell apoptosis in patients with acute herpesvirus infections: Loss of costimulatory molecules CD28, CD5 and CD6 but relative maintenance of Bax and Bcl-X expression. *Immunology* 88:508–15.

Borucinska JD, Harshbarger JC, Reimschuessel R, and Bogicevic T. 2004. Gingival neoplasms in a captive sand tiger shark, Carcharias taurus (Rafinesque), and a wild-caught blue shark, Prionace glauca (L). *J Fish Diseases* 27:185–91.

Bosari S, Lee AK, DeLellis RA, Wiley BD, Heatley GJ, and Silverman ML. 1992. Microvessel quantitation and prognosis in invasive breast carcinoma. *Hum Pathol* 23:755–61.

Bouroncle BA. 1994. Thirty-five years in the progress of hairy cell leukemia. *Leuk Lymphoma* 14 (Suppl 1):1–12.

Boveri T. 1907. *Zellen-Studien. Heft 6. Die Entwicklung dispermer Seeigel-Eier. Ein Beitrag zur Befruchtungslehre und zur Theorie des Kerns.* Jena: Gustav Fischer.

1914. *Zur Frage der Erstehung Maligner Tumoren.* Jena: Gustav Fischer (English translation by M Boveri). 1929. *The Origin of Malignant Tumors.* Baltimore, MD: Williams and Wilkins).

Bowen ID and Bowen SM. 1990. *Programmed Cell Death in Tumors and Tissues.* New York: Chapman and Hall.

Bowman T, Broome MA, Sinibaldi D, Wharton W, Pledger WJ, Sedivy JM, Irby R, Yeatman T, Courtneidge SA, and Jove R. 2001. Stat3-mediated Myc expression is required for Src transformation and PDGF-induced mitogenesis. *Proc Natl Acad Sci USA* 98:7319–24.

Boyle P, Zaridze DG, and Smans M. 1985. Descriptive epidemiology of colorectal cancer. *Int J Cancer* 36:9–18.

Brandom WF, Saccomanno G, Archer VE, Archer PG, and Bloom AD. 1978. Chromosome aberrations as a biological dose-response indicator of radiation exposure in uranium miners. *Rad Res* 76:59–171.

Braun AC. 1956. The activation of two growth-substance systems accompanying the conversion of normal to tumor cells in crown gall. *Cancer Res* 16:53–6.

1972. The usefulness of plant tumor systems for studying the basic cellular mechanisms that underlie neoplastic growth generally. In: R Harris, P Allin, and D Viza (eds.). *Cell Differentiation*, pp. 115–18. Copenhagen: Munksgaard.

1974. *The Biology of Cancer*. Reading, MA: Addison-Wesley Publishing Company.

1981. An epigenetic model for the origin of cancer. *Q Rev Biol* 56:33–60.

Breinholt V, Hendricks J, Pereira C, Arbogast D, and Bailey G. 1995. Dietary chlorophyllin is a potent inhibitor of aflatoxin B1 hepatocarcinogenesis in rainbow trout. *Cancer Res* 55: 57–62.

Brentnall TA, Haggitt RC, Rabinovitch PS, Kimmey MB, Bronner MP, Levine DS, Kowdley KV, Stevens AC, Crispin DA, Emond M et al. 1996. Risk and natural history of colonic neoplasia in patients with primary sclerosing cholangitis and ulcerative colitis. *Gastroenterology* 110:331–8.

Bridges CB. 1916. Non-disjunction as proof of the chromosome theory of heredity. *Genetics* 1: 107–63.

Briggs R and King TJ. 1952. Transplantation of living nuclei from blastula cells into enucleated frogs' eggs. *Proc Natl Acad Sci USA* 38:455–63.

Brinster R. 1993. Stem cells and transgenic mice in the study of development. *Int J Dev Biol* 37:89–99.

Brinster RL. 1974. Effects of cells transferred into the mouse blastocyst on subsequent development. *J Exp Med* 140:1049–56.

Brinster RL, Chen HY, Messing AA, Van Dyke T, Levine AJ, and Palmiter RD. 1984. Transgenic mice harboring SV40 T-antigen genes develop characteristic brain tumors. *Cell* 37:367–79.

Bromberg J and Darnell JE Jr. 2000. The role of STATs in transcriptional control and their impact on cellular function. *Oncogene* 19:2468–73.

Bromley M, Rew D, Becciolini A, Balzi M, Chadwick C, Hewitt D, Li YQ, and Potten CS. 1996. A comnparison of proliferation markers (BrdUrd, Ki-67, PCNA) determined at each cell position in the crypts of normal human colonic mucosa. *Eur J Histochem* 40:89–100.

Brown DR, Fife KH, Wheeler CM, Koutsky LA, Lupinacci LM, Railkar R, Suhr G, Barr E, Dicello A, Li W et al. 2004. Early assessment of the efficacy of a human papillomavirus type 16 virus-like particle vaccine. *Vaccine* 22:2936–42.

Bucher NLR and Malt RA. 1971. *Regeneration of the Liver and Kidney*. Boston, MA: Little, Brown.

Buchkovich K, Duffy LA, and Harlow E. 1989. The retinoblastoma protein is phosphorylated during specific phases of the cell cycle. *Cell* 58:1097–1105.

Buday L and Downward J. 1993. Epidermal growth factor regulates p21 *ras* through the formation of a complex of receptor, Grb2 adapter protein, and Sos nucleotide exchange factor. *Cell* 73:611–20.

Budillon A, Cereseto A, Kondrashin A, Nesterova M, Merlo G, Clair T, and Cho-Chung YS. 1995. Point mutation of the autophosphorylation site or in the nuclear location signal causes protein kinase A RII beta regulatory subunit to lose its ability to revert transformed fibroblasts. *Proc Natl Acad Sci USA* 92:10634–8.

Bullough WS. 1962. The control of mitotic activity in adult mammalian tissues. *Biol Rev* 37: 301–42.

Bunn PA Jr and Ridgway EC. 1993. Paraneoplastic syndromes. In: VT DeVita, Jr., S. Hellman, and SA Rosenberg (eds.). *Cancer Principles and Practice of Oncology*, 4th ed., pp. 2026–71. Philadelphia, PA: JB Lippincott.

Burke TG, Morin MJ, Sartorelli AC, Lane PE, and Tritton TR. 1987. Function of the anthracycline amino group in cellular transport and cytotoxicity. *Mol Pharmacol* 31:552–6.

Burkitt DP. 1958. A sarcoma involving the jaws of African children. *Br J Surg* 46:218–25.

——— 1971. Epidemiology of cancer of the colon and rectum. *Cancer* 28:3–13.

——— 1975. Large-bowel cancer: An epidemiological jigsaw puzzle. *J Natl Cancer Inst* 54:3–6.

Burmer GC and Loeb LA. 1989. Mutations in the KRAS2 oncogene during progressive stages of human colon carcinoma. *Proc Natl Acad Sci USA* 86:2403–7.

Butel JS and Lednicky JA. 1999. Cell and molecular biology of simian virus 40: implications or human infections and disease. *J Natl Cancer Inst* 91:119–34.

Buters JT, Sakai S, Richter T, Pineau T, Alexander DL, Savas U, Doehmer J, Ward JM, Jefcoate CR, and Gonzalez FJ. 1999. Cytochrome P450 CYP1B1 determines susceptibility to 7, 12-dimethylbenz[a]anthracene-induced lymphomas. *Proc Natl Acad Sci USA* 96:1977–82.

Butler TP and Gullino PM. 1975. Quantitation of cell shedding into efferent blood of mammary adenocarcinoma. *Cancer Res* 35:512–16.

Butlin HJ. 1892. Cancer of the scrotum in chimney-sweeps and others: II. Why foreign sweeps do not suffer from scrotal cancer. *Br Med J* 2:1–6.

Butrum RR, Clifford CK, and Lanza E. 1988. NCI dietary guidelines: Rationale. *Am J Clin Nutr* 48:888–95.

Byers T, Nestle M, McTiernan A, Doyle C, Currie-Williams A, Gansler T, and Thun M. 2002. American Cancer Society guidelines on nutrition and physical activity for cancer prevention: Reducing the risk for cancer with healthy food choices and physical activity. *CA Cancer J Clin* 52:92–119.

Calabresi P and Parks RE Jr. 1980. Antiproliferative agents and drugs used for immunosuppression. In: AG Gilman, LS Goodman, and A Gilman (eds.). *The Pharmacological Basis of Therapeutics*, pp. 1256–1313. New York: Macmillan.

Calabresi P and Schein PS (eds.). 1993. *Medical Oncology: Basic Principles and Clinical Management of Cancer.* New York: McGraw-Hill.

Calabresi P and Welch AD. 1962. Chemotherapy of neoplastic diseases. *Annu Rev Med* 13:147–202.

Calman KC. 1992. Cachexia. In: JO'D McGee, PG Isaacson, and NA Wright (eds.). *Oxford Textbook of Pathology*, vol 1, pp. 715–17. Oxford, UK: Oxford University Press.

Cameron IL, Hardman WE, and Skehan P. 1990. Regulation of growth in normal and neoplastic cell populations by a tissue sizer mechanism: Therapeutic implications. *Prog Clin Biol Res* 354A:61–79.

Campen CA, Jordan VC, and Gorski J. 1985. Opposing biological actions of antiestrogens in vitro and in vivo: Induction of progesterone receptor in the rat and mouse uterus. *Endocrinology* 116:2327–36.

Carlson DB and Perdew GH. 2002. A dynamic role for the Ah receptor in cell signaling? Insights from a diverse group of Ah receptor interacting proteins. *J Biochem Mol Toxicol* 16:317–25.

Carroll KK and Khor HT. 1975. Dietary fat in relation to tumorigenesis. *Prog Biochem Pharmacol* 10:308–53.

Carson DA, Kaye J, and Wasson DB. 1980. Differences in deoxyadenosine metabolism in human and mouse lymphocytes. *J Immunol* 124:8–12.

Carson, DA, Wasson DB, Kaye J, Ullman B, Martin DW Jr, Robins RK, and Montgomery JA. 1980. Deoxycytidine kinase-mediated toxicity of deoxyadenosine analogs toward malignant human lymphoblasts in vitro and toward murine L1210 leukemia in vivo. *Proc Natl Acad Sci USA* 77:6865–9.

Case RA. 1969. Some environmental carcinogens. *Proc R Soc Med* 62:1061–6.

Case RAM and Hosker ME. 1954. Tumours of the urinary bladder as an occupational disease in the rubber industry in England and Wales. *Br J Prev Soc Med* 8:39–50.

Case RA, Hosker ME, McDonald DB, and Pearson JT. 1954. Tumours of the urinary bladder in workmen engaged in the manufacture and use of certain dyestuff intermediates in the British chemical industry: Part I. The role of aniline, benzidine, alpha-naphthylamine and beta-napthylamine. *Br J Ind Med* 11: 75–104.

Cerhan JR, Potter JD, Gilmore JM, Janney CA, Kushi LH, Lazovich D, Anderson KE, Sellers TA, and Folsom AR. 2004. Adherence to the AICR cancer prevention recommendations and subsequent morbidity and mortality in the Iowa women's health study cohort. *Cancer Epidemiol Biomarkers Prev* 13:1114–20.

Cerutti P. 1985. Prooxidant states and tumor promotion. *Science* 227:375–81.

Chabner BA. 1990a. Cytidine analogues. In: BA Chabner and JM Collins (eds.). *Cancer Chemotherapy*, pp. 154–79. Philadelphia, PA: JB Lippincott.

    1990b. Bleomycin. In: BA Chabner and JM Collins (eds.). *Cancer Chemotherapy*, pp. 341–55. Philadelphia, PA: JB Lippincott.

    1990c. Clinical strategies for cancer treatment: The role of drugs. In: BA Chabner and JM Collins (eds.). *Cancer Chemotherapy*, pp. 1–15. Philadelphia, PA: JB Lippincott.

Chang RL, Huang MT, Wood AW, Wong CQ, Newmark HL, Yagi H, Sayer JM, Jerind DM, and Conney AH. 1985. Effect of ellagic acid and hydroxylated flavonoids on the tumorigenicity of benzo(a)pyrene and $\pm$-7,8-dihydroxy-9,10-epoxy-7,8,9,10-tetrahydrobenzo(a)pyrene on mouse skin and in the newborn mouse. *Carcinogenesis* 6:1127–33.

Chang S, Khoo C, and DePinho RA. 2001. Modeling chromosomal instability and epithelial carinogenesis in the telomerase-deficient mouse. *Semin Cancer Biol* 11:227–39.

Chaplin DJ and Trotter MJ. 1990. The nature of tumor hypoxia: Implications for therapy. *Prog Clin Biol Res* 354B:81–92.

Charbit A, Malaise EP, and Tubiana M. 1971. Relation between the pathological nature and the growth rate of human tumors. *Eur J Cancer* 7:307–15.

Chardin P, Camonis JH, Gale NW, van Aelst L, Schlessinger J, Wigler MH, and Bar-Sagi D. 1993. Human Sos1: A guanine nucleotide exchange factor for Ras that binds to GRB2. *Science* 260:1338–43.

Charlson ME. 1985. Delay in the treatment of carcinoma of the breast. *Surg Gynecol Obstet* 160:393–9.

Charron J, Malynn BA, Fisher P, Stewart V, Jeannotte L, Goff SP, Robertson EJ, and Alt FW. 1992. Embryonic lethality in mice homozygous for a targeted disruption of the N-myc gene. *Genes Dev* 6:2248–57.

Chau BN, Borges HL, Chen TT, Masselli A, Hunton IC, and Wang JY. 2002. Signal-dependent protection from apoptosis in mice expressing caspase-resistant Rb. *Nat Cell Biol* 4:757–65.

Chau BN and Wang JY. 2003. Coordinated regulation of life and death by RB. *Nat Rev Cancer* 3:130–8.

Chen LL, Trent JC, Wu EF, Fuller GN, Ramdas L, Zhang W, Raymond AK, Prieto VG, Oyedeji CO, Hunt KK et al. 2004. A missense mutation in kit kinase domain 1 correlates with imatinib resistance in gastrointestinal stromal tumors. *Cancer Res* 64:5913–19.

Chin KV, Pastan I, and Gottesman MM. 1993. Function and regulation of the human multidrug resistance gene. *Adv Cancer Res* 60:157–80.

Chiu R, Boyle WJ, Meek J, Smeal T, Hunter T, and Karin M. 1988. The c-fos protein interacts with c-jun/AP-1 to stimulate transcription of AP-1 responsive genes. *Cell* 54:541–52.

Chuang TY, Popecsh A, Su WPD, and Chute CG. 1990. Basal cell carcinoma. *J Am Acad Dermatol* 22:413–17.

Chumas P, Tyagi A, Livingston J. 2001. Hydrocephalus – what's new? *Arch Dis Child Fetal Neonatal Ed.* 85:F149–54.

Clarke R, Lippman ME, and Dickson RB. 1990. Mechanisms of hormone and cytotoxic drug interactions in the development and treatment of breast cancer. *Prog Clin Biol Res* 322: 243–78.

Cleary ML, Smith SD and Sklar J. 1986. Cloning and structural analysis of cDNAs for bcl-2 and a hybrid bcl-2/immunoglobulin transcript resulting from the t(14;18) translocation. *Cell* 47: 19–28.

Cleaver JE. 1968. Defective repair replication of DNA in xeroderma pigmentosum. *Nature* 218: 652–6.

Cleaver JE. 1994. It was a good year for DNA repair. *Cell* 76:1–4.

Cleaver JE and Crowley E. 2002. UV damage, DNA repair and skin carcinogenesis. *Front Biosci* 7:d1024–43.

Cleaver JE and Kraemer KH. 1989. Xeroderma pigmentosum. In: CR Scriver, AL Beaudet, WS Sly, and D Valle (eds.). *The Metabolic Basis of Inherited Disease*, 6th ed., pp. 2949–71. New York: McGraw-Hill.

Cohen SM and Ellwein LB. 1991. Genetic errors, cell proliferation, and carcinogenesis. *Cancer Res* 51:6493–6505.

Collaborative Group on Hormonal Factors in Breast Cancer. 2002. Breast cancer and breastfeeding: collaborative reanalysis of individual data from 47 epidemiological studies in 30 countries, including 50302 women with breast cancer and 96973 women without the disease. *Lancet* 360:187–95.

Colombo N. 1994. Controversial issues in the management of early epithelial ovarian cancer: Conservative surgery and the role of adjuvant therapy. *NIH Consensus Development Conference on Ovarian Cancer: Screening, Treatment, and Followup* April 5–7, Bethesda, MD: NIH.

Coman DR. 1944. Decreased mutual adhesiveness, a property of cells from squamous cell carcinomas. *Cancer Res* 4:625–9.

Conney AH. 1982. Induction of microsomal enzymes by foreign chemicals and carcinogenesis by polycyclic aromatic hydrocarbons. *Cancer Res* 42:4875–4917.

Conrad RA, Dobyns BM, and Sutow WW. 1970. Thyroid neoplasia as a late effect of exposure to radioactive iodine in fallout. *J Am Med Assoc* 214:316–24.

Cook JW, Hewett CL, and Hieger I. 1933. The isolation of a cancer-producing hydrocarbon from coal tar Parts I, II, and III. *J Chem Soc* Issue 395–405.

Cook SJ and McCormick F. 1993. Inhibition by cAMP of Ras-dependent activation of Raf *Science* 262:1069–72.

Cooper DY, Levin SS, Narrasimhulu S, Rosenthal O, and Estabrook RW. 1965. Photochemical action spectrum of the terminal oxidase of mixed function oxidase systems. *Science* 147:400–2.

Corcos L and Lagadic-Gossmann D. 2001. Gene induction by Phenobarbital: an update on an old question that receives key novel answers. *Pharmacol Toxicol* 89:113–22.

Cornwell MM, Tsuruo T, Gottesman MM, and Pastan I. 1987. ATP-binding properties of P-glycoprotein from multidrug-resistant KB cells. *FASEB J* 1:51–4.

Cory S, Vaux DL, Strasser A, Harris AW, and Adams JM. 1999. Insights from Bcl-2 and Myc: malignancy involves abrogation of apoptosis as well as sustained proliferation. *Cancer Res* 59(7 Suppl):1685s–1692s.

Cotran RS, Kumar V, and Robbins SL. 1994. *Pathological Basis of Cancer*, 5th ed. Philadelphia, PA: WB Saunders.

Coultas L and Strasser A. 2003. The role of the Bcl-2 protein family in cancer. *Semin Cancer Biol* 13:115–23.

Counts JL, Sarmiento JI, Harbison ML, Downing JC, McClain RM, and Goodman JL. 1996. Cell proliferation and global methylation status changes in mouse liver after phenobarbital and/or choline-devoid, methionine-deficient diet administration. *Carcinogenesis* 17: 1251–7.

Courtneidge SA. 2003. Cancer: Escape from inhibition. *Nature* 422:827–8.

Coutard H. 1932. Roentgenotherapy of epitheliomas of the tonsillar region, hypopharynx and larynx from 1920 to 1926. *Roentgenol Radium Ther* 28:313–31.

Cox AD and Der CJ. 2002. Ras family signaling: therapeutic targeting. *Cancer Biol Ther.* 1:599–606. 2003. The dark side of Ras: regulation of apoptosis. *Oncogene* 22:8999–9006.

Cox EC. 1976. Bacterial mutator genes and the control of spontaneous mutation. *Annu Rev Genet* 10:135–56.

Crum CP. 2002. The beginning of the end for cervical cancer? *N Engl J Med* 347:1703–5.

Culver K, Cornetta K, Morgan R, Morecki S, Aerbersold T, Kasid A, Lotv M, Rosenberg SA, Anderson WF, and Blaese RM. 1991. Lymphocyte as cellular vehicles for gene therapy in mouse and man. *Proc Natl Acad Sci USA* 88:3155–9.

Cummings SR, Eckert S, Krueger KA, Grady D, Powles TJ, Cauley JA, Norton L, Nickelsen T, Bjarnason NH, Morrow M et al. 1999. The effect of raloxifene on risk of breast cancer in postmenopausal women: results from the MORE randomized trial. Multiple Outcomes of Raloxifene Evaluation. *JAMA* 281:2189–97.

Cunningham M and Matthews HB. 1995. Cell proliferation as a determining factor for the carcinogenicity of chemicals: studies with mutagenic carcinogens and mutagenic noncarcinogens. *Toxicol Lett* 82–83:9–14.

Cunningham D, Humblet Y, Siena S, Khayat D, Bleiberg H, Santoro A, Bets D, Mueser M, Harstrick A, Verslype C et al. 2004. Cetuximab monotherapy and cetuximab plus irinotecan in irinotecan-refractory metastatic colorectal cancer. *N Engl J Med* 351:337–45.

Curry SJ, Byers T, and Hewitt M. (eds). 2003. *Cancer Prevention and Early Detection*. Washington, DC: The National Academies Press.

Cushing H and Wolback SB. 1927. The transformation of a malignant paravertebral neuroblastoma into a benign ganglioneuroma. *Am J Pathol* 3:203–16.

Daenen S, Vellenga E, van Dobbenburgh OA, and Halie MR. 1986. Retinoic acid as antileukemic therapy in a patient with acute promyelocytic leukemia and Aspergillus pneumonia. *Blood* 67:559–61.

Dalla Favera R, Bregni M, Erikson J, Patterson D, Gallo RC, and Croce CM. 1982. Assignment of the c-*myc* oncogene to the region of chromosome 8 which is translocated in Burkitt lymphoma cells. *Proc Natl Acad Sci USA* 79:7824–7.

Davesa SS, DJ Grauman, WJ Blot, G Pennello, RN Hoover, and JF Fraumeni Jr. 1999. Atlas of Cancer Mortality in the United States, 1950–94. Washington, DC: US Govt Print Off (NIH) 99–4564.

Davidson JR. 1934. Attempt to inhibit development of tar-carcinoma in mice: Preliminary note. *Can Med Assoc J* 31:486–7.

Davies BR, Platt-Higgins AM, Schmidt G, and Rudland PS. 1999. Development of hyperplasias, preneoplasias, and mammary tumors in MMTV-c-erbB-2 and MMTV-TGFalpha transgenic rats. *Am J Pathol* 155:303–14.

Davies H, Bignell GR, Cox C, Stephens P, Edkins S, Clegg S, Teague J, Woffendin H, Garnett MJ, and Bottomley W et al. 2002. Mutations of the BRAF gene in human cancer. *Nature* 417:949–54.

Davies MA, Lu Y, Sano T, Fang X, Tang P, LaPushin R, Koul D, Bookstein R, Stokoe D, Yung WK et al. 1998. Adenoviral transgene expression of MMAC/PTEN in human glioma cells inhibits Akt activation and induces anoikis. *Cancer Res* 58:5285–90.

Davis JMG, Beckett ST, Bolton RE, Collins P, and Middleton AP. 1978. Mass and number of fibres in the pathogenesis of asbestos-related lung disease in rats. *Br J Cancer* 37:673–88.

Davison AJ. 2002. Evolution of the herpesviruses. *Vet Microbiol* 86:69–88.

Davison AJ, Sauerbier W, Dolan A, Addison C, and McKinnell RG. 1999. Genomic studies of the Lucké tumor herpesvirus (RaHV-1). *J Cancer Res Clin Oncol* 125:232–8.

Dawe CJ. 1964. An appreciation of Dr. Mearl F Stanton. In: Use of small fish species in carcinogenecity testing. NCI Monograph 65:1–2.

    1969. Phylogeny and oncogeny. In: CJ Dawe and JC Harshbarger (eds.). *Neoplasms and Related Disorders in Invertebrate and Lower Vertebrate Animals*. Bethesda, MD: National Cancer Institute.

Dawe CJ. 1969. Phylogeny and oncogeny. In:*Neoplasms and Related Disorders in Invertebrate and Lower Vertebrate Animals*. CJ Dawe and JC Harshbarger, eds, Bethesda: Natl Cancer Inst Monogr 31:1–40.

Dawe CJ, Harshbarger JC, Kondo S, Sugimura T, and Takayama S (eds.). 1981. *Phyletic Approaches to Cancer*. Tokyo: Japan Scientific Societies Press.

Dean C, Chetty U, and Forrest APM. 1983. Effects of immediate breast reconstruction on psychosocial morbidity after mastectomy. *Lancet* 1:459–62.

DeCaprio JA, Ludlow JW, Figge J, Shew J-Y, Huang C-M, Lee W-H, Marsillo E, Paucha E, and Livingston DM. 1988. SV40 large tumor antigen forms a specific complex with the product of the retinoblastoma susceptibility gene. *Cell* 54:275–83.

DeGregori J, Leone G, Miron A, Jakoi L, and Nevins JR. 1997. Distinct roles for E2F proteins in cell growth control and apoptosis. *Proc Natl Acad Sci USA* 94:7245–50.

de Gruijl FR, Longstreth J, Norval M, Cullen AP, Slaper H, Kripke ML, Takizawa Y, and van der Leun JC. 2003. Health effects from stratospheric ozone depletion and interactions with climate change. *Photochem Photobiol Sci* 2:16–28.

Deininger MW, Goldman JM, Lydon N, and Melo JV. 1997. The tyrosine kinase inhibitor CGP57148B selectively inhibits the growth of BCR-ABL-positive cells. *Blood* 90: 3691–8.

De Jong PJ, Gorsovsky AJ, and Glickman BW. 1988. Spectrum of spontaneous mutation at the APRT locus of Chinese hamster ovary cells: An analysis at the DNA sequence level. *Proc Natl Acad Sci USA* 85:3499–3503.

de Kernion JB and Berry D. 1980. The diagnosis and treatment of renal cell carcinoma. *Cancer* 45:1947–56.

Demetri GD. 2002. Identification and treatment of chemoresistant inoperable or metastatic GIST: experience with the selective tyrosine kinase inhibitor imatinib mesylate (STI571). *Eur J Cancer* 38 Suppl 5:S-52–59.

Demetri GD, von Mehren M, Blanke CD, Van den Abbeele AD, Eisenberg B, Roberts PJ, Heimrich MC, Tuveson DA, Singer S, and Janicek M et al. 2002. Efficacy and safety of imatinib mesylate in advanced gastrointestinal stromal tumors. *N Engl J Med* 347:472–80.

Demunter A, Stas M, Degreef H, De Wolf-Peeters C, and van den Oord JJ. 2001. Analysis of N- and K-ras mutations in the distinctive tumor progression phases of melanoma. *J Invest Dermatol* 117:1483–9.

Deng T and Karin M. 1994. C-Fos transcriptional activity stimulated by H-ras-activated protein kinase distinct from JNK and ERK. *Nature* 371:171–5.

Der CJ, Finkel T, and Cooper GM. 1986. Biological and biochemical properties of human *ras*H genes mutated at codon 61. *Cell* 44:167–76.

Dérijard B, Hibi M, Wu I-H, Barret T, Su B, Deng T, Karin M, and Davis RJ. 1994. JNK1: A protein kinase stimulated by UV light and Ha-Ras that binds and phosphorylates the c-Jun activation domain. *Cell* 76:1025–37.

Devesa SS, Graman DJ, Blot WJ, Pennello G, Hoover RN, and Fraumeni Jr JF. 1999. *Atlas of Cancer Mortality in the United States, 1954–94.* Washington DC: US Govt Print Off (NIH) 99–4564.

DeVilliers E-M. 1989. Heterogeneity of the human papillomavirus group. *J Virol* 53:4898–4903.

de Vita VT Jr, Hellman S, and Rosenberg SA, eds. 2004. *Cancer: Principles and Practice of Oncology,* 7th ed., Philadelphia, PA: Lippincott Williams and Wilkins.

Dexter DL, Oki T, and Heidelberger C. 1973. Fluorinated pyrimidines. Xlii. Effect of 5-trifluoromethyl-1-2′-deoxyuridine on transcription of vaccinia viral messenger ribonucleic acid. *Mol Pharmacol* 9:283–96.

DiBerardino MA. 1997. *Genomic Potential of Differentiated Cells.* New York: Columbia University Press.

DiBerardino MA and McKinnell RG. 2004. The pathway to animal cloning and beyond – Robert Briggs (1911–1983) and Thomas J. King (1921–2000). *J Exp Zool* 301A:275–79.

DiBerardino MA, King TJ, and McKinnell RG. 1963. Chromosome studies of a frog renal adenocarcinoma line carried by intraocular transplantation. *J Natl Cancer Inst* 31:769–89.

Dickson MA, Hahn WC, Ino Y, Ronfard V, Wu JY, Weinberg RA, Louis DN, Li FP, and Rheinwald JG. 2000. Human keratinocytes that express hTERT and also bypass a p16(INK4a)-enforced mechanism that limits life span become immortal yet retain normal growth and differentiation characteristics. *Mol Cell Biol* 20: 1436–47.

Diffey BL. 1987. Cosmetic solaria and malignancies of the skin. *Photodermatology* 4:273–6.

Di Fiore PP, Pierce JH, Kraus MH, Segatto O, King CR, and Aaronson SA. 1987. *erb*B-2 is a potent oncogene when overexpressed in NIH3T3 cells. *Science* 237:178–82.

Dipple A, Khan QA, Page JE, Ponten I, and Szeliga J. 1999. DNA reactions, mutagenic action and stealth properties of polycyclic aromatic hydrocarbon carcinogens (review). *Int J Oncol* 14:103–11.

Diwan BA, Ohshima M, and Rice JM. 1989. Promotion by sodium barbital of renal cortical and transitional cell tumors, but not intestinal tumors, in F344 rats given methylacetoxymethylnitrosamine, and lack of effect of phenobarbital, amobarbital, or barbituric acid on development of either renal or intestinal tumors. *Carcinogenesis* 10:183–8.

Dixon FJ and Moore RA. 1952. Tumors of the male sex organs. *Atlas of tumor pathology*, section viii, Fascicle 31b and 32. Washington, DC: Armed Forces Institute of Pathology.

Dobson JM, Samuel S, Milstein H, Rogers K, and Wood JL. 2002. Canine neoplasia in the UK: estimates of incidence rates from a population of insured dogs. *J Small Anim Pract* 43: 240–6.

Dohner K, Tobia K, Ulrich R, Frohling S, Benner A, Schlenk RF, and Dohner H. 2002. Prognostic significance of partial tandem duplications of the MLL gene in adult patients 16 to 60 years old with acute myeloid leukemia and normal cytogenetics: a study of the Acute Myeloid Leukemia Study Group Ulm. *J Clin Oncol* 20:3254–61.

Doll R. 1967. *Prevention of Cancer: Pointers from Epidemiology.* London: The Nuffield Provincial Hospitals Trust.

1992. Health and the environment in the 1990s. *Am J Public Health* 82:933–41.

Doll R and Bradford-Hill A. 1950. Smoking and carcinoma of the lung. *Br Med J* 2:739–48.

Doll R and Peto R. 1981. The causes of cancer. *J Natl Cancer Inst* 66:1191–1308.

Doll R, Peto R, Boreham J, and Sutherland I. 2004. Mortality in relation to smoking: 50 years' observations on male British doctors. *BMJ* 328:1519–28.

Doll R, Peto R, Wheatley K, Gray R, and Sutherland I. 1994. Mortality in relation to smoking: 40-years' observations on male British doctors. *BMJ* 309:901–11.

Donaldson KL, Goolsby GL, and Wahl AF. 1994. Cytotoxicity of the anticancer agents cisplatin and Taxol during cell proliferation and the cell cycle. *Int J Cancer* 57:847–55.

Donehower LA, Harvey M, Slagle BL, McArthur MJ, Montgomery Jr CA, Butel JS, and Bradley A. 1992. Mice deficient for p53 are developmentally normal but susceptible to spontaneous tumours. *Nature* 356:215–21.

Dong J-T, Lamb PW, Rinker-Schaeffer CW, Vukanovic J, Ichikawa T, Isaacs JT, and Barrett JC. 1995. KAI1, a metastasis suppressor gene for prostate cancer on human chromosome 11p112. *Science* 268:884–6.

Dong S, Geng JP, Tong JH, Wu Y, Cai JR, Sun GL, Chen SR, Wang ZY, Larsen CJ, Berger R et al. 1993. Breakpoint clusters of the pml gene in acute promyelocytic leukemia: Primary structure of the reciprocal products of the pml-rara gene in a patient with t(15;17). *Genes Chromosomes Cancer* 6:133–9.

Doniach I and Pelc SR. 1950. Autoradiographic techniques. *Br J Radiol* 23:184–92.

Doniger J, Jacobson ED, Krell K, and DiPaolo JA. 1981. Ultraviolet light action spectra for neoplastic transformation and lethality of Syrian hamster embryo cells correlate with spectrum for pyrimidine dimer formation in cellular DNA. *Proc Natl Acad Sci USA* 78:2378–82.

Doonan J and Hunt T. 1996. Why don't plants get cancer? *Nature* 380:481–2.

Dorr RT and Fritz WL. 1980. *Cancer Chemotherapy Handbook* New York: Elsevier.

Dorr RT and Von Hoff D. 1994. *Cancer Chemotherapy Handbook.* Norwalk, CT: Appleton and Lange.

Dotto GP, Parada LF, and Weinberg RA. 1985. Specific growth response of *ras*-transformed fibroblasts to tumor promoters. *Nature* 318:472–5.

Downward J. 2003. Targeting RAS signalling pathways in cancer therapy. *Nat Rev Cancer* 3: 11–22.

Drenou B, Fardel O, Amiot L, and Fauchet R. 1993. Detection of P glycoprotein activity on normal and leukemic CD34+ cells. *Leuk Res* 17:1031–5.

Driessens MH, Stroeken PJ, Erena NF, van der Valk MA, van Rijthoven EA, and Roos E. 1995. Targeted disruption of CD44 in MDAY-D2 lymphosarcoma cells has no effect on subcutaneous growth or metastatic capacity. *J Cell Biol* 131:1849–55.

Drinkwater NR. 1990. Experimental models and biological mechanisms for tumor promotion. *Cancer Cells* 2:8–14.

Druckrey H, Ivankovic S, and Preussmann R. 1966. Teratogenic and carcinogenic effects in the offspring after single injection of ethylnitrosourea to pregnant rats. *Nature* 210:1378–9.

Druker BJ, Tamura S, Buchdunger E, Ohno S, Segal GM, Fanning S, Zimmermann J, and Lydon NB. 1996. Effects of a selective inhibitor of the Abl tyrosine kinase on the growth of Bcr-Abl positive cells. *Nat Med* 2:561–6.

Du C, Fang M, Li Y, Li L, and Wang X. 2000. Smac, a mitochondrial protein that promotes cytochrome c-dependent caspase activation by eliminating IAP inhibition. *Cell* 102: 33–42.

Dumaz N and Marais R 2003. Protein kinase A blocks Raf-1 activity by stimulating 14–3-3 binding and blocking Raf-1 interaction with Ras. *J Biol Chem* 278:29819–23.

Dungan CF, Hamilton RM, Hudson KL, McCollough CB, and KS Reece. 2002. Two epizootic diseases in Chesapeake Bay commercial clams, Mya arenaria and Tagelus plebeius. *Dis Aquat Organ* 50:67–78.

Dunn JE. 1975. Cancer epidemiology in populations of the United States – with emphasis on Hawaii and California – and Japan. *Cancer Res* 35:3240–5.

Dupont E, Falardeau P, Mousa SA, Dimitriadou V, Pepin MC, Wang T, and Alaoui-Jamali MA. 2002. Antiangiogenic and antimetastatic properties of Neovastat (AE-941) an orally active extract derived from cartilage tissue. *Clin Exp Metastasis* 19:145–53.

Dutreix J, Tubiana M, Wambersie A, and Malaise, E. 1971. The influence of cell proliferation in tumours and normal tissues during fractionated radiotherapy. *Eur J Cancer* 7:205–13.

Dworkin HJ, Meier DA, and Kaplan M. 1995. Advances in the management of patients with thyroid disease. *Semin Nucl Med* 25:205–20.

Eardley AM, Heller G, and Warrell RP Jr. 1994. Morbidity and costs of remission induction therapy with all-trans retinoic acid compared with standard chemotherapy in acute promyelocytic leukemia. *Leukemia* 8:934–9.

Earle J. 1808. *Chirurgical works of Percivall Pott.* A new edition with his last corrections to which are added a short account of the life of the author, a method of curing the hydrocele by injection and occasional notes and observations. London: J Johnson.

Easton J, Wei T, Lahti JM, and Kidd VJ. 1998. Disruption of the cyclin D/cyclin-dependent kinase/INK4/retinoblastoma protein regulatory pathway in human neuroblastoma. *Cancer Res* 58:2624–32.

Eferl R and Wagner EF. 2003. AP-1: a double-edged sword in tumorigenesis. *Nat Rev Cancer* 3:859–68.

Egeblad M and Werb Z. 2002. New functions for the matrix metalloproteinases in cancer progression. *Nature Rev Cancer* 2:161–74.

Egner PA, Wang JB, Zhu YR, Zhang BC, Wu Y, Zhang QN, Qian GS, Kuang SY, Gange SJ, Jacobson LP et al. 2001. Chlorophyllin intervention reduces aflatoxin-DNA adducts in individuals at high risk for liver cancer. *Proc Natl Acad Sci USA* 98:14601–6.

Ehrhardt A, Ehrhardt GRA, Guo X, and Schrader JW. 2002. Ras and relatives – job sharing and networking keep an old family together. *Exp Hematol* 30:1089–1106.

Eichner ER. 1987. Exercise, lymphokines, calories, and cancer. *Physician Sportsmed* 15:109–16.

Einfeld DA, Brown JP, Valentine MA, Clark EA, and Ledbetter JA. 1988. Molecular cloning of the human b cell cd20 receptor predicts a hydrophobic protein with multiple transmembrane domains. *EMBO J* 7:711–17.

el-Deiry WS, Harper JW, O'Connor PM, Velculescu VE, Canman CE, Jackman J, Pietenpol JA, Burrell M, Hill DE, and Wang Y. 1994. WAF1/CIP1 is induced in p53-mediated G1 arrest and apoptosis. *Cancer Res* 54:1169–74.

Elion GB and Hitchings GH. 1965. Metabolic basis for the actions of analogs of purines and pyrimidines. *Adv Chemother* 2:91–177.

Elwood JM, Lee JAH, Walter SD, Mo T, and Green AES. 1974. Relationship of melanoma and other skin cancer mortality to latitude and ultraviolet radiation in the United States and Canada. *Int J Epidemiol* 3:325–32.

Encarnacion CA, Ciocca DR, McGuire WL, Clark GM, Fuqua SA, and Osborne CK. 1993. Measurement of steroid hormone receptors in breast cancer patients on tamoxifen. *Breast Cancer Res Treat* 26:237–46.

Engell HC. 1955. Cancer cells in the circulating blood. *Acta Chir Scand* (Suppl 201):1–70.

English HF, Santner SJ, Levine HB, and Santen RJ. 1986. Inhibition of testosterone production with ketoconazole alone and in combination with a gonadotropin releasing hormone analogue in the rat. *Cancer Res* 46:38–42.

Ennemoser O, Ambach W, Auer T, Brunner P, Schneider P, Oberaigner W, Purtscheller F, and Stingl V. 1994. High indoor radon concentrations in an alpine region of western Tyrol. *Health Phys* 67:151–4.

Ennemoser O, Ambach W, Brunner P, Schneider P, and Oberaigner W. 1993. High domestic and occupational radon exposures: A comparison. *Lancet* 342:47.

Enterline HT and Coman DR. 1950. The amoeboid motility of human and animal neoplastic cells. *Cancer* 3:1033–8.

Epner DE, Partin AW, Schalken JA, Isaacs JT, and Coffey DS. 1993. Association glyceraldehyde-3-phosphate dehydrogenase expression with cell motility and metastatic potential of rat prostatic adenocarcinoma. *Cancer Res* 53:1995–7.

Eskes R, Desagher S, Antonsson B, and Martinou JC. 2000. Bid induces the oligomerization and insertion of Bax into the outer mitochondrial membrane. *Mol Cell Biol* 20:929–35.

Essigmann JM, Croy RG, Bennett RA, and Wogan GN. 1982. Metabolic activation of aflatoxin B_1: Patterns of DNA adduct formation, removal, excretion in relation to carcinogenesis. *Drug Metab Rev* 13:581–602.

Esteller M and Herman JG. 2002. Cancer as an epigenetic disease: DNA methylation and chromatin alterations in human tumours. *J Pathol* 196:1–7.

Evans MJ and Kaufman MH. 1981. Establishment in culture of pluripotential cells from mouse embryos cultured in medium conditioned by teratocarcinoma cells. *Nature* (Lond) 292:154–6.

Ewen ME. 1994. The cell cycle and the retinoblastoma protein family. *Cancer Metastasis Rev* 13: 45–66.

Ewing J. 1916. Pathological aspects of some problems of experimental cancer research. *J Cancer Res* 1:71–86.

———. 1928. *Neoplastic Diseases*, 3rd ed. Philadelphia, PA: WB Saunders.

Eyre H, Kahn R, Robertson RM, and ACS/ADA/AHA Collaborative Writing Committee. 2004. Preventing cancer, cardiovascular disease, and diabetes: A common agenda for the American Cancer Society, the American Diabetes Association, and the American Heart Association. *CA Cancer J Clin* 54:190–207.

Fabre I, Fabre G, and Goldman ID. 1984. Polyglutamylation, an important element in methotrexate cytotoxicity and selectivity in tumor versus murine granulocytic progenitor cells in vitro. *Cancer Res* 44:3190–5.

Falk A and Sachs L. 1980. Clonal regulation of the induction of macrophage- and granulocyte-inducing proteins for normal and leukemic myeloid cells. *Int J Cancer* 26:595–601.

Fankhauser G. 1945. The effects of changes in chromosome number on amphibian development. *Q Rev Biol* 20:20–78.

Fantl WJ, Escobedo JA, Martin GA, Turck CW, del Rosario M, McCormick F, and Williams LT. 1992. Distinct phosphoproteins on a growth factor receptor bind to specific molecules that mediate different signaling pathways. *Cell* 69:413–23.

Farina KL, Wyckoff JB, Rivera J, Lee H, Segall JE, Condeelis JS, and Jones JG.1998. Cell motility of tumor cells visualized in living intact primary tumors using green fluorescent protein. *Cancer Res* 58:2528–32.

Fattman CL, Delach SM, Dou QP, and Johnson DE. 2001. Sequential two-step cleavage of the retinoblastoma protein by caspase-3/-7 during etoposide-induced apoptosis. *Oncogene* 20: 2918–26.

Fausto N and Webber EM. 1993. Control of liver growth. *Crit Rev Eukaryot Gene Expr* 3:316–73.

Fearon ER, Vogelstein B. 1990. A genetic model for colorectal tumorigenesis. *Cell* 61:759–767.

Feig DI, Reid TM, and Loeb LA. 1994. Reactive oxygen species in tumorigenesis. *Cancer Res* (Suppl 54):1890s–1894s.

Feinberg AP and Williams BR. 2003. Wilms' tumor as a model for cancer biology. *Methods Mol Biol* 222:239–48.

Fenaux P. 1993. The role of all-trans-retinoic acid in the treatment of acute promyelocytic leukemia. *Acta Haematol* 89(Suppl 1):22–7.

Fenaux P, Le Deley MC, Castaigne S, Archimbaud E, Chomienne C, Link H, Guerci A, Duarte M, Daniel MT, Bowen D et al. 1993. Effect of all transretinoic acid in newly diagnosed acute promyelocytic leukemia. Results of a multicenter randomized trial. European apl 91 group. *Blood* 82:3241–9.

Ferguson DJ and Anderson TJ. 1981a. Ultrastructural observations on cell death by apoptosis in the "resting" human breast. *Virchows Arch A Pathol Anat Histol* 393:193–203.

    1981b. Morphological evaluation of cell turnover in relation to the menstrual cycle in the "resting" human breast. *Br J Cancer* 442:177–81.

Ferguson LR and Harris PJ. 2003. The dietary fibre debate: more food for thought. *Lancet* 361:1487–8.

Feschenko MS, Stevenson E, Nairn AC, and Sweadner KJ. 2002. A novel cAMP-stimulated pathway in protein phosphatase 2A activation. *J Pharmacol Exp Ther* 302:111–18.

Fidler IJ. 2003. The pathogenesis of cancer metastasis: the 'seed and soil' hypothesis revisited. *Nat Rev Cancer* 6:453–8.

Fidler IJ and Ellis LM. 1994. The implications of angiogenesis for the biology and therapy of cancer metastasis. *Cell* 79:185–8.

Fidler IJ and Hart IR. 1982. Biological diversity in metastatic neoplasms: Origins and implications. *Science* 217:998–1003.

Fidler IJ, Kumar R, Bielenberg DR, and Ellis LM. 1998. Molecular determinants of angiogenesis in cancer metastasis. *Cancer J Sci Am* 4 Suppl 1:S58–66.

Figlin RA, Holmes EC, Petrovich Z, and Sarna GP. 1990. Lung cancer. In: CM Haskell (ed.). *Cancer Treatment*, 3rd ed., pp. 165–88. Philadelphia, PA: WB Saunders.

Figg WD, McLeod HL (eds.). 2004. *Handbook of anticancer pharmacokinetics and pharmacodynamics.* Totowa, NJ: Humana Press.

Fisher B and Fisher ER. 1967. Experimental evidence in support of the dormant tumor cell. *Science* 130:918–19.

Fisher B, Bauer M, Margolese R, Poisson R, Pilch Y, Redmond C, Fisher ER, Wolmark N, Deutsch M, Montague E et al. 1985a. Five-year results of a randomized clinical trial comparing total mastectomy and segmental mastectomy with or without radiation in the treatment of breast cancer. *N Engl J Med* 312:665–73.

Fisher B, Gunduz N, and Saffer EA. 1983. Influence of the interval between primary tumor removal and chemotherapy on kinetics and growth of metastases. *Cancer Res* 43:1488–92.

Fisher B, Gunduz N, Coyle J, Rudock C, and Saffer E. 1989a. Presence of a growth-stimulating factor in serum following primary tumor removal in mice. *Cancer Res* 49:1996–2001.

Fisher B, Montague E, Redmond C, Barton B, Borland D, Fisher ER, Deutsch M, Schwarz G, Margolese R, Donegan W et al. 1977. Comparison of radical mastectomy with alternative treatments for primary breast cancer. *Cancer* 39:2827–39.

Fisher B, Osborne CK, Margolese R, and Bloomer W. 1993. Neoplasms of the breast. In: JF Holland, E Frei III, RC Bast, Jr, DW Kufe, DL Morton, and RR Weichselbaum (eds.). *Cancer Medicine*, vol 2, pp. 1706–74. Philadelphia, PA: Lea and Febiger.

Fisher B, Redmond C, Fisher ER, Bauer M, Wolmark N, Wickerham DL, Deutsch M, Montague E, Margolese R, and Foster R. 1985b. Ten-year results of a randomized clinical trial comparing radical mastectomy and total mastectomy with or without radiation. *N Engl J Med* 312: 674–81.

Fisher B, Redmond C, Poisson R, Caplan R, Wickerham DL, Wolmark N, Fisher ER, Deutsch M, Margolese R, Pitch Y et al. 1989b. Eight-year results of a randomized clinical trial comparing total mastectomy and lumpectomy with or without radiation in the treatment of breast cancer. *N Engl J Med* 320:822–8.

1989b. Effect of local or systemic treatment prior to primary tumor removal on the production and response to a serum growth stimulating factor in mice. *Cancer Res* 49:2002–4.

Fisher B, Wolmark N, Fisher ER, and Deutsch M. 1985c. Lumpectomy and axillary dissection for breast cancer: Surgical, pathological, and radiation considerations. *World J Surg* 9:692–8.

Flynn PJ, Miller WJ, Weisdorf DJ, Arthur DC, Brunning R, and Branda RF. 1983. Retinoic acid treatment of acute promyelocytic leukemia: In vitro and in vivo observations. *Blood* 62:1211–17.

Folberg R, Hendrix MJ, and Maniotis AJ. 2000. Vasculogenic mimicry and tumor angiogenesis. *Am J Pathol* 156:361–81.

Folkman J. 1971. Tumor angiogenesis: Therapeutic implications. *N Engl J Med* 285:1182–6.

1985. Tumor angiogenesis. *Adv Cancer Res* 43:175–203.

2002. Role of angiogenesis in tumor growth and metastasis. *Semin Oncol* 29:15–18.

2003a. Angiogenesis and apoptosis. *Semin Cancer Biol* 13:159–67.

2003b. Angiogenesis inhibitors: a new class of drugs. *Cancer Biol Ther* 2(4 Suppl 1):S127–S133.

2003c. Fundamental concepts of the angiogenic process. *Curr Mol Med* 3:643–51.

Folkman J, Merler E, Abernathy C, and Williams G. 1971. Isolation of a tumor factor responsible for angiogenesis. *J Exp Med* 133:275–88.

Forbes JF. 1990. Surgery, kinetics and biological considerations in planning adjuvant therapy protocols. *Prog Clin Biol Res* 354A:133–46.

Fornander T, Rutqvist LE, and Wilking N. 1991. Effects of tamoxifen on the female genital tract. *Ann NY Acad Sci* 622:469–76.

Forrester K, Almoguera C, Han K, Grizzle WE, and Perucho M. 1987. Detection of high incidence of K-*ras* oncogenes during human colon tumorigenesis. *Nature* 327:298–303.

Foss FM. 2002. Immunologic mechanisms of antitumor activity. *Semin Oncol* 29:5–11.

Fosså SD, Gunderson R, and Moe B. 1990. Recombinant interferon-alpha combined with prednisone in metastatic renal cell carcinoma. *Cancer* 65:2451–4.

Foty RA and Steinberg MA. 1997. Measurement of tumor cell cohesion and suppression of invasion by E- or P-cadherin. *Cancer Res* 57:5033–6.

Foulds L. 1965. Multiple etiologic factors in neoplastic development. *Cancer Res* 25:1339–47.

1969. *Neoplastic Development*, vol 1. New York: Academic Press.

Fowler RG, Schaaper RM, and Glickman BW. 1986. Characterization of mutational specificity within the lacI gene for a mutD5 mutator strain of Escherichia coli defective in 3′–5′ exonuclease proofreading activity. *J Bacteriol* 167:130–7.

Fox H and Buckley CH. 1992. The female genital tract and ovaries. In: JO'D McGee, PG Isaacson, and NA Wright (eds.). *Oxford Textbook of Pathology*, pp. 1563–1639. Oxford, UK: Oxford University Press.

Fraumeni JF, Hoover RN, DeVesa SS, and Kinlen LJ. 1989. Epidemiology of cancer. In: VT DeVita, S Hellman, and SA Rosenberg (eds.). *Cancer: Principles and Practice of Oncology*, 4th ed., pp. 150–81. Philadelphia, PA: JB Lippincott.

Fredriksson M, Bengtsson NO, Hardell L, and Axelson O. 1989. Colon cancer, physical activity, and occupational exposures. *Cancer* 63:1838–42.

Freeman B. 2003. The active migration of germ cells in the embryos of mice and men is a myth. *Reproduction* 125:635–43.

Freije JM, MacDonald NJ, and Steeg PS. 1996. Differential gene expression in tumor metastasis: Nm23. *Curr Top Microbiol Immunol* 213:215–32.

Frieben A. 1902. Cancroid des rechten Handruckens nach lang dauerden Einwirkung von Roentgenstrahlen. *Fortschr Geb Rontgenstr* 6:106–8.

Friedberg EC. 2001. How nucleotide excision repair protects against cancer. *Nat Rev Cancer* 1:22–33.

Friedwald WF and Rous P. 1950. The pathogenesis of deferred cancer: A study of the after effects of methylcholanthrene upon the rabbit skin. *J Exp Med* 91:459–84.

Friend SH, Bernards R, Rogelj S, Weinberg RA, Rapaport JM, Albert DM, and Dryja TP. 1986. A human DNA segment with properties of the gene that predisposes to retinoblastoma and osteosarcoma. *Nature* 323:643–6.

Friend SH, Horowitz JM, Gerber MR, Wang X-F, Bogenmann E, Li FP, and Weinberg RA. 1987. Deletions of a DNA sequence in restinoblastomas and mesenchymal tumors: Organization of the sequence and its encoded protein. *Proc Natl Acad Sci USA* 84:9059–63.

Frisch RE, Wyshak G, Albright NL, Albright FE, Schiff I, Witschi J, and Margullo M. 1987. Lower lifetime occurrence of breast cancer and cancers of the reproductive system among former college athletes. *Am J Clin Nutr* 45:328–35.

Frisch SM and Francis H. 1994. Disruption of epithelial cell-matrix interactions induces apoptosis. *J Cell Biol* 124:619–26.

Frixen UH, Behrens J, Sachs M, Eberle G, Voss B, Warda A, Lochner D, and Birchmeier W. 1991. E-Cadherin-mediated cell–cell adhesion prevents invasiveness of human carcinoma cells. *J Cell Biol* 113:173–85.

Frost P and Levin B. 1992. Clinical implications of metastatic process. *Lancet* 339:1458–61.

Fuchs DA and Johnson RK. 1978. Cytologic evidence that Taxol, an antineoplastic agent from *Taxus brevifolia*, acts as a mitotic spindle poison. *Cancer Treat Rep* 62:1219–22.

Fuchs IB, Landt S, Bueler H, Kuehl U, Coupland S, Kleine-Tebbe A, Lichtenegger W, and Schaller G. 2003. Analysis of her2 and her4 in human myocardium to clarify the cardiotoxicity of trastuzumab (herceptin). *Breast Cancer Res Treat* 82:23–8.

Fujimoto J, Ichigo S, Hirose R, Sakaguchi H, and Tamaya T. 1997. Expression of E-cadherin and alpha- and beta-catenin mRNAs in uterine cervical cancers. *Tumor Biol* 18:206–12.

Furth J. 1953. Conditioned and autonomous neoplasms: A review. *Cancer Res* 13:477–92.

Galaktionov K, Lee AK, Eckstein J, Draetta G, Meckler J, Loda M, and Beach D. 1995. CDC25 phosphatases as potential human oncogenes. *Science* 269:1575–7.

Gale NW, Kaplan S, Lowenstein EJ, Schlessinger J, and Bar-Sagi D. 1993. Grb2 mediates the EGF-dependent activation of guanine nucleotide exchange on Ras. *Nature* 363:88–92.

Gallagher RE, Said F, Pua I, Papenhausen PR, Paietta E, and Wiernik PH. 1989. Expression of retinoic acid receptor-alpha mRNA in human leukemia cells with variable responsiveness to retinoic acid. *Leukemia* 3:789–95.

Gamba-Vitalo C, Blair OC, Tritton TR, Lane PA, Carbone R, and Sartorelli AC. 1987. Cytotoxicity and differentiating actions of adriamycin in wehi-3b d+ leukemia cells. *Leukemia* 1: 188–97.

Gambacorti-Passerini C, le Coutre P, Mologni L, Fanelli M, Bertazzoli C, Marchesi E, Di Nicola M, Biondi A, Corneo GM, Belotti D et al. 1997. Inhibition of the ABL kinase activity blocks the proliferation of BCR/ABL+ leukemic cells and induces apoptosis. *Blood Cells Mol Dis* 23: 380–94.

Gambarotta G, Pistoi S, Giordano S, Comoglio PM, and Santoro C. 1994. Structure and inducible regulation of the human MET promoter. *J Biol Chem* 269:12852–7.

Garabrant DH, Peters JM, Mack TM, and Bernstein L. 1984. Job activity and colon cancer risk. *Am J Epidemiol* 119:1005–14.

Gardner RJM and Sutherland GR. 2004. *Chromosome abnormalities and genetic counseling*, 3rd Ed. Oxford and New York: Oxford University Press.

Garfinkel D. 1958. Studies on pig liver microsomes: 1 Enzymic and pigment composition of different microsomal fractions. *Arch Biochem Biophys* 77:493–509.

Garland FC, White MR, Garland CF, Shaw E, and Gorham ED. 1990. Occupational sunlight exposure and melanoma in the US Navy. *Arch Environ Health* 45:261–7.

Garnick MB 1994. The dilemmas of prostate cancer. *Scientific American* 1994 (April):72–81.

Gateff E. 1981. Malignancies of genetic origin in *Drosophila*. In: CJ Dawe, JC Harshbarger, S Kondo, T Sugimura, and S Takayama (eds.). *Phyletic Approaches to Cancer*, pp. 311–18. Tokyo: Japan Scientific Societies Press.

Gaudet F, Hodgson JG, Eden A, Jackson-Grusby L, Dausman J, Gray JW, Leonhardt H, and Jaenisch R. 2003. Induction of tumors in mice by genomic hypomethylation. *Science* 300:489–92.

Gautier J, Solomon MJ, Booher RN, Bazan JF, and Kirschner MW. 1991. cdc25 is a specific tyrosine phosphatase that directly activates p34cdc2. *Cell* 67:197–211.

Gefeller O and Pfahlberg A. 2002. Sunscreen use and melanoma: a case of evidence-based prevention? *Photodermatol Photoimmunol Photomed* 18:153–6.

Gehring W. 1968. The stability of the differentiated state in cultures of imaginal discs in *Drosophila*. In: H Ursprung (ed.). *The Stability of the Differentiated State*, p. 136. Berlin: Springer-Verlag.

Gelfand MJ. 1993. Meta-iodobenzylguanidine in children. *Semin Nucl Med* 23:231–42.

Geng Y, Yu Q, Sicinska E, Das M, Schneider JE, Bhattacharya S, Rideout WM, Bronson RT, Gardner H, and Sicinski P. 2003. Cyclin E ablation in the mouse. *Cell* 114:431–43.

Geraminejad P, Memar O, Aronson I, Rady PL, Hengge U, Tyring SK. 2002. Kaposi's sarcoma and other manifestations of human herpesvirus 8. *J Am Acad Dermatol* 47:641–55.

Gerhardsson M, Norell SE, Kiviranta H, Pederson NL, and Ahlbom A. 1986. Sedentary jobs and colon cancer. *Am J Epidemiol* 123:775–80.

Gerwitz G and Yallow RS. 1974. Ectopic ACTH production in carcinoma of the lung. *J Clin Invest* 53:1022–32.

Gilbert SF. 1991. *Developmental Biology*, 3rd ed Sunderland, MA: Sinauer.

Gilman A and Philips FS. 1946. The biological actions and therapeutic applications of the chloroethylamines and sulfides. *Science* 103:409–15.

Gingras D, Renaud A, Mousseau N, Beaulieu E, Kachra Z, and Beliveau R. 2001. Matrix proteinase inhibition by AE-941, a multifunctional antiangiogenic compound. *Anticancer Res* 21:145–55.

Giocanti N, Hennequin C, Rouillard D, Defrance R, and Favaudon V. 2004. Additive interaction of gefitinib ('Iressa', ZD1839) and ionising radiation in human tumour cells in vitro. *Br J Cancer* 91:2026–33.

Giovannucci E. 2002. A review of epidemiologic-studies of tomatoes, lycopene, and prostate cancer. *Exp Biol Med* 227:852–9.

   2003. Diet, body weight, and colorectal cancer: a summary of the epidemiological evidence. *J Womens Health* 12:173–82.

Girard F, Strausfeld U, Fernandez A, and Lamb NJ. 1991. Cyclin A is required for the onset of DNA replication in mammalian fibroblasts. *Cell* 67:1169–79.

Glazer RI and Rohlff C. 1994. Transcriptional regulation of multidrug resistance in breast cancer. *Breast Cancer Res Treat* 31:263–71.

Gleraminejad P, Memar O, Aronson I, Rady PL, Hengge U, and Tyring SK. 2002. Kaposi's sarcoma and other manifestations of human herpesvirus 8. *J Am Acad Dermatol* 47:641–55.

Goel SC. 1983. Role of cell death in the morphogenesis of the amniote limbs. In: JF Fallon and AL Caplan (eds.). *Limb Development and Regeneration, Part A,* pp. 175–82. New York: Alan R Liss.

Going JJ, Anderson TJ, Battersby S, and Macintyre CCA. 1988. Proliferative and secretory activity in human breast during natural and artificial menstrual cycles. *Am J Pathol* 130:193–204.

Golay J, Manganini M, Rambaldi A, and Introna M. 2004. Effect of alemtuzumab on neoplastic b cells. *Haematologica* 89:1476–83.

Goldfarb M, Shimizu K, Perucho M, and Wigler M. 1982. Isolation and preliminary characterization of a human transforming gene from T24 bladder carcinoma cells. *Nature* 296:404–9.

Goldhirsch A and Gelber RD. 1996. Endocrine therapies of breast cancer. *Semin Oncol* 23: 494–505.

Goldie JH and Coldman AJ. 1979. A mathematical model for relating the drug sensitivity of tumors to their spontaneous mutation rate. *Cancer Treat Rep* 63:1727–33.

Goldin A and Schabel FM. 1981. Cancer concepts derived from animal chemotherapy studies. *Cancer Treat Rep* 65(Suppl):11–19.

Goldsby RE, Hays LE, Chen X, Olmsted EA, Slayton WB, Spangrude GJ, and Preston BD. 2002. High incidence of epithelial cancers in mice deficient for DNA polymerase delta proofreading. *Proc Natl Acad Sci USA* 99:15560–5.

Gombert W, Borthwick NJ, Wallace DL, Hyde H, Bofill M, Pilling D, Beverley PC, Janossy G, Salmon M, and Akbar AN. 1996. Fibroblasts prevent apoptosis of il-2-deprived t cells without inducing proliferation: A selective effect on bcl-xl expression. *Immunology* 89:397–404.

Gong YL, Koplan JP, Feng W, Chen CHC, Zheng P, and Harris JR. 1995. Cigarette smoking in China. *JAMA* 274:1232–4.

Gonzalez FJ and Kimura S. 2003. Study of P450 function using gene knockout and transgenic mice. *Arch Biochem Biophys* 409:153–8.

Goodman LS, Wintrobe MM, Dameshek W, Goodman MJ, Gilman A, and McLennan M. 1946. Nitrogen mustard therapy: Use of methylbis (β-chloroethyl) amino hydrochloride for Hodgkin's disease, lymphosarcoma, leukemia, and certain allied and miscellaneous disorders. *JAMA* 132:126–32.

Goodrich LV, Milenkovic L, Higgins KM, and Scott MP. 1997. Altered neural cell fates and medulloblastoma in mouse patched mutants. *Science* 277:1109–13.

Götz J, Probst A, Ehler E, Hemmings B, and Kues W. 1998. Delayed embryonic lethality in mice lacking protein phosphatase 2A catalytic subunit Calpha. *Proc Natl Acad Sci USA* 95: 12370–5.

Gootwine E, Webb CG, and Sachs L. 1982. Participation of myeloid leukemia cells injected into embryos in haematopoietic differentiation in adult mice. *Nature* 299:63–5.

Goss PE, Ingle JN, Martino S, Robert NJ, Muss HB, Piccart MJ, al. e. 2003. A randomized trial of letrozole in postmenopausal women after five years of tamoxifen therapy for early-stage breast cancer. *N Eng J Med* 349:1793–1802.

Gottesman MM, Hrycyna CA, Schoenlein PV, Germann UA, and Pastan I. 1995. Genetic analysis of the multidrug transporter. *Ann Rev Genet* 29:607–49.

Grafstrom RH, Pan W, and Hoess RH. 1999. Defining the substrate specificity of cdk4 kinase-cyclin D1 complex. *Carcinogenesis* 20:193–8.

Graham CF and Wareing PF. 1976. *The Developmental Biology of Plants and Animals*. Philadelphia, PA: WB Saunders.

Gramzinski RA, Parchment RE, and Pierce GB. 1990. Evidence linking programmed cell death in the blastocyst to polyamine oxidation. *Differentiation* 43:59–65.

Greaves MH, Hariri G, Newman RA, Sutherland DR, Ritter MA, and Ritz J. 1983. Selective expression of the common acute lymphoblastic leukemia gp100 antigen on immature lymphoid cells and their malignant counterparts. *Blood* 61:628–39.

Green MD, Koelbl H, Baselga J, Galid A, Guillem V, Gascon P, Siena S, Lalisang RI, Samonigg H, Clemens MR et al. 2003. A randomized double-blind multicenter phase III study of fixed-dose single-administration pegfilgrastim versus daily filgrastim in patients receiving myelosuppressive chemotherapy. *Ann Oncol* 14:29–35.

Greenberg ER, Baron JA, Tosteson TD, Freeman DH Jr, Beck GJ, Bond JH, Colacchio TA, Coller JA, Frankl HD, Haile RW et al. 1994. A clinical trial of antioxidant vitamins to prevent colorectal adenoma. *N Engl J Med* 331:141–7.

Greenberger LM, Cohen D, and Horwitz SB. 1994. In vitro models of multiple drug resistance. *Cancer Treat Res* 73:69–106.

Greene HS N. 1957. Heterotransplantation of tumors. *Ann NY Acad Sci* 69:818–29.

Grem JL. 1990. Fluorinated pyrimidines. In: BA Chabner and JM Collins (eds.). *Cancer Chemotherapy*, pp. 180–224. Philadelphia, PA: JB Lippincott.

Grenman SE, Roberts JA, England BG, Gronroos M, and Carey TE. 1988. In vitro growth regulation of endometrial carcinoma cells by tamoxifen and medroxyprogesterone acetate. *Gynecol Oncol* 30:239–50.

Grenman S, Shapira A, and Carey TE. 1988. In vitro response of cervical cancer cell lines caski, hela, and me-180 to the antiestrogen tamoxifen. *Gynecol Oncol* 30:228–38.

Grever MR and Grieshaber CK. 1993. Toxicology by organ system. In: JF Holland, E Frei III, RC Bast Jr, DW Kufe, DL Morton, and RR Weichselbaum (eds.). *Cancer Medicine*, vol 1, pp. 683–97. Philadelphia, PA: Lea and Febiger.

Griffiths JD, McKinna JA, Rowbotham HD, Tsolakidis P, and Salsbury AJ. 1973. Carcinoma of the colon and rectum: Circulating malignant cells and five-year survival. *Cancer* 31:226–36.

Grimmer G and Misfeld J. 1983. Environmental carcinogens: A risk for man? Concept and strategy of the identification of carcinogens in the environment. In: G Grimmer (ed.). *Environmental Carcinogens: Polycyclic Aromatic Hydrocarbons*, pp. 1–26. Boca Raton, FL: CRC Press.

Grisham MB, Palombella VJ, Elliott PJ, Conner EM, Brand S, Wong HL, Pien C, Mazzola LM, Destree A, Parent L et al. 1999. Inhibition of nf-kappa b activation in vitro and in vivo: Role of 26s proteasome. *Methods Enzymol* 300:345–63.

Griswold DP Jr, Schabel FM Jr, Wilcox WS, Simpson-Herren L, and Skipper HE. 1968. Success and failure in the treatment of solid tumors: I Effects of cyclophosphamide (NSC-26271) on primary and metastatic plasmacytoma in the hamster. *Cancer Chemother Rep* 52: 345–87.

Grobstein C and Zwilling E. 1953. Modification of growth and differentiation of chorio-allantoic grafts of chick blastoderm pieces after culture of a glass slot interface. *J Exp Zool* 122:259–84.

Gronroos M, Maenpaa J, Kangas L, Erkkola R, Paul R, and Grenman S. 1987. Steroid receptors and response of endometrial cancer to hormones in vitro. *Ann Chir Gynaecol Suppl* 202:76–9.

Groopman JD and Kensler TW. 1999. The light at the end of the tunnel for chemical-specific biomarkers: daylight or headlight? *Carcinog* 20:1–11.

Gross L. 1970. *Oncogenic Viruses.* Oxford, UK: Pergamon Press.

Gunduz N, Fisher B, and Saffer EA. 1979. Effect of surgical removal on the growth and kinetics of residual tumor. *Cancer Res* 39:3861–5.

Guo QM. 2003. DNA microarray and cancer. *Curr Opin Oncol* 115:36–43.

Gupta RC and Lutz WK. 1999. Background DNA damage from endogenous and unavoidable exogenous carcinogens: a basis for spontaneous cancer incidence. *Mutat Res* 424:1–8.

Gutterman JU. 1994. Cytokine therapeutics: Lessons from interferon alpha. *Proc Natl Acad Sci USA* 91:1198–1205.

Haddow A and Blake I. 1933. Neoplasms in fish: A report of six cases with a summary of the literature. *J Pathology Bacteriology* 36:41–7.

Haenszel W. 1963. Variations in skin cancer incidence within the United States. *Natl Cancer Inst Monogr* 10:225–43.

Hagag N, Halegua S, and Viola M. 1986. Inhibition of growth factor induced differentiation of PC12 cells by micro-injection of antibody to *ras* p21. *Nature* 319:680–2.

Hagemann C and Rapp UR. 1999. Isotype-specific functions of Raf kinases. *Exp Cell Res* 253:34–46.

Hahn WC, Stewart SA, Brooks MW, York SG, Eaton E, Kurachi A, Beijersbergen RL, Knoll JH, Meyerson M, and Weinberg RA. 1999. Inhibition of telomerase limits the growth of human cancer cells. *Nat Med* 5:1164–70.

Hai T and Curran T. 1991. Cross-family dimerization of transcription factors fos/jun and ATF/CREB alters DNA-binding specificity. *Proc Natl Acad Sci USA* 88:3720–4.

Hait WN and Aftab DT. 1992. Rational design and preclinical pharmacology of drugs for reversing multidrug resistance. *Biochem Pharmacol* 43:103–7.

Hall JM, Lee MK, Newman B, Morrow JE, Anderson LA, Huey B, and King M-C. 1990. Linkage of early-onset familial breast cancer to chromosome 17q21. *Science* 250:1684–9.

Hall M and Peters G. 1996. Genetic alterations of cyclins, cyclin-dependent kinases, and Cdk inhibitors in human cancer. *Adv Cancer Res* 68:67–108.

Halloran MC and Berndt JD. 2003. Current progress in neural crest motility and migration and future prespects for the zebrafish model system. *Dev Dyn* 228:497–513.

Halsted WS. 1894–1895. The results of operations for the cure of cancer of the breast at the Johns Hopkins Hospital from 1889–1894. *Johns Hopkins Hosp Rep* 4:297–350.

Hammond MEH and Taube SE. 2002. Issues and barrier to development of clinically useful tumor markers: a developmental pathway proposal. *Semin Oncol* 29:213–21.

Hanks GE, Myers CE, and Scardino PT. 1993. Cancer of the prostate. In: VT DeVita Jr, S Hellman, and SA Rosenberg (eds.). *Cancer: Principles and Practice of Oncology,* 4th ed., pp. 1073–1113. Philadelphia, PA: JB Lippincott.

Hannon GJ and Beach D. 1994. p15INK4 is a potential effector of TGF-beta-induced cell cycle arrest. *Nature* 371:257–61.

Hantschel O and Superti-Furga G. 2004. Regulation of the c-Abl and Bcr-Abl tyrosine kinases. *Nat Rev Mol Cell Biol* 5:33–44.

Haque A and Kanz MF. 1988. Asbestos bodies in children's lungs. *Arch Pathol Lab Med* 112:514–18.

Harbour DA, Gunn-Moore DA, Gruffydd-Jones TJ, Caney SM, Bradshaw J, Jarrett O, and Wiseman A. 2002. Protection against oronasal challenge with virulent feline leukemia virus lasts for at least 12 months following a primary course of immunisation with Leukocell 2 vaccine. *Vaccine* 20:2866–72.

Harbour JW and Dean DC. 2000. The Rb/E2F pathway: expanding roles and emerging paradigms. *Genes Dev* 14:2393–2409.

Harbour JW, Lai S-L, Whang-Peng J, Gazdar AF, Minna JD, and Kaye FJ. 1988. Abnormalities in structure and expression of the human retinoblastoma gene in SCLC. *Science* 241:353–7.

Harley CB, Futcher AB, and Greider CW. 1990. Telomeres shorten during ageing of human fibroblasts. *Nature* 345:458–60.

Harris BE, Carpenter JT, and Diasio RB. 1991. Severe 5-fluorouracil toxicity secondary to dihydropyrimidine dehydrogenase deficiency. A potentially more common pharmacogenetic syndrome. *Cancer* 68:499–501.

Harris H. 1988. The analysis of malignancy in cell fusion: The position in 1988. *Cancer Res* 48:3302–6.

Harris RWC, Key TJA, Silcocks PB, Bull D, and Wald NJ. 1991. A case control study of dietary carotene in men with lung cancer and men with other epithelial cancers. *Nutr Cancer* 15:63–8.

Harshbarger JC. 2002. Marine turtle fibroma and fibropapilloma cases in the registry of tumors in lower animals. In: RG McKinnell and DL Carlson (eds.). *Proceedings of the Sixth International Symposium on the Pathology of Reptiles and Amphibians*, pp. 129–45. Saint Paul, MN: University of Minnesota Printing Services.

Harshbarger JC and Ostrander GK. 2000. Cancer in sharks, skates, rays and other lower fishes. *Proc Am Assoc Cancer Res* 41:762 (abstract).

Harshbarger JC, Charles AM, and Spero PM. 1981. Collection and analysis of neoplasms in sub-homeothermic animals from a phyletic point of view. In: CJ Dawe, JC Harshbarger, S Kondo, T Sugimura, and S Takayama (eds.). *Phyletic Approaches to Cancer*, pp. 357–84. Tokyo: Japan Scientific Societies Press.

Hartwell L. 1992. Defects in a cell cycle checkpoint may be responsible for the genomic instability of cancer cells. *Cell* 71:543–6.

Haseltine WA. 1983. Ultraviolet light repair and mutagenesis revisited. *Cell* 33:13–17.

Haskell CM. 1990. Principles and practice of cancer chemotherapy. In: CM Haskell (ed.). *Cancer Treatment*, 3rd ed., pp. 21–43. Philadelphia, PA: WB Saunders.

Haskell CM (ed.). 1990. *Cancer Treatment*, 3rd ed. Philadelphia, PA: WB Saunders.

Haskell CM and Berek JS. 2001. *Cancer Treatment*. 5th edition WB Saunders.

Haskell CM, Giuliano AE, Thompson RW, and Zarem HA. 1990. Breast cancer. In: CM Haskell (ed.). *Cancer Treatment*, 3rd ed., pp. 123–64. Philadelphia, PA: WB Saunders.

Haskell CM, Selch MT, and Ramming KP. 1990a. Colon and rectum. In: CM Haskell (ed.). *Cancer Treatment*, 3rd ed., pp. 232–54. Philadelphia, PA: WB Saunders.

1990b. Stomach. In: CM Haskell (ed.). *Cancer Treatment*, 3rd ed., pp. 217–31. Philadelphia, PA: WB Saunders.

Hassan HT and Zander A. 1996. Stem cell factor as a survival and growth factor in human normal and malignant hematopoiesis. *Acta Haematologica* 95:257–62.

Hawley-Nelson P, Vousden KH, Hubbert NL, Lowry DR, and Schiller JT. 1989. HPV16 E6 and E7 proteins cooperate to immortalize human foreskin keratinocytes. *EMBO J* 8:3905.

Hayashi Y, Nagao M, Sugimura T, Takayama S, Tomatis L, Wattenberg LW, and Wogan GN, eds. 1986. Preface *Int Symp Princess Takamatsu Cancer Res Fund* 16:vii.

Hayes DF, Van Zyl JA, Hacking A, Goedhals L, Bezwoda WR, Mailliard JA, Jones SE, Vogel CL, Berris RF, Shemano I et al. 1995. Randomized comparison of tamoxifen and two separate doses of toremifene in postmenopausal patients with metastatic breast cancer. *J Clin Oncol* 13:2556–66.

Hayflick L and Moorhead PS. 1961. The serial cultivation of human diploid cell strains. *Exp Cell Res* 25:585–621.

Hecker E. 1967. Phorbol esters from croton oil: Chemical nature and biological activities. *Naturwissenschaften* 54:282–4.

Heckler MM. 1984. Quoted in a news article in the *NIH Record* 36:1–12.

Hei TK, Piao CQ, He ZY, Vannais D, and Waldren CA. 1992. Chrysotile fiber is a strong mutagen in mammalian cells. *Cancer Res* 52:6305–9.

Hejna M, Raderer M, and Zielinski CC. 1999. Inhibition of metastases by anticoagulants. *J Natl Cancer Inst* 91:22–36.

Held W, Acha-Orbea H, MacDonald HR, and Waanders GA. 1994. Superantigens and retroviral infection: Insights from mouse mammary tumor virus. *Immunol Today* 15:184–90.

Helgason S, Wilking N, Carlstrom K, Damber MG, and von Schoultz B. 1982. A comparative study of the estrogenic effects of tamoxifen and 17 beta-estradiol in postmenopausal women. *J Clin Endocrinol Metab* 54:404–8.

Heller I. 1930. Occupational cancers. *J Ind Hyg* 12:169–97.

Hemminki K (ed.). 1994. DNA adducts: Identification and biological significance. *IARC Scientific Publication 125*, Lyon, FR: International Agency for Research on Cancer.

Hemminki K and Mutanen P. 2001. Genetic epidemiology of multistage carcinogenesis. *Mutat Res* 473:11–21.

Hemminki K, Koskinen M, Rajaniemi H, and Zhao C. 2000. DNA adducts, mutations, and cancer 2000. *Regul Toxicol Pharmacol* 32:264–75.

Henderson BE, Bernstein L, and Ross R. 1997. Etiology of cancer: hormonal factors. In: VT DeVita Jr, S Hellman, and SA Rosenberg (eds.). *Cancer, Principles and Practice of Oncology*, 5th ed., pp. 219–29. Philadelphia: Lippincott-Raven.

Henderson CJ, Smith AG, Ure J, Brown K, Bacon EJ, and Wolf CR. 1998. Increased skin tumorigenesis in mice lacking pi class glutathione S-transferases. *Proc Natl Acad Sci USA* 95:5275–80.

Hendricks JD, Meyers TR, and Shelton DW. 1984. Histological progression of hepatic neoplasia in rainbow trout (Salmo gairdneri). In: Use of small fish species in carcinogenicity testing. *National Cancer Institute Monograph* 65:321–36.

Hendrix MJ, Seftor EA, Seftor RE, Kirschmann DA, Gardner LM, Boldt HC, Meyer M, Pe'er J, and Folberg R. 1998. Regulation of uveal melanoma interconverted phenotype by hepatocyte growth factor/scatter factor (hgf/sf). *Am J Pathol* 152:855–63.

Hendry JH, Potten CS, Chadwick C, and Bianchi M. 1982. Cell death (apoptosis) in the mouse small intestine after low doses: effects of dose rate, 14.7 Me V neutrons, and 600 MeV (maximum energy) neutrons. *Int J Radiat Biol Relat Stud Phys Chem Med* 42:611–20.

Hengartner MO and Horvitz HR. 1994. Programmed cell death in *Caenorhabditis elegans. Curr Opin Genet Dev* 4:581–6.

Hennekens CH, Buring JE, Manson JE, Stampfer M, Rosner B, Cook NR, and Belanger C. 1996. Lack of effect of long-term supplementation with beta carotene on the incidence of malignant neoplasms and cardiovascular disease. *N Engl J Med* 334:1145–9.

Hennessy C, Henry JA, May FEB, Westley BR, Angus B, and Lennard TWJ. 1991. Expression of the antimetastatic gene *nm23* in human breast cancer: An association with good prognosis. *J Natl Cancer Inst* 83:281–5.

Heppner GH. 1982. Tumor subpopulation interactions. In: AH Owens, DS Coffey, and SB Baylin (eds.). *Tumor Cell Heterogeneity: Origins and Implications*, pp. 225–36. New York: Academic Press.

Hernandez S, Hernandez L, Bea S, Pinyol M, Nayach I, Bellosillo B, Nadal A, Ferrer A, Fernandez PL, Montserrat E, Cardesa A, and Campo E. 2000. cdc25a and the splicing variant cdc25b2, but not cdc25B1, -B3 or -C, are over-expressed in aggressive human non-Hodgkin's lymphomas. *Int J Cancer* 89:148–52.

Hersey P and Zhang XD. 2003. Overcoming resistance of cancer cells to apoptosis. *J Cell Physiol* 196:9–18.

Herzig M and Christofori G. 2002. Recent advances in cancer research: mouse models of tumorigenesis. *Biochim Biophys Acta* 1602:97–113.

Hesdorffer C, Antman K, Bank A, Fetell M, Mears G, and Begg M. 1994. Human MDR gene transfer in patients with advanced cancer. *Hum Gene Ther* 5:1151–60.

Hess AR, Seftor EA, Gardner LM, Carles-Kinch K, Schneider GB, Seftor RE, Kinch MS, and Hendrix MJ. 2001. Molecular regulation of tumor cell vasculogenic mimicry by tyrosine phosphorylation: Role of epithelial cell kinase (eck/epha2). *Cancer Res* 61:3250–5.

Higginson J, Muir CS, Muñoz N. 1992. *Human Cancer: Epidemiology and Environmental Causes*. Cambridge: Cambridge University Press.

Higinbotham KG, Rice JM, Reed CD, Watatani M, Enomoto T, Anderson LM, and Perantoni AO. 1996. Variant mutational activation of the K-ras oncogene in renal mesenchymal tumors induced in newborn F344 rats by methyl(methoxymethyl)nitrosamine. *Carcinogenesis* 17: 2625–30.

Hillion J, Mecucci C, Aventin A, Leroux D, Wlodarska I, Van Den Berghe H, and Larsen CJ. 1991. A variant translocation t(2;18) in follicular lymphoma involves the 5′ end of bcl-2 and Ig kappa light chain gene. *Oncogene* 6:169–72.

Hirayama T. 1979. Diet and cancer. *Nutr Cancer* 1:67–81.

Hjalgrim H, Askling J, Rostgaard K, Hamilton-Dutoit S, Frisch M, Zhang J-S, Madsen M, Rosdahl N, Konradsen HB, Storm HH et al. 2003. Characteristics of Hodgkin's lymphoma after infectious mononucleosis. *New England J Med* 349:1324–32.

Hochedlinger K and Jaenisch R. 2002. Nuclear transplantation: lessons from frogs and mice. *Current Opinion in Cell Biology* 14:741–8.

Hochedlinger K, Blelloch R, Brennan C, Yamada Y, Kim M, Chin L, and Jaenisch R. 2004. Reprogramming of a melanoma genome by nuclear transplantation. *Genes Dev* 18:1875–85.

Hoelzer D, Ganser A, Anger B, Seifried E, and Heimpel H. 1984. Low-dose Ara-C in the treatment of acute leukemia. Cytotoxicity or differentiation induction? *Blut* 48:233–8.

Hoglund M, Gisselsson D, Hansen GB, Sall T, and Mitelman F. 2003. Ovarian carcinoma develops through multiple modes of chromosomal evolution. *Cancer Res* 63:3378–85.

Hohl R. 2005. *Concise Clinical Pharmacology: Cancer Therapeutics*. New York: McGraw-Hill Professional.

Holaday DA. 1969. History of the exposure of miners to radon. *Health Phys* 16:547–52.

Holman CDJ, Armstrong BK, and Heenan PJ. 1986. Relationship of cutaneous malignant melanoma to individual sunlight exposure habits. *J Natl Cancer Inst* 76:403–14.

Holmes EC, Livingston R, and Turrisi A III. 1993. Neoplasms of the thorax. In: JF Holland, E Frei III, RC Bast Jr, DW Kufe, DL Morton, and RR Weichselbaum (eds.). *Cancer Medicine*, vol 2, pp. 1285–1337. Philadelphia, PA: Lea and Febiger.

Holmes FA, Jones SE, O'Shaughnessy J, Vukelja S, George T, Savin M, Richards D, Glaspy J, Meza L, Cohen G et al. 2002. Comparable efficacy and safety profiles of once-per-cycle pegfilgrastim and daily injection filgrastim in chemotherapy-induced neutropenia: A multicenter dose-finding study in women with breast cancer. *Ann Oncol* 13:903–9.

Holmes FA, O'Shaughnessy JA, Vukelja S, Jones SE, Shogan J, Savin M, Glaspy J, Moore M, Meza L, Wiznitzer I et al. 2002. Blinded, randomized, multicenter study to evaluate single administration pegfilgrastim once per cycle versus daily filgrastim as an adjunct to chemotherapy in patients with high-risk stage ii or stage iii/iv breast cancer. *J Clin Oncol* 20:727–31.

Honegger AM, Schmidt A, Ullrich A, and Schlessinger J. 1990. Evidence for epidermal growth factor (EGF)-induced intermolecular autophosphorylation of the EGF receptors in living cells. *Mol Cell Biol* 10:4035–44.

Honore S, Kamath K, Braguer D, Horwitz SB, Wilson L, Briand C, and Jordan MA. 2004. Synergistic suppression of microtubule dynamics by discodermolide and paclitaxel in non-small cell lung carcinoma cells. *Cancer Res* 64:4957–64.

Hooft van Huijsduijnen R. 1998. Protein tyrosine phosphatases: counting the trees in the forest. *Gene* 225:1–8.

Hopwood D and Levison DA. 1976. Atrophy and apoptosis in the cyclical human endometrium. *J Pathol* 119:159–66.

Horita M, Andreu EJ, Benito A, Arbona C, Sanz C, Benet I, Prosper F, and Fernandez-Luna JL. 2000. Blockade of the Bcr-Abl kinase activity induces apoptosis of chronic myelogenous leukemia cells by suppressing signal transducer and activator of transcription 5-dependent expression of Bcl-xL. *J Exp Med* 191:977–84.

Horowitz JM, Yandell DW, Park SH, Canning S, Whyte P, Buchkovich K, Harlow E, Weinberg RA, and Dryja TP. 1989. Point mutational activation of the retinoblastoma antioncogene. *Science* 243:937–40.

Horwitz KB, Koseki Y, and McGuire WL. 1978. Estrogen control of progesterone receptor in human breast cancer: Role of estradiol and antiestrogen. *Endocrinology* 103:1742–51.

Howard EB, Britt JO Jr, and Simpson JG. 1983. Neoplasms in marine mammals. In: EB Howard (ed.). *Pathobiology of Marine Mammal Diseases*, vol 2, p. 146. Boca Raton, FL: CRC Press.

Howley PM, Ganem D, and Kieff E. 1997. Etiology of cancer: viruses. In: VT DeVita Jr., S Hellman, SA Rosenberg (eds.). *Cancer: Principles and Practice of Oncology*, 5th ed., pp. 168–184. Philadelphia: Lippencott-Raven.

Howley PM and Münger K. 1999. Human papillomaviruses and squamous cell carcinomas. In: J Parsonnet and SJ Horning (eds.). *Microbes and Malignancy: Infection as a Cause of Human Cancer.* pp. 157–79. New York:Oxford University Press.

Hsieh JK, Chan FS, O'Connor DJ, Mittnacht S, Zhong S, and Lu X. 1999. RB regulates the stability and the apoptotic function of p53 via MDM2. *Mol Cell* 3:181–93.

Hsu IC, Metcalf RA, Sun T, Welsh J, Wang NJ, and Harris CC. 1991. P53 gene mutational hotspot in human hepatocellular carcinomas from Qidong, China. *Nature* 350:427–8.

Hua JR and Muschel RJ. 1996. Inhibition of matrix metalloproteinase-9 expression by a ribozyme blocks metastasis in a rat sarcoma model system. *Cancer Res* 56:5279–84.

Huang E, Cheng SH, Dressman H, Pittman J, Tsou MH, Horng CF, Bild A, Iversen ES, Liao M, Chen CM, West M, Nevins JR, and Huang AT. 2003. Gene expression predictors of breast cancer outcomes. *Lancet* 361:1590–6.

Huang E, Ishida S, Pittman J, Dressman H, Bild A, Kloos M, D'Amico M, Pestell RG, West M, and Nevins JR. 2003. Gene expression phenotypic models that predict the activity of oncogenic pathways. *Nat Genet* 34:226–30.

Huang HJS, Lee JK, Shew JY, Chen PL, Bookstein R, Friedmann T, Lee EY-HP, and Lee W-H. 1988. Suppression of the neoplastic phenotype by replacement of the RB gene in human cancer cells. *Science* 242:1563–6.

Huang JC, Svoboda DL, Reardon JT, and Sancar A. 1992. Human nucleotide excision nuclease removes thymine dimers from DNA by incising the 22nd phosphodiester bond $5'$ and the 6th phosphodiester bond $3'$ to the photodimer. *Proc Natl Acad Sci USA* 89:3664–8.

Huang ME, Ye YC, Chen SR, Chai JR, Lu JX, Zhoa L, Gu LJ, and Wang ZY. 1988. Use of all-trans retinoic acid in the treatment of acute promyelocytic leukemia. *Blood* 72:567–72.

Huang YZ, Zang M, Xiong WC, Luo Z, and Mei L. 2003. Erbin suppresses the MAP kinase pathway. *J Biol Chem* 278:1108–14.

Huff JE, McConnell EE, Haseman JK, Boorman GA, Eustis SL, Schwetz BA, Rao GN, Jameson CW, Hart LG, and Rall DP. 1988. Carcinogenesis studies: Results of 398 experiments on 104 chemicals from the US National Toxicology Program. *Ann NY Acad Sci* 534:1–30.

Huggins CB and Hodges CV. 1941. Studies on prostate cancer: I. The effect of castration, of estrogen and of androgen injection on serum phosphatases in metastatic carcinoma of the prostate. *Cancer Res* 1:293–7.

Huggins CB, Grand LC, and Brillantes FP. 1961. Mammary cancer induced by a single feeding of polynuclear hydrocarbons and its suppression. *Nature* 189:204–7.

Hunter BR, Tweedell K, and McKinnell RG. 1990. PNKT4B cells exhibit motility at invasion-restrictive temperatures. *Proc Am Assoc Cancer Res* 31:66.

Hunter T. 1991. Cooperation between oncogenes. *Cell* 64:249–70.

Hunter T and Sefton BM. 1980. The transforming gene product of Rous sarcoma virus phosphorylates tyrosine. *Proc Natl Acad Sci USA* 77:1311–15.

Hurley PM. 1998. Mode of carcinogenic action of pesticides inducing thyroid follicular cell tumors in rodents. *Environ Health Perspect* 106:437–45.

Huxley J. 1958. *Biological Aspects of Cancer.* New York: Harcourt, Brace.

Hyde JN. 1906. On the influence of light in the production of cancer of the skin. *Am J Med Sci* 131:1–22.

IARC (International Agency for Research on Cancer). 2001. *Sunscreens,* IARC Handbooks on Cancer Prevention, Volume 5. Lyon, FR: World Health Organization.

IARC Monographs on the Evaluation of Carcinogenic Risks to Humans: *Solar and Ultraviolet Radiation.* 1992. Volume 5, Lyon, FR: IARC.

IARC Working Group 2001. Ionizing radiation, part 2: some internally deposited radionuclides. *IARC Monogr Eval Carcinog Risks Hum.* 78:1–559.

Ikeda K, Sasaki K, Tasaka T, Nagai M, Kawanishi K, Takahara J, and Irino S. 1994a. Detection of pml-retinoic acid receptor-alpha fusion transcripts in acute promyelocytic leukemia with trisomy 8 but without t(15;17). *Am J Hematol* 45:212–16.

    1994b. PML-PAR alpha fusion transcripts by RNA PCR in acute promyelocytic leukemia in remission and its correlation with clinical out come. *Int J Hematol* 60:197–205.

Ikenaga M. 1994. Cytotoxic action of alkylating agents in human tumor cells and its relationship to apoptosis. *Gann Tokyo Kagaku Ryoho* 21:596–601.

Ionov Y, Peinado MA, Malkhosyan S, Shibata D, and Perucho M. 1993. Ubiquitous somatic mutations in simple repeated sequences reveal a new mechanism for colonic carcinogenesis. *Nature* 363:558–61.

Iqbal S and Lenz HJ. 2004. Integration of novel agents in the treatment of colorectal cancer. *Cancer Chemother Pharmacol* 54 Suppl 1:S32–9.

Irby RB, Mao W, Coppola D, Kang J, Loubeau JM, Trudeau W, Karl R, Fujita DJ, Jove R, and Yeatman TJ. 1999. Activating SRC mutation in a subset of advanced human colon cancers. *Nat Genet* 21:187–90.

Isaacs JT and Coffey DS. 1981. Adaptation versus selection as the mechanism responsible for the relapse of prostatic cancer to androgen ablation therapy as studied in the Dunning R-3327-H adenocarcinoma. *Cancer Res* 41:5070–5.

Isaacs JT, Heston WD, Weissman RM, and Coffey DS. 1978. Animal models of the hormone-sensitive and -insensitive prostatic adenocarcinomas, Dunning R-3327-H, R-3327-HI, and R-3327-AT. *Cancer Res* 38:4353–9.

Isaacs JT, Wake N, Coffey DS, and Sandberg AA. 1982. Genetic instability coupled to clonal selection as a mechanism for tumor progression in the Dunning R-3327 rat prostatic adenocarcinoma system. *Cancer Res* 42:2353–71.

Ishikawa T, Masahito P, Nemoto N, Matsumoto J, Shima A. 1986. Spontaneous neurinoma in an African Lungfish, Protopterus annectens, and DNA repair studies on normal and neoplastic tissues. *J Natl Cancer Inst* 77:521–8.

Isobe M, Emanuel BS, Givol D, Oren M, and Croce CM. 1986. Localization of gene for human p53 tumour antigen to band 17p13. *Nature* 320:84–5.

Iyer L, King CD, Whitington PF, Green MD, Roy SK, Tephly TR, Coffman BL, and Ratain MJ. 1998. Genetic predisposition to the metabolism of irinotecan (cpt-11). Role of uridine diphosphate glucuronosyltransferase isoform 1a1 in the glucuronidation of its active metabolite (sn-38) in human liver microsomes. *J Clin Invest* 101:847–54.

Jacks T, Fazeli A, Schmitt EM, Bronson RT, Goodell MA, and Weinberg RA. 1992. Effects of an Rb mutation in the mouse. *Nature* 359:295–300.

Jackson P and Baltimore D. 1989. N-terminal mutations activate the leukemogenic potential of the myristoylated form of c-*abl. EMBO J* 8:6649–53.

Jacobs K, Shoemaker C, Rudersdorf R, Neill SD, Kaufman RJ, Musson A, Seehra J, Jones SS, Hewick R, Fritsch EF et al. 1985. Isolation and characterization of genomic and cDNA clones of human erythropoietin. *Nature* 313:806–10.

Jain RK. 2002. Tumor angiogenesis and accessibility: Role of vascular endothelial growth factor. *Semin Oncol* 29:3–9.

Janne PA, Gurubhagavatula S, Yeap BY, Lucca J, Ostler P, Skarin AT, Fidias P, Lynch TJ, and Johnson BE. 2004. Outcomes of patients with advanced non-small cell lung cancer treated with gefitinib (ZD1839, "Iressa") on an expanded access study. *Lung Cancer* 44:221–30.

Janse JG, Mohn GR, Vrieling H, van Teijlingen CM, Lohman PH, and van Zeeland AA. 1994. Molecular analysis of hprt gene mutations in skin fibroblasts of rats exposed in vivo to N-methyl-N-nitrosourea or N-ethyl-N-nitrosourea. *Cancer Res* 54:2478–85.

Jaurand MC. 1989. Particulate-state carcinogenesis: A survey of recent studies on the mechanisms of action of fibres. In: J Bignon, J Peto, and R Saraccik (eds.). *Nonoccupational Exposure to Mineral Fibres*, pp. 54–73. IARC Scientific Publication 90. Lyon, FR: International Agency for Research on Cancer.

Jehn U, De Bock R, and Haanen C. 1984. Clinical trial of low-dose Ara-C in the treatment of acute leukemia and myelodysplasia. *Blut* 48:255–61.

Jelkmann W and Metzen E. 1996. Erythropoietin in the control of red cell production. *Anat Anz* 178:391–403.

Jemal A, Tiwari RC, Murry T, Ghafoor A, Samuels S, Ward E, Feuer EJ, and Thun MJ. 2004. Cancer Statistics, 2004. *CA Cancer J Clin* 534:8–29.

Jemal A, Murray T, Ward E, Samuels A, Tiwari RC, Ghafoor A, Feuer EJ, Thun MJ. 2005. Cancer statistics 2005. *CA Cancer J Clin* 55:10–30.

Jewett HJ. 1970. The case for radical perineal prostatectomy. *J Urol* 103:195–9.

Jhappan C, Stahle C, Harkins RN, Fausto N, Smith GH, and Merlino GT. 1990. TGF overexpression in transgenic mice induces liver neoplasia and abnormal development of the mammary gland and pancreas. *Cell* 61:1137–46.

Johnson EF, Hsu MH, Savas U, and Griffin KJ. 2002. Regulation of P450 4A expression by peroxisome proliferator activated receptors. *Toxicology* 181–182:203–6.

Johnson IS, Armstrong JG, Gorman M, and Burnett JP, Jr. 1963. The vinca alkaloids: A new class of oncolytic agents. *Cancer Res* 23:1390–1427.

Johnson IS, Wright HF, Svoboda GH, and Vlantis J. 1960. Antitumor principles derived from vinca rosea linn. I. Vincaleukoblastine and leurosine. *Cancer Res* 20:1016–22.

Johnson RL, Rothman AL, Xie J, Goodrich LV, Bare JW, Bonifas JM, Quinn AG, Myers RM, Cox DR, Epstein EH, Jr, and Scott MP. 1996. Human homolog of patched, a candidate gene for the basal cell nevus syndrome. *Science* 272:1668–71.

Johnston TP, McCaleb GS, and Montgomery JA. 1963. The synthesis of antineoplastic agents: XXXII. N-nitroso-ureas. *J Med Chem* 122:669–81.

Jonasson J, Povey S, and Harris H. 1977. The analysis of malignancy by cell fusion: VII. Cytogenetic analysis of hybrids between malignant and diploid cells and of tumours derived from them. *J Cell Sci* 24:217–54.

Jones SN, Roe AE, Donehower LA, and Bradley A. 1995. Rescue of embryonic lethality in Mdm2-deficient mice by absence of p53. *Nature* 378:206–8.

Jordan VC. 1982. Metabolites of tamoxifen in animals and man: Identification, pharmacology, and significance. *Breast Cancer Res Treat* 2:123–38.

    1988. Long-term tamoxifen therapy to control or to prevent breast cancer: Laboratory concept to clinical trials. *Prog Clin Biol Res* 262:105–23.

Jordan VC and Robinson SP. 1987. Species-specific pharmacology of antiestrogens: Role of metabolism. *Fed Proc* 46:1870–4.

Jordan VC, Gottardis MM, and Satyaswaroop PG. 1991. Tamoxifen-stimulated growth of human endometrial carcinoma. *Ann NY Acad Sci* 622:439–46.

Josephy PD. 2002. Genetically-engineered bacteria expressing human enzymes and their use in the study of mutagens and mutagenesis. *Toxicology* 181–182:255–60.

Kabbinavar F, Hurwitz HI, Fehrenbacher L, Meropol NJ, Novotny WF, Lieberman G, Griffing S, and Bergsland E. 2003. Phase II, randomized trial comparing bevacizumab plus fluorouracil (FU)/leucovorin (LV) with FU/LV alone in patients with metastatic colorectal cancer. *J Clin Oncol* 21:60–5.

Kadlubar FF, Miller JA, and Miller CE. 1977. Hepatic microsomal N-glucuronidation and nucleic acid binding of N-hydroxy arylamines in relation to urinary bladder carcinogenesis. *Cancer Res* 37:805–14.

Kamijo T, Bodner S, van de Kamp E, Randle DH, and Sherr CJ. 1999. Tumor spectrum in ARF-deficient mice. *Cancer Res* 59:2217–22.

Kammula US, Ghossein R, Bhattacharya S, and Coit DG. 2004. Serial follow-up and the prognostic significance of reverse transcriptase-polymerase chain reaction–staged sentinel lymph nodes from melanoma patients. *J Clin Oncol* 22:3989–96.

Kaplan HS. 1981. Progress in radiation therapy of cancer. In: JH Burchenal and HF Oettgen (eds.). *Cancer: Achievements, Challenges, and Prospects for the 1980s*, vol 2, pp. 829–57. New York: Grune and Stratton.

Kashishian A, Kazlauskas A and Cooper JA. 1992. Phosphorylation sites in the PDGF receptor with different specificities for binding GAP and PI3 kinase in vivo. *EMBO J* 11:1373–82.

Katzenellenbogen BS, Norman MJ, Eckert RL, Peltz SW, and Mangel WF. 1984. Bioactivities, estrogen receptor interactions, and plasminogen activator-inducing activities of tamoxifen and hydroxy-tamoxifen isomers in MCF-7 human breast cancer cells. *Cancer Res* 44: 112–19.

Kauffmann-Zeh A, Rodriguez-Viciana P, Ulrich E, Gilbert C, Coffer P, Downward J, and Evan G. 1997. Suppression of c-Myc-induced apoptosis by Ras signalling through PI(3)K and PKB. *Nature* 385:544–8.

Kawai S and Hanafusa H. 1971. The effects of reciprocal changes in temperature on the transformed state of cells infected with a Rous sarcoma virus mutant. *Virology* 46:470–9.

Kelly MG, O'Gara RW, Adamson RH, and Gadekar K. 1966. Induction of hepatic cell carcinomas in monkeys with N-nitrosodiethylamine. *J Natl Cancer Inst* 36:323–51.

Kelly ML, Winge P, Heaney JD, Stephens RE, Farell JH, Van Beneden RJ, Reinisch CL, Lesser MP, and Walker CW. 2001. Expression of homologues for p53 and p73 in the softshell clam (Mya arenaria), a naturally-occurring model for human cancer. *Oncogene* 20:748–58.

Kemp G, Rose P, Lurain J, Berman M, Manetta A, Roullet B, Homesley H, Belpomme D, and Glick J. 1996. Amifostine pretreatment for protection against cyclophosphamide-induced and cisplatin-induced toxicities: Results of a randomized control trial in patients with advanced ovarian cancer. *J Clin Oncol* 14:2101–12.

Kennaway EL. 1924. On the cancer-producing factor in tar. *Br Med J* 1:564–7.

Kenny PA and Bissell MJ. 2003. Tumor reversion: correction of malignant behavior by microenvironmental cues. *Int J Cancer* 107:688–95.

Keohavong P, DeMichele MA, Melacrinos AC, Landreneau RJ, Weyant RJ, and Siegfried JM. 1996. Detection of K-ras mutations in lung carcinomas: relationship to prognosis. *Cancer Res* 2:411–18.

Kern SE, Vogelstein B. 1991. Genetic alterations in colorectal tumors. In J Brugge, T Curran, E Harlow, F McCormick, (eds.). *Origins of Human Cancer*, pp 577–85. Plainview, NY: Cold Spring Harbor Press.

Kerr A. 1980. Biological control of crown gall through production of agrocin 84. *Plant Dis* 4:24–30.

Kerr JF and Searle J. 1973. Deletion of cells by apoptosis during castration-induced involution of the rat prostate. *Virchows Arch B* 13:87–102.

Kerr JF R, Wyllie AH, and Currie AR. 1972. Apoptosis: A basic biological phenomenon with wide-ranging implications in tissue kinetics. *Br J Cancer* 26:239–57.

Kessler II. 1976. Human cervical cancer as a venereal disease. *Cancer Res* 36:783–91.

Khanna KK and Jackson SP. 2001. DNA double-strand breaks: signaling, repair and the cancer connection. *Nat Genet* 27:247–54.

Kim B. 1996. Biological therapy: Interferons, interleukins, and monoclonal antibodies. In: R Pazdur, LR Coia, WJ Hoskins, and LD Wagman (eds.). *Cancer Management: A Multidisciplinary Approach*, pp. 581–92. Huntington, NY: PRR.

Kim JY, Ferrell JE, Chae SK, and Lee KJ. 2000. Inhibition of progesterone-induced Xenopus oocyte maturation by Nm 23. *Cell Growth Diff* 11:485–90.

Kim NW, Piatyszek MA, Prowse KR, Harley CB, West MD, Ho PL, Coviello GM, Wright WE, Weinrich SL, and Shay JW. 1994. Specific association of human telomerase activity with immortal cells and cancer. *Science* 266:2011–15.

Kim SJ, Uehara H, Karashima T, McCarty M, Shih N, and Fidler IJ. 2001. Expression of interleukin-8 correlates with angiogenesis, tumorigenicity, and metastasis of human prostate cancer cells implanted orthotopically in nude mice. *Neoplasia* 3:33–42.

Kimbrough RD. 1983. Determining exposure and biochemical effects in human population studies. *Environ Health Perspect* 48:77–9.

Kimura N, Shimazada N, Nomura K, and Watanabe K. 1990. Isolation and characterization of cDNA clone encoding rat nucleoside diphosphate kinase. *J Biol Chem* 265:15744–9.

King M-C, Rowell S, and Love SM. 1993. Inherited breast and ovarian cancer. What are the risks? What are the choices? *JAMA* 269:1975–80.

King RW, Deshaies RJ, Peters JM, and Kirschner MW. 1996. How proteolysis drives the cell cycle. *Science* 274:1652–9.

King TJ and McKinnell RG. 1960. An attempt to determine the developmental potentialities of the cancer cell nucleus by means of transplantation. *Cell Physiology of Neoplasia*, pp. 591–617. Austin, TX: University of Texas Press.

Kinzler KW and Vogelstein B. 1996. Lessons from hereditary colorectal cancer. *Cell* 87:159–70.

Kinzler KW, Nilbert MC, Su LK, Vogelstein B, Bryan TM, Levy DB, Smith KJ, Preisinger AC, Hedge P, McKechnie D et al. 1991. Identification of FAP locus genes from chromosome 5q21. *Science* 253:661–5.

Kisker O, Becker CM, Prox D, Fannon M, D'Amato R, Flynn E, Fogler WE, Sim BK, Allred EN, Pirie-Shepherd SR et al. 2001. Continuous administration of endostatin by intraperitoneally implanted osmotic pump improves the efficacy and potency of therapy in a mouse xenograft tumor model. *Cancer Res* 61:7669–74.

Kizaka-Kondoh S, Matsuda M, and Okayama H. 1996. CrkII signals from epidermal growth factor to Ras. *Proc Natl Acad Sci USA* 93:12177–82.

Klein EA. 1996. Hormone therapy for prostate cancer: A topical perspective. *Urology* 47(1A Suppl):3–12; discussion 29–32.

Kleinsmith LJ and Pierce GB. 1964. Multipotentiality of single embryonal carcinoma cells. *Cancer Res* 24:1544–51.

Klingenberg M. 1958. Pigments of rat liver microsomes. *Arch Biochem Biophys* 75:376–86.

Knight JM, Kirincich AN, Farmer ER, and Hood AF. 2002. Awareness of the risks of tanning lamps does not influence behavior among college students. *Arch Dermatol* 138:1311–15.

Knight LA, DiNicolantonio F, Whitehouse P, Mercer S, Sharma S, Glaysher S, Johnson P, and Cree IA. 2004. The in vitro effect of gefitinib ("Iressa") alone and in combination with cytotoxic chemotherapy on human solid tumours. *BMC Cancer* 4:83.

Knoepfler PS, Cheng PF, and Eisenman RN. 2002. N-myc is essential during neurogenesis for the rapid expansion of progenitor cell populations and the inhibition of neuronal differentiation. *Genes Dev* 16:2699–2712.

Knudson AG. 1985. Hereditary cancer, oncogenes, and antioncogenes. *Cancer Res* 45:1437–43.

2002. Cancer genetics. *Am J Med Genetics* 111:96–102.

Knudson AG and Strong LC. 1972. Mutation and cancer: a model for Wilms' tumor of the kidney. *J Natl Cancer Inst* 48:313–24.

Kohn EC and Liotta LA. 1995. Molecular insights into cancer invasion: Strategies for prevention and intervention. *Cancer Res* 55:1856–62.

Koizumi S, Curt GA, Fine RL, Griffin JD, and Chabner BA. 1985. Formation of methotrexate polyglutamates in purified myeloid precursor cells from normal human bone marrow. *J Clin Invest* 75:1008–11.

Konishi N, Hiasa Y, Matsuda H, Tao M, Tsuzuki T, Hayashi I, Kitahori Y, Shiraishi T, Yatani R, Shimazaki J et al. 1995. Intratumor cellular heterogeneity and alterations in ras oncogene and p53 tumor suppressor gene in human prostate carcinoma. *Am J Pathol* 147: 1112–22.

Korinek V, Barker N, Morin PJ, Van Wichen D, de Weger R, Kinzler KW, Vogelstein B, and Clevers H. 1997. Constitutive transcriptional activation by a beta-catenin-Tcf complex in APC-/- colon carcinoma. *Science* 275:1784–7.

Kottler MJ. 1974. From 48 to 46: Cytological technique, preconception, and the counting of human chromosomes. *Bull Hist Med* 48:465–502.

Koury ST, Bondurant MC, and Koury MJ. 1988. Localization of erythropoietin synthesizing cells in murine kidneys by in situ hybridization. *Blood* 71:524–7.

Koury ST, Koury MJ, and Bondurant MC. 1988. Morphological changes in erythroblasts during erythropoietin-induced terminal differentiation in vitro. *Exp Hematol* 16: 758–63.

Koutsky LA, Ault KA, Wheeler CM, Brown DR, Barr E, Alvarez FB, Chiacchierini LM, and Jansen KU. 2002. A Controlled trial of a human papillomavirus type 16 vaccine. *N Engl J Med* 347:1645–51.

Kraemer KH, Lee MM, and Scotto J. 1987. Xeroderma pigmentosum. Cutaneous, ocular, and neurologic abnormalities in 830 published cases. *Arch Dermatol* 123:241–50.

Kraemer KH, Lee MM, Andrews AD, and Lambert WC. 1994. The role of sunlight and DNA repair in melanoma and nonmelanoma skin cancer. The xeroderma pigmentosum paradigm. *Arch Dermatol* 130:1018–21.

Kraljic I, Kovacic K, and Tarle M. 1994. Serum TPS, PSA, and PAP values in relapsing stage D2 adenocarcinoma of the prostate. *Urol Res* 22:329–32.

Kraus RS, Daston DL, Caspary WJ, and Eling TE. 1986. Peroxidase-mediated metabolic activation of the thyroid carcinogen amitrole. *Proc Am Assoc Cancer Res* 27:112.

Krek W, Xu G, and Livingston DM. 1995. Cyclin A-kinase regulation of E2F-1 DNA binding function underlies suppression of an S phase checkpoint. *Cell* 83:1149–58.

Kris-Etherton PM, Hecker KD, Bonanome A, Coval SM, Binkoski AE, Hilpert KF, Griel AE, and Etherton TD. 2002. Bioactive compounds in foods: their role in the prevention of cardiovascular disease and cancer. *Am J Med* 113 Suppl 9B:71S–88S.

Krontiris TG and Cooper GM. 1981. Transforming activity of human tumor DNAs. *Proc Natl Acad Sci USA* 78:1181–4.

Kuchino Y, Mori F, Kasai H, Inoue H, Iwai S, Miura K, Ohtsuka E, and Nishimura S. 1987. Misreading of DNA templates containing 8-hydroxydeoxyguanosine at the modified base and at adjacent residues. *Nature* 327:77–9.

Kufe DW, Pollock RE, Weichselbaum RR, Bast RC, and Gansler TS. 2003. Holland-Frei Cancer Medicine. 6th edition. BC Decker.

Kullendorff CM, Soller M, Wiebe T, and Mertens F. 2003. Cytogenetic findings and clinical course in a consecutive series of Wilms tumors. *Cancer Genet Cytogenet* 140:82–7.

Kumar R, Sukumar S, and Barbacid M. 1990. Activation of ras oncogenes preceding the onset of neoplasia. *Science* 248:1101–4.

Kumar V, Cotran RS, and Robbins SL. 1992. *Robbins Pathologic Basis of Disease*, 5th ed. Philadelphia, PA: WB Saunders.

Kuniyasu H, Ohmori H, Sasaki T, Sasahira T, Yoshida K, Kitadai Y, and Fidler IJ. 2003. Production of interleukin 15 by human colon cancer cells is associated with induction of mucosal hyperplasia, angiogenesis, and metastasis. *Clin Cancer Res* 9:4802–10.

Kurzik-Dumke U, Phannavong B, Gundacker D, and Gateff E. 1992. Genetic, cytogenetic and developmental analysis of the *Drosophila melanogaster* tumor suppressor gene *lethal(2)tumorous imaginal discs (1(2)tid).Differentiation* 51:91–104.

Kurzrock R, Gutterman JU, and Talpaz M. 1988. The molecular genetics of Philadelphia chromosome–positive leukemias. *N Engl J Med* 319:990–8.

Kyprianou N, English HF, Davidson NE, and Isaacs JT. 1991. Programmed cell death during regression of the MCF-7 human breast cancer following estrogen ablation. *Cancer Res* 51:162–6.

Labandeira CC and Phillips TL. 1996. A carboniferous insect gall: Insight into early ecologic history of the Holometabola. *Proc Natl Acad Sci USA* 93:8470–4.

Labandeira CC, Johnson KR, and Wilf P. 2002. Impact of the terminal Cretaceous event on plant-insect associations. *Proc Natl Acad Sci USA* 99:2061–6.

Lacombe C, DaSilva L, Bruneval P, Fournier J-G, Wendling S, Casadevall N, Camilleri J-P, Bariety J, Varet B, and Tambourin P. 1988a. Peritubular cells are the site of erythropoietin synthesis in the murine hypoxic kidney. *J Clin Invest* 81:620–3.

Lacombe C, Da Silva JL, Bruneval P, Camilleri JP, Bariety J, Tambourin P, and Varet B. 1988b. Identification of tissues and cells producing erythropoietin in the anemic mouse. *Contrib Nephrol* 66:17–24.

Lambert JM, Lambert QT, Reuther GW, Malliri A, Siderovski DP, Sondek J, Collard JG, and Der CJ. 2002. Tiam1 mediates Ras activation of Rac by a PI(3)K-independent mechanism. *Nat Cell Biol* 4:621–5.

Lampe JW and Peterson S. 2002. Brassica, biotransformation and cancer risk: genetic polymorphisms alter the preventive effects of cruciferous vegetables. *J Nutr* 132:2991–4.

Land H, Parada LF, and Weinberg RA. 1983. Tumorigenic conversion of primary embryo fibroblasts requires at least two cooperating oncogenes. *Nature* 304:596–602.

Landrigan PJ. 1996. The prevention of occupational cancer. *CA Cancer J Clin* 46:67–9.

Lane P, Vichi P, Bain DL, and Tritton TR. 1987. Temperature dependence studies of adriamycin uptake and cytotoxicity. *Cancer Res* 47:4038–42.

Langen P. 1985. Chalones and other endogenous inhibitors of cell proliferation. In: PF Torrence (ed.). *Biological Response Modifiers, New Approaches to Disease Intervention*, pp. 265–91. Orlando, FL: Academic Press.

Lapidot T, Pflumio F, Doedens M, Murdoch B, Williams DE, and Dick JE. 1992. Cytokine stimulation of multilineage hematopoiesis from immature human cells engrafted in SCID mice. *Science* 255:1137–41.

Laster WR Jr, Mayo JG, Simpson-Herren L, Griswold DP Jr, Lloyd HH, Schabel FM Jr, and Skipper HE. 1969. Success and failure in the treatment of solid tumors. II: Kinetic parameters and "cell cure" of moderately advanced carcinoma 755. *Cancer Chemother Rep* 53:169–88.

Lawlor DA and Ness AR. 2003. The rough world of nutritional epidemiology: Does dietary fibre prevent large bowel cancer? *Int J Epidemiology* 32:239–43.

Leav I, Merk FB, Ofner P, Goodrich G, Kwan PW, Stein BM, Sar M, and Stumpf WE. 1978. Bipotentiality of response to sex hormones by the prostate of castrated or hypophysectomized dogs. Direct effects of estrogen. *Am J Pathol* 93:69–92.

LeDoux SP, Thangada M, Bohr VA, and Wilson GL. 1991. Heterogeneous repair of methylnitrosourea-induced alkali-labile sites in different DNA sequences. *Cancer Res* 51:775–9.

Lee CH, Olson P, and Evans RM. 2003. Minireview: lipid metabolism, metabolic diseases, and peroxisome proliferator-activated receptors. *Endocrinology* 144:2201–7.

Lee EY-HP, To H, Shew J-Y, Bookstein R, Scully P, and Lee W-H. 1988. Inactivation of the retinoblastoma susceptibility gene in human breast cancers. *Science* 241:218–21.

Lee JAH and Strickland D. 1980. Malignant melanoma: Social status and outdoor work. *Br J Cancer* 41:757–63.

Leevers SJ, Paterson HF, and Marshall CJ. 1994. Requirement for Ras and Raf activation is overcome by targeting Raf to the plasma membrane. *Nature* 369:411–14.

Lehman JM, Speers WC, Swartzendruber DE, and Pierce GB. 1974. Neoplastic differentiation: characteristics of cell lines derived from a murine teratocarcinoma. *J Cell Physiol* 84:13–27.

Lehvaslaiho H, Lehtola L, Sistonen L, and Alitalo K. 1989. A chimeric EGF-R-neu proto-oncogene allows EGF to regulate neu tyrosine kinase and cell transformation. *EMBO J* 8:159–66.

Lemen RA, Dement JM, and Wagoner JK. 1980. Epidemiology of asbestos-related diseases. *Environ Health Perspect* 20:1–21.

Lenihan J. 1993. *The Good News About Radiation*, pp. 1–173. Madison: Cogito Books.

Lennon SV, Martin SJ, and Cotter TG. 1991. Dose-dependent induction of apoptosis in human tumour cell lines by widely divergent stimuli. *Cell Prolif* 24:203–14.

Lenormand P, Sardet C, Pages G, L'Allemain G, Brunet A, and Pouyssegur J. 1993. Growth factors induce nuclear translocation of MAP kinases (p42mapk) but not of their activator MPA kinase kinase (p45mapkk) in fibroblasts. *J Cell Biol* 122:1079–88.

Levine EL, Davidson SE, Roberts SA, Chadwick CA, Potten CS, and West CM. 1994. Apoptosis as predictor of response to radiotherapy in cervical carcinoma. *Lancet* 344:472.

Levine EL, Renehan A, Gossiel R, Davidson SE, Roberts SA, Chadwick C, Wilks DP, Potten CS, Hendry JH, Hunter RD et al. 1995. Apoptosis, intrinsic radiosensitivity and prediction of radiotherapy response in cervical carcinoma. *Radiother Oncol* 37:1–9.

Lewandoski M. 2001. Conditional control of gene expression in the mouse. *Nat Rev Genet* 2:743–55.

Lewison EE, ed. 1976. Conference on Spontaneous Regression of Cancer. *Natl Cancer Inst Monogr* 44:99–102.

Leyden M, Manoharan A, Boyd A, Cheng ZM, and Sullivan J. 1984. Low dose cytosine arabinoside: partial remission of acute myeloid leukaemia without evidence of differentiation induction. *Br J Haematol* 57:301–7.

Li FP and Fraumeni JF. 1969. Soft-tissue sarcomas, breast cancer, and other neoplasms. A familial syndrome. *Ann Intern Med* 71:747–52.

Li J, Simpson L, Takahashi M, Miliaresis C, Myers MP, Tonks N, and Parsons R. 1998. The PTEN/MMAC1 tumor suppressor induces cell death that is rescued by the AKT/protein kinase B oncogene. *Cancer Res* 58:5667–72.

Li L and Dixon JE. 2000. Form, function, and regulation of protein tyrosine phosphatases and their involvement in human diseases. *Semin Immunol* 12:75–84.

Li Y, Dowbenko D and Lasky LA. 2002. AKT/PKB phosphorylation of p21Cip/WAF1 enhances protein stability of p21Cip/WAF1 and promotes cell survival. *J Biol Chem* 277:11352–61.

Liang J, Zubovitz J, Petrocelli T, Kotchetkov R, Connor MK, Han K, Lee JH, Ciarallo S, Catzavelos C, Beniston R et al. 2002. PKB/Akt phosphorylates p27, impairs nuclear import of p27 and opposes p27-mediated G1 arrest. *Nat Med* 8:1153–60.

Liang R, Chow WS, Chiu E, Chan TK, Lie A, Kwong YL, and Chan LC. 1993. Effective salvage therapy using all-trans retinoic acid for relapsed and resistant acute promyelocytic leukemia. *Anti-Cancer Drugs* 4:339–40.

Lichtenstein P, Holm NV, Verkasalo PK, Iliadou A, Kaprio J, Koskenvuo M, Pukkala E, Skytthe A, and Hemminki K. 2000. Environmental and heritable factors in the causation of cancer. *N Eng J Med* 343:78–85.

Liebermann D, Hoffman-Liebermann B, and Sachs L. 1982. Regulation and role of different macrophage- and granulocyte-inducing proteins in normal and leukemic myeloid cells. *Int J Cancer* 29:159–61.

Liedtke C, Groger N, Manns MP, and Trautwein C. 2003. The human caspase-8 promoter sustains basal activity through SP1 and ETS-like transcription factors and can be up-regulated by a p53-dependent mechanism. *J Biol Chem* 278:27593–27604.

Lijinsky W. 1999. N-Nitroso compounds in the diet. *Mutat Res* 443:129–38.

Lijinsky W and Shubik P. 1964. Benzo[a]pyrene and other polynuclear hydrocarbons in charcoal-broiled meat. *Science* 145:53–5.

Lindahl T and Nyberg B. 1972. Rate of depurination of native deoxyribonucleic acid. *Biochemistry* 11:3610–18.

   1974. Heat-induced deamination of cytosine residues in deoxyribonucleic acid. *Biochemistry* 13:3405–10.

Lingner J, Hughes TR, Shevchenko A, Mann M, Lundblad V, and Cech TR. 1997. Reverse transcriptase motifs in the catalytic subunit of telomerase. *Science* 276:561–7.

Link BK, Budd GT, Scott S, Dickman E, Paul D, Lawless G, Lee MW, Fridman M, Ford J, and Carter WB. 2001. Delivering adjuvant chemotherapy to women with early-stage breast carcinoma: Current patterns of care. *Cancer* 92:1354–67.

Liotta LA, Abe S, Robey PG, and Martin GR. 1979. Preferential digestion of basement membrane collagen by an enzyme derived from a metastatic murine tumor. *Proc Natl Acad Sci USA* 76: 2268–72.

Liotta LA, Rao CN, and Barsky SH. 1983. Tumor invasion and the extracellular matrix. *Lab Invest* 49:636–49.

Livant DL, Linn S, Markwart S, and Shuster J. 1995. Invasion of selectively permeable sea urchin embryo basement membrane by metastatic tumor cells, but not by their normal counterparts. *Cancer Res* 55:5085–93.

Loeb LA, Loeb KR, and Anderson JP. 2003. Multiple mutations and cancer. *Proc Natl Acad Sci USA* 100:776–81.

Lollini PL and Forni G. 2002. Antitumor vaccines: Is it possible to prevent a tumor? *Cancer Immunol Immunother* 51:409–16.

Lombardi D, Crivellari D, Scuderi C, Magri MD, Spazzapan S, Sorio R, DiLauro V, Scalone S, Veronesi A. 2004. Long-term, weekly one-hour infusion of paclitaxel in patients with metastastic breast cancer: A phase II monoinstitutional study. *Tumori* 90:285–8.

Longnecker DS and Curphey TJ. 1975. Adenocarcinoma of the pancreas in azaserine-treated rats. *Cancer Res* 35:2249–58.

Loser R, Seibel K, and Eppenberger U. 1985. No loss of estrogenic or anti-estrogenic activity after demethylation of roloxifene 3-OH-tamoxifen. *Int J Cancer* 36:701–3.

Loser R, Seibel K, Roos W, and Eppenberger U. 1985. In vivo and in vitro antiestrogenic action of 3-hydroxytamoxifen, tamoxifen and 4-hydroxytamoxifen. *Eur J Cancer Clin Oncol* 21: 985–90.

Lotem J and Sachs L. 1982. Mechanisms that uncouple growth and differentiation in myeloid leukemia cells: Restoration of requirement for normal growth-inducing protein without restoring induction of differentiation-inducing protein. *Proc Natl Acad Sci USA* 79:4347–51.

——— 1992. Hematopoietic cytokines inhibit apoptosis induced by transforming growth factor-beta1 and cancer chemotherapy compounds in myeloid leukemic cells. *Blood* 80:1750–7.

Loury DL, Goldsworthy TL, and Butterworth BE. 1987. The value of measuring cell replication as a predictive index of tissue-specific tumorigenic potential. In: BE Butterworth and TJ Slaga (eds.). Banbury Report 25: *Nongenotoxic mechanisms in carcinogenesis*, pp. 119–36. New York: Cold Spring Harbor Press.

Lowenstein EJ, Daly AG, Batzer AG, Li W, Margolis B, Lammers R, Ullrich A, Skolnik EY, Bar-Sagi D, and Schlessinger J. 1992. The SH2 and SH3 domain-containing protein GRB2 links receptor tyrosine kinases to ras signaling. *Cell* 70:431–42.

Luande J, Henschke CI, and Mohammed N. 1985. The Tanzanian human albino skin. *Cancer* 55:1823–8.

Lucké B. 1934. A neoplastic disease of the kidney of the frog, *Rana pipiens. Am J Cancer* 20:352–79.

——— 1938. Carcinoma in the leopard frog: Its probable causation by a virus. *J Exp Med* 68:457–68.

——— 1939. Characteristics of frog carcinoma in tissue culture. *J Exp Med* 70:269–76.

Lunn RM, Zhang YJ, Wang LY, Chen CJ, Lee PH, Lee CS, Tsai WY, and Santella RM. 1997. p53 mutations, chronic hepatitis B virus infection, and aflatoxin exposure in hepatocellular carcinoma in Taiwan. *Cancer Res.* 57:3471–7.

Luo J, Manning BD, and Cantley LC. 2003. Targeting the PI3K-Akt pathway in human cancer: Rationale and promise. *Cancer Cell* 4:257–62.

Lust JM, Carlson DL, Kowles R, Rollins-Smith L, Williams JW III, and McKinnell RG. 1991. Allografts of tumor nuclear transplantation embryos: Differentiation competence. *Proc Natl Acad Sci USA* 88:6883–7.

Lyman GH, Dale DC and Crawford J. 2003. Incidence and predictors of low dose-intensity in adjuvant breast cancer chemotherapy: A nationwide study of community practices. *J Clin Oncol* 21:4524–31.

Lyman SD and Jordan VC. 1985. Metabolism of tamoxifen and its uterotrophic activity. *Biochem Pharmacol* 34:2787–94.

Lynch DK, Ellis CA, Edwards PA, and Hiles ID. 1999. Integrin-linked kinase regulates phosphorylation of serine 473 of protein kinase B by an indirect mechanism. *Oncogene* 18:8024–32.

Lynch HT. 1985a. Classics in oncology. Alfred Scott Warthin, MD, PhD (1866–1931). *CA Cancer J Clin* 35:345–7.

——— 1985b. Hereditary colon cancer: Polyposis and nonpolyposis variants. *CA Cancer J Clin* 35:95–114.

Lynch HT and de la Chapelle A. 2003. Hereditary colorectal cancer. *N Engl J Med* 348: 919–32.

Lynch HT, Lynch PM, Albano WA, and Lynch JF. 1981. The cancer syndrome: A status report. *Dis Colon Rectum* 24:311–22.

Lynch TJ, Bell DW, Sordella R, Gurubhagavatula S, Okimoto RA, Brannigan BW, Harris PL, Haserlat SM, Supko JG, Haluska FG et al. 2004. Activating mutations in the epidermal growth factor receptor underlying responsiveness of non–small-cell lung cancer to gefitinib. *N Engl J Med* 350:2129–39.

Mørk S, Laerum OD, and de Ridder L. 1984. Invasiveness of tumours of the central nervous system. In: MM Mareel and KC Calman (eds.). *Invasion, Experimental and Clinical Implications*, pp. 79–125. Oxford, UK: Oxford University Press.

Macaluso M, Russo G, Cinti C, Bazan V, Gebbia N, and Russo A. 2002. Ras family genes: an interesting link between cell cycle and cancer. *J Cell Physiol* 192:125–30.

MacDonald AD and MacDonald JC. 1987. *Asbestos-Related Malignancy*, KH Antman and J Aisner, eds., pp. 31–55. Orlando, FL: Grune and Stratton.

MacLachlan TK and El-Deiry WS. 2002. Apoptotic threshold is lowered by p53 transactivation of caspase-6. *Proc Natl Acad Sci USA* 99:9492–7.

Maclure M, Katz RBA, Bryant MS, Skipper PL, and Tannenbaum SR. 1989. Elevated blood levels of carcinogens in passive smokers. *Am J Public Health* 79:1381–4.

Madigan MP, Ziegler RG, Benichou J, Byrne C, and Hoover RN. 1995. Proportion of breast cancer cases in the United States explained by well-established risk factors. *J Natl Cancer Inst* 87:1681–5.

Magee PN and Barnes JM. 1956. The production of malignant primary hepatic tumours in the rat by feeding dimethylnitrosamine. *Br J Cancer* 10:114–22.

Maiese WM, Lechevalier MP, Lechevalier HA, Korshalla J, Kuck N, Fantini A, Wildey MJ, Thomas J, and Greenstein M. 1989. Calicheamicins, a novel family of antitumor antibiotics: Taxonomy, fermentation and biological properties. *J Antibiot (Tokyo)* 42:558–63.

Malaise EP, Chavaudra N, and Tubiana M. 1973. The relationship between growth rate, labelling index and histological type of human solid tumours. *Eur J Cancer* 9:305–12.

Malaise EP, Chavaudra N, Charbit A, and Tubiana M. 1974. Relationship between the growth rate of human metastases, survival and pathological type. *Eur J Cancer* 10:451–9.

Malkin D. 1993. p53 and Li-Fraumeni syndrome. *Cancer Genet Cytogenet* 66:83–92.

Malkin D, Li FP, Strong LC, Fraumeni JF, Nelson CE, Kim DH, Kassel J, Gryka MA, Bischoff FZ, Tainsky MA, et al. 1990. Germ line p53 mutations in a familial syndrome of breast cancer, sarcomas, and other neoplasms. *Science* 250:1233–8.

Mamane Y, Petroulakis E, Rong L, Yoshida K, Ler LW, and Sonenberg N. 2004. eIF4E – from translation to transformation. *Oncogene* 23:3172–9.

Mancuso M, Pazzaglia S, Tanori M, Hahn H, Merola P, Rebessi S, Atkinson MJ, Di Majo V, Covelli V, and Saran A. 2004. Basal cell carcinoma and its development: insights from radiation-induced tumors in Ptch1-deficient mice. *Cancer Res* 64:934–41.

Mandahl N, Gustafson P, Mertens F, Akerman M, Baldetorp B, Gisselsson D, Knuutila S, Bauer HC, Larsson O. 2002. Cytogenetic aberrations and their prognostic impact in chondrosarcoma. *Genes Chromosomes Cancer* 33:188–200.

Maniotis AJ, Folberg R, Hess A, Seftor EA, Gardner LM, Pe'er J, Trent JM, Meltzer PS, and Hendrix MJ. 1999. Vascular channel formation by human melanoma cells in vivo and in vitro: Vasculogenic mimicry. *Am J Pathol* 155:739–52.

Manni A and Arafah BM. 1981. Tamoxifen-induced remission in breast cancer by escalating the dose to 40 mg daily after progression on 20 mg daily: A case report and review of the literature. *Cancer* 48:873–5.

Marchenko ND, Zaika A, and Moll UM. 2000. Death signal-induced localization of p53 protein to mitochondria. A potential role in apoptotic signaling. *J Biol Chem* 275:16202–12.

Mareel MM, DeBaetselier P, and Van Roy FM. 1991. *Mechanisms of Invasion and Metastasis*. Boca Raton, FL: CRC Press.

Mareel MM, Storme GA, DeBruyne GK, and Van Cauwenberge RM. 1982. Vinblastine, vincristine, and vindesine: Antiinvasive effect on MO4 mouse fibrosarcoma cells in vitro. *Eur J Cancer Clin Oncol* 18:199–210.

Mareel MM, Van Roy FM, Bracke ME. 1993. How and when do tumor cells metastasize. *Crit Rev Oncogenesis* 4:559–94.

Margolese R, Poisson R, Shibata H, Pilch Y, Lerner H, and Fisher B. 1987. The technique of segmental mastectomy (lumpectomy) and axillary dissection. *Surgery* 102:828–34.

Margolis B. 1999. The PTB Domain: The Name Doesn't Say It All. *Trends Endocrinol Metab* 10:262–7.

Marjon PL, Bobrovnikova-Marjon EV, and Abcouwer SF. 2004. Expression of the pro-angiogenic factors vascular endothelial growth factor and interleukin-8/cxcl8 by human breast carcinomas is responsive to nutrient deprivation and endoplasmic reticulum stress. *Mol Cancer* 3:4.

Marks R. 1996a. Prevention and control of melanoma: The public health approach. *CA Cancer J Clin* 46:199–216.

— 1996b. Squamous cell carcinoma. *Lancet* 347:735–8.

Marnett LJ. 2000. Oxyradicals and DNA damage. *Carcinogenesis* 21:361–70.

Martin GR. 1981. Isolation of a pluripotential cell line from early mouse embryos cultured in medium conditioned by teratocarcinoma stem cells. *Proc Natl Acad Sci USA* 78:7634–8.

Martineau D, Lemberger K, Dallaire A, Labelle P, Lipscomb TP, Michel P, and Mikaelian I. 2002. Cancer in wildlife, a case study: beluga from the St Lawrence estuary, Quebec, Canada. *Environ Health Perspect* 110:285–92.

Martinelli PT and Tyring SK. 2002. Human herpesvirus 8. *Dermatol Clin* 20:307–14.

Martland HS. 1931. The occurrence of malignancy in radioactive persons. *Am J Cancer* 15:2435–2516.

Masahito et al. 1994. Frequent development of pancreatic carcinomas in the Rana nigromaculata group. *Proc 8th Int Conf Int Soc Diff*, pp. 183–6, Hiroshima, Japan: International Society of Differentiation, Inc.

Masahito P, Ishikawa T, and Sugano H. 1988. Fish tumors and their importance to cancer research. *Jpn J Cancer Res (Gann)* 79:545–55.

Masahito P, Ishikawa T, and Sugano H. 1989. Pigment cells and pigment cell tumors in fish. *J Invest Dermatol* 92:266S–270S.

Masahito P, Ishikawa T, Sugano H, Uchida H, Yasuda T, Inaba T, Hirosaki Y, and Kasuga A. 1986. Spontaneous hepatocellular carcinomas in lungfish. *J Natl Cancer Inst* 77:291–8.

Masahito, P, Ishikawa T, and Takayama S. 1984. Spontaneous spermatocytic seminoma in African lungfish, *Protopterus aethiopicus*, Heckel. *J Fish Dis* 7:169–72.

Masahito P, Ishikawa T, Takayama S, and Sugimura H. 1984. Gonadal neoplasms in largemouth bass, Micropterus salmoides and Japanese Dace (Ugui), Tribolodon hakonensis. *Gann* 75:776–83.

Masahito P, Nishioka M, Ueda H, Kato Y, Yamazaki I, Nomura K, Sugano H, and Kitagawa T. 1995. Frequent development of pancreatic carcinomas in the *Rana nigromaculata* group. *Cancer Res* 55:3781–4.

Masahito P, Nishioka M, Kondo Y, Yamazaki I, Nomura K, Kato Y, Sugano H, and Kitagawa T. 2003. Polycystic kidney and renal cell carcinoma in Japanese and Chinese toad hybrids. *Int J Cancer* 103:1–4.

Masuda A and Takahashi T. 2002. Chromosome instability in human lung cancers: possible underlying mechanisms and potential consequences in the pathogenesis. *Oncogene* 21:6884–97.

Masui Y and Markert CL. 1971. Cytoplasmic control of nuclear behavior during meiotic maturation in frog oocytes. *J Exp Zool* 177:129–46.

Matsumoto J, Akiyama T, Nemoto N, Masahito P, and Ishikawa T. 1993. Appearance of tumorous phenotypes in goldfish erythrophores transfected with ras, src, and myc oncogenes and spontaneous differentiation of the transformants in vitro. *J Invest Dermatol* 100:214S–221S.

Matsumura Y and Tarin D. 1992. Significance of CD44 gene products for cancer diagnosis and disease evaluation. *Lancet* 340:1053–8.

Mayo JG, Laster WR Jr, Andrews CM, and Schabel FM Jr. 1972. Success and failure in the treatment of solid tumors 3: "Cure" of metastatic Lewis lung carcinoma with methyl-CCNU (NSC-95442) and surgery-chemotherapy. *Cancer Chemother Rep* Part 1, 56:183–95.

Mayo LD and Donner DB. 2001. A phosphatidylinositol 3-kinase/Akt pathway promotes translocation of Mdm2 from the cytoplasm to the nucleus. *Proc Natl Acad Sci USA* 98:11598–11603.

McBurney MW. 1993. P19 embryonal carcinoma cells. *Int J Dev Biol* 37:135–40.

McCague R, Parr IB, Leclercq G, Leung OT, and Jarman M. 1990. Metabolism of tamoxifen by isolated rat hepatocytes. Identification of the glucuronide of 4-hydroxytamoxifen. *Biochem Pharmacol* 39:1459–65.

McCarthy JB, Basara ML, Palm SL, Sas DF, and Furcht LT. 1985. The role of cell adhesion proteins – laminin and fibronectin – in the movement of malignant and metastatic cells. *Cancer Metastasis Rev* 4:125–52.

McCarthy NJ, Smith CA, and Williams GT. 1992. Apoptosis in the development of the immune system: Growth factors, clonal selection and bcl-2. *Cancer Metastasis Rev* 11:157–78.

McCormack JJ and Johns DG. 1990. Purine and purine nucleoside antimetabolites. In: BA Chabner and JM Collins (eds.). *Cancer Chemotherapy*, pp. 234–52. Philadelphia, PA: JB Lippincott.

McCullough ML, Robertson AS, Rodriguez C, Jacobs EJ, Chao A, Carolyn J, Calle EE, Willett WC, and Thun MJ. 2003. Calcium, vitamin D, dairy products, and risk of colorectal cancer in the cancer prevention study II nutrition cohort (United States). *Cancer Causes Control* 14: 1–12.

McDonnell TJ, Deane N, Platt FM, Nunez G, Jaeger U, McKearn JP, and Korsmeyer SJ. 1989. Bcl-2-immunoglobulin transgenic mice demonstrate extended B cell survival and follicular lymphoproliferation. *Cell* 57:79–88.

McKay JA, Loane JF, Ross VG, Ameyaw MM, Murray GI, Cassidy J, and McLeod HL. 2002a. C-erbb-2 is not a major factor in the development of colorectal cancer. *Br J Cancer* 86:568–73.

McKay JA, Murray LJ, Curran S, Ross VG, Clark C, Murray GI, Cassidy J, and McLeod HL. 2002b. Evaluation of the epidermal growth factor receptor (egfr) in colorectal tumours and lymph node metastases. *Eur J Cancer* 38:2258–64.

McKinnell RG. 1973. Nuclear transplantation. In *Seventh National Cancer Conference Proceedings*, pp. 65–72. Philadelphia, PA: JB Lippincott.

McKinnell RG. 1989a. Expression of differentiated function in neoplasms. In: AE Sirica (ed.). *The Pathobiology of Neoplasia*, pp. 435–60. New York, Plenum.

McKinnell RG. 1989b. Neoplastic cells: Modulation of the differentiated state. In: MA DiBeradino and LD Etkin (eds.). *Genomic Adaptability in Somatic Cell Specialization*, pp. 199–236. New York, Plenum.

McKinnell RG. 1994. Reduced oncogenic potential associated with differentiation of the Lucké renal adenocarcinoma. *In Vivo* 8:65–70.

McKinnell RG and DiBerardino MA. 1999. Animal cloning: History and rationale. *BioScience* 49:875–85.

McKinnell RG and Carlson DL. 1997. The Lucké renal adenocarcinoma, an anuran neoplasm: Studies at the interface of pathology, virology and differentiation competence. *J Cell Physiol* 173:115–18.

McKinnell RG and Tarin D. 1984. Temperature dependent metastasis of the Lucké renal carcinoma and its significance for studies on mechanisms of metastasis. *Cancer Metastasis Rev* 3:373–86.

McKinnell RG, Bruyneel EA, Mareel MM, Tweedell KS, and Mekela P. 1988. Temperature-dependent malignant invasion in vitro by frog renal carcinoma-derived PNKT4B cells. *Clin Exp Metastasis* 6:49–59.

McKinnell RG, DeBruyne GK, Mareel MM, Tarin D, and Tweedell KS. 1984. Cytoplasmic microtubules of normal and tumor cells of the leopard frog. Temperature effects. *Differentiation* 26:231–4.

McKinnell RG, Deggins BA, and Labat DD 1969. Transplantation of pluripotential nuclei from triploid frog tumors *Science* 165:394–6.

McKinnell RG, Kren BT, Bergad R, Schultheis M, Byrne T, and Schaad JW IV. 1980. Dominant lethality in *Xenopus laevis* induced with triethylene melamine. *Teratog Carcinog Mutagen* 1:283–94.

McKinnell RG, Mareel MM, Bruyneel EA, Seppanen ED, Mekala PR. 1986. Invasion in vitro by explants of Lucké renal carcinoma cocultured with normal tissue is temperature dependent. *Clin Exp Metastasis* 4:237–43.

McKinnell, RG, Sauerbier W, Lust JM, Williams JW, Williams CS, Rollins-Smith LA, and Carlson DL. 1991. The Lucké renal adenocarcinoma and its herpesvirus. In 4. *Internationales Colloquium für Pathologie und Therapie der Reptilien und Amphibien*. Gissen: Deutsche Veterinärmedizinische Gesellschaft e.V., pp. 204–18.

McKinnell RG, Lust JM, Sauerbier W, Rollins-Smith L, Williams III JW, Williams CS, Carlson DL. 1993. Genomic plasticity of the Lucké carcinoma: A review. *Int J Develop Biol* 37:213–19.

McLaughlin P, Grillo-Lopez AJ, Link BK, Levy R, Czuczman MS, Williams ME, Heyman MR, Bence-Bruckler I, White CA, Cabanillas F et al. 1998. Rituximab chimeric anti-cd20 monoclonal antibody therapy for relapsed indolent lymphoma: Half of patients respond to a four-dose treatment program. *J Clin Oncol* 16:2825–33.

McTiernan A and Thomas DB. 1986. Evidence for a protective effect of lactation on risk of breast cancer in young women: Results from a case-control study. *Am J Epidemiol* 124:353–8.

Medeiros F, Corless CL, Duensing A, Hornick JL, Oliveira AM, Heinrich MC, Fletcher JA, and Fletcher CD. 2004. Kit-negative gastrointestinal stromal tumors: Proof of concept and therapeutic implications. *Am J Surg Pathol* 28:889–94.

Meinsma R, Fernandez-Salguero P, Van Kuilenburg AB, Van Gennip AH, and Gonzalez FJ. 1995. Human polymorphism in drug metabolism: Mutation in the dihydropyrimidine dehydrogenase gene results in exon skipping and thymine uracilurea. *DNA Cell Biol* 14:1–6.

Mellon I, Spivak G, and Hanawalt PC. 1987. Selective removal of transcription-blocking DNA damage from the transcribed strand of the mammalian DHFR gene. *Cell* 51:241–9.

Menard S, Casalini P, Campiglio M, Pupa S, Agresti R, and Tagliabue E. 2001. HER2 overexpression in various tumor types, focussing on its relationship to the development of invasive breast cancer. *Ann Oncol* 12 Suppl 1:S15–19.

Mendelsohn ML. 1960a. The growth fraction: A new concept applied to tumors. *Science* 132:1496. 1960b. Autoradiographic analysis of cell proliferation in spontaneous breast cancer of C3H mouse. I. Growth and survival of cells labelled with tritiated thymidine. *J Natl Cancer Inst* 25:485–500.

Mendelsohn ML, Dohan Jr. FC, and Moore Jr. HA. 1960. Autoradiographic analysis of cell proliferation in spontaneous breast cancer of C3H mouse. I. Typical cell cycle and timing of DNA synthesis. *J Natl Cancer Inst.* 25:477–84.

Menssen A and Hermeking H. 2002. Characterization of the c-MYC-regulated transcriptome by SAGE: identification and analysis of c-MYC target genes. *Proc Natl Acad Sci USA* 99:6274–9.

Meyn MS. 1995. Ataxia-telangiectasia and cellular responses to DNA damage. *Cancer Res* 55:5991–6001.

Michalewicz R, Lotem J, and Sachs L. 1984. Cell differentiation and therapeutic effect of low doses of cytosine arabinoside in human myeloid leukemia. *Leuk Res* 8:783–90.

Mikaelian I, Lapointe JM, de Lafontaine Y, Harshbarger JC, Côté RJ, Naydan DK, and Martineau D. 2000. Suprasellar germinoma in three lake whitefish (Coregonus clupeaformis). *Acta Neuropathol (Berl)* 100:228–32.

Mikhailov V, Mikhailova M, Pulkrabek DJ, Dong Z, Venkatachalam MA, and Saikumar P. 2001. Bcl-2 prevents Bax oligomerization in the mitochondrial outer membrane. *J Biol Chem* 276:18361–74.

Mikheev AM, Mikheeva SA, Liu B, Cohen P, and Zarbl H. 2004. A functional genomics approach for the identification of putative tumor suppressor genes: Dickkopf-1 as suppressor of HeLa cell transformation. *Carcinog* 25:47–59.

Miki Y, Swensen J, Shattuck-Eidens D, Futreal PA, Harshman K, Tavtigian S, Liu Q, Cochran C, Bennett LM, Ding W et al. 1994. A strong candidate for the breast and ovarian cancer susceptibility gene *BRCA1*. *Science* 266:66–71.

Milburn MV, Tong L, de Voss AM, Brunger A, Yamaizumi Z, Nishimura S, and Kim S-H. 1990. Molecular switch for signal transduction: Structural differences between active and inactive forms of protooncogenic ras proteins. *Science* 247:939–45.

Miller EC. 1978. Some current perspectives on chemical carcinogenesis in humans and experimental animals. *Cancer Res* 38:1479–96.

Miller JA. 1970. Carcinogenesis by chemicals: An overview. *Cancer Res* 30:559–76.

Miller RW. 1966. Delayed radiation effects in atomic bomb survivors. *Science* 166:569–74.

Minamoto T, Mai M, and Ronai Z. 2000. K-ras mutation: early detection in molecular diagnosis and risk assessment of colorectal, pancreas, and lung cancers – a review. *Cancer Detect Prev* 24: 1–12.

Mintz B and Illmensee K. 1975. Normal genetically mosaic mice produced from malignant teratocarcinoma cells. *Proc Natl Acad Sci USA* 72:3585–9.

Misra RP, Bonni A, Miranti CK, Rivera VM, and Sheng ZH. 1994. L-type voltage-sensitive calcium channel activation stimulates gene expression by a serum response factor-dependent pathway. *J Biol Chem* 269:25483–93.

Mitelman F. 1994. *Catalog of chromosome aberrations in cancer*, 5th ed. New York: Wiley-Liss.

Miyashita T and Reed JC. 1995. Tumor suppressor p53 is a direct transcriptional activator of the human bax gene. *Cell* 80:293–9.

Molineux G, Migdalska A, Haley J, Evans GS, and Dexter TM. 1994. Total marrow failure induced by pegylated stem-cell factor administered before 5-fluorouracil. *Blood* 3:3491–9.

Monchaux G and Morlier JP. 2002. Influence of exposure rate on radon-induced lung cancer in rats. *J Radiol Prot* 22:A81–87.

Monden Y, Hamano-Takaku F, Shindo-Okada N, and Nishimura S. 1999. Azatyrosine: Mechanism of action for conversion of transformed phenotype to normal. *Ann N Y Acad Sci* 886: 109–21.

Moodie SA and Wolfman A. 1994. The 3Rs of life: Ras, Raf and growth regulation. *Trends Genet* 10:44–8.

Moore KL and Persaud TV. 1993. *The Developing Human*, 5th ed. Philadelphia, PA: WB Saunders.

Moore MA and Owen JJ. 1967. Stem cell migration in developing myeloid and lymphoid systems. *Lancet* 2:658–9.

Morris CM, Reeve AE, Fitzgerald PH, Hollings PE, Beard ME J, and Heaton DC. 1986. Genomic diversity correlates with clinical variation in Ph1-negative chronic myeloid leukaemia. *Nature* 320:281–3.

Morrison GB, Bastian A, Dela Rosa T, Diasio RB, and Takimoto CH. 1997. Dihydropyrimidine dehydrogenase deficiency: A pharmacogenetic defect causing severe adverse reactions to 5-fluorouracil-based chemotherapy. *Oncol Nurs Forum* 24:83–8.

Morrison VA. 1994. Chronic leukemias. *CA Cancer J Clin* 44:353–77.

Morse D, Dailey RC, and Bunn J. 1976. Prehistoric multiple myeloma. In: S Jarcho (ed.). *Essays on the History of Medicine*, pp. 413–24. New York: Science History Publications/USA.

Morton DL, Antman KH, and Tepper J. 1993. Soft tissue sarcoma. In JF Holland, E Frei III, RC Bast Jr, DW Kufe, DL Morton, and RR Weichselbaum (eds.). *Cancer Medicine*, vol 2, pp. 1858–87. Philadelphia, PA: Lea and Febiger.

Morton DL, Cochran AJ, and Lazar G. 1990. Melanoma. In: CM Haskell (ed.). *Cancer Treatment*, 3rd ed., pp. 500–12. Philadelphia, PA: WB Saunders.

Morton S, Davis RJ, McLaren A, and Cohen P. 2003. A reinvestigation of the multisite phosphorylation of the transcription factor c-Jun. *EMBO J* 22:3876–86.

Mossman BT, Bignon J, Corn M, Seaton A, and Gee JB. 1990. Asbestos: Scientific developments and implications for public policy. *Science* 247:294–301.

Mossman BT, Kamp DW, and Weitzman SA. 1996. Mechanisms of carcinogenesis and clinical features of asbestos-associated cancers. *Cancer Invest* 14:466–80.

Moulton JE. 1978. *Tumors in Domestic Animals*. Berkeley, CA: University of California Press.

Mueller CB and Jeffries W. 1975. Cancer of the breast: Its outcome as measured by the rate of dying and causes of death. *Ann Surg* 182:334–41.

Muir C, Waterhouse J, Mack T, Powell J, and Whelan S, eds. 1987. *Cancer Incidence in Five Continents*, Vol V IARC Scientific Publications No 88. Lyon, FR: IARC.

Mundoz-Dorado J, Inouye M, and Inouye S. 1990. Nucleoside diphosphate kinase from *Myxococcus xanthus:* I Cloning and sequencing of the gene. *J Biol Chem* 265:2702–6.

Münger K, Werness BA, Dyson N, Phelps WC, and Howley PM. 1989. Complex suppressor gene product. *EMBO J* 8:4099–4105.

Munson L. 1987. Carcinoma of the mammary gland in a mare. *J Am Vet Med Assoc* 191:71–2.

Muraoka RS, Lenferink AE, Law B, Hamilton E, Brantley DM, Roebuck LR, and Arteaga CL. 2002. ErbB2/Neu-induced, cyclin D1-dependent transformation is accelerated in p27-haploinsufficient mammary epithelial cells but impaired in p27-null cells. *Mol Cell Biol* 22:2204–19.

Murphy CS, Langan-Fahey SM, McCague R, and Jordan VC. 1990. Structure-function relationships of hydroxylated metabolites of tamoxifen that control the proliferation of estrogen-responsive T47D breast cancer cells in vitro. *Mol Pharmacol* 38:737–43.

Murphy M, Stinnakre MG, Senamaud-Beaufort C, Winston NJ, Sweeney C, Kubelka M, Carrington M, Brechot C, and Sobczak-Thepot J. 1997. Delayed early embryonic lethality following disruption of the murine cyclin A2 gene. *Nat Genet* 15:83–6.

Murray L, Chen B, Galy A, Chen S, Tushinski R, Uchida N, Negrin R, Tricot G, Jagannath S, Vesole D, et al. 1995. Enrichment of human hematopoietic stem cell activity in the CD34+Thy-1+Lin-subpopulation from mobilized peripheral blood. *Blood* 85:368–78.

Murray L, DiGusto D, Chen B, Chen S, Combs J, Conti A, Galy A, Negrin R, Tricot G, and Tsukamoto A. 1994. Analysis of human hematopoietic stem cell populations. *Blood Cells* 20:364–9; discussion 369–70.

Muss HB, Smith LR, and Cooper MR. 1987. Tamoxifen rechallenge: Response to tamoxifen following relapse after adjuvant chemohormonal therapy for breast cancer. *J Clin Oncol* 5:1556–8.

Myers CE and Chabner BA. 1990. Anthracyclines. In: CA Chabner and JM Collins (eds.). *Cancer Chemotherapy*, pp. 356–81. Philadelphia, PA: JB Lippincott.

Myers GH Jr, Fehrenbaker LG, and Kelalis PP. 1968. Prognostic significance of renal vein invasion by hypernephroma. *J Urol* 100:420–3.

Naghshineh R, Hagdoost IS, and Mokhber-Dezfuli MR. 1991. A retrospective study of the incidence of bovine neoplasms in Iran. *J Comp Pathol* 105:235–9.

Nakamura J and Swenberg JA. 1999. Endogenous apinic/apyrimidinic sites in genomic DNA of mammalian tissues. *Cancer Res* 59:2522–6.

Nakayama K, Nakayama K-I, Negishi I, Kuida K, Sawa H, and Loh DY. 1994. Targeted disruption of Bcl-2 alpha beta in mice: occurrence of gray hair, polycystic kidney disease, and lymphocytopenia. *Proc Natl Acad Sci USA* 91:3700–4.

Nakayama K, Ishida N, Shirane M, Inomata A, Inoue T, Shishido N, Horii I, Loh DY, and Nakayama K. 1996. Mice lacking p27(Kip1) display increased body size, multiple organ hyperplasia, retinal dysplasia, and pituitary tumors. *Cell* 85:707–20.

Nakazawa H, English D, Randell PL, Nakazawa K, Martel N, Armstrong BK, and Yamasaki H. 1994. UV and skin cancer: Specific p53 gene mutation in normal skin as a biologically relevant exposure measurement. *Proc Natl Acad Sci USA* 91:360–4.

National Council on Radiation Protection and Measurements. 1987. *Ionizing radiation exposure of the population of the United States.* NCRP Report 93. Bethesda, MD: National Council on Radiation Protection and Measurements.

National Research Council. 1982. *Diet, Nutrition and Cancer.* Washington, DC: National Academy Press.

National Research Council, Committee on Health Effects of Exposure to Low Levels of Ionizing Radiations (Beir VII) 1998. *Health Effects of Exposure to Low Levels of Ionizing Radiation: Time for Reassessment?* Washington, DC: National Academies Press, pp. 5–32.

Needham J. 1942. *Biochemistry and Morphogenesis.* Cambridge, UK: Cambridge University Press.

Neel JV, Schull WJ, Awa AA, Satoh C, Kato H, Otake M, and Yoshimoto Y. 1990. The children of parents exposed to atomic bombs: Estimates of the genetic doubling dose of radiation for humans. *Am J Hum Genet* 46:1053–72.

Nelson DR and Strobel HW. 1987. Evolution of cytochrome P-450 protein. *Mol Biol Evol* 4:572–93.

Nelson DR, Kamataki T, Waxman DJ, Guengerich FP, Estabrook RW, Feyereisen R, Gonzalez FJ, Coon MJ, Gunsalus IC, Gotoh O et al. 1993. The P450 superfamily: Update on new sequences, gene mapping, accession numbers, early trivial names of enzymes and nomenclature. *DNA Cell Biol* 12:1–51.

Nelson MA, Futscher BW, Kinsella T, Wymer J, and Bowden GT. 1992. Detection of mutant Ha-*ras* genes in chemically initiated mouse skin epidermis before the development of benign tumors. *Proc Natl Acad Sci USA* 89:6398–6402.

Nelson WG, De Marzo AM, Isaacs WB. 2003. Prostate cancer. *N Engl J Med* 349:366–81.

Nemoto N, Kodama K, Tazawa A, Masahito P, and Ishikawa T. 1986. Extensive sequence homology of the goldfish *ras* gene to mammalian *ras* genes. *Differentiation* 32:17–23.

Nesnow S, Miyazaki T, Khwaja T, Meyer RB Jr., and Heidelberger C. 1973. Pyridine nucleosides related to 5-fluorouracil and thymine. *J Med Chem* 16:524–8.

Neuberger JS. 1992. Residential radon exposure and lung cancer: An overview of ongoing studies. *Health Phys* 63:503–9.

Neuberger JS and Gesell TF. 2002. Residential radon exposure and lung cancer: risk in nonsmokers. *Health Phys* 83:1–18.

Neuman E, Flemington EK, Sellers WR, and Kaelin WG Jr. 1994. Transcription of the E2F-1 gene is rendered cell cycle dependent by E2F DNA-binding sites within its promoter. *Mol Cell Biol* 14:6607–15.

Neuwirth H and De Kermion JB. 1990. Prostate. In: CM Haskell (ed.). *Prostate Cancer Treatment*, 3rd ed., pp. 737–49. Philadelphia, PA: WB Saunders.

Neuwirth H, Figlin RA, and de Kernion JB. 1990. Kidney. In: CM Haskell (ed.). *Cancer Treatment*, 3rd ed., pp. 769–78. Philadelphia, PA: WB Saunders.

Newton AC. 1995. Protein kinase C: Structure, function, and regulation. *J Biol Chem* 270:28495–8.

Nikitin AY, Ballering LAP, and Rajewsky MF. 1991. Early mutatins of the *neu* (*erb*B-2) gene during ethylnitrosourea-induced oncogenesis in the rat Schwann cell lineage. *Proc Natl Acad Sci USA* 88:9939–43.

Nilsson T, Hoglund M, Lenhoff S, Rylander L, Turesson I, Westin J, Mitelman F, and Johansson B. 2003. A pooled analysis of karyotypic patterns, breakpoints and imbalances in 783 cytogenetically abnormal multiple myelomas reveals frequently involved chromosome segments as well as significant age- and sex-related differences. *Br J Haematol* 120:960–9.

Nishizuka Y. 1992. Intracellular signaling by hydrolysis of phospholipids and activation of protein kinase C. *Science* 258:607–14.

Noble RL, Beer CT, and Cutts JH. 1958. Further biological activities of vincaleukoblastine – an alkaloid isolated from *Vinca rosea* (L). *Biochem Pharmacol* 1:347–8.

Noel AC, Calle A, Emonard HP, Nusgens BV, Simar L, Foidart J, Lapiere CM, and Foidart JM. 1991. Invasion of reconstituted basement membrane matrix is not correlated to the malignant metastatic cell phenotype. *Cancer Res* 51:405–14.

Noguchi S, Miyauchi K, Nishizawa Y, and Koyama H. 1988. Induction of progesterone receptor with tamoxifen in human breast cancer with special reference to its behavior over time. *Cancer* 61:1345–9.

Norris MD, Burkhart CA, Marshall GM, Weiss WA, and Haber M. 2000. Expression of N-myc and MRP genes and their relationship to N-myc gene dosage and tumor formation in a murine neuroblastoma model. *Med Pediatr Oncol* 35:585–9.

Norton L. 1979. Thoughts on a role for cell kinetics in cancer chemotherapy. In: HJ Tagnon and MJ Staquet (eds.). *Controversies in Cancer: Design of Trials and Treatment*, pp. 105–15. New York: Masson.

1990. Biology of residual breast cancer after therapy: A kinetic interpretation. *Prog Clin Biol Res* 354A:109–32.

Norton L and Simon R. 1986. The Norton-Simon hypothesis revisited. *Cancer Treat Rep* 70:163–9.

Novick D, Cohen B, and Rubinstein M. 1992. Soluble interferon-a receptor molecules are present in body fluids. *Fed Exp Biol Soc Lett* 314:445–8.

Nowell PC. 1976. The clonal evolution of tumor cell populations. *Science* 194:23–5.

2002. Tumor progression: a brief historical perspective. *Semin Cancer Biol* 12:261–6.

Nowell PC and Hungerford DA. 1960a. A minute chromosome in human granulocytic leukemia. *Science* 132:1497.

1960b. Chromosome studies on normal and leukemic human leukocytes. *J Natl Cancer Inst* 25:85–109.

Ochsner A. 1954. *Smoking and Cancer: A Doctor's Report.* New York: J Messner.

Ochsner A and DeBakey M. 1939. Primary pulmonary malignancy: Treatment of total pneumonectomy: Analysis of seventy-nine collected cases and presentation of seven personal cases. *Surg Gynecol Obstet* 68:435–51.

——— 1940. Surgical considerations of primary carcinoma of the lung. *Surgery* 8:992–1023.

——— 1941. Carcinoma of the lung. *Arch Surg* 42:209–58.

——— 1942. Significance of metastasis in primary carcinoma of the lungs. *J Thoracic Surgery* 11:357–87.

Oda K, Arakawa H, Tanaka T, Matsuda K, Tanikawa C, Mori T, Nishimori H, Tamai K, Tokino T, Nakamura Y, and Taya Y. 2000. p53AIP1, a potential mediator of p53-dependent apoptosis, and its regulation by Ser-46-phosphorylated p53. *Cell* 102:849–62.

Ogilvie D, McKinnell RG, and Tarin D. 1984. Temperature-dependent elaboration of collagenase by the renal adenocarcinoma of the leopard frog, *Rana pipiens. Cancer Res* 44:3438–41.

Okey AB. 1990. Enzyme induction in the cytochroms P-450 system. *Pharmacol Ther* 45:241–98.

Okey AB, Roberts EA, Harper PA, and Denison MS. 1986. Induction of drug-metabolizing enzymes: Mechanisms and consequences. *Clin Biochem* 19:132–41.

Olivier M, Eeles R, Hollstein M, Khan MA, Harris CC, and Hainaut P. 2002. The IARC TP53 database: new online mutation analysis and recommendations to users. *Hum Mutat* 19:607–14.

Omenn GS, Goodman GE, Thornquist MD, Balmes J, Cullen MR, Glass A, Keogh JP, Meyskens FL, Valanis B, Williams JH et al. 1996. Effects of a combination of beta carotene and vitamin A on lung cancer and cardiovascular disease. *N Engl J Med* 334:1150–5.

Oro AE, Higgins KM, Hu Z, Bonifas JM, Epstein EH Jr, and Scott MP. 1997. Basal cell carcinomas in mice overexpressing sonic hedgehog. *Science* 276:817–21.

Orr LC, Fleitz J, McGavran L, Wyatt-Ashmead J, Handler M, Forman NK. 2002. Cytogenetics in pediatric low-grade astrocytomas. *Med Pediatr Oncol* 38:173–7.

Ortega S, Prieto I, Odajima J, Martin A, Dubus P, Sotillo R, Barbero JL, Malumbres M, and Barbacid M. 2003. Cyclin-dependent kinase 2 is essential for meiosis but not for mitotic cell division in mice. *Nat Genet* 35:25–31.

Ortner DJ and Putschar WG. 1981. *Identification of pathological conditions in human skeletal remains.* Washington, DC: Smithsonian Institution Press.

Orzechowski A, Schrenk D, and Bock KW. 1992. Metabolism of 1- and 2-naphthylamine in isolated rat hepatocytes. *Carcinogenesis* 13:2227–32.

Osborne CK, Coronado-Heinsohn EB, Hilsenbeck SG, McCue BL, Wakeling AE, McClelland RA, Manning DL, and Nicholson RI. 1995. Comparison of the effects of a pure steroidal antiestrogen with those of tamoxifen in a model of human breast cancer. *J Natl Cancer Inst* 87:746–50.

Osborne CK, Yochmowitz MG, Knight WA, and McGuire WL. 1980. The value of estrogen and progesterone receptors in the treatment of breast cancer. *Cancer* 46 (Suppl 12):2884–8.

Ozes ON, Klein SB, Reiter Z, and Taylor MW. 1993. An interferon resistant variant of the hairy cell leukemic cell line, Eskol: Biochemical and immunological characterization. *Leuk Res* 17: 983–90.

Paez JG, Janne PA, Lee JC, Tracy S, Greulich H, Gabriel S, Herman P, Kaye FJ, Lindeman N, Boggon TJ et al. 2004. EGFR mutations in lung cancer: correlation with clinical response to gefitinib therapy. *Science* 304:1497–1500.

Paffenbarger RS, Hyde RT, and Wing AL. 1987. Physical activity and incidence of cancer in diverse populations: A preliminary report. *Am J Clin Nutr* 45:312–17.

Paffenbarger RS, Hyde RT, Wing AL, and Hsieh CC. 1986. Physical activity, all-cause mortality, and longevity of college alumni. *N Engl J Med* 314:605–13.

Paget S. 1889. The distribution of secondary growths in cancer of the breast. *Lancet* 1:571–3.

Papadopoulos N, Nicolaides NC, Wei Y-F, Ruben SM, Carter KC, Rosen CA, Haseltine WA, Fleischmann RD, Fraser CM, Adams MD et al. 1994. Mutation of a *mutL* homolog in hereditary colon cancer. *Science* 263:1625–9.

Parchment RE. 1993. The implications of a unified theory of programmed cell death, polyamines, oxyradicals and histogenesis in the embryo. *Int J Dev Biol* 37:75–83.

     1998. Alternative testing systems for evaluating noncarcinogenic, hematologic toxicity. *Environ Health Perspect* 106:541–57.

Parchment RE and Pierce GB. 1989. Polyamine oxidation, programmed cell death, and regulation of melanoma in the murine limb. *Cancer Res* 49:6680–6.

Parchment RE, Gordon M, Grieshaber CK, Sessa C, Volpe D, and Ghielmini M. 1998. Predicting hematological toxicity (myelosuppression) of cytotoxic drug therapy from in vitro tests. *Ann Oncol* 9:357–64.

Parczyk K and Schneider MR. 1996. The future of antihormone therapy: Innovations based on an established principle. *J Cancer Res Clin Oncol* 122:383–96.

Park EJ and Pezzuto JM. 2002. Botanicals in cancer chemoprevention. *Cancer Metastasis Rev* 21:231–55.

Parker RG. 1990. Principles of radiation oncology. In: CM Haskell (ed.). *Cancer Treatment*, 3rd ed, pp. 15–21. Philadelphia, PA: WB Saunders.

Parker SL, Tong T, Bolden S, and Wingo PA. 1997. Cancer statistics, 1997. *CA Cancer J Clin* 47:5–27.

Parkin DM, Pisani P, Muñoz N, and Ferlay J. 1999. The global health burden of infection associated cancers. *Cancer Surv* Suppl. 33:5–34.

Patel JD, Bach PB, and Kris MG. 2004. Lung cancer in US women: A contemporary epidemic. *JAMA* 291:1763–8.

Patel NH and Rothenberg ML. 1994. Multidrug resistance in cancer chemotherapy. *Invest New Drugs* 12:1–13.

Patterson BH and Block G. 1988. Food choices and the cancer guidelines. *Am J Public Health* 78:282–6.

Payne DM, Rossomando AJ, Martino P, Erickson AK, Her J-H, Shananowitz J, Hunt DF, Weber MJ, and Sturgill TW. 1991. Identification of the regulatory phosphorylation sites in pp42/mitogen-activated protein kinase (MAP kinases). *EMBO J* 10:885–92.

Pearl R. 1922. New data on the influence of alcohol on the expectation of life in man. *Am J Hygiene* 2:463–6.

Pecorelli S. 1994. Management of the symptomatic patient: Surgery *NIH Consensus Development Conference on Ovarian Cancer*. Bethesda, MD: National Institutes of Health.

Peltomäki P, Aaltonen LA, Sistonen P, Pylkkänen L, Mecklin J-P, Järvinen H, Green JS, Jass JR, Weber JL, Leach FS et al. 1993. Genetic mapping of a locus predisposing to human colorectal cancer. *Science* 260:810–12.

Pendergast AM, Quilliam LA, Cripe LD, Bassing CH, Dai Z, Li N, Batzer A, Rabun KM, Der CJ, Schlessinger J, and Gishizky ML. 1993. BCR-ABL-induced oncogenesis is mediated by direct interaction with the SH2 domain of the GRB-2 adaptor protein. *Cell* 75:175–85.

Peraino DR, Fry RJ, and Staffeldt E. 1971. Reduction and enhancement by phenobarbital of hepatocarcinogenesis induced in the rat by 2-acetylaminofluorene. *Cancer Res* 31:1506–12.

Perantoni AO, Rice JM, Reed CD, Watatani M, and Wenk ML. 1987. Activated neu oncogene sequences in primary tumors of the peripheral nervous system induced in rats by transplacental exposure to ethylnitrosourea. *Proc Natl Acad Sci USA* 84:6317–21.

Perantoni AO and JM Rice. 1999. Mutation patterns in non-ras oncogenes and tumour suppressor genes in experimentally induced tumours. *IARC Sci Publ* 146:87–122.

Perantoni AO, Turusov VS, Buzard GS, and Rice JM. 1994. Infrequent transforming mutations in the transmembrane domain of the neu oncogene in spontaneous rat schwannomas. *Mol Carcinog* 9:230–5.

Perri RT, Weisdorf DJ, and Oken, MM. 1985. Low-dose ARA-C fails to enhance differentiation of leukaemic cells. *Br J Haematol* 59:697–701.

Perris R and Bronner-Fraser M. 1989. Recent advances in defining the role of the extracellular matrix in neural crest development. *Commun Dev Neurobiol* 1:61–83.

Pesenti-Barili B, Ferdandi E, Mosti M, and Degli-Innocenti F. 1991. Survival of Agrobacterium radiobacter K84 on various carriers for crown gall control. *Appl Environ Microbiol* 57:2047–51.

Pestell KE, Ducruet AP, Wipf P, and Lazo JS. 2000. Small molecule inhibitors of dual specificity protein phosphatases. *Oncogene* 19:6607–12.

Peter M and Herskowitz I. 1994. Joining the complex: Cyclin-dependent kinase inhibitory proteins and the cell cycle. *Cell* 79:181–4.

Phelps WC, Yee CL, Munger K, and Howley PM. 1988. The human papillomavirus type 16 E7 gene encodes transactivation and transformation functions similar to adenovirus E1a. *Cell* 53:539–47.

Phillips DH, Carmichael PL, Hewer A, Cole KJ, Hardcastle IR, Poon GK, Keogh A, and Strain AJ. 1996. Activation of tamoxifen and its metabolite alpha-hyroxytamoxifen to DNA-binding products: Comparisons between human, rat and mouse hepatocytes. *Carcinogenesis* 17:89–94.

Pickle LW, Mason TJ, Howard N, Hoover R, and Fraumeni JF. 1987. *Atlas of US Cancer Mortality Among Whites.* Bethesda, MD: Public Health Service, National Institutes of Health.

Pickle LW, Mungiole M, Jones GK, and White AA. 1996. *Atlas of United States Mortality.* Hyattsville, MD: US Department of Health and Human Services.

Pierce GB. 1974. Neoplasms, differentiation, mutations. *Am J Pathol* 77:103–18.

     1983. The cancer cell and its control by the embryo. *Am J Pathol* 113:117–24.

Pierce GB and Dixon FJ. 1959. The demonstration of teratogenesis by metamorphosis of multipotential cells. *Cancer* 12:573–89.

Pierce GB and Parchment R. 1991. Progression and teratocarcinoma. In: O Sudilovsky, H Pitot, and L Liotta (eds.). *Boundaries Between Promotion and Progression During Carcinogenesis*, pp. 71–81. Collection of Basic Life Science. New York: Plenum Press.

Pierce GB and Speers WC. 1988. Tumors as caricatures of the process of tissue renewal: prospects for therapy by directing differentiation. *Cancer Res* 48:1996–2004.

Pierce GB and Wallace C. 1971. Differentiation of malignant to benign cells. *Cancer Res* 31:127–34.

Pierce GB, Dixon FJ, and Verney EL. 1960. Teratocarcinogenic and tissue forming potentials of the cell types comprising neoplastic embryoid bodies. *Lab Invest* 9:583–602.

Pierce GB, Gramzinski RA, and Parchment RE. 1990. Amine oxidases, programmed cell death, and tissue renewal. *Philos Trans R Soc Lond Biol* 327:67–74.

Pierce GB, Lewellyn AL, and Parchment RE. 1989. Mechanisms of programmed cell death in the blastocyst. *Proc Natl Acad Sci USA* 86:3654–8.

Pierce GB, Shikes RH, and Fink LM. 1978. *Cancer: A problem of developmental biology.* Englewood Cliffs, NJ: Prentice-Hall.

Pierce GB, Stevens LC, and Nakane PK. 1967. Ultrastructural analysis of the early development of teratocarcinomas. *J Natl Cancer Inst* 39:755–73.

Pierce DA., Shimizu Y, Preston DL, Vaeth M, and Mabuchi K. 1996. Studies of the mortality of atomic bomb survivors. Report 12, Part I. Cancer: 1950–1990. *Radiat Res* 146:1–27.

Pines, J 1994. The cell cycle kinases. *Semin Cancer Biol* 5:305–13.

Pinkney AE, Harshbarger JC, May EB, and Melancon MJ. 2001. Tumor prevalence and biomarkers of exposure in brown bullheads (Ameiurus nebulosus) from the tidal Potomac River, USA, watershed. *Environ Toxicol Chem* 20:196–205.

Pirila E, Sharabi A, Salo T, Quaranta V, Tu H, Heljasvaara R, Koshikawa N, Sorsa T, and Maisi P. 2003. Matrix metalloproteinases process the laminin-5 gamma 2-chain and regulate epithelial cell migration. *Biochem Biophys Res Commun* 303:1012–17.

Pittillo RF, Schabel FM Jr, and Skipper HE. 1970. The "sensitivity" of resting and dividing cells. *Cancer Chemother Rep* 54:137–42.

Plasterk RH. 1992. Genetic switches: Mechanisms and function. *Trends Genet* 8:403–6.

Platanias LC, Pfeffer LM, Barton KP, Vardiman JW, Golomb HM, and Colamonici OR. 1992. Expression of the IFNa receptor in hairy cell leukaemia. *Br J Haematology* 82:541–6.

Pluk H, Dorey K and Superti-Furga G. 2002. Autoinhibition of c-Abl. *Cell* 108:247–59.

Podesta AH, Mullins J, Pierce GB, and Wells RS. 1984. The neurula stage mouse embryo in control of neuroblastoma. *Proc Natl Acad Sci USA* 81:7608–11.

Poland A, Glover E, and Kende AS 1976. Stereospecific high affinity binding of 2,3,7,8-tetrachlorodibenzo-p-dioxin by hepatic cytosols. Evidence that the binding species is a receptor for induction of aryl hydrocarbon hydroxylase. *J Biol Chem* 251:4936–46.

Polednak, AP 1976. College athletics, body size, and cancer mortality. *Cancer* 38:382–7.

Polek TC and Weigel NL. 2002. Vitamin D and prostate cancer. *J. Androl* 23:9–17.

Poste G and Paruch L. 1989. Stephen Paget, MD, FRCS (1855–1926): A retrospective. *Cancer and Metastasis Reviews* 8:93–7.

Pott P. 1775. *Chirurgical observations relative to the cataract, the polypus of the nose, the cancer of the scrotum, the different kinds of ruptures and the mortification of the toes and feet.* London: Hawkes, Clarke, and Collins.

Potten CS and Chadwick CA. 1994. Small intestinal regulatory factors extracted by simple diffusion from intact irradiated intestine and tested in vivo. *Growth Factors* 10:63–75.

Potten CS, Chadwick C, Ijiri K, Tsubouchi S, and Hanson WR. 1984. The recruitability and cell-cycle state of intestinal stem cells. *Int J Cell Cloning* 2:126–40.

Potten CS, Booth C, Chadwick CA, and Evans GS. 1994. A potent stimulator of small intestinal cell proliferation extracted by simple diffusion from intact irradiated intestine: In vitro studies. *Growth Factors* 10:53–61.

Potten CS, Owen G, Hewitt D, Chadwick CA, Hendry H, Lord BI, and Woolford LB. 1995. Stimulation and inhibition of proliferation in the small intestinal crypts of the mouse after in vivo administration of growth factors. *Gut* 36: 864–73.

Potten CS, Watson RJ, Williams GT, Tickle S, Roberts SA, Harris M, and Howell A. 1988. The effect of age and menstrual cycle upon proliferative activity of normal human breast. *Br J Cancer* 58:163–70.

Potter CJ, Turenchalk GS and Xu T. 2000. Drosophila in cancer research: an expanding role. *Trends in Genetics* 16:33–9.

Press OW, Appelbaum F, Ledbetter JA, Martin PJ, Zarling J, Kidd P, and Thomas ED. 1987. Monoclonal antibody 1f5 (anti-cd20) serotherapy of human b cell lymphomas. *Blood* 69: 584–91.

Pruitt K and Der CJ. 2001. Ras and Rho regulation of the cell cycle and oncogenesis. *Cancer Lett* 171:1–10.

Pulciani S, Santos E, Lauver AV, Long LK, Robbins KC, and Barbacid M. 1982. Oncogenes in human tumor cell lines: Molecular cloning of a transforming gene from human bladder cell carcinoma cells. *Proc Natl Acad Sci USA* 79:2845–9.

Purchio AF, Erikson E, Brugge JS, and Erikson RL. 1978. Identification of a polypeptide encoded by the avian sarcoma virus *src* gene. *Proc Natl Acad Sci USA* 75:1567–71.

Quaranta V. 2002. Motility cues in the tumor environment. *Differentiation* 70:590–8.

Quintanilla M, Brown K, Ramsden M, and Balmain A. 1986. Carcinogen-specific mutation and amplification of Ha-*ras* during mouse skin carcinogenesis. *Nature* 322:78–80.

Rabinowitz Y and Wilhite BA. 1969. Thymidine salvage pathway in normal and leukemic leukocytes with effects of ATP on enzyme control. *Blood* 33:759–71.

Rader K. 2002. The origin of mouse genetics: beyond the Bussey Institution. II Defining the problem of mouse supply: the 1928 National Research Council Committee on Experimental Plants and Animals. *Mamm Genome* 13:2–4.

Radonski JL. 1979. The primary aromatic amines, their biological properties and structure-activity relationships. *Annu Rev Pharmacol Toxicol* 19:129–57.

Rameh LE and Cantley LC. 1999. The role of phosphoinositide 3-kinase lipid products in cell function. *J Biol Chem.* 274:8347–50.

Rampino N, Yamamoto H, Ionov Y, Li Y, Sawai H, Reed JC, and Perucho M. 1997. Somatic frameshift mutations in the BAX gene in colon cancers of the microsatellite mutator phenotype. *Science* 275:967–9.

Rauh MJ, Blackmore V, Andrechek ER, Tortorice CG, Daly R, Lai VK, Pawson T, Cardiff RD, Siegel PM, and Muller WJ. 1999. Accelerated mammary tumor development in mutant polyomavirus middle T transgenic mice expressing elevated levels of either the Shc or Grb2 adapter protein. *Mol Cell Biol* 1912:8169–79.

Raz E. 2003. Primordial germ-cell development: the zebrafish perspective. *Nat Rev Genet* 4:690–700.

Rebello P, Cwynarski K, Varughese M, Eades A, Apperley JF, and Hale G. 2001. Pharmacokinetics of campath-1h in bmt patients. *Cytotherapy* 3:261–7.

Recamier JC. 1829. *Recherches sur le traitement du cancer, par la compression methodique simple ou combinée et sur l'histoire générale de la même maladie*, vol II. Paris: Gabon.

Rechsteiner M and Rogers SW. 1996. PEST sequences and regulation by proteolysis. *Trends Biochem Sci* 21:267–71.

Recny MA, Scoble HA, and Kim Y. 1987. Structural characterization of natural human urinary and recombinant DNA-derived erythropoietin. *J Biol Chem* 262:17156–63.

Reddel RR, Alexander IE, Koga M, Shine J, and Sutherland RL. 1988. Genetic instability and the development of steroid hormone insensitivity in cultured T 47D human breast cancer cells. *Cancer Res* 48:4340–7.

Reed E and Kohn KW. 1990. Platinum analogues. In: BA Chabner and JM Collins (eds.). *Cancer Chemotherapy*, pp. 465–90. Philadelphia, PA: JB Lippincott.

Rees E, Young RD, and Evans DJ. 2003. Spatial and temporal contribution of somatic myoblasts to avian limb muscles. *Dev Biol* 253:264–78.

Regaud C. 1930. Sur les principes radiophysiologiques de la radiotherapie des cancers. *Acta Radiol* 11:456–86.

Regaud C and Ferroux R. 1927. Discordance des effets des rayons X, d'une part dans la peau, d'autre part dans le testicule, par le fractionnement de la dose; diminution de l'efficacité dans le peau, maintien de l'efficacité dans le testicule. *C R Soc Biol* 97:431–4.

Rehn L. 1895. Bladder tumours in Fuchsine workers. *Arch Klin Chir* 50:588–600.

Reiss M, Gamba-Vitalo C, and Sartorelli AC. 1986. Induction of tumor differentiation as a therapeutic approach: Preclinical models for hematopoietic and solid neoplasms. *Cancer Treat Rep* 70:201–18.

Remmer H. 1962. Drug tolerance. In: JL Mongar and AVS De Reuck (eds.). *Ciba Foundations Symposium on Enzymes and Drug Action*, pp. 276–98. Boston, MA: Little, Brown.

Ren R. 2002. The molecular mechanism of chronic myelogenous leukemia and its therapeutic implications: studies in a murine model. *Oncogene* 21:8629–42.

Rendina GM, Donadio C, and Giovannini M. 1982. Steroid receptors and progestinic therapy in ovarian endometrioid carcinoma. *Eur J Gynaecol Oncol* 3:241–6.

Repesh LA, Drake SR, Warner MC, Downing SW, Jyring R, Seftor EA, Hedrix MJ, and McCarthy JB. 1993. Adriamycin-induced inhibition of melanoma cell invasion is correlated with decreases in tumour cell motility and increases in focal contact formation. *Clin Exp Metastasis* 11:91–102.

Reynisdottir I, Polyak K, Iavarone A, Massague J. 1995. Kip/Cip and Ink4 Cdk inhibitors cooperate to induce cell cycle arrest in response to TGF-beta. *Genes Dev* 9:1831–45.

Rideout WM, III, Coetzee GA, Olumi AF, Jones PA. 1990. 5-Methylcytosine as an endogenous mutagen in the human LDL receptor and p53 genes. *Science* 249:1248–50.

Rigel DS. 1996. Malignant melanoma: Perspectives on incidence and its effects on awareness, diagnosis, and treatment. *CA Cancer J Clin* 46:195–8.

———. 2002. The effect of sunscreen on melanoma risk. *Dermatol Clin* 20:601–6.

Rigel DS, Friedman RJ, Dzubow LM, Reintgen DS, Bystryn J-C, Marks R. (eds.). 2005. *Cancer of the skin*. Philadelphia: Elsevier Saunders.

Rijke RP, Plaisier HM, and Langendoen NJ. 1979. Epithelial cell kinetics in the descending colon of the rat. *Virchows Arch B Cell Pathol Incl Mol Pathol* 30:85–94.

Rizzino A and Crowley C. 1980. The growth and differentiation of the embryonal carcinoma cell line, F9, in defined media. *Proc Natl Acad Sci USA* 77:457–61.

Roberts JM and Sherr CJ. 2003. Bared essentials of CDK2 and cyclin E. *Nat Genet* 35:9–10.

Robertson AM, Bird CC, Waddell AW, and Currie AR. 1978. Morphological aspects of glucocorticoid-induced cell death in human lymphoblastoid cells. *J Pathol* 126:181–7.

Roninson I. B. 2002. Oncogenic functions of tumour suppressor p21(Waf1/Cip1/Sdi1): association with cell senescence and tumour-promoting activities of stromal fibroblasts. *Cancer Lett* 179:1–14.

Rose C. 2003. A comparison of the efficacy of aromatase inhibitors in second-line treatment of metastatic breast cancer. *Am J Clin Oncol* 26: S9–16.

Rose FL and Harshbarger JC. 1977. Neoplastic and possibly related skin lesions in neotenic tiger salamanders from a sewage lagoon. *Science* 298:270–9.

Rosen PR, Groshen S, Saigo PE, Kinne DW, Hellman S. 1989. A long-term follow-up study of survival in stage I (T1N0M0) and stage II (T1N1M0) breast carcinoma. *J Clin Oncol* 7:355–66.

Rosenbaum H, Webb E, Adams JM, Cory S, and Harris AW. 1989. N-myc transgene promotes B lymphoid proliferation, elicits lymphomas and reveals cross-regulation with c-myc. *EMBO J* 8:749–55.

Rosenberg B, VanCamp L, Trosko JE, and Mansour VH. 1965. Inhibition of cell division in Escherichia coli by electrolysis products from a platinum electrode. *Nature* 205:698–9.

Rosenberg B, VanCamp L, Trosko JE, and Mansour VH. 1969. Platinum compounds: A new class of potent anti-tumour agents. *Nature* 222:385–6.

Roskoski Jr R. 2004. The ErbB/HER receptor protein-tyrosine kinases and cancer. *Biochem Biophys Res Commun* 319:1–11.

Rostami M, Tateyama S, Uchida K, Naitou H, Yamaguchi R, and Ostuka H. 1994. Tumors in domestic animals examined during a ten-year period 1980 to 1990 at Miyazaki University. *J Vet Med Sci* 56:403–5.

Rous P. 1911a. A sarcoma of the fowl transmissible by an agent separable from the tumor cells. *J Exp Med* 13:397–411.

1911b. Transmission of a malignant new growth by means of a cell-free filtrate. *JAMA* 56:198–208.

Rovero S, Amici A, Carlo ED, Bei R, Nanni P, Quaglino E, Porcedda P, Boggio K, Smorlesi A, Lollini PL et al. 2000. DNA vaccination against rat her-2/Neu p185 more effectively inhibits carcinogenesis than transplantable carcinomas in transgenic BALB/c mice. *J Immunol* 165: 5133–42.

Rowley JD. 1973. A new consistent chromosomal abnormality in chronic myelogenous leukaemia identified by quinacrine fluorescence and Giemsa staining. *Nature* 243:290–3.

1977. Mapping of human chromosomal regions related to neoplasia: Evidence from chromosomes 1 and 17. *Proc Natl Acad Sci USA* 74:5729–33.

Roy HK, Olusola BF, Clemens DL, Karolski WJ, Ratashak A, Lynch HT, and Smyrk TC. 2002. AKT proto-oncogene overexpression is an early event during sporadic colon carcinogenesis. *Carcinogenesis* 23:201–5.

Rubin P and Cooper RA. 1993. Statement of the clinical oncologic problem. In: P Rubin (ed.). *Clinical Oncology: A Multidisciplinary Approach for Physicians and Students*, 7th ed., pp. 8–11. Philadelphia, PA: WB Saunders.

Ruddon RW. 1995. *Cancer Biology*, 3rd ed. New York: Oxford University.

Rudkin GT, Hungerford DA and Nowell PC. 1964. DNA contents of chromosome Ph1 and chromosome 21 in human granulocytic leukemia. *Science* 144:1229–32.

Ruggeri B, Zhang Y, Caamano J, DiRado M, Flynn SD, and Klein-Szanto AJ. 1992. Human pancreatic carcinomas and cell lines reveal frequent and multiple alterations in the p53 and Rb-1 tumor-suppressor genes. *Oncogene* 7:1503–11.

Ruggero D, Montanaro L, Ma L, Xu W, Londei P, Cordon-Cardo C, and Pandolfi PP. 2004. The translation factor eIF-4E promotes tumor formation and cooperates with c-Myc in lymphomagenesis. *Nat Med* 10:484–6.

Ruiz i Altaba A, Sanchez P, and Dahmane N. 2002. Gli and hedgehog in cancer: tumours, embryos and stem cells. *Nat Rev Cancer* 2:361–72.

Ruley HE. 1990. Transforming collaborations between ras and nuclear oncogenes. *Cancer Cells* 2:258–68.

Sachs L. 1987. Cell differentiation and bypassing genetic defects in the suppression of malignancy. *Cancer Res* 47:1981–6.

1993. Regulation of normal development and tumor suppression. *Int J Dev Biol* 37:51–9.

Safai B, Diaz B, and Schwartz J. 1992. Malignant neoplasms associated with human immunodeficiency virus infections. *CA Cancer J Clin* 42:74–95.

Salerno M, Ouatas T, Palmieri D, and Steeg PS. 2003. Inhibition of signal transduction by the nm23 metastasis suppressor: Possible mechanisms. *Clin Exptl Metastasis* 20:3–10.

Salih HR and Kiener PA. 2004. Alterations in Fas (CD 95/Apo-1) and Fas ligand (CD178) expression in acute promyelocytic leukemia during treatment with ATRA. *Leuk Lymphoma* 45:55–9.

Salonen JT. 1986. Selenium and human cancer. *Ann Clin Res* 18:18–21.

Salsbury AJ. 1975. The significance of the circulating cancer cell. *Cancer Treat Rev* 2:55–72.

Saltz LB, Meropol NJ, Loehrer PJ, Sr., Needle MN, Kopit J, and Mayer RJ. 2004. Phase II trial of cetuximab in patients with refractory colorectal cancer that expresses the epidermal growth factor receptor. *J Clin Oncol* 22:1201–8.

Samet JM and Eradze GR. 2000. Radon and lung cancer risk: taking stock at the millenium. *Environ Health Perspect* 108 Suppl 4:635–41.

Samuels ML, Weber MJ, Bishop JM, and McMahon M. 1993. Conditional transformation of cells and rapid activation of the mitogen-activated protein kinase cascade by an estradiol-dependent human *raf*-1 protein kinase. *Mol Cell Biol* 13:6241–52.

Samuels-Lev Y, O'Connor DJ, Bergamaschi D, Trigiante G, Hsieh JK, Zhong S, Campargue I, Naumovski L, Crook T, and Lu X. 2001. ASPP proteins specifically stimulate the apoptotic function of p53. *Mol Cell* 8:781–94.

Sandford NL, Searle JW, and Kerr, JF. 1984. Successive waves of apoptosis in the rat prostate after repeated withdrawal of testosterone stimulation. *Pathology* 164:406–10.

Sandow BA, West NB, Norman RL, and Brenner RM. 1979. Hormonal control of apoptosis in hamster uterine luminal epithelium. *Am J Anat* 156:15–35.

Santen RJ, Manni A, and Harvey H. 1986. Gonadotropin releasing hormone (GnRH) analogs for the treatment of breast and prostatic carcinoma. *Breast Cancer Res Treat* 7:129–45.

Santoro M, Carlomagno F, Romano A, Bottaro DP, Dathan NA, Grieco M, Fusco A, Vecchio G, Matoskova B, Kraus MH, and Di Fiore PP. 1995. Activation of RET as a dominant transforming gene by germline mutations of MEN2A and MEN2B. *Science* 267:381–3.

Santos E and Nebreda AR. 1989. Structural and functional properties of *ras* proteins. *Fed Am Soc Exp Biol J* 3:2151–63.

Saracci R. 1987. The interactions of tobacco smoking and other agents in cancer etiology. *Epidemiol Rev* 9:175–93.

Sarasin A. 1999. The molecular pathways of ultraviolet-induced carcinogenesis. *Mutat Res* 428: 5–10.

Sargent JD and DiFranza JR. 2003. Tobacco control for clinicians who treat adolescents. *CA Cancer J Clin* 53:102–23.

Sasaki M. 1982. Current status of cytogenetic studies in animal tumors with special reference to nonrandom chromosome changes. *Cancer Genet Cytogenet* 5:153–72.

Sasaoka T, Langlois WJ, Bai F, Rose DW, Leitner JW, Decker SJ, Saltiel A, Gill GN, Kobayashi M, Draznin B et al. 1996. Involvement of ErbB2 in the signaling pathway leading to cell cycle progression from a truncated epidermal growth factor receptor lacking the C-terminal autophosphorylation sites. *J Biol Chem* 271:8338–44.

Sato S, Fujita N, and Tsuruo T. 2004. Involvement of PDK1 in the MEK/MAPK signal-transduction pathway. *J Biol Chem* 279:33759–67.

Sauerbier W, Rollins-Smith LA, Carlson DL, Williams CS, Williams JW III, and McKinnell RG. 1995. Sizing of the Lucké tumor herpesvirus genome by field inversion gel electrophoresis and restriction analysis. *Herpetopathologia* 2:137–43.

Scarpelli DG. 1975. Neoplasia in poikilotherms. In: FF Becker (ed.). *Cancer 4, Biology of Tumors*, pp. 375–410. New York: Plenum.

Scartozzi M, Bearzi I, Berardi R, Mandolesi A, Fabris G, and Cascinu S. 2004. Epidermal growth factor receptor (EGFR) status in primary colorectal tumors does not correlate with EGFR expression in related metastatic sites: Implications for treatment with egfr-targeted monoclonal antibodies. *J Clin Oncol* 22:4772–8.

Schaaper RM and Dunn RL. 1987. Spectra of spontaneous mutations in Escherichia coli strains defective in mismatch repair correction: The nature of in vivo DNA replication errors. *Proc Natl Acad Sci USA* 84:6220–4.

Schabel FM Jr 1969. The use of tumor growth kinetics in planning "curative" chemotherapy of advanced solid tumors. *Cancer Res* 29:2384–9.

 1975. Concepts for systemic treatment of micrometastases. *Cancer* 35:15–24.

1976. Concepts for treatment of micrometastases developed in murine systems. *Am J Roentgenol* 126:500–11.

Schabel FM Jr, Johnston TP, Mc CG, Montgomery JA, Laster WR, and Skipper HE. 1963. Experimental evaluation of potential anticancer agents VIII. Effects of certain nitrosoureas on intracerebral L1210 leukemia. *Cancer Res* 23:725–33.

Schabel FM Jr, Skipper HE, Trader MW, and Wilcox WS. 1965. Experimental evaluation of potential anticancer agents: XIX Sensitivity of nondividing leukemic cell populations to certain classes of drugs in vivo. *Cancer Chemother Rep* 48:17–30.

Schabel FM Jr, Skipper HE, Trader MW, Laster WR Jr, Corbett TH, and Griswold DP Jr. 1980. Concepts for controlling drug-resistant tumor cells. *Eur J Cancer* Suppl 1:199–211.

Schaeffer HJ and Weber MJ. 1999. Mitogen-activated protein kinases: specific messages from ubiquitous messengers. *Mol Cell Biol* 19:2435–44.

Scheffner M, Werness BA, Huibregste JM, Levine AJ, and Howley PM. 1990. The E6 oncoprotein encoded by human papillomavirus types 16 and 18 promotes the degradation of p53. *Cell* 63:1129–36.

Scheijen B and Griffin JD. 2002. Tyrosine kinase oncogenes in normal hematopoiesis and hematological disease. *Oncogene* 21:3314–33.

Schiffer CA. 1993. Acute lymphocytic leukemia in adults. In: JF Holland, E Frei III, RC Bast Jr, DW Kufe, DL Morton, and RR Weichselbaum (eds.). *Cancer Medicine*, vol 2, pp. 1946–55. Philadelphia, PA: Lea and Febiger.

Schlegel R. 1990. Papilloma viruses and human cancer. *Semin Virol* 1:297–306.

Schlessinger J. 2000. Cell signaling by receptor tyrosine kinases. *Cell* 103:211–25.

Schlieman MG, Fahy BN, Ramsamooj R, Beckett L, and Bold RJ. 2003. Incidence, mechanism and prognostic value of activated AKT in pancreas cancer. *Br J Cancer* 89:2110–15.

Schmidt GH, Winton DJ, and Ponder BA. 1988. Development of the pattern of cell renewal in the crypt-villus unit of chimaeric mouse small intestine. *Development* 103:785–90.

Schmitt CA. 2003. Senescence, apoptosis and therapy – cutting the lifelines of cancer. *Nat Rev Cancer* 3:286–95.

Schönthal AH. 2001. Role of serine/threonine protein phosphatase 2A in cancer. *Cancer Lett* 170:1–13.

Schröck E, du Manoir S, Veldman T, Schoell B, Wienberg J, Ferguson-Smith MA, Ning Y, Ledbetter DH, Bar-Am I, Soenksen D et al. 1996. Multicolor spectral karyotyping of human chromosomes. *Science* 273:494–7.

Schuchardt A, D'Agati V, Larsson-Blomberg L, Costantini F, and Pachnis V. 1994. Defects in the kidney and enteric nervous system of mice lacking the tyrosine kinase receptor Ret. *Nature* 367:380–3.

Schuster SJ, Wilson JH, Erslev AJ, and Caro J. 1987. Physiologic regulation and tissue localization of renal erythropoietin messenger RNA. *Blood* 70:316–18.

Schwartsmann G, Brondani da Rocha A, Berlinck RG, and Jimeno J. 2001. Marine organisms as a source of new anticancer agents. *Lancet Oncol* 2:221–5.

Schwartz MK. 1993. Cancer markers. In: VT DeVita, Jr, S Hellman, and SA Rosenberg (eds.). *Cancer: Principles and Practice of Oncology*, 4th ed, pp. 531–42. Philadelphia, PA: JB Lippincott.

Sears R, Leone G, DeGregori J, and Nevins JR. 1999. Ras enhances Myc protein stability. *Mol Cell* 3:169–79.

Sears RC and Nevins JR. 2002. Signaling networks that link cell proliferation and cell fate. *J Biol Chem* 277:11617–20.

Seger R, Ahn NG, Posada J, Manur ES, Jensen AM, Cooper JA, Cobb MH, and Krebs EG. 1992. Purification and characterization of mitogen-activated protein kinase activator(s) from epidermal growth factor-stimulated A431 cells. *J Biol Chem* 267:14373–81.

Selikoff IJ, Seidman H, and Hammond EC. 1980. Mortality effects of cigarette smoking among amosite asbestos factory workers. *J Natl Cancer Inst* 65:507–13.

Sell S (ed.). 1980. *Cancer Markers Diagnostic and Developmental Significance.* Clifton, NY: Humana Press.

Sell S and Leffert HL. 1982. An evaluation of cellular lineages in the pathogenesis of hepatocellular carcinomas. *Hepatology* 2:77–86.

Sell S and Pierce GB. 1994. Maturation arrest of stem cell differentiation is a common pathway for the cellular origin of teratocarcinomas and epithelial cancers. *Lab Invest* 70:6–22.

Semenza GL. 2003. Targeting HIF-1 for cancer therapy. *Nat Rev Cancer* 3:721–32.

Sen P, Mukherjee S, Ray D, and Raha S. 2003. Involvement of the Akt/PKB signaling pathway with disease processes. *Mol Cell Biochem* 253:241–6.

Seppanen ED, McKinnell RG, Rollins-Smith LA, and Hanson W. 1984. Temperature-dependent dissociation of Lucké renal adenocarcinoma cells. *Differentiation* 26:227–30.

Serra E, Ars E, Ravella A, Sanchez A, Puig S, Rosenbaum T, Estivill X, and Lazaro C. 2001. Somatic NF1 mutational spectrum in benign neurofibromas: mRNA splice defects are common among point mutations. *Hum Genet* 108:416–29.

Serrano M, Hannon GJ, and Beach D. 1993. A new regulatory motif in cell-cycle control causing specific inhibition of cyclin D/CDK4. *Nature* 366:704–7.

Serrano M, Lee H-W, Chin L, Cordon-Cardo C, Beach D, and DePinho RA. 1996. Role of the INK4a locus in tumor suppression and cell mortality. *Cell* 85:27–37.

Setlow RB. 1974. The wavelength of sunlight effective in producing skin cancer: A theoretical analysis. *Proc Natl Acad Sci USA* 71:3363–6.

Shackney SE, McCormack GW, and Cuchural GJ Jr. 1978. Growth rate patterns of solid tumors and their relation to responsiveness to therapy: An analytical review. *Ann Int Med* 89:107–21.

Shapiro DM and Fugmann RA. 1957. A role for chemotherapy as an adjunct to surgery. *Cancer Res* 17:1098–1101.

Shaulian E and Karin M. 2002. AP-1 as a regulator of cell life and death. *Nat Cell Biol* 4:E131–6.

Shaw PE, Schroter H, and Nordheim A. 1989. The ability of a ternary complex to form over the serum response element correlates with serum inducibility of the human c-*fos* promoter. *Cell* 56:563–72.

Shelby MD and Zeiger E. 1990. Activity of human carcinogens in the Salmonella and rodent bone-marrow cytogenetics tests. *Mutat Res* 34:257–61.

Sherr CJ. 1996. *Cancer cell cycles. Science* 274:1672–7.

2001. The INK4a/ARF network in tumour suppression. *Nat Rev Mol Cell Biol* 2:731–7.

Sherr CJ and McCormick F. 2002. The RB and p53 pathways in cancer. *Cancer Cell* 2:103–12.

Shevde LA and Welch DR. 2003. Metastasis suppressor pathways – an evolving paradigm. *Cancer Lett* 198:1–20.

Shibata MA, Liu ML, Knudson MC, Shibata E, Yoshidome K, Bandey T, Korsmeyer SJ, and Green JE. 1999. Haploid loss of bax leads to accelerated mammary tumor development in C3(1)/SV40-TAg transgenic mice: reduction in protective apoptotic response at the preneoplastic stage. *EMBO J* 18:2692–2701.

Shibutani S, Takeshita M, and Grollman AP. 1991. Insertion of specific bases during DNA synthesis past the oxidation-damaged base 8-oxodG. *Nature* 349:431–4.

Shields SE, Ogilvie DJ, McKinnell RG, and Tarin, D. 1984. Degradation of basement membrane collagens by metalloproteases released by human, murine, and amphibian tumours. *J Pathol* 143:193–7.

Shih C and Weinberg RA. 1982. Isolation of a transforming sequence from a human bladder carcinoma cell line. *Cell* 29:161–9.

Shih C, Padhy LC, Murray M, and Weinberg RA. 1981. Transforming genes of carcinomas and neuroblastomas introduced into mouse fibroblasts. *Nature* 290:261–4.

Shimizu Y, Kato H, and Schull WJ. 1990. Studies of the mortality of A-bomb survivors; mortality, 1950–1985: Part 2. Cancer mortality based on the recently revised doses DS86. *Radiat Res* 121:120–41.

Shimizu Y, Kato H, Schull WJ, Preston DL, Fujita S, and Pierce DA. 1989. Studies of the mortality of A-bomb survivors; mortality, 1950–1985: Part 1. Comparison of risk coefficients for site-specific cancer mortality based on the DS86 and T65DR shielded kerma and organ doses. *Radiat Res* 118:502–24.

Shimizu Y, Nakatsuru Y, Ichinose M, Takahashi Y, Kume H, Mimura J, Fujii-Kuriyama Y, and Ishikawa T 2000. Benzo[a]pyrene carcinogenicity is lost in mice lacking the aryl hydrocarbon receptor. *Proc Natl Acad Sci USA* 97:779–82.

Shimkin MB. 1977. *Contrary to Nature*. Pub NIH 76–720. Washington, DC: US Dept Health, Education and Welfare.

Sieweke MH, Thompson NL, Sporn MB, and Bissell MJ. 1990. Mediation of wound-related Rous sarcoma virus tumorigenesis by TGF-beta. *Science* 248:1656–60.

Sifri R, Gangadharappa S, and Acheson LS. 2004. Identifying and testing for hereditary susceptibility to common cancers. *CA Cancer J Clin* 54:309–26.

Sikic BI. 1993. Modulation of multidrug resistance – at the threshold. *J Clin Oncol* 11: 1629–35.

Sills RC, Boorman GA, Neal JE, Hong HL, and Devereux TR. 1999. Mutations in ras genes in experimental tumours of rodents. *IARC Sci Publ* 146:55–86.

Silvennoinen O, Ihle JN, Schlessinger J, and Levy DE. 1993. Interferon-induced nuclear signalling by Jak protein tyrosine kinases. *Nature* 366:583–5.

Simon R, Nocito A, Hubscher T, Bucher C, Torhorst J, Schraml P, Bubendorf L, Mihatsch MM, Moch H, Wilber K, Schotzau A, Kononen J, and Sauter G. 2001. Patterns of her-2/neu amplification and overexpression in primary and metastatic breast cancer. *J Natl Cancer Inst* 93: 1141–6.

Simpson L and Parsons R. 2001. PTEN: life as a tumor suppressor. *Exp Cell Res* 264:29–41.

Simpson-Herren L and Noker PE. 1990. Effects of the initial chemotherapy on subsequent therapy. *Prog Clin Biol Res* 354A:21–9.

Sincock AM, Delhanty JD, and Casey GA. 1982. A comparison of the cytogenetic response to asbestos and glass fibers in CHO cell lines. *Mutat Res* 101:257–68.

Sinha AA, Gleason DF, Staley NA, Wilson MJ, Sameni M, and Sloane B. 1995. Cathepsin B in angiogenesis of human prostate: An immunohistochemical and immunoelectron microscope analysis. *Anat Rec* 241:353–36.

Sinha AA, Quast BJ, Korkowski JC, Wilson MJ, Reddy PK, Ewing SL, Sloane BF, and Gleason DF. 1999. The relationship of cathepsin B and Stefin A mRNA localization identifies a potentially aggressive variant of human prostate cancer within a Gleason histologic score. *Anticancer Res* 19:2821–9.

Sinha AA, Quast BJ, Wilson MF, Fernandes ET, Reddy PK, Ewing SL, Sloane BF, and Gleason DF. 2001. Ratio of cathepsin B to stefin A identifies heterogeneity within Gleason histologic scores for human prostate cancer. *Prostate* 48:274–84.

Sinha AA, Quast BJ, Wilson MF, Fernandes ET, Reddy PK, Ewing SL, and Gleason DF. 2002. Prediction of pelvic lymph node metastasis by the ratio of cathepsin B to stefin A in patients with prostate carcinoma. *Cancer* 94:3141–9.

Sioud M. 2002. How does autoimmunity cause tumor regression? A potential mechanism involving cross-reaction through epitope mimicry. *Mol Med* 8:115–19.

Sisskin EE, Gray T, and Barrett JC. 1982. Correlation between sensitivity to tumor promotion and sustained epidermal hyperplasia of mice and rats treated with 12-O-tetradecanoylphorbol-13-acetate. *Carcinogenesis* 3:403–7.

Sivridis E and Alison MR. 2003. Tissue-based stem cells: ABC transporter protein take center stage. *J Pathol* 200:547–50.

Skipper HE and Perry S. 1970. Kinetics of normal and leukemic leukocyte populations and relevance to chemotherapy. *Cancer Res* 30:1883–97.

Skipper HE, Schabel FM Jr, and Lloyd HH. 1978. Experimental therapeutics and kinetics: Selection and overgrowth of specifically and permanently drug-resistant tumor cells. *Semin Hematol* 15:207–19.

Skipper HE, Schabel FM Jr, and Wilcox WS. 1967. Experimental evaluation of potential anticancer agents: XXI Scheduling of arabinosylcytosine to take advantage of its S-phase specificity against leukemia cells. *Cancer Chemother Rep* 51:125–65.

Skipper HE, Schabel FM Jr, Mellett LB, Montgomery JA, Wilkoff LJ, Lloyd HH, and Brockman RW. 1970. Implications of biochemical, cytokinetic, pharmacologic, and toxicologic relationships in the design of optimal therapeutic schedules. *Cancer Chemother Rep* 54: 431–50.

Slack JM. 1983. *From Egg to Embryo: Determinative Events in Early Development*, pp. 136–61. Cambridge, London: Cambridge University Press.

Slaga T, Bowden GT, and Boutwell RK. 1975. Acetic acid, a potent stimulator of mouse epidermal macromolecular synthesis and hyperplasia but weak promoting activity. *J Natl Cancer Inst* 55:983–7.

Slamon D and Pegram M. 2001. Rationale for trastuzumab (herceptin) in adjuvant breast cancer trials. *Semin Oncol* 28:13–19.

Slamon DJ, Clark GM, Wong SG, Levin WJ, Ullrich A, and McGuire WL. 1987. Human breast cancer: Correlation of relapse and survival with amplification of the HER2/*neu* oncogene. *Science* 235:177–82.

Slamon DJ, Leyland-Jones B, Shak S, Fuchs H, Paton V, Bajamonde A, Fleming T, Eiermann W, Wolter J, Pegram M et al. 2001. Use of chemotherapy plus a monoclonal antibody against her2 for metastatic breast cancer that overexpresses her2. *N Engl J Med* 344:783–92.

Slee EA, Harte MT, Kluck RM, Wolf BB, Casiano CA, Newmeyer DD, Wang HG, Reed JC, Nicholson DW, Alnemri ES et al. 1999. Ordering the cytochrome c-initiated caspase cascade: hierarchical activation of caspases-2, -3, -6, -7, -8, and -10 in a caspase-9-dependent manner. *J Cell Biol* 144:281–92.

Slichenmyer WJ, Rowinsky EK, Grochow LB, Kaufmann SH, and Donehower RC. 1994. Camptothecin analogues: Studies from the Johns Hopkins Oncology Center. *Cancer Chemother Pharmacol* 34 Suppl:S53–7.

Sloan DA, Schwartz RW, McGrath PC, and Kenady DE. 1996. Diagnosis and management of adrenal tumors. *Curr Opin Oncol* 8:30–6.

Smith EF and Townsend CO. 1907. A plant-tumor of bacterial origin. *Science* 25:671–3.

Smith JR, Freije D, Carpten JD, Grönberg H, Xu J, Isaacs SD, Brownstein MJ, Bova GS, Guo H, Bujnovszky P et al. 1996. Major susceptibility locus for prostate cancer on chromosome 1 suggested by a genome-wide search. *Science* 274:1371–4.

Smith MW and Jarvis LG. 1980. Use of differential interference contrast microscopy to determine cell renewal times in mouse intestine. *J Micros* 118:153–9.

Smith MW, Jarvis LG, and King IS. 1980. Cell proliferation in follicle-associated epithelium of mouse Peyer's patch. *Am J Anat* 159:157–66.

Smith RB and Haskell CM. 1990. Testis. In: CM Haskell (ed.). *Cancer Treatment*, 3rd ed., pp. 779–97. Philadelphia, PA: WB Saunders.

Smith-Hicks CL, Sizer K C, Powers J F, Tischler A S, and Costantini F. 2000. C-cell hyperplasia, pheochromocytoma and sympathoadrenal malformation in a mouse model of multiple endocrine neoplasia type 2B. *EMBO J* 19:612–22.

Smolev JK, Coffey DS, and Scott WW. 1977. Experimental models for the study of prostatic adenocarcinoma. *J Urol* 118:216–20.

Smolev JK, Heston WD, Scott WW, and Coffey DS. 1977. Characterization of the Dunning R3327H prostatic adenocarcinoma: An appropriate animal model for prostatic cancer. *Cancer Treat Rep* 61:273–87.

Snell RS. 1978. *Clinical and Functional Histology for Medical Students*. Boston, MA: Little, Brown.

Snellwood RA, Kuper SWA, Payne PM, and Burn JI. 1969. Factors affecting the finding of cancer cells in the blood. *Br J Surg* 56:649–52.

Snustad DP and Simmons MJ. 2003. *Principles of Genetics*, 3rd ed. New York: John Wiley & Sons, Inc.

Song MJ, Reilly AA, Parsons DF, and Hussain M. 1986. Patterns of blood-vessel invasion by mammary tumor cells. *Tissue Cell* 18:817–25.

Songyang Z, Shoelson SE, Chaudhuri M, Gish G, Pawson T, Haser WG, King F, Roberts T, Ratnofsky S, Lechleider RJ et al. 1993. SH2 domains recognize specific phosphopeptide sequences. *Cell* 72:767–78.

Sonoda J, Rosenfeld JM, Xu Lu, Evans RM, and Xie W. 2003. A nuclear receptor-mediated xenobiotic response and its implication in drug metabolism and host protection. *Curr Drug Metab* 4:59–72.

Sora M, Einisto P, Husgafvel-Pdursiainen K, Jarventaus H, Kivisto H, Peltonen Y, Tuomi T, and Valkonen S. 1985. Passive and active exposure to cigarette smoke in a smoking experiment. *J Toxicol Environ Health* 16:523–34.

Sordella R, Bell DW, Haber DA, and Settleman J. 2004. Gefitinib-sensitizing EGFR mutations in lung cancer activate anti-apoptotic pathways. *Science* 305:1163–7.

Span M, Moerkerk PT, De Goeij AF, and Arends JW. 1996. A detailed analysis of K-ras point mutations in relation to tumor progression and survival in colorectal cancer patients. *Int J Cancer* 69:241–5.

Speers WC. 1982. Conversion of malignant murine embryonal carcinomas to benign teratomas by chemical induction of differentiation in vivo. *Cancer Res* 42:1843–9.

Spinelli G, Bardazzi N, Citernesi A, Fontanarosa M, and Curiel P. 1991. Endometrial carcinoma in tamoxifen-treated breast cancer patients. *J Chemother* 3:267–70.

Spratt NT. 1946. Formation of the primitive streak in the explanted chick blastoderm marked with carbon particles. *J Exp Zool* 103:259–304.

Stanglmaier M, Reis S, and Hallek M. 2004. Rituximab and alemtuzumab induce a nonclassic, caspase-independent apoptotic pathway in b-lymphoid cell lines and in chronic lymphocytic leukemia cells. *Ann Hematol* 83:634–45.

Stanton BR, Perkins AS, Tessarollo L, Sassoon DA, and Parada LF. 1992. Loss of N-*myc* function results in embryonic lethality and failure of the epithelial component of the embryo to develop. *Genes Dev* 6:2235–47.

Stark MB. 1918. An hereditary tumor in the fruit fly, Drosophila. *J Cancer Res* 3:279–301.

Stayner LT, Dankovic DA, and Lemen RA. 1996. Occupational exposure to chrysotile asbestos and cancer risk: A review of the amphibole hypothesis. *Am J Public Health* 86:179–86.

Steeg PS, Bevilacqua G, Kopper L, Thorgeirsson UP, Talmadge JE, Liotta LA, and Sobel ME. 1988. Evidence for a novel gene associated with low tumor metastatic potential. *J Natl Cancer Inst* 80:200–4.

Steelman LS, Pohnert SC, Shelton JG, Franklin RA, Bertrand FE, and McCubrey JA. 2004. JAK/STAT, Raf/MEK/ERK, PI3K/Akt and BCR-ABL in cell cycle progression and leukemogenesis. *Leukemia* 18:189–218.

Stehelin D, Varmus HE, Bishop JM, and Vogt PK, 1976. DNA related to the transforming gene(s) of avian sarcoma virus is present in normal avian DNA, *Nature* 260:170–3.

Steinbach G, Lynch PM, Phillips RK, Wallace MH, Hawk E, Gordon GB, Wakabayashi N, Saunders B, Shen Y, Fujimura T et al. 2000. The effect of celecoxib, a cyclooxygenase-2 inhibitor, in familial adenomatous polyposis. *N Engl J Med* 342:1946–52.

Steinmetz KA and Potter JD. 1991. Vegetables, fruit, and cancer: II Mechanisms. *Cancer Causes Control* 2:427–42.

Stellman JM and Stellman SD. 1996. Cancer and the workplace. *CA Cancer J Clin* 46:70–92.

Stevens LC. 1967. Origin of testicular teratomas from primordial germ cells. *J Natl Cancer Inst* 38:549–52.

1970. The development of transplantable teratocarcinomas from intertesticular grafts of pre- and post-implantation mouse embryos. *Dev Biol* 21:364–82.

1984. Germ cell origin of testicular and ovarian teratomas. *Transplant Proc* 16:502–4.

Stewart BW and Kleihues P. (eds.). 2003. *World Cancer Report*, Lyon, FR: International Agency for Research on Cancer.

Stoker AW, Hatier C, and Bissell MJ. 1990. The embryonic environment strongly attentuates v-*src* oncogenesis in mesenchymal and epithelial tissues but not in endothelia. *J Cell Biol* 111:217–28.

Strasser A, Harris AW, Bath ML, and Cory S. 1990. Novel primitive lymphoid tumours induced in transgenic mice by cooperation between *myc* and *bcl*-2. *Nature* 348:331–3.

Strasser A, Harris AW and Cory S. 1993. E mu-bcl-2 transgene facilitates spontaneous transformation of early pre-B and immunoglobulin-secreting cells but not T cells. *Oncogene* 8:1–9.

Strauli P and Haemmerli G. 1984. Cancer cell locomotion: Its occurrence during invasion. In: MM Mareel and KC Calman (eds.). *Invasion: Experimental and Clinical Implications*, pp. 252–74. Oxford, UK: Oxford University Press.

Straume T, Rugel G, Marchetti AA, Ruhm W, Korschinek G, McAninch JE, Carroll K, Egbert S, Faestermann T, Knie K et al. 2003. Measuring fast neurons in Hiroshima at distances relevant to atomic-bomb survivors. *Nature* 424:539–42.

Strickland S, Smith KK, and Marotti KR. 1980. Hormonal induction of differentiation in teratocarcinoma stem cells. Generation of endoderm with retinoic acid and dibuteryl cAMP. *Cell* 21:347–55.

Studzinski GP and Moore DC. 1995. Sunlight – Can it prevent as well as cause cancer? *Cancer Res* 55:4014–22.

Sugerbaker EV. 1981. Patterns of metastasis in human malignancies. *Cancer Biol Rev* 2:235–78.

Sugio K, Kishimoto Y, Virmani AK, Hung JY, and Gazdar AF. 1994. K-*ras* mutations are a relatively late event in the pathogenesis of lung carcinomas. *Cancer Res* 54:5811–15.

Sutherland BM, Delihas NC, Oliver RP, and Sutherland JC. 1981. Action spectra for ultraviolet light-induced transformation of human cells to anchorage-independent growth. *Cancer Res* 41:2211–14.

Suzuki K, Hattori Y, Uraji M, Ohta N, Iwata K, Murata K, Kato A, and Yoshida K. 2000. Complete nucleotide sequence of a plant tumor-inducing T_i plasmid. *Gene* 242:331–6.

Sweeney DC and Johnston GS. 1995. Radioiodine therapy for thyroid cancer. *Endocrinol Metab Clin North Am* 24:803–39.

Swift CH, 1914. Origin and early history of the primordial germ cells in the chick, *Am J Anat* 15:483–516.

Symonds G and Sachs L. 1982a. Autoinduction of differentiation in myeloid leukemic cells: Restoration of normal coupling between growth and differentiation in leukemic cells that constitutively produce their own growth-inducing protein. *EMBO J* 1:1343–6.

  1982b. Cell competence for induction of differentiation by insulin and other compounds in myeloid leukemic clones continuously cultured in serum-free medium. *Blood* 60:208–12.

  1982c. Modulation of cell competence for induction of differentiation in myeloid leukemic cells. *J Cell Physiol* 111:9–14.

Symonds H, Krall L, Remington L, Saenz-Robles M, Lowe S, Jacks T, and Van Dyke T. 1994. p53-dependent apoptosis suppresses tumor growth and progression in vivo. *Cell* 78:614–17.

Szekely L, Salivanova G, Magnusson KP, Klein G, and Wiman KG. 1993. EBNA-5, an Epstein-Barr virus-encoded nuclear antigen, binds to the Rb and p53 proteins. *Proc Natl Acad Sci USA* 90:5455–9.

Takanami I and Takeuchi K. 2003. Autocrine motility factor-receptor gene expression in lung cancer. *Jpn J Thorac Cardiovasc Surg* 51:368–73.

Takayama S, Ishikawa T, Masahito P, and Matsumoto J. 1981. Overview of biological characterization of tumors in fish. In: CJ Dawe et al. (eds.). pp. 3–17. *Phyletic Approaches to Cancer*. Tokyo: Japan Sci Soc Press.

Takimoto R and El-Deiry WS. 2000. Wild-type p53 transactivates the KILLER/DR5 gene through an intronic sequence-specific DNA-binding site. *Oncogene* 19:1735–43.

Tallman MS. 1994. All-trans-retinoic acid in acute promyelocytic leukemia and its potential in other hematologic malignancies. *Semin Hematol* 31 (Suppl 5):38–48.

Tamimi RM, Lagiou P, Adami HO, and Trichopoulos D. 2002. Prospects for chemoprevention of cancer. *J Intern Med* 251:286–300.

Tamura M, Gu J, Matsumoto K, Aota S, Parsons R, and Yamada K M. 1998. Inhibition of cell migration, spreading, and focal adhesions by tumor suppressor PTEN. *Science* 280:1614–17.

Tamura M, Gu J, Tran H, and Yamada KM. 1999. PTEN gene and integrin signaling in cancer. *J Natl Cancer Inst* 91:1820–8.

Tamura K, Fukuoka M. 2005. Gefitinib in non-small cell lung cancer. *Expert Opin Pharmacother* 6:985–93.

Tanaka K, Oshimura M, Kikuchi R, Seki M, Hayashi T, and Miyaki M. 1991. Suppression of tumorigenicity in human colon carcinoma cells by introduction of normal chromosome 5 or 18. *Nature* 349:340–2.

Tarin D. 1992. Tumour metastasis. In: J O'D McGee, PG Isaacson, and NA Wright (eds.). *Oxford Textbook of Pathology*, pp. 607–33. Oxford, Oxford University Press.

——— 1997. Prognostic markers and mechanisms of metastasis. In: PP Anthony, RNM MacSween, and D Lowe (eds.). *Recent Advances in Histopathology 17:14–45*. Edinburgh: Churchill Livingstone.

Tarin D, Price JE, Kettlewell MG, Souter RG, Vass AC, and Crossley B. 1984. Clinicopathological observations on metastasis in patients treated with peritoneovenus shunts. *Br Med J (Clin Res Ed)* 288:749–51.

Tarkowski AK and Wroblewska J. 1967. Development of blastomeres of mouse eggs isolated at the 4–8 cell stage. *J Embryol Exp Morph* 18:155–80.

Taub R, Kirsch I, Morton C, Lenoir G, Swan D, Tronick S, Aaronson S, and Leder P. 1982. Translocation of the c-myc gene into the immunoglobulin heavy chain locus in human Burkitt lymphoma and murine plasmacytoma cells. *Proc Natl Acad Sci USA* 79:7837–41.

Tedder TF, Boyd AW, Freedman AS, Nadler LM, and Schlossman SF. 1985. The b cell surface molecule b1 is functionally linked with b cell activation and differentiation. *J Immunol* 135:973–9.

Teicher BA, Ara G, Herbst R, Palombella VJ, and Adams J. 1999. The proteasome inhibitor ps-341 in cancer therapy. *Clin Cancer Res* 5:2638–45.

Terranova VP and Maslow DE. 1991. Interactions of tumor cells with basement membrane. In: FW Orr, MR Buchanan, and L Weiss (eds.). *Microcirculation in Cancer Metastasis*, pp. 23–44. Boca Raton, FL: CRC Press.

Thomas RK, Re D, Zander T, Wolf J, and Diehl V. 2002. Epidemiology and etiology of Hodgkin's lymphoma. *Ann Oncol* 13 (Suppl 4):147–52.

Thomas SM and Grandis JR. 2004. Pharmacokinetic and pharmacodynamic properties of EGFR inhibitors under clinical investigation. *Cancer Treat Rev* 30:255–68.

Thompson TC, Southgate J, Kitchener G, and Land H. 1990. Multistage carcinogenesis induced by *ras* and *myc* oncogenes in a reconstituted organ. *Cell* 56:917–30.

Thorgeirsson SS, Gant TW, and Silverman JA. 1994. Transcriptional regulation of multidrug resistance gene expression *Cancer Treat Res* 73:57–68.

Thorgeirsson UP, Turpeenniemi-Hujanen T, Williams JE, Westen EH, Heilman CA, Talmadge JE, and Liotta LA. 1985. NIH/3T3 cells transfected with human tumor DNA containing activated ras oncogenes express the metastatic phenotype in nude mice. *Mol Cell Bio* 5: 259–62.

Tirmarche M, Raphalen A, Allin F, Chameaud J, and Bredon P. 1993. Mortality of a cohort of French uranium miners exposed to relatively low radon concentrations. *Br J Cancer* 67: 1090–7.

Tisdale MJ. 2000. Biomedicine. Protein loss in cancer cachexia. *Science* 289:2292–4.

Tjio JH and Levan A. 1956. The chromosome number of man. *Hereditas* 42:1–6.

Toker A and Newton AC. 2000. Akt/protein kinase B is regulated by autophosphorylation at the hypothetical PDK-2 site. *J Biol Chem* 275:8271–4.

Tomatis L. 1988. The contribution of the IARC monographs program to the identification of cancer risk factors. *Ann NY Acad Sci* 534:31–8.

Tone H and Heidelberger C. 1973. Fluorinated pyrimidines. Xliv. Interaction of 5-trifluoromethyl-2'-deoxyuridine 5'-triphosphate with deoxyribonucleic acid polymerases. *Mol Pharmacol* 9:783–91.

Tonini T, Rossi F, and Claudio PP. 2003. Molecular basis of angiogenesis and cancer. *Oncogene* 22:6549–56.

Tortora G and Ciardiello F. 2003. Antisense strategies targeting protein kinase C: preclinical and clinical development. *Semin Oncol* 30(Suppl 10):26–31.

Tracy S, Mukohara T, Hansen M, Meyerson M, Johnson BE, and Janne PA. 2004. Gefitinib induces apoptosis in the EGFRL858R non-small-cell lung cancer cell line H3255. *Cancer Res* 64:7241–4.

Trask BJ. 2002. Human cytogenetics: 46 chromosomes, 46 years and counting. *Nat Rev Genet* 3:769–78.

Traver D, Akashi K, Weissman IL, and Lagasse E. 1998. Mice defective in two apoptosis pathways in the myeloid lineage develop acute myeloblastic leukemia. *Immunity* 9:47–57.

Trayner ID and Farzaneh F. 1993. Retinoid receptors and acute promyelocytic leukaemia. *Eur J Cancer* 29A:2046–54.

Treisman R. 1990. The SRE: A growth factor responsive transcription regulator. *Semin Cancer Biol* 1:47–58.

——— 1994. Ternary complex factors: Growth factor regulated transcriptional activators. *Curr Opin Genet Dev* 4:96–101.

Trichopoulos D, Petridou E, Lipworth L, and Adami H-O. 1997. Epidemiology of Cancer. In: VT DeVita Jr., S Hellman, SA Rosenberg (eds.). *Cancer: Principles and Practice of Cancer*, 5th ed., pp. 231–57. Philadelphia: Lippincott-Raven.

Trigg ME. 1993. Acute lymphoblastic leukemia in children. In: JF Holland, E Frei III, RC Bast, Jr, DW Kufe, DL Morton, RR Weichselbaum (eds.). *Cancer Medicine*, vol 2, pp. 2153–66. Philadelphia, PA: Lea and Febiger.

Trosko JE. 2001. Commentary: is the concept of "tumor promotion" a useful paradigm? *Mol Carcinog* 30:131–7.

Tsujimoto Y, Cossman J, Jaffe E, and Croce CM. 1985. Involvement of the bcl-2 gene in human follicular lymphoma. *Science* 228:1440–3.

Tsujimoto Y, Finger LR, Yunis J, Nowell PC, and Croce CM. 1984. Cloning of the chromosome break-point of neoplastic B cells with the t(14;18) chromosome translocation. *Science* 226:1097–9.

Turesky RJ. 2002. Heterocyclic aromatic amine metabolism, DNA adduct formation, mutagenesis, and carcinogenesis. *Drug Metab Rev* 34:625–50.

Turusov VS and Mohr U, eds. 1994. *Pathology of Tumours in Laboratory Animals*, II: *Tumours of the Mouse.* Lyon, FR: Int Agency Res Cancer.

Tuyns AJ. 1990. Alcohol and cancer. *Proc Nutr Soc* 49:145–51.

Tylecote FE. 1927. Cancer of the lung. *Lancet* 2:256–7.

Ueda K, Cornwell MM, Gottesman MM, Pastan I, Robinson IB, Ling V, and Riordan JR. 1986. The *mdr1* gene, responsible for multidrug-resistance, codes for P-glycoprotein. *Biochem Biophys Res Commun* 141:956–62.

Uesugi M and Sugiura Y. 1992. DNA cleaving modes in minor groove of DNA helix by esperamicin and calicheamicin antitumor antibiotics. *Nucleic Acids Symp* Ser 1–2.

Urbach F. 1993. Environmental risk factors for skin cancer. *Recent Res Cancer Res* 128:243–62.

Vadhan-Raj S. 1996. Recombinant human erythropoietin in combination with other hematopoietic cytokines in attenuating chemotherapy-induced multilineage myelo-suppression: Brief communication. *Semin Hematol* 33:16–18.

Vadhan-Raj S, Broxmeyer HE, Hittelman WN, Papadopoulos NE, Chawla SP, Fenoglio C, Cooper S, Buescher ES, Frenck Jr. R, Holian A et al. 1992. Abrogating chemotherapy-induced myelo-suppression by recombinant granulocyte-macrophage colony-stimulating factor in patients with sarcoma: Protection at the progenitor level. *Journal of Clinical Oncology* 10:1266–77.

Vainio H, Miller AB, and Bianchini F. 2000. An international evaluation of the cancer-preventive potential of sunscreens. *Int J Cancer* 88:838–42.

Valentine MA, Meier KE, Rossie S, and Clark EA. 1989. Phosphorylation of the cd20 phosphoprotein in resting b lymphocytes. Regulation by protein kinase c. *J Biol Chem* 264:11282–7.

Van Beneden RJ, Henderson KW, Blair DG, Papas TS, and Gardner HS. 1990. Oncogenes in hematopoietic and hepatic fish neoplasms. *Cancer Res* 50 Suppl:5671s–5674s.

Van Beneden RJ, Rhodes LD, and Gardner GR. 1998. Studies of the molecular basis of gonadal tumors in the marine bivalve, Mya arenaria. *Mar Environ Res* 46:209–13.

van der Bliek AM and Borst P. 1989. Multidrug resistance. *Adv Cancer Res* 52:165–203.

van der Schaft DW, Seftor RE, Seftor EA, Hess AR, Gruman LM, Kirschmann DA, Yokoyama Y, Griffioen AW, and Hendrix MJ. 2004. Effects of angiogenesis inhibitors on vascular network formation by human endothelial and melanoma cells. *J Natl Cancer Inst* 96:1473–7.

van Steenbrugge GJ, Groen M, Romijn JC, and Schroder FH. 1984. Biological effects of hormonal treatment regimens on a transplantable human prostatic tumor line (PC-82). *J Urol* 131: 812–17.

Van't Veer LJ and Weigelt B. 2003. Road map to metastasis. *Nature Medicine* 9:999–1000.

Vaughan WP, Karp JE, and Burke PJ. 1984. Two-cycle-timed sequential chemotherapy for adult acute nonlymphocytic leukemia. *Blood* 64:975–80.

Vedantham S, Gamliel H, and Golomb HM. 1992. Mechanism of interferon action in hairy cell leukemia: A model of effective cancer biotherapy. *Cancer Res* 52:1056–66.

Vermeulen K, Van Bockstaele DR, and Berneman ZN. 2003. The cell cycle: a review of regulation, deregulation and therapeutic targets in cancer. *Cell Prolif* 36:131–49.

Vigneri P and Wang JY. 2001. Induction of apoptosis in chronic myelogenous leukemia cells through nuclear entrapment of BCR-ABL tyrosine kinase. *Nat Med* 7:228–34.

Villunger A, Michalak EM, Coultas L, Mullauer F, Bock G, Ausserlechner MJ, Adams JM, and Strasser A. 2003. p53- and drug-induced apoptotic responses mediated by BH3-only proteins puma and noxa. *Science* 302:1036–8.

Virchow R. 1863. Über bewegliche thierische Zellen. *Arch Pathol Anat* 28:237–40.

Vitaliano PP. 1978. The use of logistic regression for modeling risk factors: With application to non-melanoma skin cancer. *Am J Epidemiol* 108:402–14.

Vlach J, Hennecke S, and Amati B. 1997. Phosphorylation-dependent degradation of the cyclin-dependent kinase inhibitor p27. *EMBO J* 16:5334–44.

Vogel CL. 1996. Hormonal approaches to breast cancer treatment and prevention: An overview. *Semin Oncol* 23 (4 Suppl 9):2–9.

Vogel CL, Cobleigh MA, Tripathy D, Gutheil JC, Harris LN, Fetirenbacher L, Slamon DJ, Murphy M, Norotny WF, Burchmore M, et al. 2002. Efficacy and safety of trastuzumab as a single agent in first-line treatment of HER2-overexpressing metastatic breast cancer. *J Clin Oncol* 20:719–26.

Vogelstein B and Kinzler KW. 1993. The multistep nature of cancer. *Trends Genet* 9:138–42.

Vogelstein B, Fearon ER, Hamilton SR, and Kern SE. 1988. Genetic alteration during colorectal-tumor development. *N Engl J Med* 319:525–32.

Vousden KH. 2002. Activation of the p53 tumor suppressor protein. *Biochim Biophys Acta* 1602:47–59.

Wagner JC, Newhouse ML, Corrin B, Rossiter CE R, and Griffiths DM. 1988. Correlation between fibre content of the lung and disease in east London asbestos factory workers. *Br J Ind Med* 45:305–8.

Wagner JC, Sleggs CA, and Marchand P. 1960. Diffuse pleural mesothelioma and asbestos exposure in the North Western Cape Province. *Br J Ind Med* 17:260–71.

Waldron HA, Waterhouse JA, and Tessema N. 1984. Scrotal cancer in the West Midlands 1936–1976. *Br J Ind Med* 41:473–4.

Wales JH and Sinnhuber RO. 1966. An early hepatoma epizootic in rainbow trout, Salmo gairdnerii. *Calif Fish Game* 52:85–91.

Wallet V, Mutzel R, Troll H, Barzu O, Wurster B, Veron M, and Lacombe M. 1990. Dictyostelium nucleoside diphosphate kinase highly homologous to *nm*23 and awd proteins involved in mammalian tumor metastasis and Drosophila development. *J Natl Cancer Inst* 82:1199–1202.

Walsh PC, Lepor H, and Eggleston JC. 1983. Radical prostatectomy with preservation of sexual function: Anatomical and pathological considerations. *Prostate* 4:473–85.

Walsh PC, Quinlan DM, Morton RA, and Steiner RA. 1990. Radical retropubic prostatectomy. Improved anastomosis and urinary continence. *Urol Clin North Am* 17:679–84.

Walter JB. 1982. *An Introduction to the Principles of Disease*, p. 245 Philadelphia, PA: WB Saunders.

Wang JYJ. 1994. Nuclear protein tyrosine kinases. *Trends Biochem Sci* 19:373–6.

Wang L, Patel U, Lagnajita G, and Banerjee S. 1992. DNA polymerase b mutations in human colorectal cancer. *Cancer Res* 52:4824–7.

Wang SS, Esplin ED, Li JL, Huang L, Gazdar A, Minna J, and Evans GA. 1998. Alterations of the PPP2R1B gene in human lung and colon cancer. *Science* 282:284–7.

Wang YL, Bagg A, Pear W, Nowell PC, and Hess JL. 2001. Chronic myelogenous leukemia: laboratory diagnosis and monitoring. *Genes Chromosomes Cancer* 32:97–111.

Wang Z, Atencio J, Robinson ES, and McCarrey JR. 2001. Ultraviolet B-induced melanoma in Monodelphis domestica occurs in the absence of alterations in the structure or expression of the p53 gene. *Melanoma Res* 11:239–45.

Wani MC, Taylor HL, and Wall ME. 1971. Plant antitumor agents: VI The isolation and structure of Taxol, a novel antileukemic and antitumor agent from *Taxus brevifolia. J Am Chem Soc* 93:2325–7.

Ward M, Richardson C, Pioli P, Smith L, Podda S, Goff S, Hesdorffer C, and Bank A. 1994. Transfer and expression of the human multiple drug resistance gene in human CD34+ cells. *Blood* 84:1408–14.

Warne PH, Viciana PR, and Downward J. 1993. Direct interaction of Ras and the amino-terminal region of Raf-1 *in vitro. Nature* 364:353–5.

Warren JR, Scarpelli DG, Reddy JK, and Kanwar YS. 1987. *Essentials of General Pathology.* New York: Macmillan.

Wasserman E, Myara A, Lokiec F, Goldwasser F, Trivin F, Mahjoubi M, Misset JL, and Cvitkovic E. 1997. Severe cpt-11 toxicity in patients with gilbert's syndrome: Two case reports. *Ann Oncol* 8:1049–51.

Wattenberg LW. 1985. Chemoprevention of cancer. *Cancer Res* 45:1–8.

———. 1996. Chemoprevention of cancer. *Prev Med* 25:44–5.

Weh HJ, Zschaber R, and Hossfeld DK. 1984. Low-dose cytosine-arabinoside in the treatment of acute myeloid leukemia (AML) and myelodysplastic syndrome (MDS). *Blut* 48:239–42.

Weichselbaum RR, Hallahan DE, and Chen GT. 1993. Biological and physical basis to radiation oncology. In: JF Holland, E Frei III, RC Bast, Jr, DW Kufe, DL Morton, and RR Weichselbaum (eds.). *Cancer Medicine*, vol 2, pp. 539–66. Philadelphia, PA: Lea and Febiger.

Weinberg RA. 1995. The retinoblastoma protein and cell cycle control. *Cell* 81:323–30.

Weisburger JH, Wynder EL. 1987. Etiology of colorectal cancer with emphasis on mechanisms of action and prevention. *Important Adv Oncol* 1987:197–221.

Weisenburger DD, Sanger WG, Armitage JO, and Purtilo DT. 1987. Intermediate lymphocytic lymphoma: immunophenotypic and cytogenetic findings. *Blood* 69:1617–21.

Weiss WA, Aldape K, Mohapatra G, Feuerstein BG, and Bishop JM. 1997. Targeted expression of MYCN causes neuroblastoma in transgenic mice. *EMBO J* 16:2985–95.

Werness BA, Münger K, and Howley PM. 1991. The role of the human papillomavirus oncoproteins in transformation and carcinogenic progression. In: VT DeVita, Jr, S Hellman, and SA Rosenberg (eds.). *Important Advances in Oncology*, pp. 2–18. Philadelphia, PA: JB Lippincott.

Whang-Peng J, Lee EC, and Knutsen TA. 1974. Genesis of the Ph chromosome. *J Natl Cancer Inst* 52:1035–6.

Whitehead VM, Rosenblatt DS, Vuchich MJ, and Beaulieu D. 1987. Methotrexate polyglutamate synthesis in lymphoblasts from children with acute lymphoblastic leukemia. *Dev Pharmacol Ther* 10:443–8.

Whyte P, Williamson NM, and Harlow E. 1989. Cellular targets for transformation by the adenovirus E1A proteins. *Cell* 56:67–75.

Wilbourn J, Haroun L, Heseltine E, Kaldor J, Partensky C, and Vainio H. 1986. Response of experimental animals to human carcinogens: An analysis based upon the IARC monographs programme. *Carcinogenesis* 7:1853–63.

Wilcox WS. 1966. The last surviving cancer cell: The chances of killing it. *Cancer Chemother Rep* 50:541–2.

——— 1970. The last surviving cancer cell: the chances of killing it. *CA Cancer J Clin* 20:164–7.

Wilcox WS, Griswold DP, Laster WR Jr, Schabel FM Jr, and Skipper HE. 1965. Experimental evaluation of potential anticancer agents: XVII. Kinetics of growth and regression after treatment of certain solid tumors. *Cancer Chemother Rep* 47:27–39.

Willett CG, Boucher Y, di Tomaso E, Duda DG, Munn LL, Tong RT, Chung DC, Sahani DV, Kalva SP, Kozin SV et al. 2004. Direct evidence that the VEGF-specific antibody bevacizumab has antivascular effects in human rectal cancer. *Nat Med* 10:145–7.

Williams BO, Remington L, Albert DM, Mukai S, Bronson RT, and Jacks T. 1994. Cooperative tumorigenic effects of germline mutations in Rb and p53. *Nat Genet* 7:480–4.

Williams D. 2002. Cancer after nuclear fallout: lessons from the Chernobyl accident. *Nature Rev Cancer* 2:543–9.

Williams JW III, Carlson DL, Gadson RG, Rollins-Smith L, Williams CS, and McKinnell RG. 1993. Cytogenetic analysis of triploid renal carcinoma in Rana pipiens. *Cytogenet Cell Genet* 64:18–22.

Willis RA. 1967. *Pathology of Tumors*, 4th ed. London: Butterworth.

——— 1973. *The Spread of Tumours in the Human Body*, 3rd ed. London: Butterworth.

Wilson MJ and Sinha AA. 2004. Matrix degradation in prostate cancer. In: RJ Ablin and MD Mason (eds.). *Metastasis of Prostate Cancer*, Dordrecht: Kluwer Academic Publishers (In press).

Winton DJ and Ponder BA. 1990. Stem-cell organization in mouse small intestine. *Proc R Soc Lond*, Part B 241:13–18.

Witschi E. 1948. Migrations of germ cells of human embryos from the yolk sac to the primitive gonadal folds. *Contr Embryol Carnegie Inst* 32:67–80.

Wittes RE. 1986. Adjuvant chemotherapy – Clinical trials and laboratory models. *Cancer Treat Rep* 70:87–103.

Wogan GN. 1986. Diet and nutrition as risk factors for cancer. *Int Symp Princess Takamatsu Cancer Res Fund* 16:3–10.

Wolfel T, Hauer M, Schneider J, Serrano M, Wolfel C, Klehmann-Hieb E, De Plaen E, Hankeln T, Meyer zum Buschenfelde KH, and Beach D. 1995. A p16INK4a-insensitive CDK4 mutant targeted by cytolytic T lymphocytes in a human melanoma. *Science* 269:1281–4.

Wolk A, Gridley G, Svensson M, Nyren O, McLaughlin JK, Fraumeni JF, and Adam HO. 2001. A prospective study of obesity and cancer risk (Sweden). *Cancer Causes Control* 12:13–21.

Won KA and Reed SI. 1996. Activation of cyclin E/CDK2 is coupled to site-specific autophosphorylation and ubiquitin-dependent degradation of cyclin E. *EMBO J* 15:4182–93.

Wong C, Rougier-Chapman EM, Frederick JP, Datto MB, Liberati NT, Li JM, and Wang XF. 1999. Smad3-Smad4 and AP-1 complexes synergize in transcriptional activation of the c-Jun promoter by transforming growth factor beta. *Mol Cell Biol* 19:1821–30.

Wood DW, Setubal JC, Kaul R, Monks DE, Kitajima JP, Okura VK, Zhou Y, Chen L, Wood GE, Almeida NF Jr et al. 2001. The genome of the natural genetic engineer Agrobacterium tumefaciens C58. *Science* 294:2317–28.

Woodhouse E, Hersperger E and Shearn A. 1998. Growth, metastasis, and invasiveness of Drosophila tumors caused by mutations in specific tumor suppressor genes. *Dev Genes Evol* 207: 542–50.

Woodhouse E, Hersperger E, Stetler-Stevenson WG, Liotta LA, and Shearn A. 1994. Increased type-IV collagenase in IgI-induced invasive tumors of Drosophila. *Cell Growth Differ* 5:151–9.

Wooster R and Weber BL. 2003. Breast and ovarian cancer. *N Engl J Med* 348:2339–47.

World Cancer Research Fund Panel. 1997. *Food, nutrition and the prevention of cancer: a global perspective.* Washington, DC: American Institute for Cancer Research.

Wright N, Watson A, Morley A, Appleton D, Marks, J, and Douglas A. 1973. The cell cycle time in the flat (avillous) mucosa of the human small intestine. *Gut* 14:603–6.

Wu CC, Fang TH, Yang MD, Wu TC, and Liu TJ. 1992. Gastrectomy for advanced gastric carcinoma with invasion to the serosa. *Int Surg* 77:144–8.

Wu J, Dent P, Jelinek T, Wolfman A, Weber MJ, and Sturgill TW. 1993. Inhibition of the EGF-activated MAP kinase signaling pathway by adenosine $3',5'$-monophosphate. *Science* 262:1065–9.

Wu X, Wang X, Qien X, Liu H, Ying J, Yang Z, and Yao H. 1993. Four years' experience with the treatment of all-trans retinoic acid in acute promyelocytic leukemia. *Am J Hematol* 43:183–9.

Wylie CV, Nakane PK, and Pierce GB. 1973. Degrees of differentiation in nonproliferating cells of mammary carcinoma. *Differentiation* 1:11–20.

Wyllie AH and Morris RG. 1982. Hormone-induced cell death. Purification and properties of thymocytes undergoing apoptosis after glucocorticoid treatment. *Am J Pathol* 109:78–87.

Wyllie AH, Kerr JF, and Currie AR. 1980. Cell death: The significance of apoptosis. *Int Rev Cytol* 68:251–306.

Wynder EL and Graham EA. 1950. Tobacco smoking as a possible etiological factor in bronchiogenic carcinoma. A study of six hundred and eighty-four proved cases. *JAMA* 143:329–36.

Xiao D, Srivastava SK, Lew KL, Zeng Y, Hershberger P, Johnson CS, Trump DL, and Sing SV. 2003. Allyl isothiocyanate, a constituent of cruciferous vegetables, inhibits proliferation of human prostate cancer cells by causing G(2)/M arrest and inducing apoptosis. *Carcinogenesis* 24: 891–7.

Xie J, Murone M, Luoh SM, Ryan A, Gu Q, Zhang C, Bonifas JM, Lam CW, Hynes M, Goddard A et al. 1998. Activating Smoothened mutations in sporadic basal-cell carcinoma. *Nature* 391:90–2.

Yamada KM and Araki MJ. 2001. Tumor suppressor PTEN: modulator of cell signaling, growth, migration and apoptosis. *Cell Sci* 114:2375–82.

Yamada T and McDevitt D. 1974. Direct evidence for transformation of differentiated iris epithelial cells into lens cells. *Dev Biol* 38:104–18.

Yamagawa K and Ichikawa K. 1918. Experimental study of the pathogenesis of carcinoma. *J Cancer Res* 3:1–29.

Yamaguchi A, Tamatani M, Matsuzaki H, Namikawa K, Kiyama H, Vitek MP, Mitsuda N, and Tohyama M. 2001. Akt activation protects hippocampal neurons from apoptosis by inhibiting transcriptional activity of p53. *J Biol Chem* 276:5256–64.

Yandell DW, Campbell TA, Dayton SH, Petersen R, Walton D, Little JB, McConkie-Rosell A, Buckley EG, and Dryja TP. 1989. Oncogenic point mutations in the human retinoblastoma gene: Their application to genetic counseling. *N Engl J Med* 321:1689–95.

Yang E, Lerner L, Besser D, and Darnell JE Jr. 2003. Independent and cooperative activation of chromosomal c-fos promoter by STAT3. *J Biol Chem* 278:15794–9.

Yang Q, Rasmussen SA, and Friedman SA. 2002. Mortality associated with Down's syndrome in the USA from 1983 to 1997: a population-based study. *Lancet* 359:1019–25.

Yano S, Shinohara H, Herbst RS, Kuniyasu H, Bucana CD, Ellis LM, Davis DW, McConkey DJ, and Fidler IJ. 2000. Expression of vascular endothelial growth factor is necessary but not sufficient for production and growth of brain metastasis. *Cancer Res* 60:4959–67.

Yarden Y and Schlessinger J. 1987. Epidermal growth factor induces rapid, reversible aggregation of the purified epidermal growth factor receptor. *Biochemistry* 26:1443–51.

Ye H, Liu H, Raderer M, Chott A, Ruskone-Fourmestraux A, Wotherspoon A, Dyer MJ, Chuang SS, Dogan A, Isaacson PG, and Du MQ. 2003. High incidence of t(11;18)(q21;q21) in Helicobacter pylori-negative gastric MALT lymphoma. *Blood* 101:2547–50.

Yordy JS and Muise-Helmericks RC. 2000. Signal transduction and the Ets family of transcription factors. *Oncogene* 19:6503–13.

Yoshino K, Rubin JS, Higinbotham KG, Uren A, Anest V, Plisov SY, and Perantoni AO. 2001. Secreted Frizzled-related proteins can regulate metanephric development. *Mech Dev.* 102:45–55.

Young RC. 1994. Management of early ovarian epithelial cancer. *NIH Consensus Development Conference on Ovarian Cancer: Screening, Treatment, and Followup* April 5–7, Bethesda, MD: NIH.

Yunis JJ. 1976. High resolution of human chromosomes. *Science* 191:1268–70.

Yuspa SH. 1994. The pathogenesis of squamous cell cancer: Lessons learned from studies of skin carcinogenesis. *Cancer Res* 54:1178–89.

Zacharski LR. 1984. The coagulation hypothesis of cancer dissemination. In: K Hellmann and SA Eccles (eds.). *Treatment of Metastasis: Problems and Prospects*, pp. 77–80. London: Taylor and Francis.

Zarbl H, Sukumar S, Arthur AV, Martin-Zanca D, and Barbacid M. 1985. Direct mutagenesis of Ha-*ras*-1 oncogenes by N-nitroso-N-methylurea during initiation of mammary carcinogenesis in rats. *Nature* 315:382–5.

Zech L, Haglund U, Nilsson K, and Klein G. 1976. Characteristic chromosomal abnormalities in biopsies and lymphoid-cell lines from patients with Burkitt and non-Burkitt lymphomas. *Int' J Cancer* 17:47–56.

Zeegers MP, Jellema A, and Ostrer H. 2003. Empiric risk of prostate carcinoma for relatives of patients with prostate carcinoma. *Cancer* 97:1894–1903.

Zhang X and Ren R. 1998. Bcr-Abl efficiently induces a myeloproliferative disease and production of excess interleukin-3 and granulocyte-macrophage colony-stimulating factor in mice: a novel model for chronic myelogenous leukemia. *Blood* 92:3829–40.

Zhang X, Settleman J, Kyriakis JM, Takeuchi-Suzuki E, Elledge SJ, Marshall MS, Bruder JT, Rapp UR, and Avruch J. 1993. Normal and oncogenic p21*ras* proteins bind to the amino-terminal reulatory domain of c-Raf-1. *Nature* 364:308–13.

Zhang Y and Xiong Y. 2001. A p53 amino-terminal nuclear export signal inhibited by DNA damage-induced phosphorylation. *Science* 292:1910–15.

Zong WX, Lindsten T, Ross AJ, MacGregor GR, and Thompson CB. 2001. BH3-only proteins that bind pro-survival Bcl-2 family members fail to induce apoptosis in the absence of Bax and Bak. *Genes Dev* 15:1481–6.

Zörnig M, Hueber A, Baum W and Evan G. 2001. Apoptosis regulators and their role in tumorigenesis. Biochim Biophys Acta. 1551:F1–37.

Zwart P and Harshbarger JC. 1991. A contribution to tumors in reptiles. Description of new cases. In: K Gabrisch, B Schildger, and P Zwart (eds.). *4 Internationales Colloquium für Pathologie und Therapie der Reptilien und Amphibien*, pp. 219–24. Bad Nauheiml: Deutsche Veterinärmedizinische Gesellschaft eV.

Zwickey RE and Davis KJ. 1959. Carcinogenicity screening. In *Appraisal of the Safety of Chemicals in Foods, Drugs and Cosmetics*. Austin, TX: Association of Food and Officials of the United States.

Index

Note to readers: The term "cancer" is not included in many terms because all entries in the index pertain to cancer. Further, many specific carcinogens, oncogenes, gene products and drugs are not listed if the section in which they are contained is focused primarily on another topic.

14-3-3 proteins, 165

abl, 165, 175, 184
adduct, 82
aflatoxin, 86–93, 363
Agrobacterium tumefaciens
  etiological role in crown gall, 199
  plasmid pAqK84, 200
Agrosin 84 plant protection, 200
Ah receptor, 11, 112
Akt/PKB, 161, 186
alcohol, 261
alkylating agents, 84, 85, 91, 92–95
American Cancer Society Guidelines on Nutritional
  and Physical Activity, 254
Ames test, 86, 98
androgen ablation therapy, 328
aneuploidy. *See* chromosomes
angiogenesis,
  and tumor cell growth, 68, 180
  antiangiogenesis, 320, 322
  host response, 15
angiostatin, 181
antioxidants, 115
AP-1, 164, 166, 167
APC (Adenomatous Polyposis Coli) suppressor, 187,
  192
apoptosis, 48–49, 149, 188, 189, 271–273, 279, 323
  anti-apoptotic factor, 150
  gefitinib-induced, 270
  growth regulation, 188–191
aralkylating agent, 84, 85, 91, 93
ARF (Alternate Reading Frame)/p14^{ARF}/p19^{Arf}, 186
arylhydroxylamine, 84, 85, 91, 94

asbestos, 95, 97, 98
ATF (Activating Trascription Factor), 167
autochthonous cancer cells, 78

Bak, 189, 190
basement membrane, 56
  breached by cancer cells, 61
  capillaries and dissemination, 66
  collagenase, 56
  degradation by cathepsin B, 58
  migration of cancer cells through, 68
  sea urchin, 64
Bax, 189, 190, 191
bcl-2, 150, 177, 178, 189, 191
bcr, 175
Bcr-Abl, 176, 177, 191
benign tumor
  and atrophy of normal cells, 19
  compared with malignant tumor, 18–21, 24
  diagnosis, 21–24
binary transgenic systems, 150
biomarker, 82, 93, 95, 99, 117
birds, 215–216
  Marek disease, 216
  Rous sarcoma, 215
Boveri, Theodor, and aneuploidy, 128
brain, 45
breast, 233–330, 335, 355
  breast feeding, 234
  dietary fat, 234
  family history, 235
  genes and inheritance, 141
  hyperplasia during pregnancy, 16
  geography, 234

breast (*cont.*)
    letter, 4
    living in "developed" countries, 263
    raloxifene, 116
    risk factors, 234
    Romans of first century, 127
    susceptibility, 141
    treatment, 301, 310
Burkitt, Denis Parsons, 256
Burkitt lymphoma, 134

cachexia, 33
calcium and risk for colorectal cancer, 260
cancer cells
    arrest as emboli, 67
    circulating, 67
    detected in blood, 67
    disseminated via vascular system, 65
    extravasation, 68
    loss of cohesiveness, 59
    motility, 61, 68
cancer
    cure, 279–282
    increase in prevalence by 2020, 225
    phyletic aspects, 196
carcinogen
    defined, 81
    distribution in body after uptake, 114
    inactivated, 114
    list, 87
    metabolism, 109
    oncogene activation, 178
    report, 80
carcinogenesis, 82
    chemoprevention, 115
    endogenous, 107–109
    modified, 117–120
    multistep process, 82–83
    tumor promotion, 120–122
caricature of tissue renewal, 47
caspase, 161, 188, 189, 190
cat (feline) leukemia, 219
cathepsin B, 58
Cdc25, 172
CDK (Cyclin-Dependent Kinase), 168, 169, 170,
    183
celecoxib and colorectal polyps, 116
cell cycle, 168
    arrest, 268
    radiation, 311

Masui, Yoshio, 202
    regulation, 168, 269
    tumor size, 15
cells
    hybrids (cell fusion), 181
    normal with suppression of tumor growth, 180
    surface markers and therapy, 345–349
cervical. *See* uterus
    "beginning of the end", 197, 243–244
    human papilloma virus (HPV), 106, 243
    urinary obstruction, 32
chemoprevention, 115, 116
chemotherapy, 314–320, 341, 349
Chernobyl and radiation, 101
chimera, of normal and malignant cells, 266
chromosomes
    aberrations, 128, 192
    aneuploidy, 107, 123, 127–129
    banding, 129
    chronic granulocytic leukemia, 131. *See also*
        leukemia
    diploid number in humans, 128
    failure to detect cytogenetic changes in some
        tumors, 129
    Philadelphia (a minute), 132
Cip/Kip, 161, 170, 171, 185
CKI (Cyclin-dependent Kinase Inhibitor), 170,
    171
clams (*Mya arenaria*), 202
classification and nomenclature, 27–28
collagen
    enzyme degrades type I, 56
    enzyme degrades type IV, 56
    sea urchin basement membrane, 64
collagenase, 66, 69
colorectal
    description, 359
    epidemiology, 240–241, 243
    familial, 139
    intestinal obstruction, 31
    letter, 2
    loss of suppressor sequence, 192
    polyps, 116
complete carcinogen, 83
conversion of malignant to benign, 266. *See also*
    differentiation
crown gall, 198, 199
cruciferous (brassicacous) vegetables, 261
cryotherapy, 314
cyclin, 168, 169, 170

CYPs
  carcinogen metabolism, 110–114
  gene families, 110
  induction, 111
cytochrome P-450. *See* CYPs
cytokine trophic factors, 335
cytoreductive therapies, 271–273, 279, 307, 311,
    314
cytostatic agent, 268–270
cytotoxic chemotherapy, 341

death
  cachexia, 33
  how it occurs, 30–34
  late organ failure, 30
  obstruction of hollow organs, 31–33
defenses
  chemical carcinogens, 109
  chemical uptake, 109
  cytochrome P-450 oxidases (CYPs), 110–114
  selective absorption, 110
determination, 42
dissemination, 65, 67
developmental aspects of cancer, 40–48
diagnosis, 21–24
diet, 253–255, 256, 258, 263
differentiation, 342, 349
  and dedifferention, 40
  cell surface markers and therapy, 345–349
  embryonal carcinoma cells, 46
  stem cells, 44
differentiating agents, 267
DNA
  double-strand breaks (DSB), 102
  methylation, 122
  synthesis regulation, 168
dog (canine), 218
Doll, William Richard Shaboe, 231
dormancy, 35
Drosophila melanogaster
  imaginal discs, 42
  melanotic "tumors", 201
  neuroblastomas, 201
drug resistance, 296, 297
dysplasia, 16–32

E2F, 169, 171, 183, 186
e-cadherin, 61
eIF4E (elongation initiation factor 4E), 161
elasmobranchs (sharks), 196, 206. *See also* fish

elastin, 56
electrophile, 84
Elion, Gertrude B., 318
emboli, 67
embryonic development and metastasis, 74–76
endocrine therapy, 324
endogenous carcinogenesis, 107
endometriosis, 74
endostatin, 181
Environmental Protection Agency (EPA), 124
Epstein-Barr virus (EBV), 106, 183
ERK (Extracellular signal-Regulated Kinase), 163,
    164
erythrophoroma (goldfish), 207
estrogen, 330–333, 335. *See also* breast
excision repair, 118, 119
exercise, 263
extracellular matrix, 56, 57, 58, 62, 64

familial cancer syndromes, 139–143, 187
fat, 258
federal regulations, 122–123
fibronectin, 56, 59, 62, 64
fish, 204–207
  environmental studies, 205
  erythrophoromas, 207
  fossil, 204
  germ cell tumors, 206
  oncogenes, 207
  pigment cell tumors, 206
  sentinel organisms, 196
Fisher, Bernard, 311
Food and Drug Administration (FDA), 124
fos, 164, 166, 167
fossil, human, 226
frogs
  kidney tumors in hybrids, 208
  Lucké renal adenocarcinoma, 209–212
  nuclear transplantation (cloning), 209
  pancreatic, 207
fruit fly, *See Drosophila melanogaster*

gall, insect, 197
GAP (GTPase-activating protein), 159, 162, 165
garlic (*Allium* sp.), 115, 261
GEF (Guanine nucleotide Exchange Factor), 162
gene targeting, 150, 152, 153
geography, 223
goldfish, 207
grading of tumors, *See* staging and grading

grb2, 158, 159, 162
GSK3β (glycogen synthase kinase 3β), 161, 167, 187

Halsted, William, 311
Harshbarger, John C., 214
Hartwell, Leland, 269
heart, 16–32
heart disease and cancer compared, 224
hepatitis B virus (HBV), 106
HER-2/neu, 153, 154, 158, 173, 174, 175
hereditary cancer, 136–138
herpesvirus
    Epstein-Barr, 244
    Kaposi sarcoma, 246
    RaHV-1 (frog), 209
    turkey, 216
heterogeneity, 38–40, 54, 56, 58, 77, 123
Hh (Hedgehog), 187, 188
HIF1α (Hypoxia-Inducible Factor-1α), 181
Hiroshima and Nagasaki, 101
Hitchings Jr., George H., 318
Hodgkin lymphoma, 32, 244
Huggins, Charles Brenton, 37, 240–241, 243,
    329
human papilloma virus (HPV), 106, 183. See cervical
Hunt, R. Timothy, 269
hyperplasia, 121

imatinib mesylate/Gleevec/STI-571, 177
immune surveillance, 119
immune system, 335
initiation, 36
INK4, 170, 171, 182
interferon, 335
International Agency for Research on Cancer
    (IARC), 80, 122–123
ionizing radiation, 100, 101, 102, 103

JNK (Jun N-terminal Kinase), 166
jun, 164, 166, 167

Kaposi sarcoma
    HIV infection, 245
    human herpesvirus 8, 246
    treatment, 336
kidney, 7. See also Lucké renal adenocarcinoma

laminin
    hyrolysis, 59
    sea urchin basement membrane, 64
latency, 36

leukemia
    acute promyelocytic (APL), 267
    description, 378
    chronic myelogenous, 336
    feline, 219
    letter, 5
    lymphoma, 378
    retinoic acid therapy, 342
    stem cells, 45
Li-Fraumeni Syndrome (LFS), 184
Lucké renal adenocarcinoma, 61
    cell detachment, 59
    chromosomes, 129
    differentiation, 61
lung
    description, 370
    endoscopic procedure, 310
    epidemiology, 227–233
    letter, 6
    preventable disease, 249–251
    smoking, 221
lycopene, 260
lymphoma, 134, 378
liver, 365

mammals, 216–220
mammary carcinoma (mouse), 39, 217. See also breast
malignant cells, 21–23
MAP Kinase signaling, 159
Marek disease, 216
markers, tumor, 30
Marshall Islands and thyroid tumors, 101
Masui, Yoshio, 202
matrix metalloproteinases, 57, 63
Mdm2 (Murine Double Minute 2), 186
met, 168
methylation, 173
melanoma, See skin
metabolic activation
    aflatoxin, 86–93
    chemical carcinogens, 113
    polycyclic aromatic hydrocarbons, 94
metaplasia of respiratory epithelium, 16
metastasis
    cardinal attribute of malignancy, 52
    cascade, 52, 54
    cell cohesiveness, 59
    contrasted with primary tumor, 76
    dissemination, 28
    extravasation, 68
    feared by both patient and physician, 51

non-random distribution, 52, 70, 71
    genes, 69
    Rana pipiens, 56
    Recamier, 52
    survival of "finest" cells, 78
metronomic therapy, 268
microarray, 154, 155
microarray technology, 143, 154–155
microtubules and vinca alkaloids, 63
migration, *See* motility
Miller, Elizabeth Cavert, 116
mismatch repair (mmr), 193
molecular oncology, 353
motility
    autocrine motility factor, 63
    chemoattractants, 63
    extravasation, 68
    in vitro, 61
    normal cell, 73
mouse
    mammary carcinoma, 39, 217
    teratocarcinoma, 217
mTOR (mammalian Target of Rapamycin), 161
mutagenesis
    carcinogens, 84
    "spontaneous", 136
    ultraviolet, 239
multiple myeloma, 226
mutagens, 91
mutator phenotype, 107, 108
myc, 149, 150, 151, 177, 178, 179
mycotoxins, 92

2-naphthylamine, 94, 95, 96
National Institute of Environmental Health Sciences (NIEHS), 124
National Toxicology Program (NTP), 80, 124
necrosis, 48
neoplasm, 14–18
neuroblastoma, 12, 201, 374
NF1 (Neurofibromatosis type I), 165
N-nitroso compounds, 92
Nm23, 69
nomenclature, 27–28
nonreceptor tyrosine kinases, 161, 162
normal cells
    metastasis, 73–74
    and tumor growth, 180
normal tissue from cancer cells, 199
Northern Leopard Frog, *See Rana pipiens*
Nurse, Paul, 269

obesity, *See* diet
occupational, 244–245
Occupational Safety and Health Administration (OSHA), 124
Ochsner, Edward William Alton, 229
oncogene, 145, 146, 157. *See also* proto-oncogene
    activation mechanisms, 173–178
    cell transformation, 179
    cellular functions, 145
    classification, 155, 157
    defined, 145
    fish, 207
    growth and differentiation, 149
    signal transduction, 156, 171
    suppressor genes, 145
ozone layer depletion, 239
opossum melanoma, 217

P-450 cytochromes/CYPs (mixed fuction oxidase enzymes), 91, 110, 111, 112
p53, 171, 184, 185, 186, 190
p53 suppressor gene, 184–187
Paget, Stephen, 71
PAH (polycyclic aromatic hydrocarbon), 84, 93
pancreatic cancer in frog hybrids, 207
paraneoplastic syndrome, 30, 33
pharmacology, 349–353
Phase II conjugating enzymes, 114
Philadelphia chromosome, 175
phosphatase, 172
photodynamic therapy, 310
PI3K (phosphotidylinositol 3-kinase), 159, 160, 161, 164, 172
PKA (Protein Kinase A), 165
PKC (protein kinase C), 160, 164
PKD1, 161, 173
PLC (phospholipase C), 159, 160, 164
plants
    crown galls, 198
    insect galls not true tumors, 197
plasminogen activators, 59
Pott, Percivall, 84, 244
PP2A (Protein Phosphatase 2A), 172
PPAR (peroxisome proliferator-activated receptors), 112
progression, 37, 83, 122–123
promotion, 120–122
promotor, 81, 83, 91, 120, 121, 122
prostate
    androgen ablation therapy, 328
    cathepsin B, 58

prostate (*cont.*)
   description, 358
   epidemiology, 239–241
   heredity, 142
   heterogeneity, 58
   Huggins, Charles Brenton, 37, 240-241, 243, 329
   hyperplasia, 15
   KAI1, 70
   lycopene, 260
   tissue-sparing, 310
proto-oncogene, *See also* oncogene
   classification, 155, 157
   and differentiation, 149
   growth factors, 156
   knockout mutations, 150
   nuclear signaling, 165
   serine/threonine kinases, 163
   transcriptional factors, 166
   tyrosine kinases, 161
proteoglycan, 56
PTB (posphotyrosine-binding) domain, 158
Ptch (Patched), 187, 188
PTEN (Phosphatase and TENsin homolog deleted
      on chromosome 10), 172, 173, 182
pyknosis, 48

Rac/Rho, 164
radon, 103, 104, 105
Raf, 163, 164
raloxifene and breast cancer, 116
Ramazzini, Bernardino, 233
Rana pipiens
   chromosomes, 129
   herpesvirus, 196, 209
   Lucké renal adenocarcinoma, 56, 209–212
   nuclear transplantation (cloning), 209
radiation therapy, 311–314
ras, 149, 153, 162, 163, 174, 178, 179
   activation, 162
   differentiation, 153
   evolution, 149
   suppression, 165
RB (Retinoblastoma tumor suppressor), 169, 171,
      182, 183, 184, 185, 186
reactive oxygen species (ROS), 102
Recamier, Joseph Claude Anselme, 52
reptiles, 212
risk for cancer, 136–138, 139
ret, 149, 151, 175
retinoblastoma, 136–138, 139, 183, 374

retinoic acid therapy, 342
retroviruses, 105, 146, 147, 148
Rous sarcoma, 146, 215
RTKs (receptor tyrosine kinases), 156, 158, 161

Salmonella typhimurium, 86–93
sarcoma, 377
selection and heterogeneity, 38–40
selenium, 260
senescence, 191, 192, 269
sentinel organisms, 196
serine/threonine kinase, 156, 163
sFRP, 153
SH2 (*src* Homology 2) domain, 158, 159, 166, 176
SH3 (*src* Homology 3) domain, 176
sharks, 196, 206
Shc, 158, 159
signal transduction, 156, 171
skin
   basal cell carcinoma, 236
   epidemiology, 236
   malignant melanoma, 11, 100, 237, 309, 336, 372
   minimize risk, 252
   squamous cell carcinoma, 8, 237, 361
   sunscreen, 253
   opossum, 217
Skipper, Howard, 275
Smac/DIABLO, 189, 190
snuff (smokeless tobacco), 250
"soil and seed" theory
   distribution of metastases, 71
   Paget, 70
   primordial germ cells, 73
Sos (Son of Sevenless), 158, 159, 162
spontaneous regression, 34
squamous cell carcinoma, *See* skin, cervical
src, 161, 162
SRF (Serum Response Factor), 167
staging and grading, 25–26, 282–289, 293
   breast cancer, 356
   molecular markers, 30
STAT, 155, 166, 177
stem cells, 43–46, 267, 289, 293, 342
stomach, 10
suppressor genes
   adenomatous polyposis coli, 187, 192
   known functions, 182
   p53, 184–187
   patched (Ptch), 187
surgery, 308–310

Tamoxifen, 331
Tasmanian devil, 216
telomerase, 192
teratocarcinoma, 217, 365
testis, 9, 365
TGFβ (Transforming Growth Factor-β), 167, 180
therapeutic index, 313
therapy
  breast, 301
  cytoreductive, 311
  cytotoxic, 341
  factors for success, 282–289, 293
  failure, 293–301
  multi-modality, 282–289
TNF (Tumor Necrosis Factor), 188, 190, 191
TNM staging, 25
toxicity of therapy, 312, 341
tretinoin, 343
trophic factors
  antagonism, 325–335, 337, 340
  apoptosis, 322, 324
  chemotherapy, 341
  steroid hormones, 328
Tumor suppressor, 181, 182

ultraviolet
  basal cell carcinoma, 236
  mutagenesis, 239
  opossum melanoma, 217
  ozone layer depletion, 239
  protective clothing, 252
  radiation, 98–100
  squamous cell carcinoma, 237
  sunscreens, 98–100
  tanning, 251
  UV-A and OV-B compared, 252

ultraviolet radiation, 98, 99
uranium miners, 103–105, 107
uterine cancer, 335. *See also* cervical cancer

vaccines, 197
  cervical carcinoma, 244
  feline leukemia, 219
  Marek disease, 216
VEGF (Vascular Endothelial Growth Factor), 181
vinca alkaloids, 197
virus
  carcinogenesis, 103–105, 107, 215
  Hodgkin lymphoma, 244
  human papilloma virus (HPV), 106, 243
  milk of mice, 217
vitamins
  antioxidants, 115
  ascorbic acid (vitamin C), 115, 259
  beta-carotene (vitamin A), 115, 259, 267, 342
  calciferol (vitamin D), 260
  tocopherols (vitamin E), 115

Walsh, Patrick, 311
warfarin, 67
Wilms tumor, 137–138, 375
Withers, H. Rodney, 291
wnt, 151, 168, 187

xenobiotic metabolism, 109, 110
xenobiotics, 109, 110
xeroderma pigmentosum (XP), 99, 118, 119, 137–138
x-ray, 100, 311–314

Yamagiwa and Ichikawa, 84, 245